# Lecture Notes
# in Control and Information Sciences 250

Editor: M. Thoma

Springer-Verlag London Ltd.

# Index of Authors

# Chapter 1

# Keynotes

Two keynote presentations were made. The first (not included in this volume) was by Rodney Brooks of MIT and explored issues relating to intelligence and emotion in robots. Emotional response is being seriously explored by researchers since humans have a strong and subconscious ability to perceive emotional states, often displayed through facial expressions. For example if we are talking to somebody and perceive they are confused we would modify our words accordingly. Tapping into this communications 'channel' may make human/robot interaction more effective.

The second keynote presentation was by Mandyam Srinivasan from the Australian National University. It was wonderfully presented despite the less than ideal (for talking) environment of a boat on Sydney Harbour. He described experiments and theories that attempt to explain how simple biological automata (insects) can perform complex navigation tasks with limited sensing and computational resources.

PIC

# From Insects to Robots

M.V. Srinivasan[1], J.S. Chahl[1], K. Weber[2],
S. Venkatesh[2], S.W. Zhang[1] and M.G. Nagle[1]
[1]Centre for Visual Science, Research School of Biological Sciences
Australian National University
email M.Srinivasan@anu.edu.au
P.O. Box 475, Canberra, A.C.T. 2601, Australia
and
[2]Department of Computer Science, School of Computing
Curtin University
GPO Box U 1987, Perth, WA 6001, Australia
email svetha@cs.curtin.edu.au

May 29, 1999

## Abstract

Recent studies of insect visual behaviour and navigation reveal a number of elegant strategies that can be profitably applied to the design of autonomous robots. Bees, for example, negotiate narrow gaps by balancing the velocity of image motion in the two eyes, rather than using computationally complex stereo mechanisms to measure distance. Distance flown is gauged by integrating, over time, the optic flow that is experienced by the visual system en route to the destination. These and other visually mediated behaviours are described along with applications to robot vision.[1]

# 1   Introduction

Anyone who has tried to swat a fly, or watched it execute a flawless landing on the rim of a teacup would have noticed that insects possess a visual system that is fast, precise and reliable. This is so despite the fact that the insect brain weighs less than a tenth of a milligram and possesses four orders of magnitude fewer neurons than does the human brain. Do simple nervous systems such as these use computational "short cuts" to achieve their goals? Observation of the visual behaviour of insects suggests that this may indeed be the case.

---

[1]Corresponding author: M.V. Srinivasan, Phone: 61-6-249-2409; Fax: 61-6-249-3808; email: M.Srinivasan@anu.edu.au

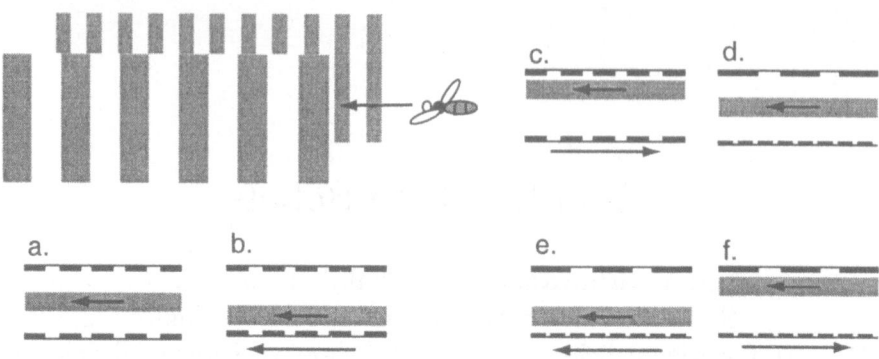

Figure 1: Illustration of an experiment which demonstrates that flying bees infer range from apparent image speed. The shaded areas represent the means and standard deviations of the positions of the flight trajectories, analysed from video recordings of several hundred flights.

Unlike vertebrates, insects have immobile eyes with fixed-focus optics. Thus, they cannot infer the distance of an object from the extent to which the directions of gaze must converge to view the object, or by monitoring the refractive power that is required to bring the image of the object into focus on the retina. Furthermore, compared with human eyes, the eyes of insects are positioned much closer together, and possess inferior spatial acuity. Therefore the precision with which insects could estimate the range of an object through binocular stereopsis would be much poorer and restricted to relatively small distances, even if they possessed the requisite neural apparatus [1]. Not surprisingly, therefore, insects have evolved alternative strategies for dealing with the problem of three-dimensional vision. Many of these strategies rely on using image motion as a significant cue. Some of them are outlined below, together with applications to robot navigation [2]. References to more complete accounts are also provided.

## 2 The centring response in bees, and its application to robot navigation

When a bee flies through a hole in a window, it tends to fly through its centre. How does the insect, lacking stereo vision, gauge and balance the distances to the sides of the opening? One possibility is that it simply balances the speeds of image motion on the two eyes. To investigate this hypothesis, bees were trained to enter an apparatus which offered a reward of sugar solution at the end of a tunnel [3]. Each side wall carried a pattern consisting of a vertical black-and-white grating (Fig. 1). The grating on one wall could be moved horizontally at any desired speed, either towards the reward or away from it. After the bees had received several rewards with the gratings stationary, they were filmed from above, as they flew along the tunnel.

When both gratings were stationary, the bees tended to fly along the midline of the tunnel, i.e. equidistant from the two walls (Fig. 1a). But when one of the gratings was moved at a constant speed in the direction of the bees' flight, thereby reducing the speed of retinal image motion on that eye relative to the other eye, the bees' trajectories shifted towards the wall with the moving grating (Fig. 1b). When the grating moved in a direction opposite to that of the bees' flight, thereby increasing the speed of retinal image motion on that eye relative to the other, the bees' trajectories shifted away from the wall with the moving grating (Fig. 1c). These findings demonstrate that when the walls were stationary, the bees maintained equidistance by balancing the apparent angular speeds of the two walls, or, equivalently, the speeds of the retinal images in the two eyes. A lower image speed on one eye was evidently taken to mean that the grating on that side was further away, and caused the bee to fly along a trajectory closer to it. A higher image speed, on the other hand, had the opposite effect. Variation of the periods of the gratings on the two walls did not change the results (Figs. 1d-f), indicating that the bees were measuring the speeds of the images of the walls independently of their spatial structure.

This biological finding offers a simple strategy for autonomous visual navigation of robots along corridors. By balancing the speeds of the images of the two side walls, one can ensure that the robot progresses along the middle of the corridor without bumping into the walls. Computationally, this method is far more amenable to real-time implementation than methods that use stereo vision to calculate the distances to the walls. Four different laboratories have now built robots that negotiate corridors successfully using the bee-derived principle of balancing lateral image motion [4], [5], [6], [7].

The design and performance of one of these robots [5] is shown in Figure 2. The robot is approximately the size of a small skateboard, with a single video camera mounted facing upwards (Fig. 2a). This camera captures views of the side walls (one of each wall) through a mirror assembly positioned above the lens. Video information from the camera is transmitted to a desktop computer, where the image velocities of the two walls, induced by the motion of the robot, are measured using a simplified version of the image-interpolation algorithm described above. The computer then issues appropriate steering commands to the robot to ensure that it stays close to the midline of the tunnel. The tunnel-following performance of the robot is illustrated by the examples shown in Figs. 2b-e. In all cases the robot reliably follows the axis of the corridor. The presence of an obstacle next to one of the walls causes the robot to go through the middle of the gap remaining between the obstacle and the other wall. Additional control algorithms have been developed for controlling the speed of the robot. Speed control is achieved by holding constant the sum of the image speed from the two walls, thereby ensuring that the robot automatically slows down to a safe speed when the corridor narrows. This strategy, again, is based on observation of bees flying though tunnels of varying widths [8].

6

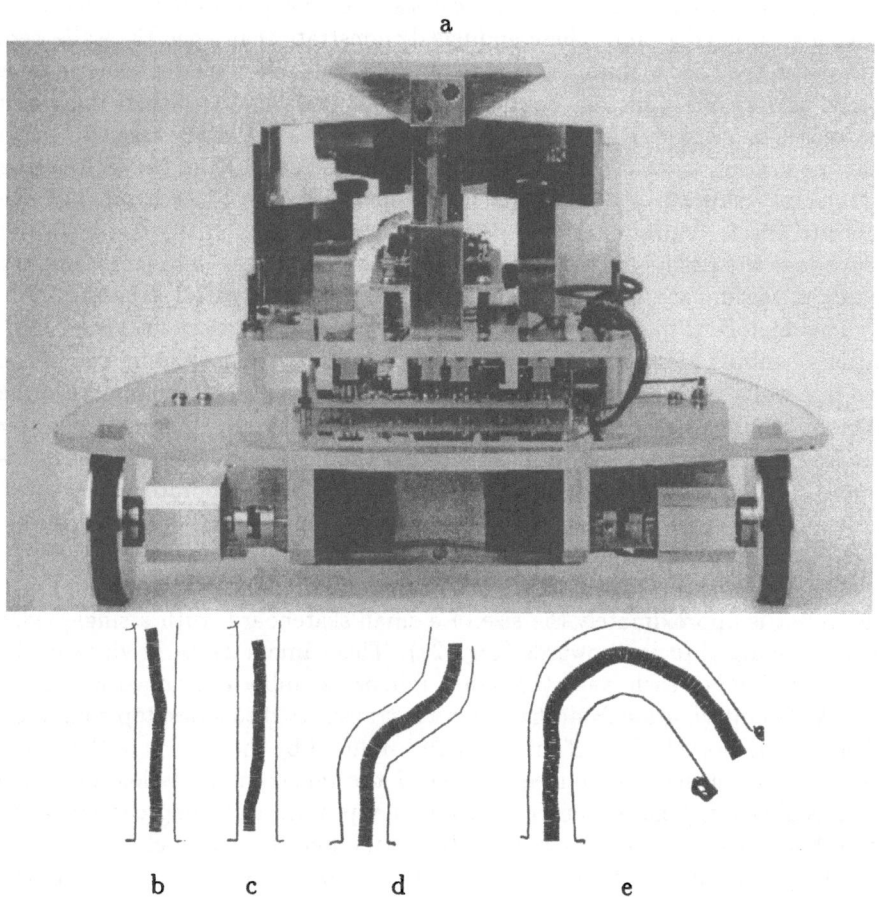

Figure 2: Upper panel: frontal view of corridor-following robot. Lower panels: performance in variously shaped corridors, of width approximately 80-100cm.

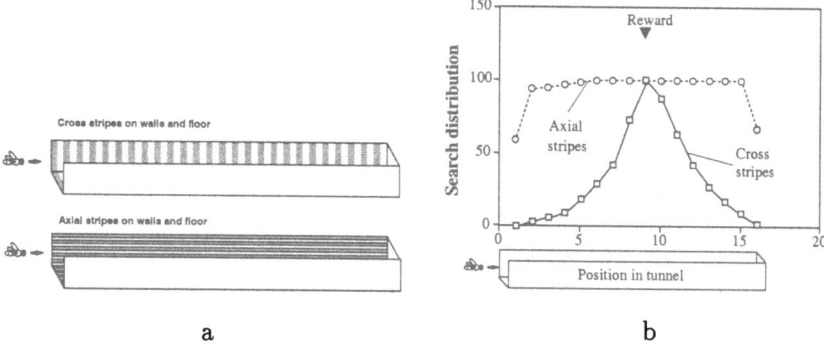

Figure 3: (a) Experimental setup for investigating visual odometry in bees. (b) Performance of bees in gauging distance travelled in the tunnel, as shown by distributions of searching locations

# 3   Visual odometry: from bees to robots

Although it is well established that foraging honeybees can gauge the distances to food sources, the mechanisms by which they do so have remained enigmatic. To elucidate these mechanisms, we recently investigated whether bees can estimate and learn short distances flown under controlled laboratory conditions [8,9]. Bees were trained to enter a tunnel and collect a reward of sugar solution at a feeder placed in the tunnel at a fixed distance from the entrance. The walls and floor of the tunnel were lined with black-and-white stripes, usually perpendicular to the tunnel's axis (Fig. 3a). During training, the position and orientation of the tunnel were changed frequently to prevent the bees from using any external landmarks to gauge their position relative to the tunnel entrance. The bees were then tested by recording their searching behaviour in a fresh tunnel which carried no reward, and was devoid of any scent cues. The training and test tunnels were covered by a transparent sheet of perspex, and subdivided into 16 sections for the purposes of analysis. In the tests, the bee's behaviour whilst searching for the reward was recorded by noting the locations of their first, second, third and fourth U-turns. From this data it was possible to estimate the mean searching location, and the extent to which the search was distributed about this mean (Fig. 3b).

Bees trained in this way showed a clear ability to search for the reward at the correct distance, indicated by the thick curve in Fig. 3b. How were the bees gauging the distance flown? A number of hypotheses were examined, as described below.

Were the bees counting landmarks en route to the goal? To examine this possibility, bees were trained in a tunnel lined with stripes of a particular spatial period and tested in a tunnel lined with stripes of a different period. The test bees searched at the correct distance from the tunnel entrance, regardless of stripe period [9]. Therefore, distance is not gauged by counting the number of stripes or other features passed whilst flying through the tunnel.

Were the bees measuring distance flown in terms of the time required to reach the goal? To examine this possibility, bees were trained as above and tested in a tunnel that presented a headwind or a tailwind, generated by a fan at the far end of the tunnel. In a headwind, bees flew slower and took longer to reach the estimated location of the reward. The opposite was true in a tailwind [9]. Therefore, distance is not estimated in terms of time of flight, or other correlated parameters such as number of wingbeats. In a headwind, bees overshot the location of the reward; in a tailwind, they undershot it. Therefore, distance flown is not measured in terms of energy consumption.

Were the bees measuring distance flown by integrating the motion of the image of the surrounding panorama as they flew to the goal? To examine this possibility, bees were trained in a tunnel of a given width and then tested in a tunnel that was narrower or wider. In the narrower tunnel, the bees searched at a shorter distance from the entrance; in the wider tunnel, they searched farther into the tunnel [9]. These results suggest that distance flown is gauged by integrating the speed of the images of the walls and floor on the eyes whilst flying through the tunnel.

To further examine the image motion hypothesis, bees were trained and tested in conditions where image motion was eliminated or reduced. In tunnels with axially oriented stripes on the walls and floor, the bees showed no ability to gauge distance travelled. In these tests, the bees searched uniformly over the entire length of the tunnel, showing no tendency to stop or turn at the former location of the reward (thin curve, Fig. 3b). These results indicate that image motion is critical to odometry in bees, and confirms the hypothesis that distance flown is measured by integrating image motion. Unlike an energy-based odometer, a visually driven odometer would not be affected by wind or by the load of nectar that the bee carries.

Inspired by the bee's visual odometer, we have designed a robot that uses image motion to determine how far it has travelled. This machine uses a version of the 'Image-Interpolation algorithm' developed in our laboratory [10]. The algorithm makes use of the fact that if a sensor captures images of the environment from two different reference positions, a known distance apart, the image corresponding to an unknown intermediate position can be approximated by a weighted linear combination of the images obtained from the two reference positions. The intermediate position of the sensor can then be determined from the coefficients of the interpolation. This technique is accurate provided that the reference images are captured at positions that are close enough to permit accurate linear interpolation of the intermediate image [11].

This concept is applied to compute egomotion as follows. Two panoramic views of the world are captured simultaneously, a fixed, known distance apart on the axis along which the robot translates (the fore-aft axis). Each view corresponds to a ring-like, horizontal slice of the environment, and is acquired by a separate video camera mounted under a cone-shaped mirror (see Fig. 4a). The separation of the two cameras is equal to the largest displacement that is expected. Let $f_0(\theta)$ and $f_1(\theta)$ denote the images captured by the rear and front cameras, respectively. Now assume that the robot translates forward by an unknown distance a expressed as a fraction of the separation between the two cameras. We capture a third image $f(\theta)$ at this location, using the rear camera. We wish to determine the translation of the robot by measuring how far the rear camera has moved. The rear camera is now located somewhere between the original locations of the front and rear cameras. We assume that the new image captured by the rear camera at this intermediate location can be approximated by $\hat{f}(\theta)$, a weighted linear combination of the reference images $f_0(\theta)$ and $f1(\theta)$ as follows:

$$\hat{f}(\theta) \cong (1 - \alpha)f_0(\theta) + \alpha f_1(\theta) \tag{1}$$

where $\alpha$ $(0 \le \alpha \le 1)$ specifies the fractional displacement of the rear camera relative to the distance between the two cameras. This equation is valid provided the local spatial gradient of the intensity profile is approximately constant between points $f_0(\theta)$ and $f_1(\theta)$. In other words, it assumes that the intensity profile is smooth enough that the image displacement at any location is not greater than half the period of the highest local spatial frequency. (In practice, this can be ensured by low-pass filtering the images). From (4) we see that when $\alpha = 0$, $f(\theta) = f_0(\theta)$ and when $\alpha = 1$, $f(\theta) = f_1(\theta)$. Thus, the approximation $\hat{f}(\theta)$ satisfies the boundary conditions.

To determine the value of $\alpha$ that gives the best approximation to $f(\theta)$, we minimise the mean square error between $f(\theta)$ and $\hat{f}(\theta)$. That is, we minimise

$$\int_0^{2\pi} [(1 - \alpha)f_0(\theta) + \alpha f_1(\theta) - f(\theta)]^2 d\theta \tag{2}$$

with respect to $\alpha$. This minimisation yields

$$\alpha = \frac{\int_0^{2\pi} [f(\theta) - f_0(\theta)][f_1(\theta) - f_0(\theta)]d\theta}{\int_0^{2\pi} [f_0(\theta) - f_1(\theta)]^2 d\theta} \tag{3}$$

Since $f_0(\theta)$, $f(\theta)$ and $f1(\theta)$ are known, $\alpha$ can be readily computed. Thus, the translatory displacement of the robot can be measured optically by capturing three images, and processing them as specified above.

An analogous approach can be used to measure rotations of the robot about the optical axis of the rear camera. In this case, the reference images can be obtained by software-rotating the image captured by the rear camera, because rotation about the optical axis causes all points in the ring image to be displaced by the same amount, irrespective of the distances of the objects in the environment.

We have embodied this technique in a mobile robot thereby enabling it to measure its egomotion in an unknown environment. A view of the robot and its environment is shown in Fig. 4a. The robot is approximately 40cm long and moves on three wheels, consisting of two drive wheels at the rear and a casting wheel at the front. The drive wheels are independently controllable, allowing the robot to either (i) translate along the long axis, (ii) turn about a point on the rear axle midway between the two rear wheels or (iii) move along a curve, i.e. translate and rotate simultaneously. Fig. 9a also shows the two camera/cone assemblies. The image captured by each camera is 344x288 pixels of 128 grey levels (Fig. 4b).The egomotion algorithm uses a ring-shaped strip of this image, of radius 140 pixels and thickness 5 pixels, centred on the axis of the cone, that is divided into an array of 250 units. These are shown as a ring of black and white circles superimposed on the grey level image in Fig. 4b. This ring corresponds to a horizontal slice of the environment at a height of 15cm above the floor, covering an elevational angle of 1 deg.

The performance of the egomotion algorithm in computing the motion of the robot is illustrated in Fig. 4c. The robot moved along three different trajectories, each encompassing 30-40 individual steps. The figure compares the path determined by integrating the motions computed for the individual steps by the egomotion algorithm, with the actual path followed by the robot, as established from the grid marked out on the floor. Clearly, the algorithm performs well.

It is important to note that the egomotion algorithm requires absolutely no knowledge of the 3- D structure of the environment. For example, there is no need to assume that the world is locally planar. Nor is there need for any object recognition, feature matching, or even image- velocity computation. The algorithm yields accurate results, provided the environment carries sufficient texture to provide the relevant visual information, and provided there is no movement of other objects in the environment. The Image-Interpolation algorithm, as applied here, uses the world *as its own model* to infer the motion of the robot. Recently, we have extended this approach to the measurement of egomotion in six degrees of freedom [12].

# 4 Conclusions

Analysis of vision in simple natural systems, such as those found in insects, can often point to novel ways of tackling tenacious problems in autonomous navigation. This is probably because insects, with their "stripped down" nervous systems, have been forced to evolve ingenious strategies to cope with visual challenges within their environment. This article has outlined some ways in which insects use motion cues to perceive their environment in three dimensions, and navigate in it. Each of these strategies has inspired a novel solution to a problem in machine vision or robotics. In constructing these robots, our aim is not to copy insect vision in the literal sense. For instance, our egomotion computing robot, with its forward and rear cameras, could hardly be described as a faithful imitation of an insect. At present we do not know enough about the actual processing mechanisms that underlie insect vision to 'produce a carbon (or silicon) copy, anyway. Our aim, instead, is to carry out behavioural experiments to reveal the cues that insects use to navigate, and to design machines that use such cues to advantage.

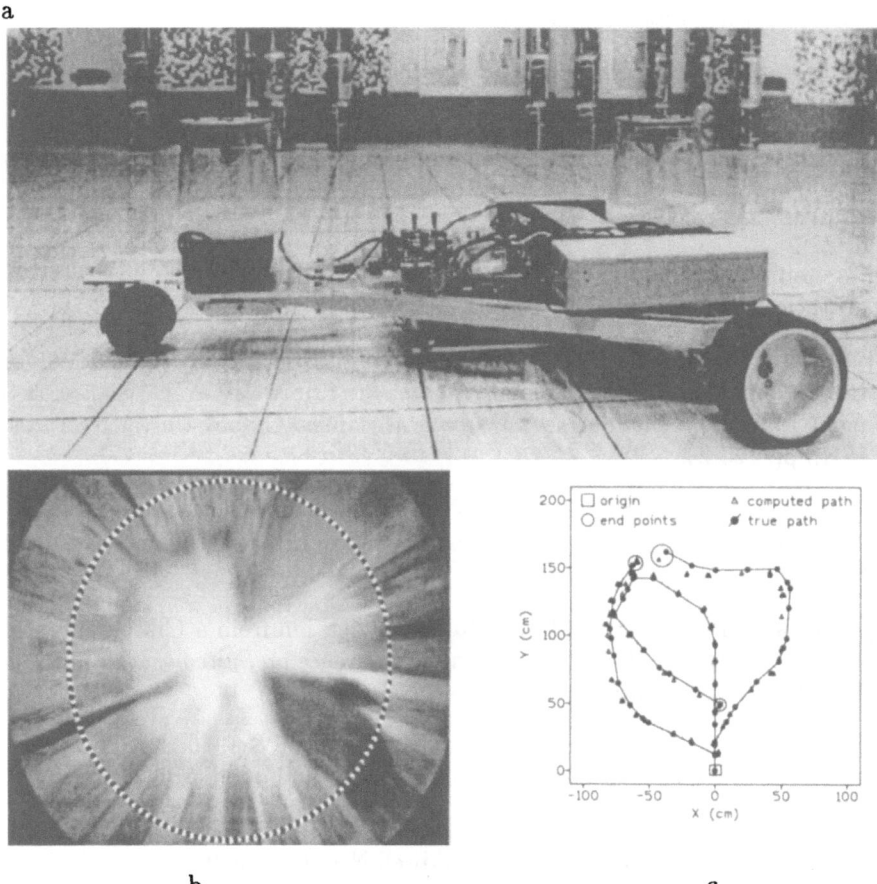

a

b                                    c

Figure 4: (a) Photograph of egomotion computing robot in the laboratory in which all experiments were undertaken, showing the grid on the floor and some of the objects in the environment. (b) An image captured by one of the video cameras via the conical reflector which is used to obtain a horizontal, panoramic strip view of the environment. The black and white annulus represents the part of the image that was decoded to form the narrow strip. Individual visual units along the strip are indicated by the alternating black and white segments. (c) Comparisons between the path determined by the integration of motion calculated by the interpolation algorithm (dotted line), and the actual path followed by the robot (solid line). Three different trajectories are shown.

# References

[1] Srinivasan M V How insects infer range from visual motion. In: Miles F A, Wallman J (eds) 1993 *Visual Motion and its Role in the Stabilization of Gaze*, Elsevier, Amsterdam, pp 139-156

[2] Srinivasan M V, and S. Venkatesh S 1997 *From Living Eyes to Seeing Machines*. Oxford University Press, U.K.

[3] Srinivasan M V, Lehrer M, Kirchner W, Zhang S W 1991 Range perception through apparent image speed in freely-flying honeybees. *Vis. Neurosci.* 6: 519-535

[4] Coombs D, Roberts K 1992 Bee-bot: using peripheral optical flow to avoid obstacles. *Proceedings SPIE* Boston, 1825:714-721

[5] Weber K, Venkatesh S, Srinivasan M V 1997 Insect inspired behaviours for the autonomous control of mobile robots. In: Srinivasan M V, Venkatesh S (eds) 1997 *From Living Eyes to Seeing Machines*. Oxford University Press, U.K., pp 226-248.

[6] Sandini G, Gandolfo F, Grosso E, Tistarelli M 1993 Vision during action.In: Aloimonos Y (ed).*Active Perception*. Lawrence Erlbaum, Hillsdale, NJ pp 151-190

[7] Duchon A P, Warren W H 1994 Robot navigation from a Gibsonian viewpoint. *Proceedings, IEEE International Conference on Systems, Man and Cybernetics, San Antonio, TX* pp 2272-2277

[8] Srinivasan M V, Zhang S W, Lehrer M, Collett T S 1996 Honeybee navigation en route to the goal: visual flight control and odometry. *J. Exp. Biol.* 199:237-244

[9] M.V. Srinivasan M V, Zhang S W, Bidwell N 1997 Visually mediated odometry in honeybees. *J. Exp. Biol.* 200:2513-2522

[10] Srinivasan M V 1994 An image-interpolation technique for the computation of optic flow and egomotion. *Biol. Cybernetics* 71:401-416

[11] Chahl J S, Srinivasan M V 1996 Visual computation of egomotion using an image interpolation technique. /em Biol. Cybernetics 74:405-411

[12] Nagle M, Srinivasan M V, Wilson D 1997 Image interpolation technique for measurement of egomotion in six degrees of freedom. *J. Opt. Soc. Am. A* 14:3233-3241

# Chapter 2

# Manipulation

Manipulation is fundamental to robotics, whether by humans, animals or machines. This chapter presents some intriguing work on manipulation, and the video presentations at the conference were certainly entertaining.

Thomas Sugar and Vijay Kumar have been building robots to handle loads cooperatively: two or more robots carry a load together even though they are both autonomous and have different mechanical arrangements. They discuss experiments with different mechanical compliance and control schemes to adjust for the inevitable differences in their movements without dropping the load or colliding. Their method builds on the work of Khatib and others, and they claim their method allows any number of robots to work as a team.

While they have not yet managed to deal with thick stacks of stapled paper, Matthew Mason and his colleagues have been playing with tiny wheeled robots with sticky paws to automate desktop tasks. They are realists in accepting the inevitable: the paperless office will always be a figment of our imagination! They show how wheels can be used to lift and fold paper or climb over and roll cardboard tubes carelessly left lying around by you or me.

Makoto Kaneko, Kensuke Harada and Toshio Tsuji have carefully studied how human hands grasp different objects and manipulate them, even though the objects cannot be seen. They have built a hand which imitates these movements and demonstrated some striking movements in their video presentation at the conference. Their presentation was most noted for the way in which they film their experimental subjects, humans, trying to pick up and manipulate different objects. The last subject, a female, was presented with a slimy octopus without warning with dramatic results!

JPT

# Control and Coordination of Multiple Mobile Robots in Manipulation and Material Handling Tasks

Thomas Sugar and Vijay Kumar
University of Pennsylvania
Philadelphia, U.S.A.
tgs@grip.cis.upenn.edu

**Abstract:** We describe a framework and control algorithms for coordinating multiple mobile robots with manipulators focusing on tasks that require grasping, manipulation, and transporting large and possibly flexible objects without special purpose fixtures. Because each robot has an independent controller and is autonomous, the coordination and synergy are realized through sensing and communication. The robots can cooperatively transport objects and march in a tightly controlled formation, while also having the capability to navigate autonomously. We describe the key aspects of the overall hierarchy and the basic algorithms, with specific applications to our experimental testbed consisting of three robots.

## 1. Introduction

We address the coordination of a small team of robots that cooperatively perform such manipulation tasks as grasping a large, flexible object and transporting it in a two-dimensional environment with obstacles. Such systems are useful in material handling applications where there are no special purpose material handling devices (for example, conveyors, fixtures or pallets).

Figure 1. *(a) Our experimental testbed of two cooperating mobile manipulators (left). The robot to the right is the Nomad XR4000, an omnidirectional platform, with a fork-lift. On the left is a TRC Labmate platform, a nonholonomic platform, equipped with a three degree-of-freedom compliant arm. (b) Three robots carry a flexible board (right). The one to the right is the TRC platform with the stiff, passive, one degree-of-freedom arm.*

We focus on tasks that simply cannot be performed by a single mobile robot. The large box and the flexible board shown in Figure 1 are large enough that one platform cannot transport the object without special purpose tooling, that is specific to the object and the task. In our approach, two or more robots may be coordinated to accomplish the task of manipulating and transporting a wide range of objects.

The physical interaction between the robots and the tight coupling required of autonomous controllers poses many challenging engineering problems. It should be possible to organize the robots differently for various tasks which means that the controllers should be independent and yet be able to function in a tightly coupled fashion. The manipulators must be capable of controlling the grasp forces and maintaining force closure in a robust fashion; and the robots must coordinate their trajectories in order to maintain a desired formation while maintaining the grasp. This means that (a) there needs to be an efficient way of communicating and sharing information in real time; and (b) the team must be robust with respect to positioning errors and modeling errors.

Our previous work [1] attempted to address some of the important issues such as controlling and maintaining the grasp forces in a robust fashion. In this paper, we describe the cooperative control of three independently controlled robots. Two critical issues are the coordination between the manipulation and the locomotion system (arm and the platform), and the coordination between the three different locomotion systems.

Others, notably [2], have developed mobile robot systems for transporting objects. One approach to the problem is to control each mobile manipulator by a computed torque scheme and let the mobile manipulators share real-time servo level information via a wireless network [2]. This approach has the disadvantage that it does not scale easily with an increase in the number of robots and it is not clear how the robots can be reconfigured to perform different tasks.

An important question in this and other related papers [3, 4] has to do with the sharing of information between the robots. The nature of communication and the information shared also determines the extent to which the load can be physically shared between the robots in the team.

In our approach, the control of each platform is decomposed into the control of the gross trajectory and the control of the grasp. The two control problems are mechanically decoupled. One or more actively controlled, compliant arms control the grasp forces in the formation allowing the robot platforms to be position controlled. The excessive forces due to platform positioning errors and odometry errors are accommodated by the compliant arms. As shown later, a designated lead robot is responsible for task planning and communicates its planned path to the follower robots. The communication between the mobile manipulators is necessary for cooperative behavior.

In the next section, we describe our control algorithms that enable forces to be applied to the object and the method for planning trajectories for the team of mobile platforms. We present experimental results to demonstrate our methods and algorithms, and to establish benchmarks for the system performance.

## 2. Controller

Our framework for controlling multiple robots is *flexible* because the robots can be organized to perform different tasks. The controllers of the robots are decentralized and they do not exchange information at servo-rates. However, they are weakly coupled because they exchange motion plans. A designated lead robot(s) plans its trajectory based on sensory information and the plan is broadcast to follower robots using a wireless Ethernet system. The follower robots are responsible for their own plans and execution. Our framework is also *scalable* in the sense that it can accommodate a larger number of robots. The number of robots is only limited by the bandwidth of the communication network. Additional robots can easily be added because each follower robot is autonomous and functions independently of the other robots. The planning and control for each mobile manipulator can be divided into three subsystems.

1. Planner: The planner receives the information broadcast from the lead robot, gets information from its sensors, and plans an appropriate reference trajectory that avoids obstacles while maintaining the tight formation. It provides setpoints for the platform controller (a reference trajectory). The trajectory must be computed in real time and is described in [5].

2. Platform Controller: The platform controller insures that the robot will follow the specified reference trajectory. Our look-ahead controller adds an error signal to the specified trajectory to compensate for errors and guarantee that the arm does not extend past its workspace boundary. For the nonholonomic platforms, a nonlinear controller based on feedback linearization is used, while the omnidirectional platform uses three decoupled linear controllers.

3. Arm Controller: The arm senses and controls the grasp forces and accommodates platform positioning errors. The arm controller is independent from the platforms and controls the internal grasp forces.

## 3. Actively Controlled Arm

In this section, we describe the actively controlled, compliant arm that controls the grasp forces in a multirobot manipulation task. The arm is a three degree-of-freedom, parallel manipulator with three limbs each consisting of a linear actuator with a spring in series. While the stiffness of the springs cannot be changed, the equilibrium position of the spring can be actively adjusted in order to control the Cartesian stiffness of the entire arm [6]. The arm is naturally compliant because of the springs and gives it the ability to control the grasp forces and counteract disturbances.

### 3.1. Force control in individual limbs

The schematic for each limb in the arm is shown in Figure 2(a), and the complete arm is shown in Figure 3. While the length $l$ is constrained by the interaction of the limb with the environment, the active control allows the regulation of the length $s$. We can control the stiffness of the limb (as shown in [6]) as well as the force applied by the link (as shown below).

18

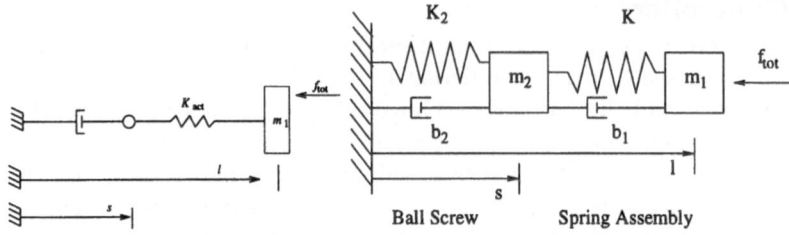

Figure 2. *(a) Diagram for one link. (b) Model of one link.*

The force applied on the mass $m_1$ is given by the physics of the spring:

$$F = -K_{act}(l - s - d) \qquad (1)$$

where $d$ equals the free length of the spring. A simple position control law, with the set point, $s_{des}$, given by:

$$s_{des} = \frac{f_{des} + lK_{act} - K_{act}d}{K_{act}} \qquad (2)$$

guarantees that the force $F$ is equal to $f_{des}$. However, this analysis does not directly address the force applied on the environment.

Figure 3. *The three degree-of-freedom, in-parallel, actively controlled arm allows forces to be applied in the $X$ and $Y$ direction as well as a moment in the $Z$ direction. Each limb has a spring attached in series to a DC motor attached to a ball screw transmission.*

To understand the behavior of each limb, an in-depth analysis of the dynamics as well as experimental measurements are presented below. A more accurate model of the system, incorporating the dynamics, is shown in Figure 2(b). In the diagram, the ball screw is modeled as a second order system with coefficients, $m_2$, $b_2$, and $K_2$. The ball screw is adjusted by a motor and controlled using a proportional and derivative controller on an embedded motor control board. The motor inertia, torque constant, and controller are lumped into the mass system. The spring assembly is modeled using the actual spring constant, $K$, a small mass on the end of the spring, $m_1$, as well as a damping constant, $b_1$, caused by the linear bearings.

The movement of the mass, $m_1$, is measured by the length, $l$, while $q$ measures the movement of the mass, $m_2$. The dynamics of the spring assembly are found by summing the forces.

$$-K(l - q) - f_{tot} = m_1\ddot{l} + b_1(\dot{l} - \dot{q}) \qquad (3)$$

where $q$ is defined: $q = s + d$, and $f_{tot}$ is the external force applied on the end effector by the environment. The simple position control law can be found from Equation (2).

$$q_{des} = f_{des}/K + l \tag{4}$$

The dynamics of the ball screw assembly are modeled using a second order system. The transfer function, $\frac{q(s)}{q_{des}(s)}$, describes the ability of the motor to move the mass, $m_2$. Because of the addition of the proportional and derivative control, the mass system is described with a spring constant as well as a damping constant.

$$\frac{q(s)}{q_{des}(s)} = \frac{K_2}{m_2 s^2 + b_2 s + K_2} \tag{5}$$

Consider the simple case in which $l$ is fixed and the desired force is varied. The equations then simplify to an easier case.

$$\frac{f_{spring}(s)}{f_{des}(s)} = \frac{K_2}{m_2 s^2 + b_2 s + K_2} \tag{6}$$

$$\frac{f_{tot}(s)}{f_{des}(s)} = \frac{\frac{b_1 K_2}{K} s + K_2}{m_2 s^2 + b_2 s + K_2} \tag{7}$$

The transfer function, $\frac{f_{tot}(s)}{f_{des}(s)}$, describes the performance of the force control system. It is worth noting that the damping constant, $b_1$, adds numerator dynamics. The friction and damping were reduced in the mechanical design to minimize the value of $b_1$.

The experimental frequency response, $\frac{f_{tot}(s)}{f_{des}(s)}$, for a single limb is shown in Figure 4 by the dashed line in the case when $l = 0$. The transfer function for actuator position versus desired actuator position, $\frac{q(s)}{q_{des}(s)}$, obtained from the actuator-spring assembly (without the mass of the end effector) is also shown on the same plot. The parameters in the system are experimentally identified: $m_2 = 1.8 \ kg$, $b_2 = 298 \ N \left(\frac{s}{m}\right)$, and $K_2 = 41220 \ \frac{N}{m}$.

The experimental frequency response for the spring force versus the desired force is shown by the solid line. The transfer functions, $\frac{f_{spring}(s)}{f_{des}(s)}$ and $\frac{q(s)}{q_{des}(s)}$, show a close match. The cutoff frequency of the system is approximately 24Hz and is basically limited by the bandwidth of the motor control subsystem.

The actual force versus the desired force, $\frac{f_{tot}(s)}{f_{des}(s)}$, (dashed line), shows a close match to the other transfer functions over a fairly wide range. There is a peak in the dashed line, which is due to the damping in the linear bearings (the term $b_1$). The experimental phase curve does not match as well and has an additional 90 degree phase lag.

## 4. Grasp Performance Measures

When multiple mobile manipulators are used to hold a grasped object, there are two, often conflicting, measures of grasp performance. First, if the robot end effectors can be accurately positioned, the system's ability to reject force disturbances is a measure of grasp stability. The grasp stiffness matrix, or a

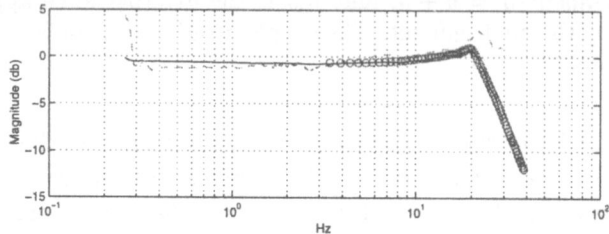

Figure 4. *Frequency response plots for $\frac{f_{tot}}{f_{des}}(i\omega)$ (dashed), $\frac{f_{spring}}{f_{des}}(i\omega)$ (solid), and $\frac{q}{q_{des}}(i\omega)$(circles).*

frame invariant measure of the minimum grasp stiffness [7], provides one choice for a performance metric. This assumption of being able to accurately position the robot end effector is generally used in the fixturing and grasping literature extensively. When there are errors in positioning and orienting the robot end effectors, because of significant platform positioning and tracking errors, it is important to choose a grasp such that the system performance is insensitive to these positioning errors. Thus, it makes sense to minimize the dependence of grasp forces on such positioning errors.

As an example, consider a symmetric grasp with two opposing, flat end effectors, with disturbances and errors along the line joining the centers of contact. A simple model for a grasp of an object of mass $m$ by robots 1 and 2 is shown in Figure 5. $K_i$ represents the effective stiffness due to the arm and contact for each robot $i$.

If the position of the grasped object is perturbed by $\Delta x$, the restoring force $\Delta F$ that restores the condition of equilibrium is maximized by maximizing the stiffness:

$$K = \frac{\Delta F}{\Delta x} = K_1 + K_2 \tag{8}$$

On the other hand, if there are positioning errors $\Delta x_i$ for each platform, the new equilibrium position for the object will be given by the position $(x + \Delta x)$ and the grasp forces $F_i + \Delta F_i$. The magnitudes of the changes in grasp forces are given by:

$$\left|\frac{\Delta F_i}{\Delta x_i}\right| = \frac{K_1 K_2}{K_1 + K_2} \tag{9}$$

The changes are minimized by minimizing the effective stiffness:

$$\kappa = \frac{K_1 K_2}{K_1 + K_2} \tag{10}$$

Thus, in the simple case of robots marching in a straight line, making both the arms (and the contacts with the grasped object) as stiff as possible will result in maximizing the measure in Equation (8), but will cause the effective stiffness in Equation (10) to increase as well. The compromise is clearly a choice of end effectors and control algorithms where either $K_1$ or $K_2$ is maximized while the other is minimized. It is clear that the combination of a compliant

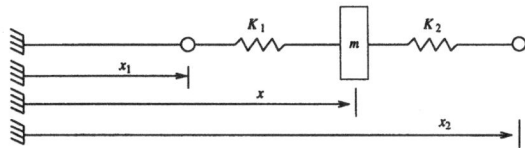

Figure 5. *Diagram for a simple spring system.*

arm with a rigid arm used in our experiments (see Figure 1) reflects this design trade-off.

There is a natural extension of these ideas from $R^1$ to $SE(2)$, the special Euclidean group of translations and rotations in the plane. First, the two stiffness measures $K$ and $\kappa$ are $(0,2)$ tensors on $SE(2)$. Because of this, it is necessary to find the smallest stiffness associated with $K$, and the largest stiffness associated with $\kappa$. Second, the determination of largest or smallest stiffness reduces to simply finding the largest or smallest eigenvalues in a generalized eigenvalue problem. As shown in [7], it is necessary to "normalize" the stiffness matrix by an inertia tensor $M$, before comparing eigenvalues. Thus, the two appropriate measures become $\lambda_{min}\left(M^{-1}K\right)$ and $\lambda_{max}\left(M^{-1}\kappa\right)$. The general problem of maximizing grasp performance is to select a grasp configuration and a combination of end effector stiffnesses that address both of these measures. It is clear that $K$ is given by an expression of the form:

$$\frac{\Delta F}{\Delta x} = M^{-1}\left(\sum_{i=1}^{n} T_i K_i T_i^T\right) \tag{11}$$

where $T_i$ is the transformation from the *ith* effector to a common reference frame. However, the expression for $\kappa$ requires a more detailed derivation. The interested reader is referred to [8].

## 5. Experimental Results

We show results from three different experiments with two robots carrying a large box as shown in Figure 1. In the first two experiments, the lead robot (robot 1) follows a desired trajectory while the rear platform (robot 2) follows while maintaining the desired formation. The arm controller is based on a stiffness law in the first experiment, while a pure force controller is used in the second experiment. In the third experiment, the role of the leader switches when the follower detects an obstacle and forces the system to change its path. This causes the follower to assume the role of the leader, and the leader becomes the follower. In the first two experiments, robot 1 is the Nomad omnidirectional robot with a stiff arm, while robot 2 is a nonholonomic platform with an actively controlled compliant arm.

The trajectories for the two platforms are shown in Figure 6. The trajectories in the first two experiments are similar and only the results from the first experiment are shown in the figure.

In the first experiment, the variation of the active arm's position and orientation is shown in Figure 7. The nominal position and orientation is

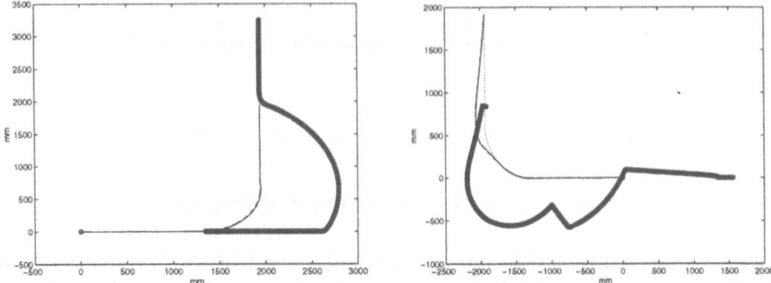

Figure 6. *Two robots carrying an object in the configuration shown in Figure 1. The actual (solid, thick) path of robot 1 and the desired (solid thin) and actual (dashed) paths of robot 2 is shown for experiment 1 (left) and experiment 3 (right). The dotted line on the right for experiment 3 gives the desired path for robot 2 when no obstacles are encountered.*

$x = 254$ *mm*, $y = 568$ *mm*, and $\theta = 0$ *degrees*. In this experiment, the stiffness control law tries to keep the arm near this home position. The actual forces and moment applied by the arm to the object are given in Figure 8.

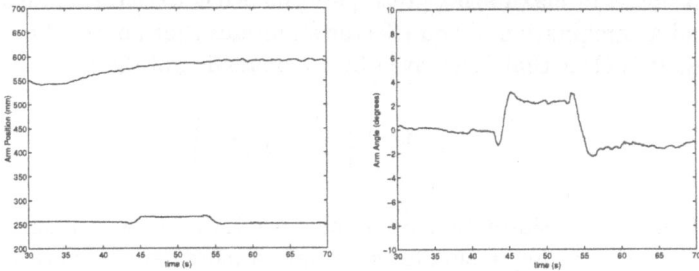

Figure 7. *Positions(left) and orientation(right) of the active arm during the first experiment using a stiffness control law.*

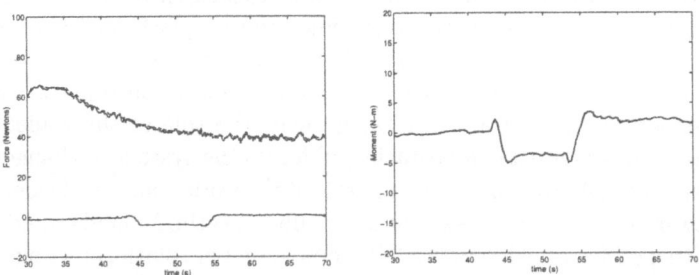

Figure 8. *Forces(left) and moment(right) applied to the object with the arm controlled using a stiffness law. Actual values are shown by a solid line while desired values are shown dashed.*

In the second experiment, the task is identical and the trajectories are therefore similar, but the arm on robot 2 is controlled using a force control law. The arm position and orientation is shown in Figure 9. Comparing this to Figure 7, it is clear that the angular variation of the arm is much greater than in the stiffness control case where restoring forces drive the arm toward the nominal orientation of zero degrees. However, from the force histories shown in Figure 10, it is evident that the forces and moment are held close to the desired values of $F_x = 0$, $F_y = 44.48N$, $M = 0$, even during the large position and orientation changes.

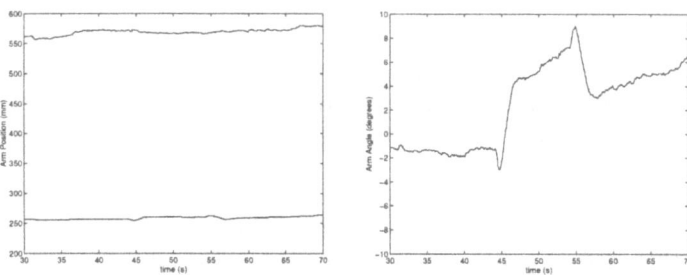

Figure 9. *Positions (left) and orientation (right) of the active arm during the second experiment using a force control law.*

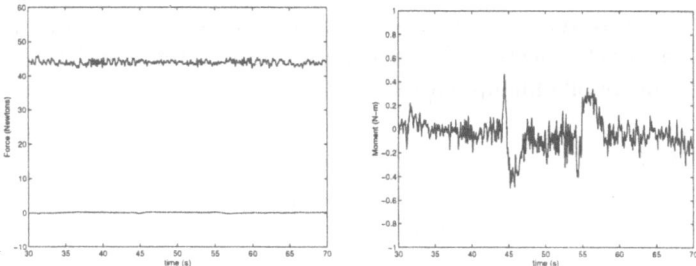

Figure 10. *Forces(left) and moment(right) applied to the object with the arm controlled using a force law.*

In the third experiment, there are three distinct stages. Robot 2 is initially the leader until robot 1, the follower, detects an obstacle that had not been accounted for by the leader. Robot 1 gains control of the team and becomes the leader, until it passes the obstacle. Then the lead platform regains control and plans a path back toward the original goal position. The flexibility in the control architecture is evident in this experiment. At all times, the arm is controlled using a stiffness control law. The trajectories are shown in Figure 6 and the forces and moment are shown in Figure 11.

## 6. Conclusion

We have developed an architecture for a team of robots which can pick up, carry, and lower different sized objects. Additional platforms can easily be added because each platform is responsible for its own safety and its precise

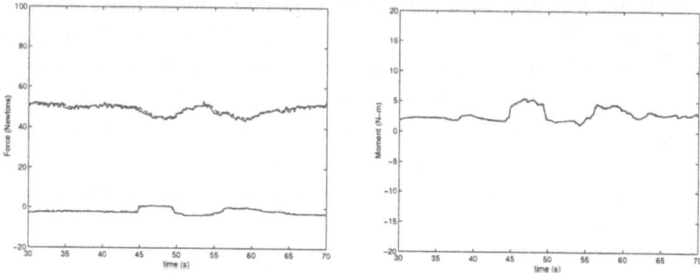

Figure 11. *Forces (left) and moment (right) applied to the object with the stiffness controlled arm on robot 2. In spite of the changes in the leader and the three transitions, the forces and moment are held relatively constant.*

position among the team of robots. A novel compliant arm is added to one platform which can easily adjust the internal forces applied to the object. It is shown that a stiff arm in conjunction with a compliant arm is beneficial for both maintaining a large stiffness as well reducing unwanted forces due to platform positioning errors. As the platforms move, the force and position control problems are decoupled by allowing the position controlled robots to follow a desired path and adding an active arm which controls the internal forces. With a systematic design of each part of the system, we are able to robustly transport objects with our experimental system consisting of two or three different mobile manipulators.

## References

[1] Sugar T, Kumar V, 1998 Decentralized control of cooperating manipulators. in *Proceedings of 1998 International Conference on Robotics and Automation*. Belgium

[2] Khatib O, Yokoi K, Chang K, Ruspini D, Holmberg R, Casal A, Baader A, 1995 Force strategies for cooperative tasks in multiple mobile manipulation systems. in *Int. Symp. of Robotics Research*. Munich

[3] Kosuge K, Oosumi T, Chiba K, 1997 Load sharing of decentralized-controlled multiple mobile robots handling a single object. in *Proceedings of 1997 International Conference on Robotics and Automation*. New Mexico

[4] Stillwell D, Bay J, 1993 Toward the development of a material transport system using swarms of ant-like robots. in *Proceedings of 1993 International Conference on Robotics and Automation*. Atlanta

[5] Sugar T, Kumar V, 1999 Multiple cooperating mobile manipulators. in *Proceedings of 1999 International Conference on Robotics and Automation*. Detroit

[6] Sugar T, Kumar V, 1998 Design and control of a compliant parallel manipulator for a mobile platform. in *Proceedings of the 1998 ASME Design Engineering Technical Conferences and Computers in Engineering Conference*. Atlanta

[7] Bruyninckx H, Demey S, Kumar V, 1998 Generalized stability of compliant grasps. in *Proceedings of 1998 International Conference on Robotics and Automation*. Belgium

[8] Sugar T, 1999 Design and Control of Cooperative Mobile Robotic Systems. PhD thesis, University of Pennsylvania, expected (August 1999)

# Experiments in Constrained Prehensile Manipulation: Distributed Manipulation with Ropes

Bruce Donald, Larry Gariepy, and Daniela Rus
Dartmouth
Hanover NH, USA
{brd,gariepy,rus}@cs.dartmouth.edu

**Abstract:** This paper describes our experiments with a distributed manipulation system. We study a system in which multiple robots cooperate to move large objects such as furniture and boxes using a constrained prehensile manipulation mode, by wrapping ropes around them. The system consists of three manipulation skills: tieing ropes around objects, affecting translations using a flossing manipulation gait, and affecting rotations using a ratcheting manipulation gait. We present experimental data and discuss the non-holonomic nature of this system.

## 1. Introduction

We are working on distributed manipulation problems, in which two or more robots apply spatially-distributed forces to rearrange the objects in their work space; the computation is distributed among the robots so as to qualify as a distributed computation. In our previous work [10, 17, 8, 9] we presented several distributed manipulation protocols for non-prehensile pushing and re-orienting of large objects, where the robots have no models of these objects, there is no explicit synchronization between their actions, and there is no explicit communication between the robots. In this paper we describe a new class of distributed manipulation algorithms in which multiple mobile robots manipulate large objects using ropes.

We are implementing a system in which multiple robots cooperate to move large objects such as furniture and boxes using a constrained prehensile manipulation mode, by wrapping ropes around them. In our system, two RWI B14 robots are linked by a long rope (5 meters) which is tied around the "waist-line" of the robots. This system can rotate and translate objects by wrapping the rope around them and pulling on the rope in a coordinated fashion. The rope acts as both coupling and indirect communication medium in this system, much like the hands of a dancing couple.

Manipulating objects with ropes has several advantages over the direct manipulation of an object using a team of robots [17, 8, 9]. First, the method allows for the parallel and synchronized manipulation of multiple objects. (That is, several objects can be wrapped together). Second, wrapping a rope around an object permits *constrained prehensile manipulation* with non-prehensile robots. The rope can conform to any object geometry on-line, in that the same wrapping protocol works for all geometries and there is no need to specify a model in advance. Finally, the rope can be viewed as a tool that allows the robot system to exert torques that are larger than the torque limit of each individual robot. Our notion of constrained prehensile manipulation is different than the classical definition of prehensile manipulation, which usually denotes a force closure grasp. In *constrained prehensile manipulation,* perturbations along any differential direction can be resisted (in our case, due to the taut rope) but the object can be moved only along certain lower dimensional degrees of freedom. Thus, constrained prehensile manipulation systems are *non-holonomic.*

## 1.1. Inspiration: Biomimetic Designs

This approach to distributed manipulation is inspired by a class of biological systems, including flagella and cilia. Many bacteria possess flagella – that is, long, protein-containing hair-like appendages that propel the bacteria through liquid media or across solid surfaces, usually toward a desired nutrient or away from a deleterious agent. In biological systems, these thin helical filaments self-assemble from 53-kd sub-units of a single protein, flagellin. An *E. coli* or *S. typhimurium* bacterium has about six flagella, which are about 150 Å in diameter and 10 $\mu$M long. Bacterial flagella are much smaller and far less complex than eucaryotic flagella and cilia. Artificial flagella and cilia could, in principle, permit robot systems to combine structure and function, and allow dynamic and automatic reconfiguration in response to environmental challenges. Such physical mechanisms allow unit modules to dynamically and automatically configure themselves into more complex robots, or for more complex tasks (for example, manipulation vs. locomotion). In particular, aggregations of flagella or cilia may be able to self-organize into the optimum configuration for a manipulation or locomotion task. Artificial *Cilia* with active control have already been demonstrated for distributed manipulation tasks on the micro scale, by our group and others [3, 4, 5]. Artificial *flagella* are more difficult to construct. As a first step towards exploring the power of distributed manipulation using artificial flagella, we have built a system of distributed robots employing synthetic *passive* flagella, which we term "ropes." These experiments allow us to "factor" the power gained in a flagellary system into two components: (1) the passive kinematics of the ropes/flagella, and (2) the additional actuated degrees of freedom gained by active control of the flagella. In this paper, we explore the power of (1), using entirely passive flagella/ropes. Note, however, that the flagella/ropes can move and effect manipulation when their "parent" robots move; however, they cannot be independently actuated. Thus, our work explores a new kind of underactuated distributed manipulation system based on a simple biomimetic model. As we will see later, this system may be encoded and analyzed as a non-holononmic system for distributed manipulation.

## 2. Distributed Manipulation with Ropes

Consider the task whose goal is to change the pose of a large box, or a group of boxes. A team of robots can use non-prehensile strategies such as those described in [17, 8, 9] to accomplish this task. A different approach is to use a constrained prehensile strategy. Because the objects we wish to move are large, it is impractical to envision robots equipped with big enough grippers to do such manipulation tasks. Instead, we focus on strategies that augment the basic architecture of mobile robots with simple tools. In this section we describe an approach where the robots can achieve a grasp on large boxes or collections of objects by wrapping the objects with a rope (independent of the overall geometry). We then explore cooperative strategies for manipulating an object wrapped with a rope. Our experimental infrastructure is described in Section 2.1. Section 2.2 describes a system that implements such strategies. Section 3 describes our experimental results with this system. Section 4 describes some interesting issues related to the geometric foundations of motion control [15] in the context of a mechanical analysis for these motions.

### 2.1. Experimental Infrastructure

Our distributed manipulation team consists of three mobile robots: two RWI B14 robots equipped with sonar, discrete contact sensors, IR, and radio modems; and one RWI Pioneer robot equipped with sonar, a radio modem, and a parallel-jaw gripper. The B14 robots are named Bonnie and Clyde and the RWI Pioneer robot is named Ben. We augmented the basic hardware of these robots by using a long rope (over 5 meters) which is tied at one end to Bonnie and at the other end to Clyde. We incorporated a tool in Ben's gripper that extends the effective width of the gripper from 2 cm. to 15 cm.

### 2.2. A Distributed System for Manipulation with Ropes

The distributed manipulation system consists of three robots: Bonnie and Clyde, which are linked by a long rope (5 meters) tied around their "waist-lines", and Ben, who is free to move and carries a rope grasping tool. When an object is wrapped and constrained by the rope, we say it is *bound*. Bonnie and Clyde are unable to bind an object by themselves; therefore Ben acts as an "enzyme" to catalyze the binding "reaction".

The distributed manipulation system developed using the three robots consists of three skills for manipulating objects with ropes: tieing a rope around a large object or a collection of objects, effecting rotations on a bound object, and two ways of effecting translations on a bound object. The control for each skill is implemented as a protocol using active and passive robots. We now detail the components of this system.

#### 2.2.1. Binding: Wrapping the rope around an object

The goal of this skill is for a team of two robots to surround an object with a rope to generate a *rope grasp*. The ends of the rope are tied along the "waistline" of each robot. The basic idea of the algorithm is to keep one robot stationary while controlling the other robot to move around the object with the rope held taut. At some point, the moving robot has to cross over the

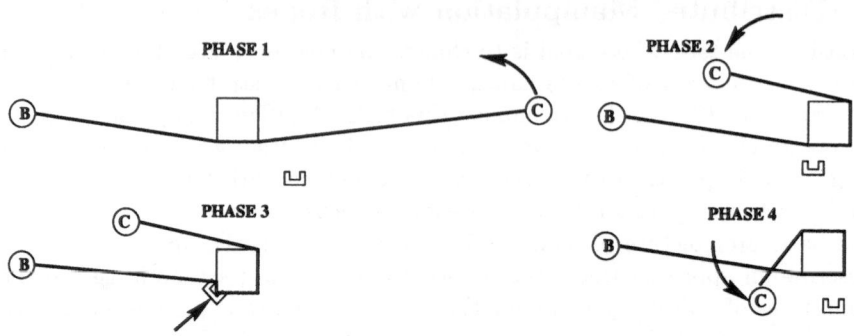

Figure 1. The 4 phases of binding (wrapping a rope around a box). The RWI robots are denoted by disks. Bonnie is denoted by B; Clyde is denoted by C. The Pioneer helper is denoted by the gripper icon. *Translation*: From the state obtained in Phase 4, translations can be effected by having Bonnie and Clyde move with parallel, synchronous velocities. For example, from Phase 4, the box can be translated 'southwest' by commanding the same motion to B and C. For pure translations, B and C should be equidistant from the box. That can be accomplished using a ratchet (regrasp) step.

rope in order to complete the grasp; this step requires coordination with the stationary robot.

We have developed a control algorithm for this skill and implemented it using two RWI B14 robots and one Pioneer robot (see Figure 1). In the first phase, the B14 robots are on the same side of the box. Clyde, the active robot (the right robot in Figure 1) moves so as to keep the rope taut around the box. Bonnie, the left robot in Figure 1, is passive and stays in place. The control for encircling the box is implemented as hybrid control of a compliant rotation. The robot starts with a calibration step to determine the level of power necessary to achieve rotational compliance to the box. The robot tries to move in a polygonal approximation to a circle. Due to the radial force along the rope, the robot's heading complies to remain normal to the radial force, thereby effecting a spiral. The robot continues this movement until it gets close to having to cross the rope. The following function implements a half-tour of the box:

```
(define (half-surround-box)
  (let* ((initial-heading 0))
    (rotate-power (calibrate 1 200))
    (set-translate-velocity! 200)
    (set! initial-heading (rotate-where))
    (let repeat ()
        (translate-relative 100)
        (if (and (= 0 (contact-mode)) ;Stop after 165 degrees
                 (< (abs (- initial-heading (rotate-where))) 165))
        (repeat)
        (stop)))))
```

The first phase ends when the encircling robot has completed a 165 degrees tour around the box. In the second phase, Clyde crosses the rope. The algorithm requires Bonnie to drop the rope to the ground to enable Clyde to drive over the rope. A simple forward movement of Bonnie would cause the rope to drop to the ground, but it will break the grasp. Furthermore, the inverse move is not guaranteed to reestablish the grasp because the rope can slide under the box. Thus, it is important to maintain the partial grasp on the box during the crossover. Our approach is to use a Pioneer helper robot, that can support the rope while the passive robot gives slack.

The algorithm for the second phase starts with Ben locating the corner of the object closest to the Bonnie. Ben uses its gripper to push the rope against the box, thus securing the rope grasp. Then Bonnie gives some slack in the rope. In the third phase, Clyde crosses over the rope. In the fourth phase, Bonnie tightens the rope. The tension in the rope signals the helper to let go and move away.

### 2.2.2. Translating a bound object

In the translation skill, two robots translate a bound object by moving with parallel, synchronous velocities (see Figure 1). The basic idea of this algorithm is to regrasp start by regrasping a bound object so that (1) the two robots are equidistant from the box along a line called *the rope line* and (2) the rope line is perpendicular to the desired direction of translation. These two properties are achieved by using the active part of wrapping algorithm: keeping the rope taut, both robots rotate counter-clockwise about the box until the rope line becomes perpendicular to the desired direction of translation. At this point, a regrasp step is necessary to ensure that both robots are equidistant along the new rope line. From this position, the robots can translate by commanding synchronous velocities.

### 2.2.3. Flossing: rotating a bound object

In the flossing skill, two robots rotate an object in place by coordinating movements of the rope between them. This type of manipulation is similar in flavor to the pusher/steerer system described in [6]. Flossing is preceded by a binding operation. We have developed a control algorithm for this skill and implemented it using two RWI B14 robots (see Figure 2). In the first phase the active robot translates forward. The passive robot follows along, feeling force $F_p$ and keeping the rope taut. This action causes the object to rotate. This phase terminates when the passive robot stops in place, thus increasing the tension in the rope. The active robot senses this impediment and stops. At this point, the robots can reverse the passive/active roles and repeat the first phase. When Bonnie is the active robot, the object rotates clockwise; when Clyde is the active robot, the object rotates counterclockwise.

### 2.2.4. Ratcheting: translating a bound object

In the ratcheting skill, two robots translate an object by coordinating movements of the rope between them. As in the flossing skill, this type of manipulation is also preceded by a binding operation. The manipulation system may also combine ratcheting manipulation with flossing manipulation.

Figure 2. The 2 phases of rotating an object by "flossing". The robots are denoted by disks. $V_a$ denotes the velocity of the active robot. $V_p$ denoted the velocity of the passive robot. $F_p$ is the force felt through the rope by the passive robot.

We have developed a control algorithm for this skill and implemented it using two RWI B14 robots (see Figure 3). The algorithm for translations iterates the following phases. In the first phase the active robot translates forward with velocity $v_a$. The passive robot follows along feeling $F_p$ and keeping the rope taut. This action causes the robot to translate. When this phase is executed for the first time, it terminates with the passive robot sensing contact with the object. This is called the *calibration step*. During a calibration step, the passive robot measures the approximate distance to the object. In subsequent repetitions of this phase, the passive robot uses dead reckoning to decide when to stop.

In the second phase the robots decrease the tension in the rope and translate together in the opposite direction. The object does not move in this phase and the rope slips against the object. The effect of this phase is to re-grasp the object so as to move the active robot closer to the object.

### 2.3. Details of Manipulation Algorithms

Please see http://www.cs.dartmouth.edu/~brd/Research/iser99/ for details.

Figure 3. The 2 phases of translating an object with a ratchet motion. The robots are denoted by disks. $V_a$ denotes the velocity of the active robot. $V_p$ denoted the velocity of the passive robot. $F_p$ is the force felt through the rope by the passive robot.

## 3. Experiments

We have implemented the distributed manipulation skills described in Section 2.2 using the robots described in Section 2.1. The RWI robots are programmed in MOBOT-SCHEME [16] and the helper Pioneer robot is programmed in C++. Figure 4 show some snapshots from a typical run.

Figure 5 shows reliability data for some recent experiments run with this system. The failures of the binding skill were related to sensor inaccuracies. The majority of the observed errors were due to the fact that Clyde, the active robot during binding, uses odometry to terminate the first phase of the algorithm. This method introduces an error of up to 3 feet in the positioning of

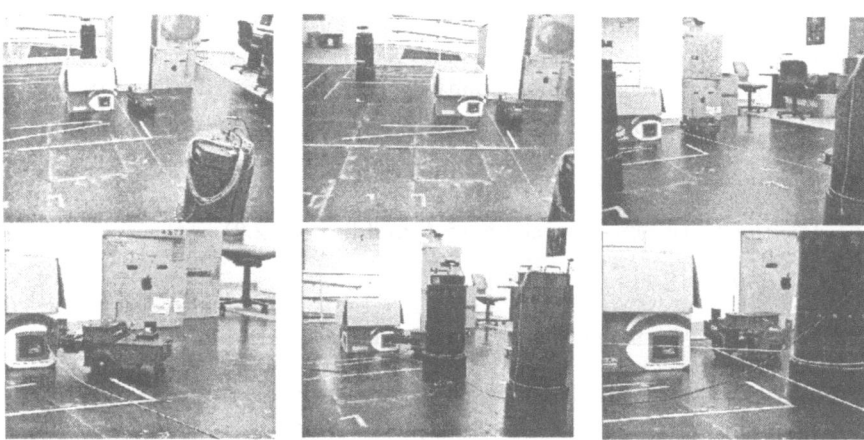

Figure 4. Six snapshots from an execution of binding with ropes. The top-left figure shows the initial configuration for the binding experiment. The top-middle figure shows the robot Clyde encircling the box. The top-right picture show that Clyde has come halfway around the box and Ben is ready to help with the knot tieing component of the experiment. The bottom-left picture shows that Ben has traveled to the box corner and is currently holding the rope. In the bottom-middle picture Clyde crosses over the rope. In the bottom-right image the box is wrapped with the rope. We have also implemented the translation motion (Figure 1), the ratchet motion (Figure 3) and the flossing motion (Figure 2) once the object is bound.

| Task | Successes | Tries | Reliability |
|------|-----------|-------|-------------|
| Binding | 29 | 40 | 72.5 % |
| Flossing | 19 | 21 | 90.5 % |
| Ratcheting | 19 | 26 | 73 % |
| Transitions | 8 | 11 | 72.7 % |

Figure 5. This table contains reliability data for our experiments with binding, and for manipulation by flossing and ratcheting. The last row shows data for transitions from binding to flossing and from binding to ratcheting.

the robot relative to the rope and leads to the failure of the rope crossing step. The rest of the errors were due to the helper robot, Ben, who uses odometry to locate the corner of the box. Occasionally, Ben misses the corner and thus does not get a good grasp of the rope. The failures in flossing and ratcheting were due to signaling errors. Occasionally, the passive robot stops prematurely. We believe that these errors are due to low battery levels during those specific experiments, but more investigation is necessary.

## 4. Analysis

In this section we provide a brief analysis of the manipulation grammar and the non-holonomic nature of the rope manipulation system.

## 4.1. Manipulation Grammar

It is possible to develop a simple manipulation grammar [3] to describe our manipulation system. The manipulation algorithms described above provide four basic skills:

B. *Binding* (Section 2.2).

T. *Translation* (Figure 1, post-Phase 4, and Section 2.2),

O. *Orientation* (Section 2.2), and

G. *Regrasping* (when the rope slips along the box in Section 2.2).

Let us call these primitives $B$, $T$, $O$, and $G$, respectively. If $O$ is an orientation step, let $O^R$ be the *reverse* step, performed by reversing the robots' direction. In this case, flossing can be written as $(O, O^R)^*$, the repeated iteration of $O$ followed by $O^R$. Note that no direct communication is required to switch between $O$ and $O^R$; the signal is transmitted through the task (the rope).

Similarly, the ratcheting operation can be written as $(O \mid T, G)^*$. Again, no direct communication is required to switch between regrasp and orientation/translation steps. To summarize:

1. *Flossing* $= (O, O^R)^*$.

2. *Ratcheting* $= (O \mid T, G)^*$.

All operations must be preceeded by a binding step $B$. For pure translations, Bonnie and Clyde should be equidistant from the box. That can be accomplished inserting a regrasp step $G$. Thus, a simple translation could be accomplished by $(B, G, T)$.

## 4.2. Non-Holonomic Manipulation Systems

In mechanics, systems may be *holonomic* or *non-holonomic*. In general, a *holonomic constraint* is a wholly integrable sub-bundle $E$ of the tangent bundle. The system outcome for a non-holonomic system is path-dependent. Non-holonomic systems have been studied in robotics. Examples include: car-like robots, tractor-trailers, bicycles, roller-blades, airplanes, submarines, satellites, and spherical fingertips rolling on a manipulandum. In robotics, a non-holonomic system is usually defined by a series of non-integrable constraints of the form $\Gamma_i(p, v) = 0$ on the tangent bundle [1, 13, 2, 11, 14]. For example, whereas holonomic kinematics can be expressed in terms of algebraic equations which constrain the internal, rotational coordinates of a robot to the absolute position/orientation of the body of interest, non-holonomic kinematics are expressible with differential relationships only. This distinction has important implications for the implementation of a control system.

It is well known (see Figure 6) that many underactuated manipulation systems are naturally modeled as non-holonomic systems. Figure 6 illustrates a *one-sided* non-holonomic constraint: $R$ can push $O$ but not pull it. Therefore

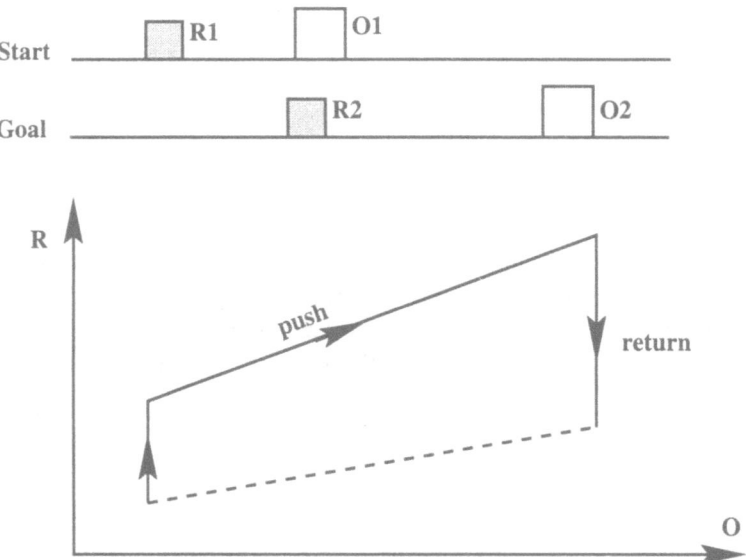

Figure 6. An underactuated manipulation system may be modeled as a non-holonomic system (simplified example from [7, pp.40-42,260-271], based on a suggestion of Koditschek). Here a robot $R$ starting at $R_1$ must push an object $O$ from configuration $O_1$ to $O_2$. The goal state of the robot is $R_2$. Both $R$ and $O$ have one-dimensional configuration spaces, so their combined configuration space is $\mathbb{R}^2$. If the system were fully actuated, then $R$ could move from $R_1$ to $R_2$ while $O$ simultaneously moved from $O_1$ to $O_2$ (dotted path). Since $O$ is not actuated, $R$ must first move to $O_1$, then push $O$ from $O_1$ to $O_2$, and finally return to $R_2$ (solid path). Not all degrees of freedom (DOF) are available to $R$ (for example, diagonal pushing paths are only possible when $R$ and $O$ are in contact). However, by a series of maneuvers along the reduced differential DOF, higher DOF strategies may be carried out. This is the essence of a non-holonomic system.

only 'northeast' diagonal pushing motions are possible. On the other hand, if $R$ could *pull* $O$, then diagonal pushing motions in either direction (northeast or southwest) are possible. This would define an *isotropic nonholonomy*.

Our system admits an isotropic non-holonomic model as follows: Let $R_1$ and $R_2$ be the positions of Bonnie and Clyde (respectively), and let $r$ be one half the diameter of the manipulandum (for a sensible definition of the *diameter* of a polygon, see [12]). Although in general $R_i$ is two dimensional, for the sake of illustration, assume that $R_1, R_2 \in \mathbb{R}$. With the rope taut, the distance between $R_1$ and $R_2$ may be closely approximated as fixed once the box is wrapped. Thus the combined, simultaneous configurations of $R_1$ and $R_2$ may be collapsed into a single parameter $d \in \mathbb{R}$, representing the *rope position*. Finally, let $\theta$ be the orientation of the box, and $x$ its translational position.

Now, let us consider a single orientation step $O$ (Figure 7-left). When the robots move to manipulate the box, it translates and rotates in a non-linear

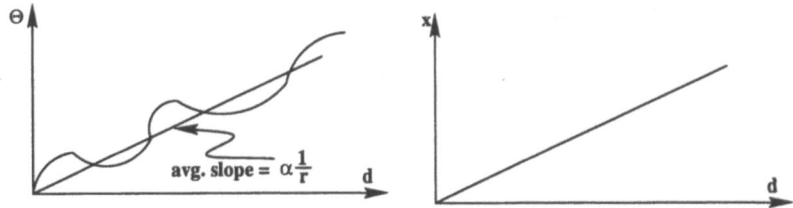

Figure 7. (Left) Non-holonomic Model of a single Orientation Step $O$. The orientation $\theta$ of the box as a function of the control parameter $d$, modeled in the non-holonomic configuration space $\mathbb{R} \times S^1$. This figure shows one half $(O)$ of an orientational flossing cycle $(O, O^R)$. (Right) Non-holonomic Model of a Translation Step $T$. $x$ represents the position of the box.

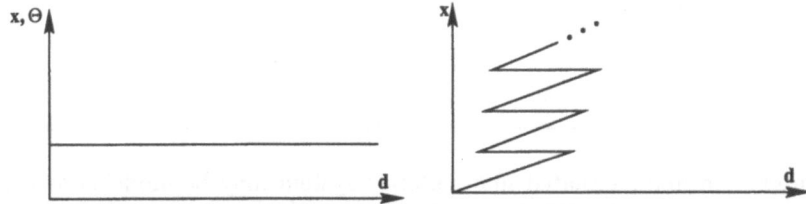

Figure 8. (Left) Non-holonomic Model of a Regrasp Step $G$. When the rope is slightly slack, it slips along the box without affecting its pose. (Right) The translational ratchet. Non-holonomic Model of interleaved translational and regrasp steps, $(T, G)^*$.

fashion that depends on the box geometry (Figure 7). For the orientation task, the progress this strategy makes is the average slope of the linearization of this curve in $d$-$\theta$ space. This slope is approximately $(2\pi r)^{-1}$.

Non-holonomic analyses can also be done for translation $(T)$, flossing $(O, O^R)^*$), translational ratcheting $(T, G)^*$, and orientational ratcheting $(O, G)^*$. (Figures 7, 8, and 9).

By interleaving regrasp steps with translation steps, the box can be translated repeatedly; the motion in configuration space is like a 'ratchet' (Figure 8). Note that not all of the degrees of freedom (DOF) in the configuration space are accessible; only a 1D subset of the DOF are differentially possible. Therefore, this kind of ratchet maneuver must be employed to access all the DOF. Note that the ratchet maneuvers are similar to parking maneuvers common in non-holonomic motion planning for wheeled vehicles [13, 2, 11, 14]. A similar picture is obtained by interleaving regrasp steps with orientation steps, except the orientation steps are non-linear (Figure 9). By interleaving all three kinds of steps (translational, orientation, and regrasp), arbitrary $(x, \theta)$ poses of the box can be achieved. This corresponds to a non-holonomic ratchet in the 3-dimensional configuration space $\mathbb{R}^2 \times S^1$ parameterized by $(d, x, \theta)$ (Figure 9). Both 2D and 3D ratchets can be viewed as manipulation gaits in a non-holonomic control space.

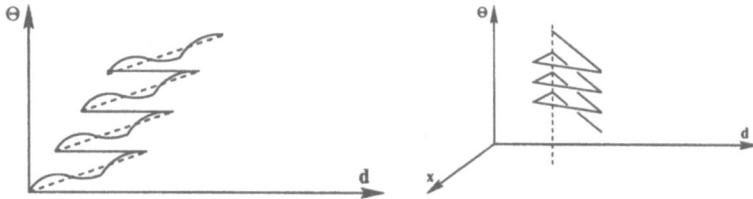

Figure 9. (Left) The orientational ratchet. Non-holonomic manipulation gait of interleaved orientational and regrasp steps $(O, G)^*$.
(Right) The 3D ratchet in the configuration space $\mathbb{R}^2 \times S^1$ is a non-holonomic manipulation gait composed of interleaved translational, rotational, and regrasp steps $(O, T, G)^*$.

## 5. Summary

In this paper we described the constrained prehensile manipulation of an object bound by a rope. We presented four algorithms for the distributed manipulation of a roped object: binding (that is, tieing a rope around an object), translations, flossing (that is, effecting rotations of a bound object), and ratcheting (that is, effecting translations of a bound object). Our algorithm use a team of three robots that do not communicate directly but use the tension in the rope as a signaling mechanism. The role of the three robots employed by our systems can be separated into: active, passive, and helper. The algorithms consist of phases that can be repeated by switching the roles of the active and passive robots. We have also discussed briefly some interesting issues related to the geometric foundations of motion control in the context of a mechanical analysis for these motions.

We implemented the three manipulation skills and experimented with a distributed system of three robots that can bind an object, rotate a bound object in place, and translate a bound object. In our system, the robots can also make transitions between these skills: a binding operation can be followed immediately by a flossing or ratcheting operation. Flossing and ratcheting can also be alternated. Our experimental results show levels of reliability between 72.5 % and 90 % for the manipulation skills.

## Acknowledgements

This paper describes research done in the Dartmouth Robotics Laboratory. Support for this work was provided in part by the following grants from the National Science Foundation: NSF CAREER IRI-9624286, NSF IRI-9714332, NSF IIS-9818299, NSF 9802068, NSF CDA-9726389, NSF EIA-9818299, NSF CISE/CDA-9805548, NSF IRI-9896020 and NSF IRI-9530785. We are grateful to Keith Kotay for his help in many aspects of this work.

## References

[1] Baillieul, Brockett, B. R. Donald, and *et al*. *Robotics*, volume 41 of *Symposia in Applied Mathematics*. American Mathematical Society Press, Providence, RI, 1990.

[2] J. Barraquand and J.-C. Latombe. Nonholonomic multibody mobile robots: Controllability and motion planning the the presence of obstacles. *Algorithmica*, 10(2):121–155, August 1993. Special Issue on Computational Robotics.

[3] K.-F. Böhringer, B. R. Donald, and N. C. MacDonald. Programmable vector fields for distributed manipulation, with applications to MEMS actuator arrays and vibratory parts feeders. *IJRR* 18(2):168–200, February 1999.

[4] K.-F. Böhringer, B. R. Donald, N. C. MacDonald, G. T. A. Kovacs, and J. W. Suh. Computational methods for design and control of MEMS micromanipulator arrays. *Computer Science and Engineering*, pages 17–29, January – March 1997.

[5] K.-F. Böhringer, J. W. Suh, B. R. Donald, and G. T. A. Kovacs. Vector fields for task-level distributed manipulation: Experiments with organic micro actuator arrays. pages 1779–1786, Albuquerque, New Mexico, Apr. 1997.

[6] K. Bohringer, R. Brown, B. Donald, J. Jennings, and D. Rus. Information Invariants for distributed manipulation: experiments in minimalism *Experimental Robotics IV*, Lecture Notes in Control and Information Sciences 223, eds. O. Khatib and K. Salisbury, pp. 11-25, Springer-Verlag, 1997.

[7] B. R. Donald. *Error Detection and Recovery in Robotics*, volume 336 of *Lecture Notes in Computer Science*. Springer Verlag, Berlin, 1989.

[8] B. Donald, J, Jennings, and D. Rus. Minimalism + Distribution = Supermodularity. *Journal of Experimental and Theoretical Artificial Intelligence*, 9(1997) 293-321, 1997.

[9] B. Donald, J. Jennings, and D. Rus. Information invariants for cooperating autonomous mobile robots. *International Journal of Robotics Research*, 16(5) 673-702, 1997.

[10] B. Donald, J. Jennings, and D. Rus. Information invariants for distributed manipulation. *The First Workshop on the Algorithmic Foundations of Robotics, eds. K. Goldberg, D. Halperin, J.-C. Latombe, and R. Wilson*, pages 431–459, 1994.

[11] K. Goldberg, J.C.-Latombe, D. Halperin, and R. Wilson, editors. *The Algorithmic Foundations of Robotics: First Workshop*. A. K. Peters, Boston, MA, 1995.

[12] K. Y. Goldberg. Orienting polygonal parts without sensing. *Algorithmica*, 10(2/3/4):201–225, August/September/October 1993.

[13] J.-C. Latombe. *Robot Motion Planning*. Kluwer Academic Press, 1991.

[14] J.-C. Latombe. Robot algorithms. In T. Kanade and R. Paul, editors, *Robotics Research: The Sixth International Symposium*, 1993.

[15] J. Ostrowski and J. Burdick. Geometric Perspectives on the Mechanics and Control of Robotic Locomotion. *Proceedings of the 1995 International Symposium on Robotics Research.*

[16] J. Rees and B. Donald. Program Mobile Robots in SCHEME. *Proceedings of the 1992 International Conference on Robotics and Automatio.* Nice, France 1992.

[17] D. Rus, B. Donald, and J. Jennings. Moving furniture with teams of autonomous robots. *Proceedings of the 1995 Conference on Intelligent Robot Systems.* 1995.

# Experiments with Desktop Mobile Manipulators

Matthew T. Mason
Carnegie Mellon University
Pittsburgh PA, USA
matt.mason@cs.cmu.edu

Dinesh K. Pai
University of British Columbia
Vancouver, BC, Canada
pai@cs.ubc.ca

Daniela Rus and Jon Howell
Dartmouth
Hanover NH, USA
rus@cs.dartmouth.edu and jonh@cs.dartmouth.edu

Lee R. Taylor and Michael A. Erdmann
Carnegie Mellon University
Pittsburgh PA, USA
lrt+@andrew.cmu.edu and me@cs.cmu.edu

**Abstract:** This paper describes our work on Desktop Robotics. The main focus is two robots that locomote and manipulate paper on a desktop. One robot uses wheels for both manipulation and locomotion. The other robot uses wheels for locomotion and a sticky foot to lift and carry pieces of paper. We outline the goals of our work on desktop robotics, describe the design of the robots built so far, present experimental data, and outline some of the issues for future work.

## 1. Introduction

Desktop robotics addresses applications of robotics in office environments: storage and retrieval of documents and other desktop paraphernalia, integration of hardcopy documents with computer files, or perhaps routine maintenance and cleaning tasks. The goal is to create small robots that can be plugged into a desktop computer with the same ease as a camera, a CD ROM, or any other peripheral device. We believe that motors, sensors, and electronic components have become small enough and inexpensive enough to build robotic systems that would be practical and useful in an office environment. Besides the possible practical value of such systems, we are interested in the fresh perspective on research issues. We believe that ideas arising from this work may be useful in a variety of applications, for example manipulating flexible objects, designing prosthetic devices, or human-robot interaction.

A particular task might be to keep a desktop organized: to store, retrieve, or discard items on demand; to stack papers neatly; to scan papers or books; and so on. This is a complex problem which may ultimately involve a large number of different sensor and motor systems. Closely related work [15] describes a system that uses a camera to capture electronically and index the contents of the papers contained on a desktop. In the future, we envision combining this previous work with a robot (perhaps some variation on the robots described in the present paper) to address the larger problems of desktop robotics.

This paper focuses on two robots for manipulating paper, and perhaps other objects, on a desktop. The first is a mobile manipulator that we call the *mobipulator*. It looks like a small car, with four wheels independently driven, none of them steered (Figure 1). It uses its wheels both for locomotion and manipulation. The second robot, called *Fiat*, can manipulate paper on a desktop, even when the surface is crowded with several sheets of paper. Fiat has a sticky foot, which it uses to lift a sheet of paper, carry it to a designated location of the desk, and place it down on the desk. Section 2 describes the Mobipulator, and Section 3 describes Fiat. Section 4 address previous work, and Section 5 is a discussion and conclusion.

## 2. The Mobipulator

The Mobipulator (Figure 1) looks like a small car, with four independently powered wheels. None of the wheels are steered. We envision several different modes of manipulation/locomotion, of which five have been demonstrated so far:

- **Translation mode.** This mode is pure locomotion. To translate forward or backward, all four wheels are driven at equal rates.

- **Dual diff drive mode.** This mode (Figure 2) combines locomotion and manipulation. To maneuver a piece of paper on a desktop, one pair of wheels drives the paper relative to the robot, while the other pair drives the robot relative to the desktop.

- **Inchworm mode.** This mode (Figure 1) is pure manipulation. With all four wheels on the paper, by quick alternation of the front wheels and rear wheels, the paper can be advanced incrementally beneath the robot.

- **Cylinder rolling mode.** This mode (Figure 2) is inspired by dung beetles. The robot's front wheels are placed on a cylinder. It uses its rear wheels to propel itself forward while its front wheels turn in the opposite direction to roll the cylinder forward.

- **Scoot mode.** With all four wheels on the paper (Figure 1) the robot accelerates hard and then decelerates hard. During the high acceleration the paper slips backward, and during the high deceleration the paper slips forward. There is a net forward motion of the paper.

While we have seen all of these techniques work in the lab, so far we have performance data only for dual diff drive mode.

Figure 1. (Left) Mobipulator I, shown in scoot mode on a sheet of US standard letter paper. (Right) Inchworm mode.

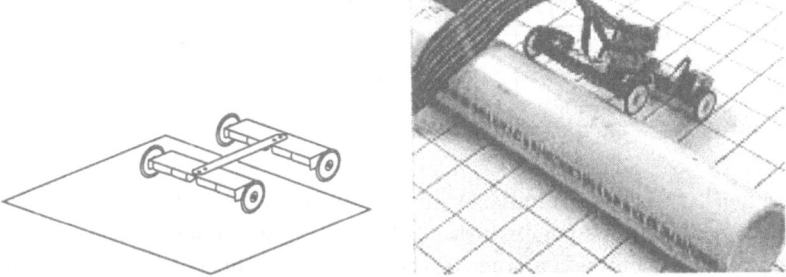

Figure 2. (Left) Dual diff drive mode. (Right) Cylinder rolling mode.

A typical scenario would switch modes freely, to avoid the limitations inherent to any single mode. The robot might begin with translation mode (and perhaps some slip-steering) to roll on to a piece of paper. Then would follow a sequence of dual diff-drive motions alternating with translations, to coarsely position the robot. Finally, the robot can use dual diff-drive mode to perform some closed-loop fine positioning of the paper. More exotic (and highly speculative) scenarios involve the robot using inchworm mode to maneuver a page onto a stack of paper, or using other modes to turn the pages of a book.

These scenarios will require research progress in a number of areas: the interaction between tire and paper, the behavior of a hump in the paper during inchworm mode, and planning techniques that integrate a variety of fundamentally different actions.

### 2.1. Experiments with the Mobipulator

The goal of our experiments was to test the basic concept of dual diff-drive mode, and specifically to see whether unwanted slip would be a problem.

The car has a square wheelbase of about 125mm along each edge. Each wheel is driven by a DC gearmotor with incremental encoder. The front half of the car is joined to the rear half by a thin piece of steel, providing a suspension to evenly distribute the load among the wheels. The tires are rubber O-rings. The "desktop" was cardboard, and the "paper" was mylar.

The mobipulator was operated in dual diff drive mode, moving the square piece of mylar along a path in the shape of a rounded square, while holding the mylar orientation constant. The motions had trapezoidal velocity profiles, with synchronized changes in acceleration. Each wheel was controlled independently by a PID servo. To record the motion, we hand-digitized data taken with a video camera positioned above the desk.

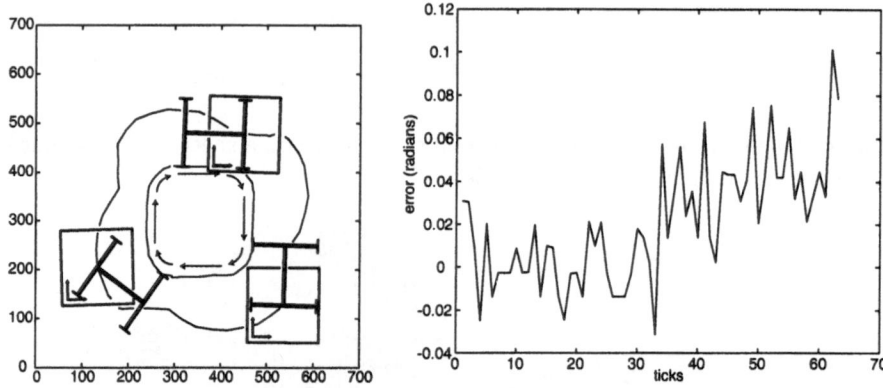

Figure 3. (Left) The mobipulator driving a rounded square, while holding the orientation of a mylar page constant. One curve is a trace of the mylar sheet center; the other curve is a trace of the right back wheel. Dimensions are mm. (Right) Control error for page orientation during the motion of the left figure. The page orientation was commanded to be a constant zero, so any motion is an error. "Ticks" is a nonuniform sampling of time. Duration of the motion was one minute, with about half that time spent at rest. After rotating $2\pi$ relative to the robot, the max error is around 0.1 radians, or about $6°$.

Our conclusions were that unwanted slip is bad enough to require additional sensors, but not bad enough to prevent effective manipulation and locomotion. We also discovered that we need a better approach to control, coupling the wheel servos and attending to the nonholonomic nature of the device.

## 3. Fiat: the paper-lifting robot

The Mobipulator can move single sheets of paper in the $x$ and $y$ directions when the papers rest directly on a desktop. To cope with areas of desktops that are crowded with stacks of paper we designed the mobile robot called *Fiat* (see Figure 4). In the future, we plan to combine the functionality of Fiat and the Mobipulator into a single mobile robot.

Fiat can lift a single piece of paper from a stack and move it to a different location on the desktop. The mechanical design for Fiat was motivated by the goal of moving paper in the vertical direction, in order to stack sheets one above the other deterministically. The manipulator of this robot is a foot made sticky with a supply of removable tape. A spool of removable adhesive tape provides a supply of "fresh stickiness" for the foot (see Figure 4).

The manipulator needs a way to attach the foot to a piece of paper and to release the paper from the foot; this is accomplished by integrating the manipulator with the locomotive design. The foot travels on an eccentric path relative to the free rear wheels. In one direction, the foot emerges below the wheels, using the weight of the robot to stick the foot to the paper. As the eccentric motion continues, the foot moves behind the wheels, and lifts up the paper. The opposite motion detaches the foot from the paper, using the rear wheels to hold down the paper while the sticky foot is freed. Notice that the wheels have been placed over the edge of the sheet of paper, holding it in place and pulling it free from the sticky foot. The result is that we can reliably grab and release a flexible sheet of paper using only a single rotary motion.

Figure 4. (Left) A picture of the paper lifting robot. (Right) Details of the sticky foot mechanism of the robot Fiat.

There are four total actuators (steering servo, drive motor, cam motor, tape winding motor). There are also two internal sensors: one shaft encoder to allow dead reckoning motion, and one optical transmissive sensor to detect the phase of the cam assembly.

The total weight of this robot is 2.7 kg (5.9 pounds), of which the batteries account for 0.6 kg (1.4 pounds). Exterior dimensions are approximately 29 cm long by 23 cm high by 17 cm wide (11.5 x 9 x 6.7 inches). Our fabrication method permits rapid redesign and prototyping. We expect that each linear dimension will be halved by the next design revision, to enable the robot to operate in much smaller spaces.

The Fiat robot contains a processor that is used for controlling the motors and future sensors for this robot. A 3Com Palm III personal organizer serves as the main processor on the robot. We chose this approach for several reasons: (1) A PalmOS/Motorola 68000 compiler is readily available for open systems; (2) Its built-in user interface makes "field tests" very convenient; (3)It has a built-

42

in infrared data port, which allows the robot to interface to an external camera hosted by a computer without a tether; (4) Its generous 4MB of memory and 16MHz processor come in a small, inexpensive package that is hard to match with custom components; (5) It consumes very little power; and (6) Its "Hot Sync" function serves as a very convenient bootstrap mechanism. Figure 6 shows the control components for this robot.

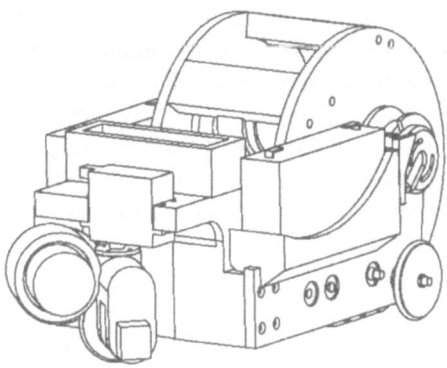

Figure 5. The CAD model of the paper-lifting robot Fiat used for fabrication.

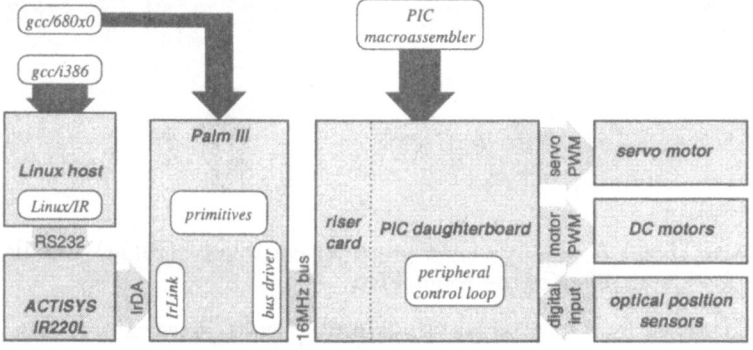

Figure 6. The schematic diagram of the paper-lifting robot Fiat. The Palm III interfaces with an expansion board that contains a 4MHz Microchip PIC 16C74A microcontroller and I/O pins through the Palm's memory bus.

## 3.1. Control and Capabilities

The basic capabilities of Fiat consist of planar movement on a desk top, lifting a paper, releasing a paper, and advancing the tape on the sticky foot. The following primitives were implemented on the Palm III to effect these capabilities:

1. `drive(float distance)` moves the robot straight for `distance` centimeters, dead reckoning with the drive wheel's shaft encoder. Speed is ramped along a trapezoidal envelope.

2. `rotate(float angle)` turns the steering wheel perpendicular to the rear wheels and rotates the robot `angle` degrees around the midpoint of the rear wheels, the location of the sticky foot when retracted.

3. `paperPickup()` winds the sticky tape spool a little, then rotates the cam out to pick up a sheet of paper. Each rear wheel should be approximately at the edge of the paper.

4. `paperDrop()` rotates the cam inward, detaching the paper from the foot.

Using these primitives, the robot can move a sheet of paper from one side of a desk to the other. Since the leading edge of the paper is at least a centimeter above the lowest point on the rear wheels, the motion of the paper is over other sheets of paper that are resting on the desk.

### 3.2. Experiments

We have run two sets of performance tests on the Fiat robot.

A first set of experiments was designed to quantify the reliability of shifting a single piece of paper on a desk top. Placement can be implemented as a sequential execution of `paperPickup()`, `rotate(float angle)`, and `paperDrop()`. A single sheet of paper was placed in front of the robot. The robot was commanded to pick up the paper, rotate it 180 degrees, and place it down. The location of the paper was then compared with the expected location to estimate the placement error. This experiment was repeated 54 times. We observed a 93% success rate for this cycle. Each experiment resulted in a reasonable location for the placement of the paper. The 7% error were due to problems in the cam alignment for the paper lifting mechanism. We believe that this problem can be solved by using a better shaft encoder. To measure the placement errors, we measured the location of one point (in terms of its $x$ and $y$ coordinates) and the orientation of the paper. We found that the standard deviations in the $(x, y)$ location of the point was $(0.47cm., 0.59cm.)$, and the standard deviation for the orientation was 3.29 degrees. Since the standard deviation numbers are low, we conclude that the experiment has a high degree of repeatability.

We also conducted an experiment to test how well the robot can shift piles of paper. The basic experimental setup consists of a pile of three sheets of paper placed at a designated location on a desktop. The robot starts at a known location and follows a pre-coded trajectory to the location of the paper stack. The robot then uses the sticky foot to lift the top page and transport it to another pre-specified location on the desktop, following a given trajectory. After depositing the paper, the robot returns to its starting location and the process is ready to be repeated. We have repeated this loop many times. We observed failures in the system due to odometry. Because the robot uses no sensors to navigate, position errors accumulate and result in a misalignment between the robot and the paper stack. We plan to enhance the architecture of the robots with sensors that will provide guidance and paper detection capabilities.

## 4. Related Work

This section reviews some previous work and its relation to the present paper.

Several preceding systems have explored the connection between manipulation and locomotion. One of the earliest influential robots, Shakey [11], was a mobile manipulator. A direct approach is to attach a manipulator to a mobile platform. The JPL Cart [17] provides an early example, while Romeo and Juliet [8] provide a current example. These projects have demonstrated effective coordination of wheels and arm joints in manipulation tasks.

One goal of the work described here is to explore the relation of manipulation and locomotion, a goal shared with the distributed manipulation work of Donald et al [4]. This work included a set of mobile robots pushing objects, as if each robot were a finger in a multi-fingered grasp. The OSU Hexapod [16] used dynamic stability analysis and algorithms quite similar to those sometimes used to coordinate or analyze dexterous manipulation with multi-fingered hands.

The Platonic Beast [12] is probably closest in spirit to the present work. It had several limbs, each of which could be used for locomotion or manipulation.

The present work can also be viewed in relation to other manipulation systems. In fact it fits naturally with work on nonprehensile manipulation. A few examples are manipulation of objects in a tilting tray [6] or on a vibrating plate [14], manipulation of parts on a conveyor belt with a single-joint robot [1], manipulation of planar objects dynamically with simple robots [10], use of a passive joint on the end of a manufacturing arm to reorient parts [13], control of an object by rolling it between two plates [3], and manipulation of planar shapes using two palms [5]. These examples are simpler than a conventional general purpose manipulator, they can manipulate a variety of parts without grasping, and they exploit elements of the task mechanics to achieve goals. In the case of the present work, each robot uses four motors to manage six freedoms (mobipulator) or seven freedoms (Fiat). Manipulation is accomplished by friction, gravity, dynamics, and adhesion, without grasping. Perhaps the most relevant previous work is the business card manipulation work of Kao and Cutkosky [7], which addressed manipulation of laminar objects by fingers pressing down from above.

We still have a great deal of work to do on analysis, planning, and control, which will depend heavily on well-established techniques for non-holonomic robots. [2, 9]

## 5. Discussion and Conclusion

Our goal is to develop reliable desktop robots, capable of performing useful and interesting tasks on a desktop, while attached as peripherals to a desktop computer. In this paper we explore two designs for robot systems that allow the robots to move paper on a desktop in the $x$, $y$, and $z$ directions. Both robots are small in scale and architecturally minimalist.

More generally, our robots explore the deep connection between locomotion and manipulation. While several authors have noted the existence of this connection (see §4) the present work seeks to take the connection even further: the robot is just one of several movable objects in the task. The job of each actuator is resolved according to the task, be it manipulation, locomotion, or something not clearly classifiable as either.

Our experiments show that these robots are capable of manipulating individual pieces of paper in the $x$, $y$, and $z$ directions on a desktop[1]. Many components of these demonstrations were hard wired. The purpose of the experiments was to demonstrate the basic capabilities of the robots to manipulate paper and other common objects on the desktop. We are currently incorporating a planning module in the system.

Our systems are very preliminary. In the future, we plan to enhance these robots by combining their two functionalities into one mobile robot and to augment their mechanical designs with sensing. Another goal is to develop automated planners able to compute complex motions for desktop tidying tasks. We plan to incorporate (1) an external vision system to guide the robot; (2) reflective optical sensors for detecting edges of contrasting color in order to help the robot align to the edge of a sheet of paper on a contrasting background; (3) an inkjet head to be used to print barcodes on new papers arriving on the desk; and (4) optical sensors that, when combined with distance information from the shaft encoder, could be used to read barcodes on pages as the robot passes over them.

## Acknowledgments

Many of these ideas arose during discussions with Illah Nourbakhsh. Stephen Moore and Jason Powell helped in the design and construction of the mobipulator. We thank Keith Kotay for many valuable discussions. This work is supported by the National Science Foundation under Grant IRI-9318496. DKP would like to thank E. P. Krotkov, MAE, and MTM for hosting his sabbatical at CMU, and NSERC for financial support. Daniela Rus has been supported under NSF Career award IRI-9624286.

## References

[1] S. Akella, W. Huang, K. M. Lynch, and M. T. Mason. Planar manipulation on a conveyor with a one joint robot. In *International Symposium on Robotics Research*, pages 265–276, 1995.

[2] J. Barraquand and J.-C. Latombe. Nonholonomic multibody mobile robots: Controllability and motion planning in the presence of obstacles. *Algorithmica*, 10:121–155, 1993.

[3] A. Bicchi and R. Sorrentino. Dexterous manipulation through rolling. In *IEEE International Conference on Robotics and Automation*, pages 452–457, 1995.

[4] B. R. Donald, J. Jennings, and D. Rus. Analyzing teams of cooperating mobile robots. In *Proceedings 1994 IEEE International Conference on Robotics and Automation*, 1994.

[5] M. A. Erdmann. An exploration of nonprehensile two-palm manipulation. *International Journal of Robotics Research*, 17(5), 1998.

---

[1] Currently, neither robot can handle thick stacks of stapled papers, which is a challenge for the future.

[6] M. A. Erdmann and M. T. Mason. An exploration of sensorless manipulation. *IEEE Transactions on Robotics and Automation*, 4(4):369–379, Aug. 1988.

[7] I. Kao and M. R. Cutkosky. Dextrous manipulation with compliance and sliding. In H. Miura and S. Arimoto, editors, *Robotics Research: the Fifth International Symposium*. Cambridge, Mass: MIT Press, 1990.

[8] O. Khatib, K. Yokoi, K. Chang, D. Ruspini, R. Holmberg, A. Casal, and A. Baader. Force strategies for cooperative tasks in multiple mobile manipulation systems. In G. Giralt and G. Hirzinger, editors, *Robotics Research: The Seventh International Symposium*, pages 333–342, 1996.

[9] J. P. Laumond. Nonholonomic motion planning for mobile robots. Technical report, LAAS, 1998.

[10] K. M. Lynch, N. Shiroma, H. Arai, and K. Tanie. The roles of shape and motion in dynamic manipulation: The butterfly example. In *Proceedings of the 1998 IEEE International Conference on Robotics and Automation*, pages 1958–1963, 1998.

[11] N. J. Nilsson. Shakey the robot. Technical Report 323, SRI International, 1984.

[12] D. K. Pai, R. A. Barman, and S. K. Ralph. Platonic beasts: a new family of multilimbed robots. In *Proceedings 1994 IEEE International Conference on Robotics and Automation*, pages 1019–1025, 1994.

[13] A. Rao, D. Kriegman, and K. Y. Goldberg. Complete algorithms for reorienting polyhedral parts using a pivoting gripper. In *IEEE International Conference on Robotics and Automation*, pages 2242–2248, 1995.

[14] D. Reznik and J. Canny. A flat rigid plate is a universal planar manipulator. In *Proceedings of the 1998 IEEE International Conference on Robotics and Automation*, pages 1471–1477, 1998.

[15] D. Rus and P. deSantis. The self-organizing desk. In *Proceedings of the International Joint Conference on Artificial Intelligence*, 1997.

[16] S.-M. Song and K. J. Waldron. *Machines that Walk: The Adaptive Suspension Vehicle*. MIT Press, 1988.

[17] A. M. Thompson. The navigation system of the jpl robot. In *Fifth International Joint Conference on Artificial Intelligence*, pages 749–757, 1977.

# Experimental Approach on Grasping and Manipulating Multiple Objects

Makoto Kaneko, Kensuke Harada, and Toshio Tsuji

Industrial and Systems Engineering
Hiroshima University
Kagamiyama, Higashi-Hiroshima 739-8527, JAPAN
{kaneko, kharada, tsuji}@huis.hiroshima-u.ac.jp

**Abstract:** This paper discusses grasping and manipulating multiple objects by an enveloping style. We formulate the motion constraint for enveloping multiple objects under rolling contacts. We further provide a sufficient condition for producing a relative motion between objects. We show that a three-fingered robot hand experimentally succeeds in lifting up two cylindrical objects from a table and in changing their relative positions within the hand.

## 1. Introduction

Multifingered robot hands have a potential advantage to perform various skillful tasks like human hands. So far, much research has been done on multifingered robot hands focusing on the grasp of an object such as the stability of the grasp, the equilibrium grasp, the force closure grasp, and the manipulation of an object by utilizing either the rolling or the sliding contact. Most of works have implicitly assumed that a multifingered hand treats only one object. In this paper, we relax the assumption of single object, and discuss the manipulation of multiple objects by a multifingered robot hand. While there have been a couple of works [1, 2, 3, 4] discussing the grasp of multiple objects, as far as we know, there has been no work on the manipulation of multiple objects within the hand.

The goal of this paper is to realize the manipulation of multiple objects by a multi-fingered robot hand. For grasping and manipulating an object, there are two grasp styles, finger-tip grasp and enveloping grasp. While we can expect a dexterous manipulation through a finger-tip grasp, it may easily fail in grasping multiple objects under a small disturbance. On the other hand, an enveloping grasp may ensure even more robustness for grasping multiple objects than a finger-tip grasp, due to a large number of distributed contacts on the grasped objects, while we can not expect a dexterous manipulation for enveloped objects. Taking advantage of the robustness of enveloping grasp, we focus on the enveloping grasp for handling multiple objects. One big problem for manipulating enveloped objects is, however, that the manipulating force cannot be obtained uniquely for a given set of torque commands and it generally spans a bounded space. As a result, each object motion is not uniquely determined either. Knowing of such basic properties of the manipulation of enveloped objects, we provide a sufficient condition for producing a relative motion between objects.

The highlight of this work is the experimental validation. For two cylindrical objects placed on a table, the specially developed robot hand first approaches and lifts up them with a simple pushing motion by finger tips. After two objects are fully enveloped by the hand, it starts the manipulation mode composed of four phases where torque commands are appropriately switched depending upon the relative position of objects. By switching torque commands step by step, we succeeded in manipulation two cylindrical objects within the hand.

## 2. Related Works

### Enveloping Grasp:

There have been a number of works concerning the enveloping grasp. Especially, Salisbury et al.[5] has proposed the Whole-Arm Manipulation(WAM) capable of treating a big and heavy object by using one arm which allows multiple contacts with an object. Bicchi[6], Zhang et al.[7] and Omata et al.[8] analyzed the grasp force of the enveloping grasp. In our previous work, we have proposed the grasping strategies for achieving enveloping grasp for cylindrical objects[9].

### Grasp and Manipulation of Multiple Objects:

Dauchez et al.[1] and Kosuge et al.[2] used two manipulators holding two objects independently and tried to apply to an assembly task. However, they have not considered that two manipulators grasp and manipulate two common objects simultaneously. Recently, Aiyama et al.[3] studied a scheme for grasping multiple box type objects stably by using two manipulators. For an assembly task, Mattikalli et al.[10] proposed a stable alignments of multiple objects under the gravitational field. While these works treated multiple objects, they have not considered any manipulation of objects.

### Grasp by Rolling Contacts:

Kerr et al.[11] and Montana[12] formulated the kinematics of manipulation of objects under rolling contacts with the fingertip. Li et al.[13] proposed a motion planning method with nonholonomic constraint. Howard et al.[14] and Maekawa et al.[15] studied the stiffness effect for the object motion with rolling. Cole et al.[16] and Paljug et al.[17] proposed a control scheme for achieving a rolling motion for a grasped object.

## 3. Modeling

Fig.1 shows the hand system enveloping $m$ objects by $n$ fingers, where finger $j$ contacts with object $i$, and additionally object $i$ has a common contact point with object $l$. $\Sigma_R$, $\Sigma_{Bi}$ $(i = 1, \cdots, m)$ and $\Sigma_{Fjk}$ $(j = 1, \cdots, n, \quad k = 1, \cdots, c_j)$ denote the coordinate systems fixed at the base, at the center of gravity of the object $i$ and at the finger link including the $k$th contact of finger $j$, respectively. Let $p_{Bi}$ and $R_{Bi}$ be the position vector and the rotation matrix of $\Sigma_{Bi}$, and $p_{Fjk}$ and $R_{Fjk}$ be those of $\Sigma_{Fjk}$, with respect to $\Sigma_R$, respectively. $^{Bi}p_{Cjk}$

and $^{Fjk}p_{Cjk}$ are the position vectors of the $k$th contact point of finger $j$ with respect to $\Sigma_{Bi}$ and $\Sigma_{Fjk}$, respectively. $^{Bi}p_{COt}(t = 1, \cdots, r)$ is the position vector of the common contact point between object $i$ and object $l$ with respect to $\Sigma_{Bi}$.

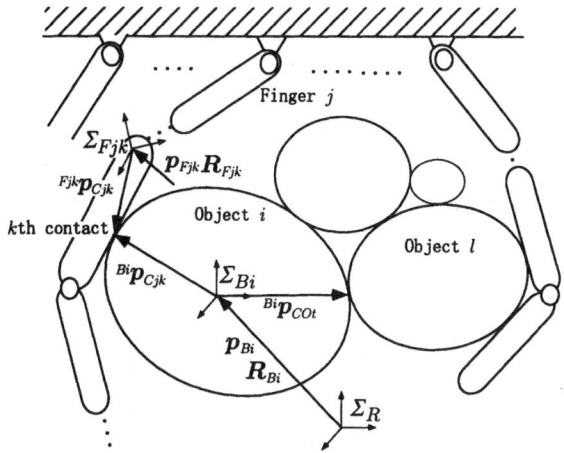

**Fig. 1.** Model of the System

### 3.1. Basic Formulation

In this subsection, we derive the constraint condition. The contact point between object $i$ and the $k$th contact of finger $j$ can be expressed by $\Sigma_{Bi}$ and $\Sigma_{Fjk}$. Similarly, the contact point between object $i$ and object $l$ can be expressed by $\Sigma_{Bi}$ and $\Sigma_{Bl}$. As a result, we have the following relationships:

$$p_{Bi} + R_{Bi}{}^{Bi}p_{Cjk} = p_{Fjk} + R_{Fjk}{}^{Fjk}p_{Cjk}, \tag{1}$$

$$p_{Bi} + R_{Bi}{}^{Bi}p_{COt} = p_{Bl} + R_{Bl}{}^{Bl}p_{COt}. \tag{2}$$

Suppose the rolling contact at each contact point. The velocity of object $i$ and the finger $j$ should be same at the contact point. Also, the velocity of two objects should be same at the common contact point between them. These constraints lead to the following equations[11]:

$$D_{Bjk}\begin{bmatrix} \dot{p}_{Bi} \\ \omega_{Bi} \end{bmatrix} = D_{Fjk}\begin{bmatrix} \dot{p}_{Fjk} \\ \omega_{Fjk} \end{bmatrix}, \tag{3}$$

$$D_{Oit}\begin{bmatrix} \dot{p}_{Bi} \\ \omega_{Bi} \end{bmatrix} = D_{Olt}\begin{bmatrix} \dot{p}_{Bl} \\ \omega_{Bl} \end{bmatrix}, \tag{4}$$

$$D_{Bjk} = [I - (R_{Bi}{}^{Bi}p_{Cjk}\times)], \tag{5}$$

$$D_{Fjk} = [I - (R_{Fjk}{}^{Fjk}p_{Cjk}\times)], \tag{6}$$

$$D_{Oit} = [I - (R_{Bi}{}^{Bi}p_{COt}\times)], \tag{7}$$

where $I$, $(R_{Bi}{}^{Bi}p_{Cjk}\times)$, $(R_{Fjk}{}^{Fjk}p_{Cjk}\times)$ and $(R_{Bi}{}^{Bi}p_{COt}\times)$ denote the identity and the skew-symmetric matrices which are equivalent to the vector product, respectively, and $\omega_{Bi}$ and $\omega_{Fjk}$ denote the angular velocity vectors of $\Sigma_{Bi}$ and $\Sigma_{Fjk}$ with respect to $\Sigma_R$, respectively. Since the velocity of the finger link including the $k$th contact point of finger $j$ can also be expressed by utilizing the joint velocity of finger $j$, we obtain the following relationships:

$$\begin{bmatrix} \dot{p}_{Fjk} \\ \omega_{Fjk} \end{bmatrix} = J_{jk}\dot{\theta}_j, \qquad (8)$$

where $J_{jk}$ is the jacobian matrix of the finger link with respect to the joint velocity. Substituting eq.(8) into eq.(3) and aggregating for $j = 1, \cdots, n$ and $k = 1, \cdots, c_j$, and aggregating eq.(4) for $t = 1, \cdots, r$, the the relationship between joint velocity and object velocity can be derived as follows:

$$D_F\dot{\theta} = D_B\dot{p}_B, \qquad (9)$$

where

$$D_F = \begin{bmatrix} D_{F1} & \cdots & 0 \\ \vdots & \ddots & \vdots \\ 0 & \cdots & D_{Fn} \\ 0 & \cdots & 0 \end{bmatrix},$$

$$\dot{\theta} = [\dot{\theta}_1^T \cdots \dot{\theta}_n^T]^T,$$
$$D_B = [D_{B1}^T \cdots D_{Bn}^T \, D_O^T]^T,$$
$$\dot{p}_B = [\dot{p}_{B1}^T \, \omega_{B1}^T \cdots \dot{p}_{Bm}^T \, \omega_{Bm}^T]^T.$$

Definition of $D_{Fj}$ ($j = 1, \cdots, n$), $D_{Bj}$ ($j = 1, \cdots, n$) and $D_O$ are shown in [4], and an example for a grasp of two objects by two fingers will be shown in the next section. In eq.(9), $D_B \in R^{(\sum_{j=1}^n 3c_j + 3r) \times 6m}$ and $D_F \in R^{(\sum_{j=1}^n 3c_j + 3r) \times \sum_{j=1}^n s_j}$, where $s_j$ shows the number of joints of finger $j$.

Note that, while the kinematics for a 3D model is discussed so far, for a 2D model, the skew-symmetric matrix equivalent to the vector product $a\times$ is redefined as $(a\times) = [-a_y \ a_x]$, and $D_B \in R^{(\sum_{j=1}^n 2c_j + 2r) \times 3m}$.

## 4. A Sufficient Condition for Producing a Relative Motion for Two Objects

Fig. 2 shows the 2D model of the grasp system, where $\tau_{ij}$, $f_{Cik}$, $p_{Cik}$, $f_{ti}$, $n_{ti}$ and $f_{CO}$ denote the torque of the $j$th joint of the $i$th finger, the $k$th contact force of the $i$th object, the total force and moment of the $i$th object, and the contact force between two objects, respectively. Let $f_{CB}$ and $\tau$ be the contact force vector from each finger to the object and the joint torque vector, where $f_{CB} = [f_{C11}^T \cdots f_{C22}^T]^T$ and $\tau = [\tau_{11} \cdots \tau_{22}]^T$. We have the following relationship between $f_{CB}$ and $\tau$

$$\tau = J^T f_{CB}, \qquad (10)$$

where $J$ is the Jacobian matrix for two fingers. Assuming a friction cone at

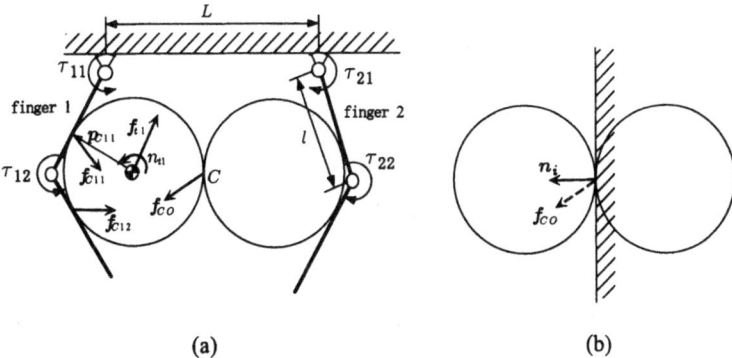

**Fig. 2.** Model of the Grasp System

each point of contact, we can express the contact force $f_{CB}$ in the following form:

$$f_{CB} = V\lambda, \quad \lambda \geq 0, \tag{11}$$

where

$$\lambda = [\lambda_{11}^1 \ \lambda_{11}^2 \ \lambda_{12}^1 \ \cdots \ \lambda_{22}^2]^T \in R^{8 \times 1},$$

$$V = \begin{bmatrix} V_{11} & 0 & \cdots & 0 \\ 0 & V_{12} & \ddots & \vdots \\ \vdots & \ddots & V_{21} & 0 \\ 0 & \cdots & 0 & V_{22} \end{bmatrix},$$

$$V_{ij} = [v_{ij}^1 \ v_{ij}^2] \in R^{2 \times 2}.$$

Substituting eq.(11) into eq.(10) yields

$$\tau = J^T V \lambda. \tag{12}$$

By solving eq.(12) with respect to $\lambda$, we finally obtain

$$f_{CB} = V \left[ (J^T V)^{\sharp} \tau + \{I - (J^T V)^{\sharp} J^T V\} k \right], \tag{13}$$

where $(J^T V)^{\sharp}$, $I$ and $k$ denote the pseudo-inverse matrix of $J^T V$, unit matrix, and an arbitrary vector whose dimension is same as $\lambda$.

In order to introduce the contact force $f_{CO}$, we first consider the equation of motion for two objects

$$M_B \ddot{p}_B + h_B = D_{CB}^T f_{CB} + D_O f_{CO}, \tag{14}$$

where $M_B = \text{diag}[m_{B1} I \ H_{B1} \ m_{B2} I \ H_{B2}]$, $\ddot{p}_B = [\ddot{p}_{B1}^T \ \ddot{\phi}_{B1} \ \ddot{p}_{B2}^T \ \ddot{\phi}_{B2}]^T$, $h_B$, $m_{Bi}$, and $H_{Bi}$ denote the inertia matrix, acceleration vector at the center of object, various forces(such as gravitational force, centrifugal and Colioris force), mass and inertia of the $i$th object, respectively. $\boldsymbol{D}_{CB}$ and $\boldsymbol{D}_O$ are given by

$$\boldsymbol{D}_{CB} = [\boldsymbol{D}_{B1}^T \ \boldsymbol{D}_{B2}^T]^T, \tag{15}$$
$$\boldsymbol{D}_{B1} = [\boldsymbol{D}_{B11} \ \mathbf{o}], \tag{16}$$
$$\boldsymbol{D}_{B2} = [\mathbf{o} \ \boldsymbol{D}_{B21}], \tag{17}$$
$$\boldsymbol{D}_O = [\boldsymbol{D}_{O11} - \boldsymbol{D}_{O21}]. \tag{18}$$

The constraint condition for keeping contact at point $C$ is expressed by

$$\boldsymbol{D}_O \dot{\boldsymbol{p}}_B = \mathbf{o}. \tag{19}$$

From both eqs.(14) and (19), we can obtain $\boldsymbol{f}_{CO}$ in the following form:

$$\boldsymbol{f}_{CO} = (\boldsymbol{D}_O \boldsymbol{M}_B^{-1} \boldsymbol{D}_O^T)^{-1}(-\boldsymbol{D}_O \boldsymbol{M}_B^{-1} \boldsymbol{D}_{CB}^T \boldsymbol{f}_{CB} + \boldsymbol{D}_O \boldsymbol{M}_B^{-1} \boldsymbol{h}_B - \dot{\boldsymbol{D}}_O \dot{\boldsymbol{p}}_B). \tag{20}$$

Note that $\boldsymbol{f}_{CO}$ cannot be uniquely determined since $\boldsymbol{f}_{CB}$ includes an arbitrary vector $\boldsymbol{k}$. As a result, $\boldsymbol{f}_{CO}$ spans a contact force space $\mathcal{F}_C$. Now, let us consider the total force $\boldsymbol{f}_{ti}$ summing up all contact forces for the $i$th object:

$$\boldsymbol{f}_{t1} = \boldsymbol{f}_{C11} + \boldsymbol{f}_{C12} + \boldsymbol{f}_{CO}, \tag{21}$$
$$\boldsymbol{f}_{t2} = \boldsymbol{f}_{C21} + \boldsymbol{f}_{C22} - \boldsymbol{f}_{CO}. \tag{22}$$

A sufficient condition for moving the $i$th object in the direction given by a unit vector $\boldsymbol{a}_i$ is shown by the following condition:

$$\boldsymbol{a}_i^T \boldsymbol{f}_{ti} > 0, \tag{23}$$
$$\boldsymbol{n}_i^T \boldsymbol{f}_{ti} > 0, \tag{24}$$

where $\boldsymbol{n}_i$ denotes the unit vector expressing the normal direction of object $i$ at $C$.

Now, let us consider an example. Two fingers with $l = L = 1.0$[m] initially grasp two cylindrical objects whose diameter and mass is $R = 0.45$[m] and $M = 1.0$[kg], respectively. Consider the problem computing the command torque, so that the object 1 and 2 may move upward and downward, respectively. In order to reduce the number of parameters, we assume the following relationship between $\tau_{i1}$ and $\tau_{i2}$.

$$\tau_{i2} = k_i \tau_{i1}, \ (i = 1, 2). \tag{25}$$

Also, we assume the maximum torque of $\tau_{max} = 3.0$[Nm]. Fig. 3 shows the simulation results where (a) $k_1 = 0.9$, $k_2 = 0.1$, and (b) $k_1 = 1.2$, $k_2 = 0.5$, respectively. When choosing the torque commands from the region computed, it is guaranteed that the object 1 and 2 moves upward and downward respectively, although we cannot specify that either rolling or sliding motion may happen.

## 5. Experiments

### 5.1. Manipulating Two Cylindrical Objects by Human

Fig. 4 shows a grasp experiment by human, where Phase 1 through 4 denote a series of continuous photos taken when the line connecting each center of objects

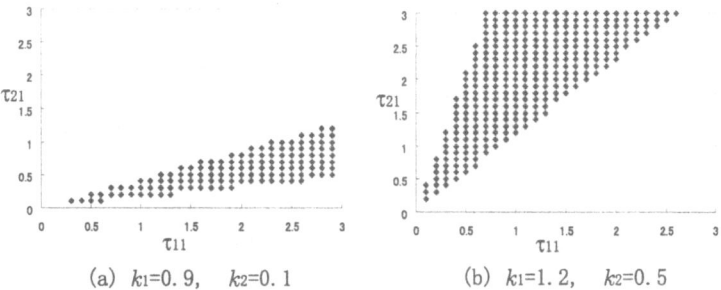

(a) $k_1=0.9$, $k_2=0.1$    (b) $k_1=1.2$, $k_2=0.5$

**Fig. 3.** Computed Torque Set

rotates in every 45[deg]. Human can achieve such a manipulating motion easily and quickly. We would note that, in most phase, a slipping contact is kept between two objects, while rolling is a dominant motion between each object and hand.

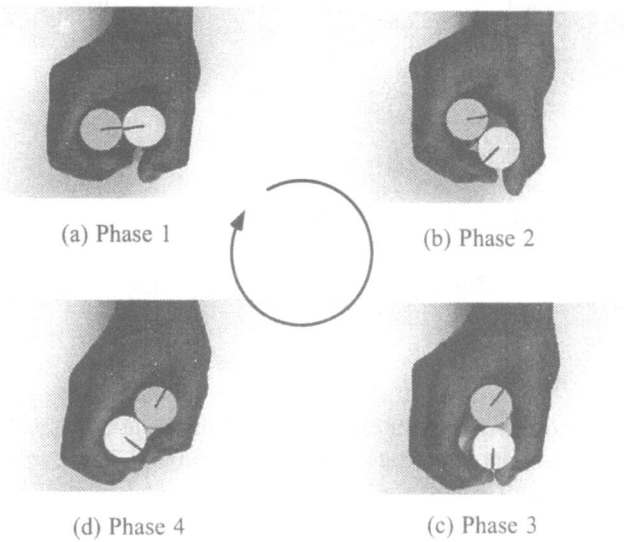

(a) Phase 1    (b) Phase 2

(d) Phase 4    (c) Phase 3

**Fig. 4.** Manipulating Two Cylinders by Human

## 5.2. Manipulating Two Cylindrical Objects by a Robot Hand

We execute a whole grasping experiment for two objects initially placed on a table, where there is no interaction between the hand and objects. For this experiment, we utilize the Hiroshima-Hand[18], where each finger unit has three rotational joints. Each joint is driven by tendon whose end is fixed at the drive pully connected to the shaft of actuator. Tension Differential type Torque sensor (TDT sensor) is mounted in each joint for joint torque control. Fig.5 shows an experimental result, where each finger first approaches and grasps two objects placed on the table, and starts to manipulate them by changing the torque commands step by step. During manipulation of two objects, we prepared four set of torque commands depending upon the relative position of objects. Each torque command is chosen so that it is enough for producing a slipping motion at the point of contact between objects. We believe that this is the first experiment of the manipulation of multiple objects within the hand.

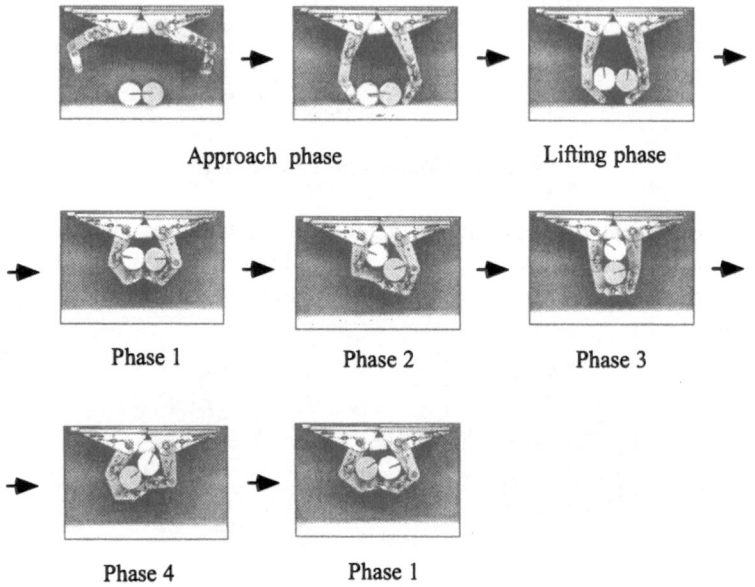

Approach phase           Lifting phase

Phase 1       Phase 2       Phase 3

Phase 4       Phase 1

**Fig. 5.** An Experimental Result

## 6. Conclusions

In this paper, we discussed the manipulation of multiple objects. We formulated the motion constraint for enveloping multiple objects under rolling contacts. We showed a sufficient condition for changing the relative position of two objects. For two cylindrical objects placed on a table, we succeeded in grasping and manipulating them within a robot hand.

This work was partly supported by the Inter University Project (The Ministry of Education, Japan). We would like to express our sincere gratitude to Mr. N. Thaiprasert, Mr. T. Shirai, and Mr. T. Mukai for their assistance in the experiments.

## References

[1] P.Dauchez and X.Delebarre 1991. Force-Controlled Assembly of two Objects with a Two-arm Robot. *Robotica, vol.9, pp.299-306.*

[2] K.Kosuge, M.Sakai, and K.Kanitani 1995 Decentralized Coordinated Motion Control of Manipulators with Vision and Force Sensors. *Proc. of 1995 IEEE Int. Conf. on Robotics and Automation, pp.2456-2462.*

[3] Y.Aiyama, M.Minami, and T.Arai 1997. Manipulation of Multiple Objects by Two Manipulators. *Proc. of 1998 IEEE Int. Conf. on Robotics and Automation, pp.2904-2909.*

[4] K.Harada and M.Kankeo 1998. Enveloping Grasp for Multiple Objects. *Proc. of 1998 IEEE Int. Conf. on Robotics and Automation, pp.2409-2415.*

[5] K.Salisbury et al. 1988 Preliminary Design of a Whole-Arm Manipulation Systems(WAMS). *Proc. of 1988 IEEE Int. Conf. on Robotics and Automation, pp.256-260.*

[6] A.Bicchi 1993 Force Distribution in Multiple Whole-Limb Manipulation. *Proc. of 1993 IEEE Int. Conf. on Robotics and Automation, pp.196-201.*

[7] X.-Y.Zhang, Y.Nakamura, K.Goda, and K.Yoshimoto 1994 Robustness of Power Grasp. *Proc. of 1994 IEEE Int. Conf. on Robotics and Automation, pp.2828-2835.*

[8] T.Omata and K.Nagata 1996 Rigid Body Analysis of the Indeterminate Grasp Force in Power Grasps. *Proc. of 1996 IEEE Int. Conf. on Robotics and Automation, pp.1787-1794.*

[9] M.Kaneko, Y.Hino, and T.Tsuji 1997 On Three Phases for Achieving Enveloping Grasps. *Proc. of 1997 IEEE Int. Conf. on Robotics and Automation.*

[10] R.Mattikalli, D.Baraff, P.Khosla, and B.Repetto 1995 Gravitational Stability of Frictionless Assemblies. *IEEE Trans. on Robotics and Automation, vol.11, no.3, pp.374-388.*

[11] J.Kerr and B.Roth 1986 Analysis of Multifingered Hands. *The Int. J. of Robotics Research, vol.4, no.4, pp.3-17.*

[12] D.J.Montana 1988 The Kinematics of Contact and Grasp. *The Int. J. of Robotics Research, vol.7, no.3, pp.17-32.*

[13] Z.Li and J.Canny 1990 Motion of Two Rigid Bodies with Rolling Constraint. *IEEE Trans. on Robotics and Automation, vol.6, no.1, pp.62-72.*

[14] W.S.Howard and V.Kumar 1995 Modeling and Analysis of Compliance and Stability of Enveloping Grasps. *Proc. of 1995 IEEE Int. Conf. on Robotics and Automation, pp.1367-1372.*

[15] H.Maekawa, K.Tanie, and K.Komoriya 1997 Kinematics, Statics and Stiffness Effect of 3D Grasp by Multifingered Hand with Rolling Contact at the Fingertip. *Proc. of 1997 IEEE Int. Conf. on Robotics and Automation, pp.78-85.*

[16] A.B.A.Cole, J.E.Hauser, and S.S.Sastry 1989 Kinematics and Control of Multifingered Hands with Rolling Contacts. *IEEE Trans. on Automatic Control, vol.34, no.4, pp.398-404.*

[17] E.Paljug, X.Yun, and V.Kumar 1994 Control of Rolling Contacts in Multi-Arm Manipulation. *IEEE Trans. on Robotics and Automation, vol.10, no.4, pp.441-452.*

[18] N. Imamura, M.Kaneko, and T.Tsuji 1998. Three-fingered Robot Hand with a New Design Concept. *Proc. of the 6th IAESTED Int. Conf. on Robotics and Manufacturing, pp. 44-49.*

# Chapter 3

# Vision

One of the oldest sensing modalities for robots, dating back decades to early block worlds work, is vision. Researchers ambitiously attempted to replicate human visual ability but were hampered by inadequate understanding of the visual process and the technology of the day. Today we have low cost cameras, laser rangefinders and abundant computing, but still no robotic vision system to rival that of the human. The papers in this chapter give a flavor of recent advances in sensors, data interpretation and control.

The paper by Vandapel et al. compares the performance of various ranging sensors in the antarctic environment. This environment is hostile not just because of temperature, but also because the environment confounds the sensors commonly used: lack of scene texture defeats stereo matching and specularly reflecting surfaces and airborne particles (snow) provide nil or spurious returns to laser rangefinders.

At the opposite extreme of temperature is the application to firefighting, a dangerous human occupation and one that has been recognized as an important application for advanced robotics. The paper by Schraft, Rust and Gehringer provides important experimental data regarding the performance of various sensors such as cameras, rangefinders, ultrasonics and radar in the presence of heat and smoke.

Traditionally monocular and stereo vision has been heavily reliant on camera calibration, but recent research has shown the significant potential of uncalibrated camera systems. The paper by Van Gool, Koch and Moons shows how multiple views from a single uncalibrated camera with unknown motion can be used to reconstruct a 3D model of the scene.

Tsubouchi et al. describe an extension to the classical ranging technique of structured lighting. They describe a very wide angle structured lighting system for a corridor navigating mobile robot capable of accepting linguistic commands.

In image-based visual servoing we aim to move image feature points from their initial configuration to a desired configuration which corresponds to a desired pose in the workspace. While it is straightforward to design a controller that is stable, its execution frequently causes the feature points to leave the camera's field of view. Morel et al. introduce an explicit trajectory generator on the image plane in order to overcome this.

PIC

# Preliminary Results on the use of Stereo, Color Cameras and Laser Sensors in Antarctica

Nicolas Vandapel,* Stewart J. Moorehead, William "Red" Whittaker
Carnegie Mellon University
5000 Forbes Avenue
Pittsburgh, PA 15213 - USA
{vandapel,sjm,red}@ri.cmu.edu

Raja Chatila
LAAS-CNRS
7, av. du Colonel Roche
31077 Toulouse Cedex 4 - France
raja@laas.fr

Rafael Murrieta-Cid
Stanford University
Computer Science Department Gates 133
Stanford, CA 94305 - USA
murrieta@Robotics.Stanford.EDU

**Abstract:** In November of 1998, an expedition from Carnegie Mellon University travelled to the Patriot Hills, Antarctica. The purpose of the expedition was to demonstrate autonomous navigation and robotic classification of meteorites and the characterization of various robotics technologies in a harsh, polar setting. This paper presents early results of experiments performed on this expedition with CCD cameras and laser range finders. It evaluates the ability of these sensors to characterize polar terrain. The effect of weather on this characterization is also analyzed. The paper concludes with a discussion on the suitability of these sensors for Antarctic mobile robots.

## 1. Introduction

Antarctica is a unique area on Earth for meteorite search. The flow of blue ice causes meteorites to concentrate on stranding surfaces, often near mountains. The cold and relatively dry environment in Antarctica helps to protect meteorites against significant weathering. Also, as meteorites commonly occur as

⁰This work - supported in part under NASA Ames Grant NAG2-1233, "Accurate Localization from Visual Features" - was performed at Carnegie Mellon University as Visiting Student Scholar from LAAS-CNRS.

dark objects against a lighter background, they are more easily spotted. Meteorites are typically 3cm in diameter and search is performed by humans on foot or skidoo. The search and initial identification is done using vision alone, and up to now no practical method exists to detect meteorites buried in ice or snow.

Systematic search and reliable identification of meteorites in Antarctica can be difficult for humans [1]. The development of robotic capabilities can help scientists find buried meteorites as proposed in [2] and in surface detection and classification as performed currently at Carnegie Mellon University (CMU) [3]. Robots can also be used to support human activities in Antarctica with logistical applications and could be a powerful tool to deploy instruments for Antarctic exploration and study as proposed in [4].

### 1.1. Expedition Overview

As part of the three year Robotic Antarctic Meteorite Search program [3] at CMU, the rover Nomad, designed for the Atacama Desert Trek [5], was deployed at Patriot Hills (80S,81W) in Ellsworth Land, Antarctica in collaboration with the University of Pittsburgh and the NASA Ames Research Center. The deployment was for 35 days in November and December 1998.

The expedition demonstrated autonomous navigation in polar terrain [7] and meteorite detection and classification [8]. Experiments were also performed on systematic patterned search [9], ice and snow mobility, landmark based navigation and millimeter wave radar. Foot search by the expedition found two meteorites [1].

### 1.2. Robotic activities in Antarctica

Previously underwater vehicles like ROBY and SARA from Italy [10] and TROV from NASA Ames [11] have explored the sea near coastal bases. On land, the walking robot Dante explored Mt. Erebus [12]. Italy has also conducted mobile robot research for robotic applications in Antarctica with the RAS project as detailed in [13]. We can also note the design of a chassis, see [14], for an Antarctic rover lead at LAAS.

## 2. Camera and Stereo Results

Vision, and in particular stereo vision, is a common sensing modality for robots. Antarctica provides many challenges to the use of vision for navigation and other tasks. The cold temperatures mean that the cameras must be heated and kept in sealed enclosures to prevent snow melting on the warm camera. Ice may form on the lenses, distorting and obstructing the view (Figure 1). Further, the nature of the terrain - large, featureless plains of white snow and blue ice, make it difficult for stereo matching.

This section presents the results of experiments performed using color cameras mounted on a tripod and Nomad's stereo cameras. Nomad has two pairs of B&W stereo cameras mounted on a sensor yard 1.67m above the ground (Figure 9).

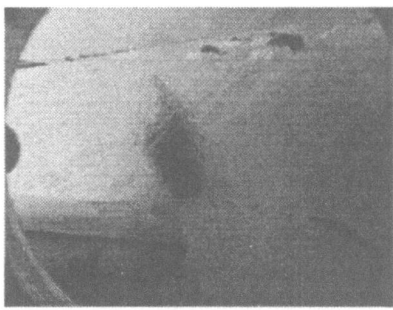

Figure 1. Image with ice on lens

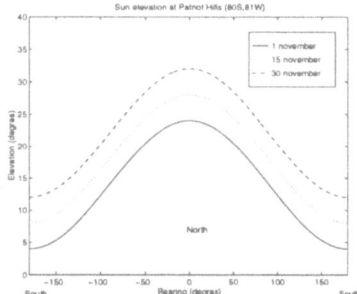

Figure 2. Sun elevation

## 2.1. Sun Influence

In the austral summer, the sun is above the horizon 24 hours a day but always remains low on the horizon as seen in Figure 2. The terrain in the Patriot Hills region is also highly reflective, consisting largely of snow and ice, and so a high level of glare was expected. It was anticipated that this glare, combined with low sun elevation, would cause high light levels producing pixel saturation and blooming in the CCD cameras.

In practice sun influence was not an issue. Conditions were bright on sunny days but the camera iris and shutter were sufficient to regulate the ambient light and produce good pictures. As seen in Figure 3 glare was present in the images taken on blue ice fields. The use of linear polarizing filters reduced the glare but had little effect on the number of pixels matched in stereo processing. Direct viewing of the sun by the stereo cameras was also not a problem. The

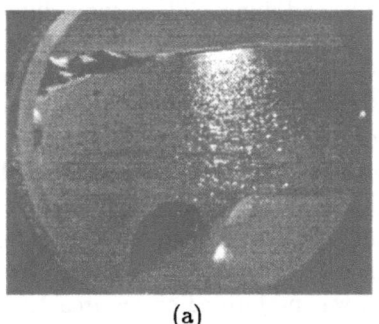

(a)

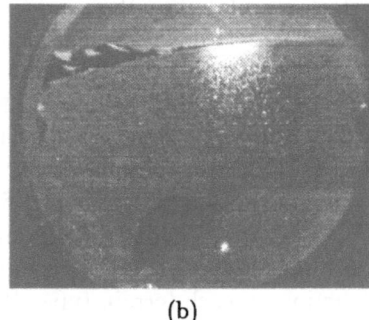

(b)

Figure 3. (a) Image without any filter (b) Image with linear polarizing filter.

local topography of the Patriot Hills helped to reduce this problem. Since the Patriot and behind them the Independence Hills occupied the entire southern horizon, the sun was behind them when it was at its lowest point - at midnight. To demonstrate the effect of sun position on stereo matching image sequences were captured as Nomad drove in circles of 4.0m in radius. The number of pixels matched in each image pair are shown in Figure 4. The graphs show minor dependence on sun position. The sun was in front of Nomad in images 0 to 7 and 34 to 42 in Figure 4(a) and 13 to 25 in Figure 4(b).

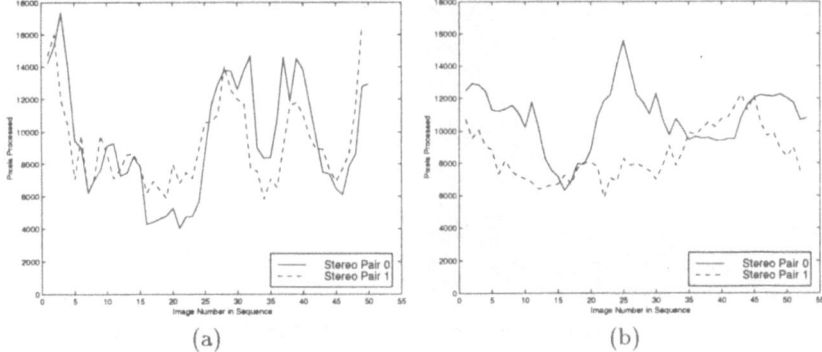

(a)  (b)

Figure 4. (a) Pixels processed for stereo pairs 0 and 1 while driving in 4.0m radius circle on snow. (b) On blue ice.

| Terrain Type | Av. Num. of Pixels | |
|---|---|---|
| | Pair 0 | Pair 1 |
| Snow | 9526 | 9231 |
| Ice | 10800 | 8602 |
| Moraine | 9460 | 10637 |

Table 1. Effects of Terrain on Stereo

| Weather | Av. Num. of Pixels | |
|---|---|---|
| | Pair 0 | Pair 1 |
| Sunny | 9526 | 9231 |
| Overcast | 4973 | 5960 |
| Blowing Snow | 14223 | 9917 |

Table 2. Effects of Weather on Stereo

## 2.2. Terrain Effects

Three common Antarctic terrain types - snow, blue ice and moraine - can be found in the Patriot Hills region. The snow fields consist of hard packed snow which is sculpted by the wind to resemble small sand dunes. These snow dunes are referred to as sastruggis and can be obstacles for Nomad. The blue ice forms large plains that seldom have features to impede a robot. The surface of the ice is pock marked with many small depressions (5cm diameter) called sun cups. A moraine is a large collection of various sized rocks on ice. The rocks at the Patriot Hills moraine were very sparsely distributed. Blue ice with small snow patches made up the terrain between the rocks.

The ability to model the terrain in sufficient detail to permit navigation is an important factor. Several images were taken to compare stereo's ability to function on each terrain type. Pixel matching was performed on an area 1.5m to 750m in front of the robot (308x480 pixels). The average number of pixels matched on each terrain, for image sequences to 50 images, is presented in Table 1.

The results indicate that the terrain type had little effect on the number of pixels matched. For comparison a typical disparity map from each terrain type and one of the corresponding stereo images can be found in Figures 5, 6 and 7. Unfortunately, the low density of matched pixels in the images meant that stereo was insufficient for navigation. Determining if the low pixel count is due to poor calibration or low image texture is still to be done.

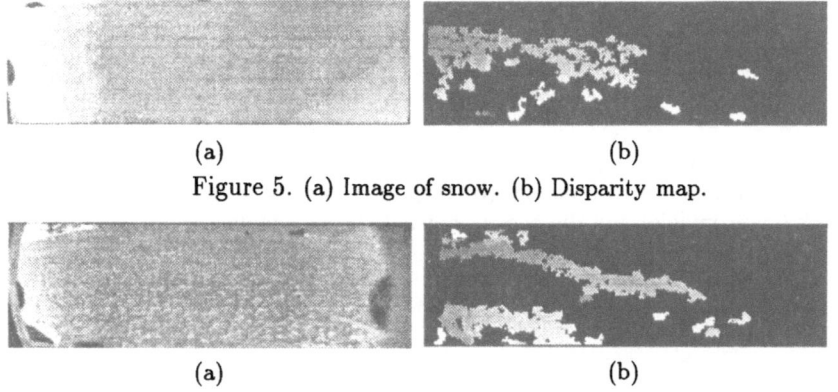

(a)            (b)

Figure 5. (a) Image of snow. (b) Disparity map.

(a)            (b)

Figure 6. (a) Image of ice. (b) Disparity map.

## 2.3. Weather Effects

Another thing which can affect the results of stereo vision is the weather. Images were collected under three types of weather conditions - sunny, overcast and blowing snow. The average number of pixels processed under these conditions is presented in Table 2. These sequences were taken on snow terrain.

It is interesting to note that blowing snow has very little effect on the stereo results. In fact, it is difficult to see the blowing snow in the images. Overcast weather creates diffuse lighting conditions which cause the terrain to have no contrast and makes it impossible, even for a human, to see depth. The first two authors both managed to fall into holes left from dug out tents without seeing them under these types of conditions, so it is not surprising that stereo works poorly.

## 2.4. Color Segmentation

We use the method presented in [15] to segment[1] color images in order to extract skyline and ground features such as rocks - or meteorites - and sastruggi shadows. The segmentation method is a combination of two techniques: clustering and region growing. Clustering is performed by using a non supervised color classification method, this strategy allows the growing process to work independently of the beginning point and the scanning order of the adjacent regions. Figure 8 presents an example of image segmentation on a snow field.

On a snowy flat area - the Twin-Otter runway - we took images of a collection of rocks of different sizes - 3cm to 5cm - at several distances - 2m to 11m. Unfortunately, we were not able to extract 3cm rocks - the standard size of a meteorite - with our 4.8mm lenses in a reliable fashion because they were too small.

## 3. Laser Results

Two different lasers (Figure 9) were used in Antarctica to perform a set of common and complementary experiments dealing with obstacle detection, ground classification, weather influence and range capability (Table 3).

---

[1]This method was developed at LAAS-CNRS by R. Murrieta-Cid during his Ph.D thesis

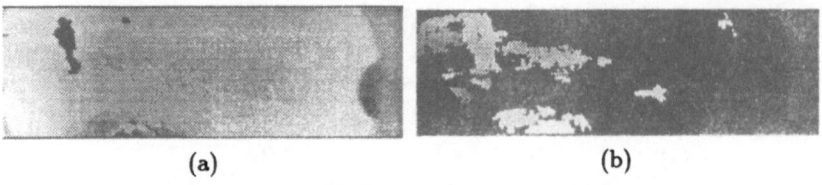

(a)  (b)

Figure 7. (a) Image of rock on ice. (b) Disparity map. Note the good disparity on the rock (lower middle) and the person, but not the surrounding ice.

(a)  (b)

Figure 8. segmentation examples : (a) original (b) segmented image

## 3.1. Ground Classification

On blue ice fields snow patches can hide crevasses, so differentiating between snow and ice may be the first step towards crevass detection. Towards that goal, an attempt to classify the ground based on the energy return of a laser is presented. Using the Riegl, the energy return and distance were recorded for measurements of rocks, blue ice and snow at close distances. This data is plotted in Figure 10. The "rock phg" was taken on Antarctic rocks brought back to Pittsburgh. This graph shows that for a given distance, snow usually has a higher energy return than blue ice and rocks. Thus it appears possible to distinguish snow from blue ice and rocks but a better model of the laser is needed to decorrelate the influence of temperature, distance and photon noise before implementing a classification process using, for example, Bayesian rules.

| Characteristics | Riegl laser LD-3100HS | SICK laser LMS-220 |
|---|---|---|
| Type | telemeter | laser stripe |
| Wavelength | 900±100 nm | 920 nm |
| Beam divergence | 3 mrad max. | 7 mrad |
| Information recorded | Distance (12 bits) Energy returned (8 bits) Temperature | Distance (13 bits) |
| Theoretical range | 50 to 150 m | 50 m |
| Accuracy (mean value) | ±20mm | ±20mm in single scan mode |
| Operation Temperature | $-30.. + 50°C$ | $-30.. + 70°C$ |

Table 3. Laser characteristics

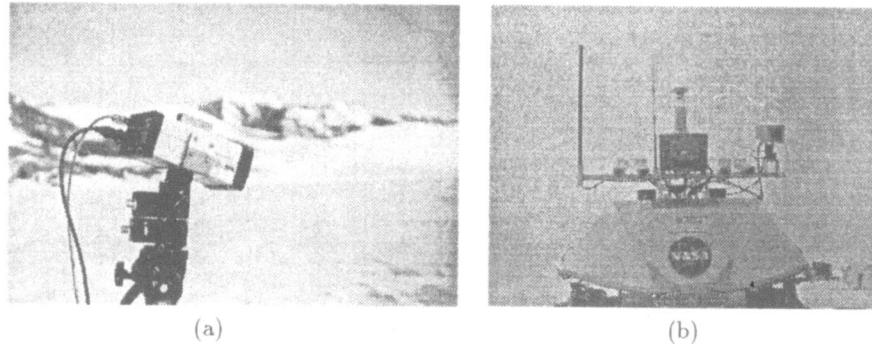

(a)           (b)

Figure 9. Lasers used in Antarctica (a) Riegl (b) SICK

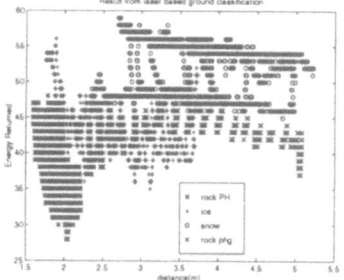     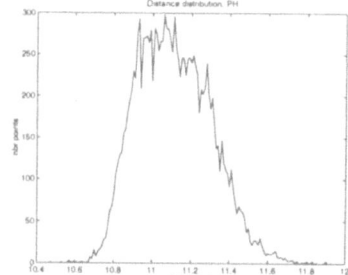

Figure 10. Ground Classification      Figure 11. Distance Distribution

## 3.2. Weather Influence

Figures 11-14 show that both the Riegl and SICK lasers were affected by the presence of blowing snow. Figure 11 shows the distribution of distance returns of the Riegl laser pointed at a static object 11m away during a period of heavy blowing snow. The snow has two effects on the distance measurements. First, the standard deviation of the data is much greater with blowing snow than for clear conditions. Second, the presence of a large spike of data at a range of 0m (Figure 12(b)). This corresponds to readings which had no return. Figures 13(a) and 14(a) show the distribution of distance returns for a single direction (straight ahead) of the SICK laser scanner mounted on the Nomad robot. They correspond to conditions of moderate blowing snow and heavy blowing snow respectively (this data was taken at the same time as the stereo blowing snow data reported in Table 2). Under clear conditions the SICK laser reports the distance ±2cm. Two differences with the Riegl data are apparent. First, a no return is interpreted as having a distance measurement of infinity (1000 cm) instead of 0. Secondly, the SICK laser will produce readings with very short distances at times, the Riegl will not.

For navigation in blowing snow it is useful to think of the distance measurement as confirmation that nothing lies between the laser and that distance, not that an object exists at that distance [16]. Using this interpretation the SICK laser data was filtered to attempt to remove the effects of blowing snow. Three

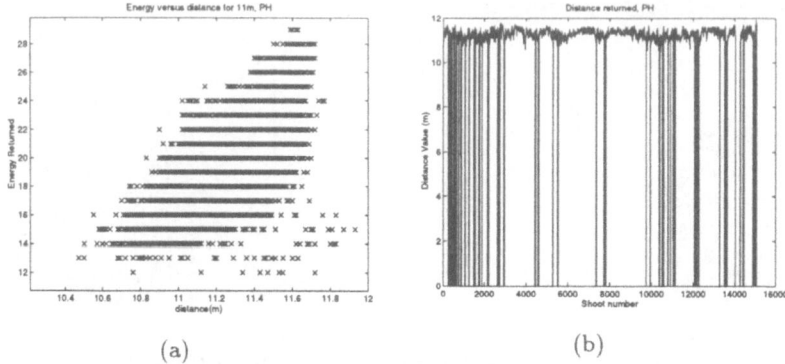

(a)                                    (b)

Figure 12. (a) Energy versus distance (b) Distance signal over time

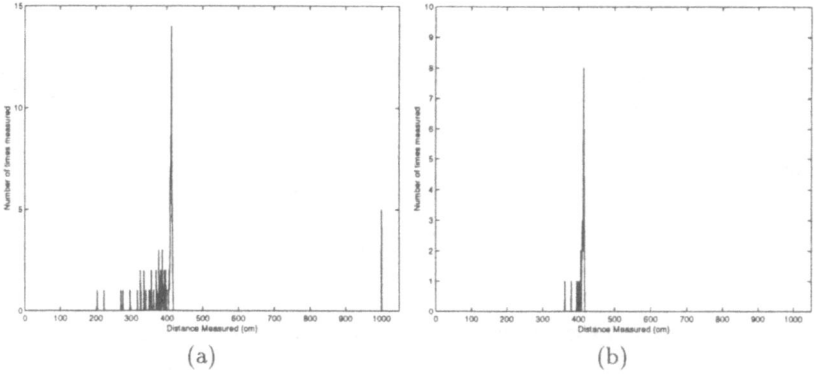

(a)                                    (b)

Figure 13. Distance Distribution Medium Snow - (a) Raw data (b) Filtered

measurements of the scene were taken. Measurements of infinity were discarded unless all three readings were infinite. The largest remaining measurement was used as correct. This filter was applied to the data from Figures 13(a) and 14(a). The results are shown in Figures 13(b) and 14(b). For a moderate amount of blowing snow the filter does a very good job of reducing the error in distance. However, for heavy blowing snow, we can see that the filter is still not adequate for navigation purposes since none of the raw data contains the proper distance of approximately 4m.

## 3.3. Maximum Range

The Riegl laser was tested to detect the maximum useful range on snow terrain. A flat area with slight undulations and sastruggis less than 30cm in height was chosen near the main Chilean Patriot Hills camp. The laser was mounted 2m above the ground and the incidence angle was changed until the signal was lost. A distance of 55m was reached with a reliable signal. It was possible to get readings from as far as 64m but not reliably. The experiment was repeated on another snow field approximately 8km away and produced similar results.

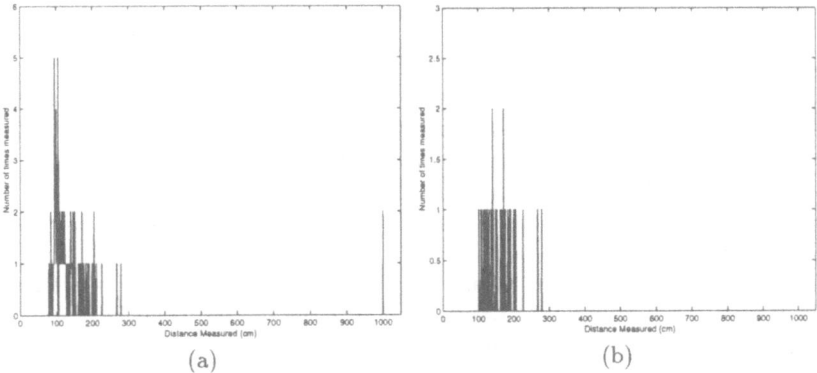

(a)                                                          (b)

Figure 14. Distance Distribution Heavy Snow - (a) Raw Data (b) Filtered

## 4. Conclusion

This paper presents early results from camera and laser tests performed in Antarctica. As this was the first deployment of an autonomous mobile robot to Antarctica very little was know beforehand on the performance of these sensors.

The results presented show that stereo vision was largely unaffected by the terrain type but performed very poorly on cloudy days. Polarizing filters were able to reduce glare from the sun on blue ice fields but in general the low sun angles and high light levels did not present a problem for stereo. Color segmentation was able to segment out sastruggi on snow fields.

The laser sensors worked well on both snow, ice and rocks. By measuring the energy returned and distance it is possible to distinguish snow from blue ice and rocks. This is important since it could allow a robot to find crevasses as well as areas where a visual meteorite search will not find meteorites. The main limitation of the laser sensors was during periods of blowing snow. The snow reflects the laser resulting in bad distance measurements.

Since stereo failed on cloudy days and laser on blowing snow, it is possible that a rover equipped with both sensors for navigation could operate under a wide range of Antarctic weather and terrain conditions.

**Acknowledgement**

N. Vandapel's participation was funded by the French Polar Institute (IFRTP) and supported by the Région Midi-Pyrénées. The Robotic Antarctic Meteorite Search program is funded by the Telerobotics Program NAG5-7707 of NASA's Office of Space Science Advanced Technology & Mission Studies Division.

A special thanks to Martial Hebert and Dimitrios Apostolopoulos for their support during this expedition and its preparation.

We would like to thank Riegl for their technical support during the preparation of the expedition and Sick Optic-Electronic for providing the LMS 220.

We would also like to thank the National Science Foundation, the Fuerza Area de Chile, the Instituto Chileno Antarctico, the 1998 RAMS expedition team members the people involved in this project at CMU and LAAS.

68

# References

[1] P. Lee et al., 1999, Search for Meteorites at Martin Hills and Pirrit Hills, *Lunar and Planetary Society Conference.*

[2] J.Y. Prado, G. Giralt and M. Maurette, January 1997, Systematic Meteorite Search in Antarctica Using a Mobile Robot, *LAAS-CNRS Technical Report 97030.*

[3] http://www.frc.ri.cmu.edu/projects/meteorobot/

[4] R. Chatila, S. Lacroix and N. Vandapel, December 1997, Preliminary Design of Sensors and of a Perceptual System for a Mobile Robot in Antarctica, *LAAS-CNRS Technical Report 97523, In French.*

[5] W. Whittaker, D. Bapna, M. Maimone, E. Rollins, July 1997, Atacama Desert Trek: A Planetary Analog Field Experiment, *i-SAIRAS, Tokyo, Japan.*

[6] http://www.laas.fr/~vandapel/antarctica/

[7] S.J. Moorehead, R. Simmons and W. Whittaker, 1999, Autonomous Navigation Field Results of a Planetary Analog Robot in Antarctica, *to appear i-SAIRAS.*

[8] L. Pedersen, D. Apostolopoulos, W. Whittaker, W. Cassidy, P. Lee and T. Roush, 1999, Robotic Rock Classification Using Visible Light Reflectance Spectroscopy : Preliminary results from the Robotic Antarctic Meteorite Search program, *Lunar and Planetary Society Conference.*

[9] K. Shillcutt, D. Apostolopoulos and W. Whittaker, 1999, Patterned Search Planning and Testing for the Robotic Antarctic Meteorite Search, *Meeting on Robotics and Remote Systems for the Nuclear Industry.*

[10] B. Papalia et. al., 1994, An Autonomous Underwater Vehicle for Researches in Antarctica, *Oceans '94 Conference.*

[11] C. Stroker, 1995, From Antarctica to Space: use of telepresence and virtual reality in control of a remote underwater vehicle, *SPIE Proceedings 1995, vol 2352, Mobile Robots IX.*

[12] D. Wettergreen, C. Thorpe and W. Whittaker, 1993, Exploring Mount Erebus by Walking Robot, *Robotics and Autonomous Systems, 11(3-4).*

[13] P. Fichera, J. Manzano and C. Moriconi, October 1997, RAS - A Robot for Antactica Surface Exploration : A project Review, *IARP Workshop.*

[14] G. Giralt, July 1996, Study of the RISP-CONCORDIA robot for Antarctica, *LAAS-CNRS Technical Report 96301, In French.*

[15] R. Murrieta-Cid, M. Briot and N. Vandapel, 1998, Landmark identification and tracking in natural environment, *IROS.*

[16] H.P. Moravec and A. Elfes, 1985, High Resolution Maps from Wide Angle Sonar, *ICRA.*

# Sensors for the object detection in environments with high heat radiation and photoresist or diffuse visibility conditions

Rolf Dieter Schraft, Hendrik Rust, Hagen R. Gehringer
Fraunhofer Institute Manufacturing Engineering and Automation (IPA)
Nobelstraße 12, D-70569 Stuttgart, Germany
phone: ++49 (0)711 970 1043
fax.: ++49 (0)711 970 1008
rust@ipa.fhg.de

**An application for teleoperated fire fighting shall be equipped with a sensor system which offers the operator the required data of the robot's environment so that he is able to guide and control it safely. The environmental perception is based on a sensor for two-dimensional pictures and a sensor for the measurement of the one-dimensional depth. In several live tests a number of sensors were investigated and their data were analysed. Suitable sensors were chosen according to the criteria evaluating the overall quality of sensor data acquisition and integrated in a sensor system for the robot.**

## 1   Introduction

Mobile robots are more and more used for service applications. They are necessary to fulfil tasks in areas which are inaccessible for human beings for any reason (for example: toxic air, radioactive radiation, narrow or small workplaces).

Fire fighting robots will perform such tasks in the near future – as they deliver fire extinguishing systems to the scene of fire. Because these mobile robots will be teleoperated by an operator, it is necessary to acquire data of the environment of the robot so that the operator is able to control and guide it safely to its workplace avoiding any collision with obstacles. For the acquisition of geometric data of the environment, a sensor system is needed. Due to the environmental conditions of the robot while heading to the scene of fire, the sensors must detect objects through smoke at high heat radiation.

Currently there is no reasonable sensor system available on the market which fulfils these tasks sufficiently.

## 2   Requirements and boundary conditions

### 2.1   Requirements for the robot system

Due to the experiences of the past with fire fighting robots, a new concept has been developed [1]. A small chain-driven remote-controlled vehicle is used as a tractor to pull a wheeled hose reels with an extinguishing monitor (fig. 1).

The tractor is a tracked vehicle called MF4 and is manufactured by telerob (Germany) with the following technical data:

| | |
|---|---|
| dimensions: | 1300·850·400 mm (L·W·H) |
| speed: | 50 m/min |
| total weight: | 350,0 kg |
| umbilical: | sensor data transfer control signals |

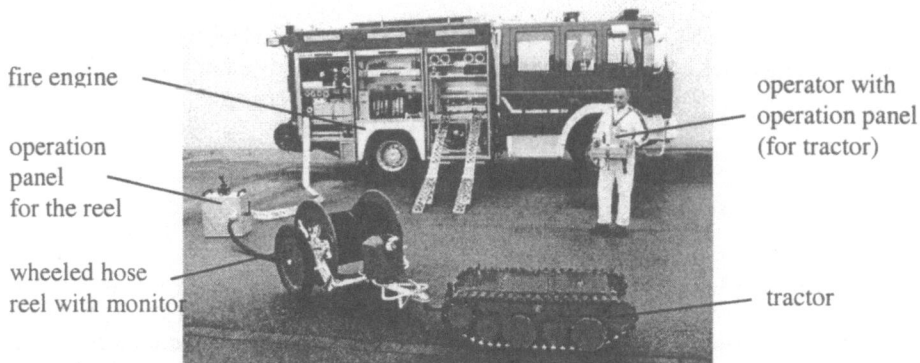

fire engine

operation panel for the reel

wheeled hose reel with monitor

operator with operation panel (for tractor)

tractor

Fig. 1: Modular concept of the fire fighting robot

The robot will be used for fire fighting in plants or apartment houses. In that case the tractor drives into the building, navigates through bad visibility conditions caused by heavy smoke, pulls the reel nearby the fire, drops it and drives back to the starting point. The reel aligns the monitor automatically in the direction of the fire to extinguish it with water. In special cases the robot is equipped with a manipulator to be capable to perform manipulation tasks. All processes are teleoperated by operators outside the building. [2]

To perform this task, the tractor needs a sensor system to provide sufficient information for the operator. Therefore it must be equipped with a data processing system, a data transfer system, and a man-machine-interface (e. g. an operation panel).

Concerning the different environmental conditions with different smoke/fog densities, four density classes (A-D) were defined (fig. 2).

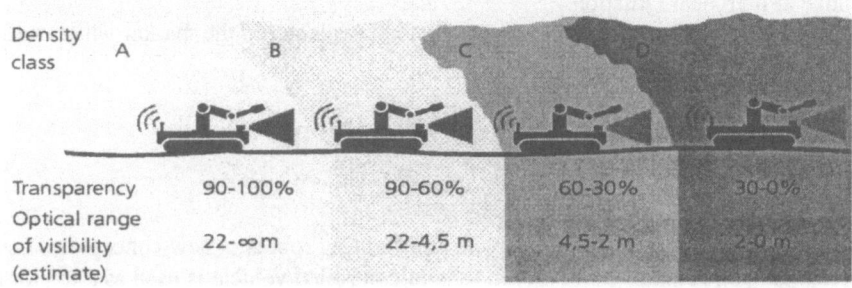

| Density class | A | B | C | D |
|---|---|---|---|---|
| Transparency | 90-100% | 90-60% | 60-30% | 30-0% |
| Optical range of visibility (estimate) | 22-∞ m | 22-4,5 m | 4,5-2 m | 2-0 m |

Fig. 2: The four smoke density classes

The possible temperature in the environment of the robot was divided into three zones: I: <20°C; II: 20-50°C; III: >50°C.

## 2.2 Requirements for the operation

To fulfil its tasks, the robot has to perform different manoeuvres. Therefore it is proposed to define different surveillance areas in the periphery of the robot (fig.3):

(1) Area I is the collision belt around the vehicle where objects shall be detected which are within a range of approx. 1,5 meters.
(2) Area II is the section in front of the vehicle's driving direction. Appearing holes in the ground, stairs, shafts or objects in this area have to be detected up to a range of 1-3 meters.
(3) In area III space structures, far objects (3-6 meters or further) and work items shall be detected or surveyed. Therefore a sensor system is needed to carry out sensor data acquisition providing depth measurement and two-dimensional pictures.
(4) Area IV is the working zone of the application module (for example a manipulator) of the teleoperated vehicle which must be surveyed in a range of 1-3 meters.

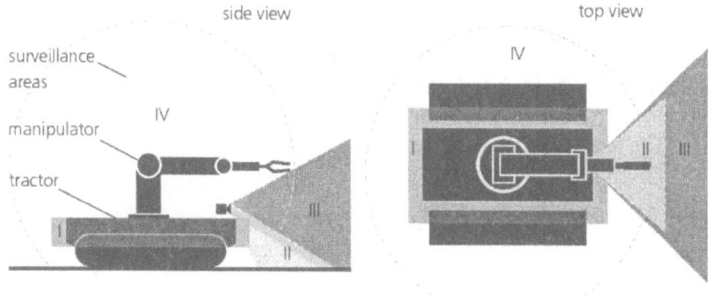

Fig. 3: Relevant areas for the sensor concept

To survey the areas, different sensors are used to gain the necessary information.

## 2.3 Boundary conditions

Typical boundary conditions for applications at the scene of fire are:
- temperature:     10-500°C          - smoke density:     10-100%
- smoke colour:    white/grey/black

Furthermore the system has to be robust, water and heat resistant. It should be easy to integrate it in existing platforms and the costs of the system must recline less than 80.000 US $.

## 2.4 Reference task

As a reference task the tractor has to pull the reel through the doorway and the hall into room 2 and to position it in front of the fire in this room. After that it drives back to the starting point (fig. 4):

72

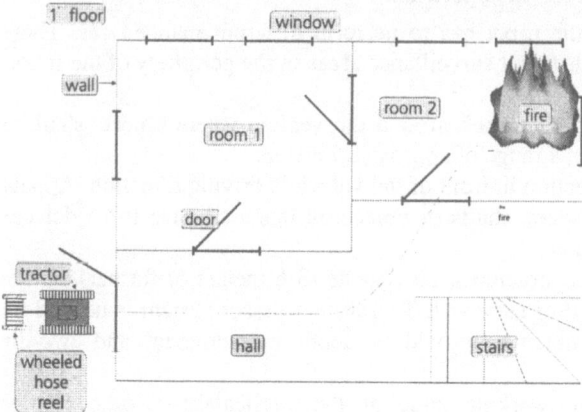

Fig. 4: Reference task of the robot

# 3 Sensors

## 3.1 Principles

The sensor principles shown in fig. 5 were chosen as generally suitable for sensor data acquisition of the geometric conditions in smoke-contaminated environments.

| Physical Principle | Measuring Principle | Sensors | |
|---|---|---|---|
| Optical | Laser | Distance sensor | Distance of a point strongly focused |
| | | 2D scanner | Distance of points within a measuring plane |
| | | 3D scanner | Distance of points within a measuring space |
| | Image processing | 3D greyscale image | Scene analysis by means of the model of the environment |
| | | stereo image | Depth information by picture overlay |
| | | structured illumination | Depth information by light cuts |
| Acoustic | Ultrasonic | Distance sensor | Distance of the next point within the measuring cone |
| | | 2D scanner | Distance of the next point within the measuring cone |
| | Radar | Distance sensor | Distance of the next point within the measuring cone |
| Electro-magnetic | Capacitive | Distance sensor | Distance to a plane |
| | Inductive | Distance sensor | Distance to a plane |
| Tactile | Bumper | contact sensor | Contact to a plane |

Fig. 5: Sensor principles

### 3.2 Available sensor systems

Already available systems which fulfil the preconditions are used in the military sector. These systems which are capable of measuring distances through smoke or to gain two-dimensional pictures are very expensive (over 60.000 US $ per unit) and information about the functions and modes of operation is only incomplete or not available at all.

Commercially available infrared cameras (10.000 – 50.000 US $ per unit) are used by firemen to find missing persons in smoke-contaminated buildings.

Rescue helicopters use high-tech radar in order to be able to fly in case of fog or in the night [3] (system price >100.000 US $). Low cost radar sensors are used in cars for distance measurement to the preceding vehicle.

### 3.3 Experimental experience

Reports and documents of experiments in smoke-contaminated environments and in high heat radiation are numerously available, however, the aim of these experiments was only in a few cases the testing of sensor systems. [4], [5], [6], [7]

Only once portable infrared cameras were tested under these conditions.

### 3.4 Theoretical research

The first step of our theoretical work was to examine the characteristics of smoke relating to the incendiaries. A literature research determined the size of the smoke particles with 0,5-1,6 $\mu$m [6]. According to [7], [8] and [9], sensor systems which work at a wave length which is lower than 8$\mu$m cannot penetrate the smoke.

Due to temperature influence, the ultrasonic sensors have problems with their measuring accuracy because of the changing density of the air. All sensor principles were theoretically examined in order to determine if smoke or heat bias the function or accuracy of the sensor and to proof that they are suitable for the application. Furthermore the difficulty of data processing was checked.

After theoretical analysis, several sensors were chosen to create a concept for the sensor system and for extensive live tests.

## 4    Experimental validation

### 4.1 Test bed

An experiment container (fig. 6, fig. 7) with the length of 12 meters was built for the tests. Objects with different materials, geometry and surfaces were located in the test bed in order to simulate the conditions of the reference task (fig. 4). The sensors were installed in a sealed box which is moveable and rotatable to simulate the robot's movement.

Different smoke densities (according to the density classes - fig. 2) and different temperatures were generated during the test.

The data of the sensors were recorded considering the test objects, their characteristics and the different environmental conditions. At the same time, the temperature of the object and of different air layers, the smoke density, the distance, and the relative position to the object according to the defined 3 object distance zones (differing while simulating the movement of the robot) were measured.

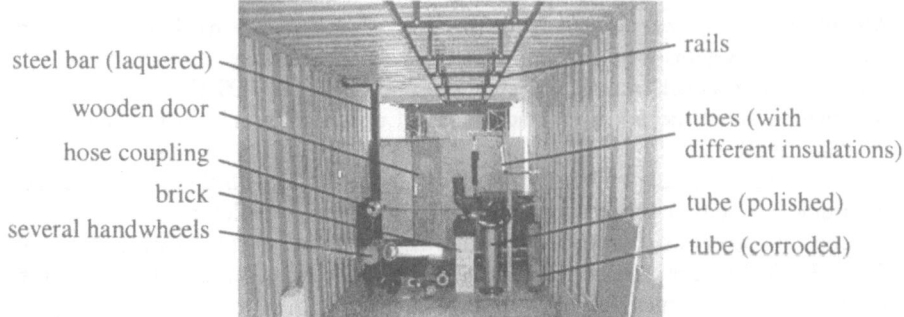

steel bar (laquered) — rails

wooden door — tubes (with different insulations)

hose coupling —

brick — tube (polished)

several handwheels — tube (corroded)

Fig. 6: Reference objects in test bed

rails

rotatable and turnable frame

sensor box — smoke density meter

reference camera —

Fig. 7: Sensor box movable on rails

## 4.2 Experiments

Different smoke types of different colours and particle concentrations in the air were used for the tests. The smoke types were produced by different incendiaries (1: white smoke bomb, 2: black smoke bomb, 3: petrol, 4: heptan, 5: plastic, 6: wood).

For the experiments the prechosen sensor types (fig. 5) were installed in the sensor box. The standard test program for each sensor was defined (fig. 8) as follows:

1. Choice of temperature zone
2. Examination of all smoke types in each temperature zone
3. Adjustment of the smoke density classes for each smoke type
4. Change of object distance (three distance classes) by moving the sensor box

According to this test program, 216 measurements (3 temperature zones·6 smoke types·4 density classes·3 object distances) have to be made.

Due to technical problems only some transducers could be tested under high temperatures. The problems were:

• temperature sensibility of the transducers and
• the missing possibility to integrate the sensors in the sensor box because of their size.

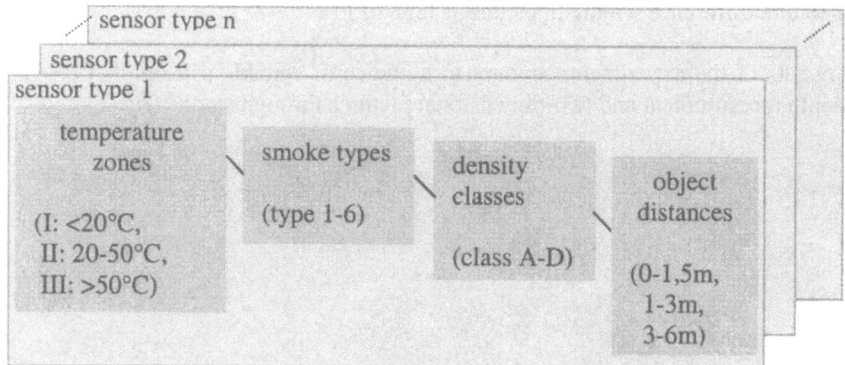

Fig. 8: Test program of sensors considering all possible combinations

# 5 Results

## 5.1 Sensors

Firstly the quality of the sensor information was examined in a range between 0 and 6 m (clustered in three range zones) without smoke (under the conditions of smoke density zone A (fig. 2)). According to these results the suitability of the sensors in the different range zones were evaluated. Secondly it was described up to which smoke density the sensor receives the same quality of information as in smoke density zone A. Thirdly the difficulty of data processing of each sensor was valued.

Especially infrared systems were examined extensively because of their supposed importance for the sensor system and because of the big number of available systems on the market. In fig. 9 and 10 two examples of the difference in picture quality of infrared cameras are shown – taken of the same scene.

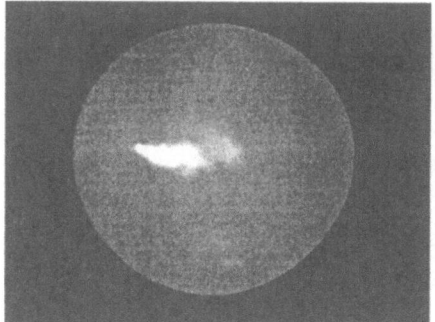 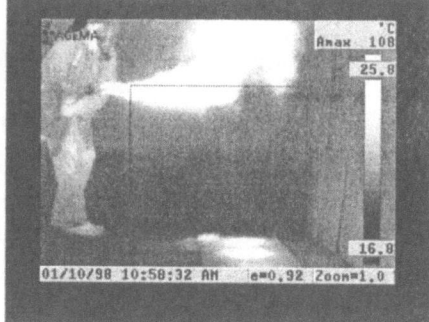

Fig. 9: Infrared camera 1             Fig. 10: Infrared camera 2
(wavelength 8-14µm)                   (wavelength 8-14µm)

It was proved that only cameras using a wavelength >8 µm are able to get information through the smoke. Some infrared systems are not able to handle high heat radiation or big temperature leaps within a picture. Due to these results an infrared camera with a wavelength of 8-14 µm is proposed for the sensor system which delivers pictures no matter if the environmental temperature or the temperature difference within an picture is high or low.

The results of the experiments amount to a choice of suitable sensors that provide the depth measurement and two-dimensional pictures through smoke (fig. 11).

| Measuring principle | Sensors | Measuring distance[1] | | | Function up to density zone | Difficulty of data processing |
|---|---|---|---|---|---|---|
| | | 0-1,5m | 1-3m | 3-6m | | |
| Laser | Distance sensor | ● | ● | ● | A, D[2] | ◐ |
| | 2D scanner | ● | ● | ● | A, D[2] | ◐ |
| | 3D scanner | ● | ● | ● | A, D[2] | ● |
| Image processing | 3D greyscale image | ◐ | ● | ◐ | B[3], D[4] | ● |
| | Stereo image | ◐ | ● | ● | B[3], D[4] | ● |
| | Structured illumination | ● | ● | ● | B[3], D[2] | ◐ |
| Ultrasonic | Distance sensor | ◐ | ● | ○ | C | ○ |
| | 2D scanner | ◐ | ● | ○ | C | ◐ |
| Radar | Distance sensor | ● | ● | ● | D | ◐ |
| Capacitive | Distance sensor | ● | ○ | ○ | D | ○ |
| Inductive | Distance sensor | ● | ○ | ○ | D | ○ |
| Bumper | Contact sensor | ● | ○ | ○ | D | ○ |

Legend: [1]in smoke density zone A
[2]with $CO_2$-laser and infrared detector/ infrared camera
[3]with CCD-camera
[4]with infrared camera

Smoke density: ● suitable; ◐ slightly suitable; ○ not suitable

Data processing: ● very difficult; ◐ difficult; ○ easy

Fig. 11: Survey of the results of the experiments

Relating to the different conditions and objects, suitable sensors of certain wave spectrums can be proposed because of the theoretical work validated by the results of the experiments. It was distinguished between sensors which are applicative for gaining one-, two- or three- dimensional information and it was tested if they are suitable for the object detection nearby or at the distance.

## 5.2  Sensor system and robot system

Corresponding to these results the sensors for the different areas (fig. 3) were defined:

(1) For the collision belt ultrasonic sensors (range 0,5-1,5m) and tactile sensors (range 0 – 0,5m) will be used.

(2) A CCD-camera will be used for the smoke density zones A and B. Ultrasonic and radar sensors are suitable for the depth measurement. Tactile sensors will be used for the collision detection in front of the vehicle, too.

(3) To gain two-dimensional pictures, a standard CCD-camera module will be used for the smoke density zones A and B. Further an infrared camera system was chosen to get pictures especially in smoke density zone B-D. We developed two new kinds of depth measuring systems which are especially configured for the requirements of measuring through smoke. [Because we are currently taking out a patent for this system, we cannot go into detail about it at the moment.] Radar is one other option to get depth information of high quality.

(4) The CCD-camera module and the infrared camera used in area III will be mounted on a pan-tilt unit so that the work space of the application module (for example a manipulator) can be surveyed.

The result of the experiments are suitable low cost sensors that provide the depth measurement and two-dimensional pictures through smoke. In several live tests a number of sensors was investigated and their data were analysed.

The concept for the sensor system provides a sensor for two-dimensional pictures and a sensor for the one-dimensional depth information of the environment.

# 6  Conclusion and future perspective

For the application of a teleoperated fire fighting robot for areas with bad visibility conditions caused by smoke, a low cost sensor system was designed which can provide sufficient information about the robot's environment to the operator. Its functionality was proven by experiments.

The design for the sensor system we develop at present will be integrated in the prototype in the fourth quarter of 1999. The full operational efficiency and practice aptitude of the prototype will be tested in a predefined check programme under real conditions of use.

The fire fighting robot will be commercially available with a standard video system in the beginning of the year 2000. The sensor system developed by Fraunhofer IPA will be available as an add-on package in the middle of the same year.

This system could be the basis for a sensor system for other applications in the field of fire fighting – for example for autonomous fire trucks at airports or for forest fires. It is suitable to acquire data through dense fog (containing water drops or drops of chemical products for example, Oleum fog) or dust as well.

Other possible use cases for the sensor system could be the integration in divers vehicles for hazardous environments (nuclear power plants, mining, chemical accidents) [10].

# 7 References

[1]   Schraft, Rolf Dieter; Schmierer, Gernot:
      Serviceroboter: Produkte, Szenarien, Visionen.
      Berlin; Heidelberg: Springer, 1998

[2]   Gehringer Hagen R.; Rust, Hendrik:
      Brandbekämpfung mit Robotern.
      Spektrum der Wissenschaft - Dossier Robotik,
      Verlag Spektrum der Wissenschaft, Heidelberg, 4/98

[3]   Kreitmair-Steck, W., A. P. Wolframm, A. Schuster:
      HeliRadar - the pilot's eye for flights at adverse weather conditions.
      Proc. of the SPIE.
      AeroSense Symposium. Orlando, April 1995

[4]   Butcher, E. G. ; Parnell, A. C.:
      Smoke Control in Fire Safety Design.
      London: E. & F. N. Spon, 1979.

[5]   Heins, Thomas; Kordina, Karl:
      Untersuchungen über die Brand- und Rauchentwicklung.
      in unterirdischen Verkehrsanlagen - Katastrophenschutz in Verkehrstunneln.
      In: Schriftenreihe „Forschung" des Bundesministers für Raumordnung,
      Bauwesen und Städtebau (1990) Heft-Nr. 481

[6]   Bankstone, C. B.:
      Aspects of the Mechanisms of Smoke Generation by Burning Materials.
      In: Combustion and Flame (1981), Nr. 41

[7]   Hinkley, P. L.:
      Rates of production of hot gases in roof venting experiments.
      In: Fire Savety Journal (1986), Vol. 10, No 3

[8]   Friedlander, S. K.:
      Smoke, Dust and Haze, Fundamentals of Aerosol Behavior.
      New York; London; Sydney; Toronto: John Wiley & Sons, 1977.

[9]   Jin, T.:
      Visibility through fire smoke. Part 2, Visibility of monochromatic
      sings through fire smoke.
      In: Report of Fire Research Institute of Japan (1971), No. 33.

[10]  Rust, Hendrik:
      Löschroboter- Brandbekämpfung der Zukunft.
      Seminar Umweltschutz.
      Daimler Chrysler Aerospace, Immenstaad, November 1998.
      November 1998

# New techniques for 3D modeling ...
# ... and for doing without

Luc Van Gool, Reinhard Koch, and Theo Moons
ESAT, Univ. of Leuven
Leuven, Belgium
Luc.VanGool@esat.kuleuven.ac.be

**Abstract:** Object recognition, visual robot guidance, and several other vision applications require models of objects or scenes. Computer vision has a tradition of building these models from inherent object characteristics. The problem is that such characteristics are difficult to extract. Recently, a pure view-based object recognition approach was proposed, that is surprisingly performant. It is based on a model that is extracted directly from raw image data. Limitations of both strands raise the question whether there is room for middle ground solutions, that combine the strengths but avoid the weaknesses. Two examples are discussed, where in each case the only input required are images, but where nevertheless substantial feature extraction and analysis are involved. These are non-Euclidean 3D reconstruction from multiple, uncalibrated views and scene description based on local, affinely invariant surface patches that can be extracted from single views. Both models are useful for robot vision tasks such as visual navigation.

## 1. Detailed 3D models or view-based representations?

One way of looking at computer vision is to consider it the art of inverse graphics. One tries to gain knowledge about the inherent characteristics of objects and scenes irrespective of the concealing effects of projection and illumination. With 'inherent' characteristics we mean physical properties of the objects themselves, in contrast to how objects may look in images, after the forementioned effects of projection and illumination. Examples are the albedo of a surface or its Gaussian and mean curvatures. Object recognition, visual robot guidance, vision for object manipulation and assembly, and scene understanding in general are areas where knowledge of such features is extremely useful. The models only contain relevant information and are compact as they can be used under different viewing conditions.

Yet, one has to consider how much effort one is willing to invest in retrieving such characteristics. This often proves very difficult. Some research directions with that goal are among the notoriously hard problems of vision. Shape-from-shading and colour constancy are cases in point. One often has to resort to an off-line modeling step, where special apparatus is used or tedious calibration procedures are involved. Although the models have been stripped of all viewing related variability, this very restriction to inherent characteristics renders their use less efficient. Viewing related variations may have been factored out from the model, but they will be present in the actual data to compare the models against. One simultaneously has to hypothesize a model and viewing conditions that together explain the image data. This may entail a painfully slow search in high-dimensional parameter space. There are also limitations to the complexity of objects that such approaches can handle. Building detailed 3D models of intricate shapes is not straightforward, for instance.

In response to the great difficulties encountered in the extraction of inherent characteristics, some remarkably successful alternatives have been proposed recently. The work of Nayar *et al.* [1] has been especially influential. Their approach can be considered an extension of plain template matching. Image data are correlated against principal components of a set of training images. Different objects are represented by different manifolds in the space spanned by a subset of the principal components. Although no on-line feature extraction is applied, the method can efficiently recognize hundreds of objects from a range of viewpoints.

Such techniques that are based on raw image data have the advantage that they require little image processing overhead. They also tend to require less or no camera, robot, and/or hand-eye calibration. The model of an object or scene is extracted directly from their images. These approaches can usually deal with complicated objects and scenes because they do not have to analyse the data. For the very same reason, they are more vulnerable to changing backgrounds and scene clutter, however. Another disadvantage is that many images might have to be taken, both for training (as allowable scene variability should be learned), as for operating the system, because reference images should densely sample the space of possible views. This is a definite disadvantage, that implies that such techniques can only be used in controled worlds, with not too many objects and not too many changes in the viewing conditions.

The pro's and con's of both strands to some extent are complementary. It stands to reason to try and combine aspects of both. In the sequel, two examples are described where we started from one strand, but then tried to include some of the properties of the other.

The first example is a system that builds 3D scene reconstructions from multiple views. In contradistinction with typical structure-from-motion approaches, no explicit camera and motion calibration is necessary, however. The 3D model is built from the input images without any further input. The price to pay is that one has to give up knowledge about the absolute scale of the objects. The example suggests that some inherent characteristics can be exchanged for increased flexibility, while the remaining ones still suffice to solve the tasks.

The second example is a system that represents objects and scenes on the basis of interest points and their surrounding colour patterns. This system is mainly view-based, but it strongly relies on geometric and photometric invariance for which substantial feature extraction is necessary. A big advantage is that this allows to cut down on the number of reference views that have to be used. These two examples are described next.

## 2. Uncalibrated 3D scene reconstruction

3D scene and object models are very useful for tasks such as recognition, grasping, and visual navigation. But how important is a full 3D description? Taking navigation as an example, one is interested in following a path in the middle of a corridor, parallel to the walls, while avoiding collisions with obstacles. These subtasks do not require a full, 'Euclidean' reconstruction of the environment. Middle-of, parallel-to, time-of-contact, etc. are all affine concepts as well. This means that if one were to produce a 3D model that is only precise up to some unknown 3D affine transformation, this model could nevertheless support several aspects of mobile robot navigation. Such model is easier to produce than a Euclidean model. Here we will discuss another example, namely when the model is allowed to have an unknown scale. In that case also relative directions can be read directly off the model. As comes out, giving up scale yields substantial advantages in return. Moreover, the scale is easy to fix. Knowledge of one length would suffice. Odometry could provide it if necessary.

Here we concisely describe a system we developed to create 3D models from a series of images for which the camera parameters and viewpoints are not known beforehand. There are a few limitations, but these are relatively minor. Typically, the baseline between the images has to be small on a pairwise basis, i.e. for every image there is another one that has been taken from a rather similar viewpoint. The reason is that finding correspondences between different views should not be made too difficult for the computer. Such images can be taken from a handheld camera or a camera that is mounted on an autonomously guided vehicle (AGV, for short), as the platform necessarily continuously moves between positions.

The first step consists of automatically tracking image features over the different views. First, a corner detector is applied to yield a limited set of initial correspondences, which enable the process to put in place some geometric constraints (e.g. the epipolar and trilinear constraints). A self-contained overview of such constraints and their mathematical foundations is given by Moons [2]. These constraints support the correspondence search for a wider set of features and in the limit, for a dense, i.e. point-wise, field of disparities between the images.

The initial, limited set of corner correspondences also yields the camera projection matrices that are necesssary for 3D reconstruction [3, 4, 5]. Both the internal and external camera parameters are determined automatically. Internal parameters include the focal length, pixel aspect ratio, angle between image rows and columns, and the coordinates of the image center. The external parameters specify the 3D position and orientation of the camera. In general, to arrive at metric structure (Euclidean except scale) — i.e. to undo any remaining projective and affine skew from the 3D reconstruction — the camera's internal parameters like the focal length etc. have to remain fixed. But even if one has limited a priori knowledge about these parameters, like the pixel aspect ratio or the fact that rows and columns in the images are orthogonal, then also focal length can be allowed to change [6, 7].

Figure 1. *Two frames of an image sequence of an Indian temple.*

Consider the example of fig. 1. It shows two images of an Indian temple from a series of 8, used for its 3D reconstruction. All images were taken from the same ground level as these two. Fig. 2 shows 3 views of the 3D reconstruction, from viewpoints different from those of the input images.

As this example shows, it is possible to generate 3D models directly from images. Hence, a robot could build 3D maps of its environment from an image sequence that it takes and it can compare it against a 3D model that has been built off-line, but with the same hardware. This approach still captures most of the inherent shape characteristics, but an advance over older methods is that it no longer requires camera calibration or precise odometry. Inherent characteristics are derived and exploited up to the point where they can still be extracted from the images alone. An affine reconstruction can be obtained from as few as two images if the camera translates in between [8]. This motion is very typical of an AGV. As mentioned earlier, such model would still be useful for navigation.

A more detailed account is given in Pollefeys *et al.* [4].

Figure 2. *Three views of the 3D reconstruction obtained for the Indian temple.*

## 3. Affinely invariant patches of interest

Over the last couple of years, important progress has been made in object recognition research. They have seen the introduction of strongly view-based approaches, such as the approach of Nayar *et al.* [1]. In the area of geometry-based methods novel invariants have been derived and tested [9].

Recently, a particularly interesting approach has been proposed by Schmid and Mohr [10]. It heavily leans towards using a view-based approach, but it does a fair amount of feature extraction in that local invariants are extracted. On the one hand, their approach inherits from the view-based school the ability to deal with objects of high complexity. On the other hand, their approach is less sensitive to scene clutter and occlusions, an advantage of using local invariants. The idea is to identify special points of interest, e.g. corners, and to extract translation and 2D rotation invariant features from the colour pattern in a circular neighbourhood. Invariance under changing illumination has been included as well. Invariance under scaling is handled by using circular neighbourhoods of several sizes. The invariants are derived from Gaussian derivatives of intensity calculated over the neighbourhoods. The invariance ensures that the number of reference images can be reduced. In particular, shifting or rotating the object in the image will have no effect on the invariants and hence a single reference image in the model suffices to cover all these cases. Recognition requires that similar points of interest are extracted from the given image, together with their invariants. The same points of interest are quickly found in the model database of reference views by comparing their neighbourhood invariants.

We have extended this framework. The goal has been to further reduce the number of required reference images. To that end, affinely invariant neighbourhoods are derived. This means that instead of being insensitive only to 2D translation and rotation, the descriptions that are generated for the neighbourhoods of the corners are now invariant under plane affine transformations. This is not a straightforward generalization, however. The crux of the matter is to define the neighbourhoods in such a way that they change with the viewpoint.

The selected neighbourhoods correspond to viewpoint dependent parallelogram or ellipse-shaped patches around corners. When changing the position of the camera the neighbourhoods in the image change their shape in order to keep on representing the same physical part of the object or scene. We will refer to these configurations of corners and neighbourhoods as "affine patches". In a way the work is akin to that of Pritchett and Zisserman [11] who look for parallelogram shapes that are explicitly present in the image to find correspondences between wide baseline stereo pairs, but we construct virtual parallelograms from surface textures.

The affine patches are subsequently characterised by allowable geometric deformations which are plane affinities now, and the changes to the colour bands are similar to those envisaged by Schmid and Mohr, i.e. different scalings and offsets for each band. The steps in the construction of the affine patches are all invariant under the forementioned geometric and photometric changes. The geometric/photometric invariants used to characterize the neighbourhoods are moment invariants. We have made complete classifications for moments of lower orders, which yield the best results. The moments in these expressions can be traditional moments or 'Generalized Colour Moments' which we have introduced to better exploit the multispectral nature of the data [12]. The latter contain powers of the image coordinates and the intensities of the different colour channels. They yield a broader set of features to build the moment invariants from and, as a result, moment invariants that are simpler and more robust.

The result of the affine patches is that the viewing angles from which an object can still be recognized based on the same points of interest and their invariant descriptions is substantially increased. Reference views that sample

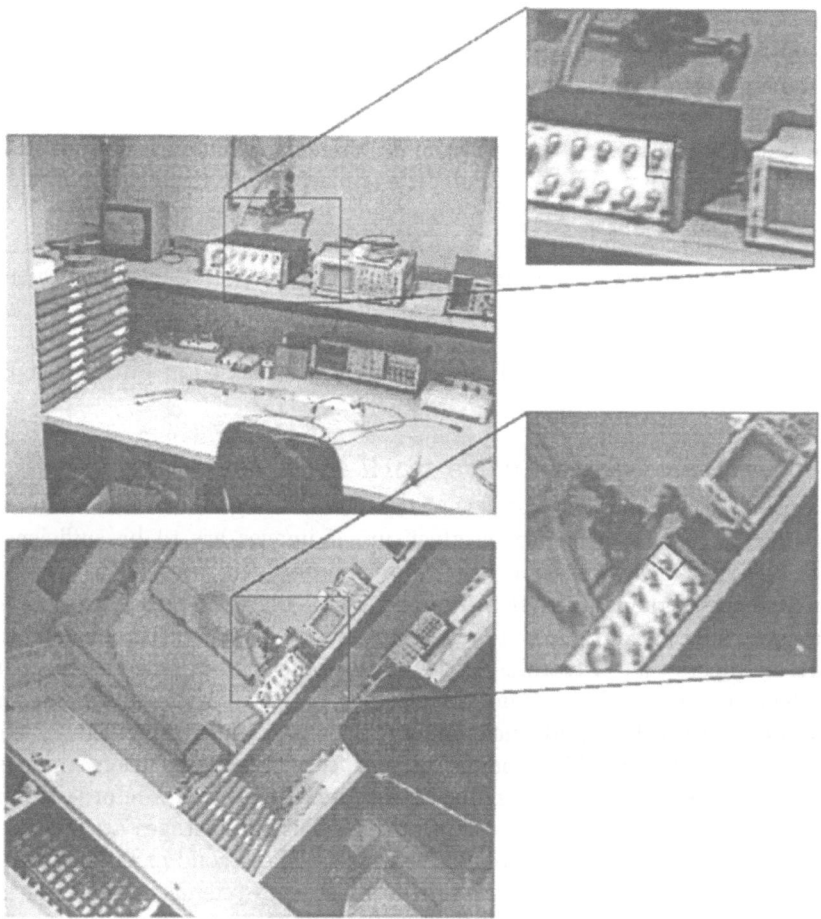

Figure 3. *Two images of an office scene, taken from very different viewpoints. The inlets show a detail and one of the many affine patches for which a correspondence was found (parallelograms).*

the viewing sphere more sparsely can now suffice for viewpoint independent recognition.

We currently investigate the use of affine patches for image database retrieval, wide baseline stereo matching, and visual guidance of a mobile robot. The example in fig. 3 shows two images of a scene, taken from substantially different viewpoints. A pair of corresponding affine patches is shown in the inlets. Parallelograms represent the same physical patches on the instrument. Their image shapes are different, but they cover the same physical part of the scene. Typically, more than enough initial correspondences are found to calculate the epipolar geometry for the image pair, such that the system can bootstrap itself to find additional correspondences. Finding epipolar geometry is also useful for navigation, as it yields the general motion direction of the robot platform. The robot can have a substantially different viewpoint from any image used to build the set of model reference views. The affine invariance ensures a high probability of matching the given image against at least one reference image. The epipolar geometry between that pair then gives general indications of how one should move to e.g. maximize the resemblance of the observed views with reference views that are taken along a learned path. The affine patches serve as natural beacons. No special markers have to be used.

Also in this second example, care has been taken that the affine patches can be picked up automatically and solely from the images. Effects of projection and illumination do not stand between the model and the features extracted in the given images.

The extraction and use of affine patches is discussed at greater length in [13].

## 4. Conclusions and future work

In this paper, we have suggested that the idea of using inherent characteristics does not necessarily have to be given up, if images are to be sufficient for both the creation of the models and the solution of a task. It is rather the completeness of the inherent descriptions that may have to be sacrificed. Taking the first example as an illustration, there we have given up the absolute scale. Several objects that only differ in their size would in that case be represented by the same model. This model therefore no longer uniquely represents those objects. On the other hand, such models can be produced from images, without special equipment or calibration procedures. With respect to AGVs, there is no need for precise odometry or regular calibration of the camera in order to build descriptions similar to the models. It seems these advantages are well worth the relatively slight degradation of the model. The second example exploited invariance as a way to cut down on the number of reference images needed. There the description is much more incomplete. The features cover only part of the surface. The affine invariance filters out inherent characteristics that change with viewing direction. What is left is an impoverished description of the shape. Yet, the locality of the features and the level of invariance in the description render the technique efficient. The locality makes sure that the assumption of patch planarity often holds to a good degree and that affine deformations model the effect of projection well. The affine invariance has the advantage that the models can be obtained directly from images. Such a technique is also more robust than techniques based on raw image data. Occlusions, partial visibility, and scene clutter can be handled due to the locality of the affine patches. Not all of them need be visible and at least a few probably are. This technique will also be immune against some changes never actually shown to the system during training. This is a problem when raw image data are used. Indeed, as the degrees of freedom in relative camera motion and changes of illumination increase, it quickly becomes absolutely impossible to collect a representative set of reference views. The invariance lowers the dimension of the problem.

Thus, it is interesting to ask how complete (or how precise) a model should be to be useful, while it can still be retrieved from images. The more completely models capture inherent object features, the more difficult they are to produce. Distant objects can e.g. not be reconstructed in 3D and hence a skyline, which is useful for navigation, will not be included in a 3D model. It is useful to consider descriptions that cover the middle ground between complete sets of inherent features and raw image data. In terms of geometric scene models, recent research in graphics and computer vision has e.g. turned towards pseudo-3D descriptions such as lightfields, scene descriptions that consist of planar layers at different depths [14], and planes with an additional relief component [15]. One can well imagine mixed descriptions with detailed 3D models for the objects in the foreground that have to be manipulated, affine patch based layer descriptions for less relevant objects and objects somewhat farther away that can be recognized or used as beacons, and a cylindrical mosaic of the skyline to determine overall orientation. The automatic generation of such composite descriptions is an issue for future research.

**Acknowledgements:** The authors gratefully acknowledge support of ACTS project AC074 'VANGUARD' and IUAP project 'Imechs', financed by the Belgian OSTC (Services of the Prime Minister, Belgian Federal Services for Scientific, Technical, and Cultural Affairs).

# References

[1] Murase H, Nayar S 1995 Visual learning and recognition of 3-D objects from appearance. *International Journal of Computer Vision* 14:5-24

[2] Moons T 1998 A guided tour through multiview relations. In: Koch R, Van Gool L (eds) 1998 *Proceedings of the SMILE Workshop — 3D Structure from Multiple Images of Large-scale Environments.* Springer-Verlag, Berlin (Lecture Notes in Computer Science No. 1506), pp. 304-346

[3] Armstrong M, Zisserman A, Beardsley P 1994 Euclidean structure from uncalibrated images. *Proc. British Machine Vision Conference (BMVC '94)* pp.

[4] Pollefeys M, Koch R, Van Gool L 1998 Self-calibration and metric reconstruction in spite of varying and unknown internal camera parameters. *Proc. International Conference on Computer Vision (ICCV '98)*, Bombay, India, pp. 90-95

[5] Triggs B 1997 The absolute quadric. *Proc. International Conference on Computer Vision and Pattern Recognition (ICCV '97)*, pp. 609-614

[6] Heyden A and Åström K 1997 Euclidean reconstruction from image sequences with varying and unknown focal length and principal point. *Proc. IEEE Conference on Computer Vision and Pattern Recognition (CVPR '97)*

[7] Pollefeys M, Van Gool L, Proesmans M 1996 Euclidean 3D reconstruction from image sequences with variable focal lengths. *Proc. European Conference on Computer Vision (ECCV '96)* pp. 31-42

[8] Moons T, Van Gool L, Proesmans M, Pauwels E 1996 Affine reconstruction from perspective image pairs with a relative object-camera translation in between. *IEEE Trans. Pattern Analysis and Machine Intelligence (T-PAMI)* 18:77-83

[9] Mundy J, Zisserman A (eds) 1992 *Applications of invariance in vision.* MIT Press, Boston

[10] Schmid C, Mohr R 1997 Local greyvalue invariants for image retrieval. *IEEE Trans. Pattern Analalysis and Machine Intelligence (T-PAMI)* 19:872-877

[11] Pritchett P, Zisserman A 1998 Wide baseline stereo matching. *Proc. International Conference on Computer Vision (ICCV '98)*, pp. 754-759

[12] Mindru F, Moons T, Van Gool L 1998 Color-based moment invariants for the viewpoint and illumination independent recognition of planar color patterns, *Proc. International Conference on Application of Pattern Recognition (ICAPR' 98)*, pp. 113-122

[13] Tuytelaars T, Van Gool L, D'Haene L, Koch R 1999 Matching affinely invariant regions for visual servoing, accepted for oral presentation at the *International Conference on Robotics and Automation*, Detroit

[14] Shade J, Gortler S, He L-W, Szeliski R 1998 Layered Depth Images. *Computer Graphics (SIGGRAPH '98)*

[15] Debevec P E, Taylor C J, Malik J 1996 Modeling and rendering architecture from photographs: A hybrid geometry- and image-based approach. *Computer Graphics (SIGGRAPH '96)*, pp. 11-20

# A Mobile Robot Navigation by Means of Range Data from Ultra Wide Angle Laser Sensor and Simple Linguistic Behavior Instructions

Takashi Tsubouchi, Yoshinobu Ando and Shin'ichi Yuta
Intelligent Robot Laboratories, Colledge of Engineering Systems
University of TSUKUBA,
1-1-1, Tennoudai, Tsukuba, Ibaraki, 305-8573 Japan
tsubo@esys.tsukuba.ac.jp

**Abstract:** A navigation scheme for a mobile robot is presented, where a set of linguistic instructions are formulated to command the robot to move toward destination in a corridor environment. This approach do not need a detailed model of the environment. The robot is equipped with ultra-wide-angle laser range sensor. Fundamental functions based on the surroundings information obtained from the laser range sensor: to move along the corridor without any collision with walls nor objects, and to find a crossings are provided on the robot to move in the corridor directed by the set of instructions.

## 1. Introduction

This research work is aimed at a mobile robot navigation in a corridor environment. In the conventional approach for indoor navigation, a map as a detailed model of workspace is provided and all the objects are represented in the map with respect to a pre-defined coordinate system. Once a start and a goal locations are provided, a path which leads the robot to the goal is generated from the model. The path is described with respect to the coordinate system and is followed by the robot. As the robot usually estimates its position by an odometry system, the robot is well controlled to follow the path. Such a navigation scheme is regarded as a "position based navigation".

In the position based navigation approach, navigation could be realized by "cheap" sensors to estimate the robot position – such as pulse encoders. When we assume the map is almost perfect and estimated position by odometry is accurate enough, the robot can reach the destination only by keeping track of the path. However, such a detailed map takes "laborious" work if the workspace for the robot spreads extremely large. In such a case, another approach emerges: necessary navigation is realized by "rich" sensors and "simple" instructions. Navigation by this approach is a principal issue of this paper.

The authors equipped the mobile robot with a ultra-wide-angle laser range sensor as a "rich" sensor which consists of laser fan beam projectors and CCD cameras (Figure 1 and 2). The laser fan beam is projected so as to be parallel

Camera   Camera   Camera

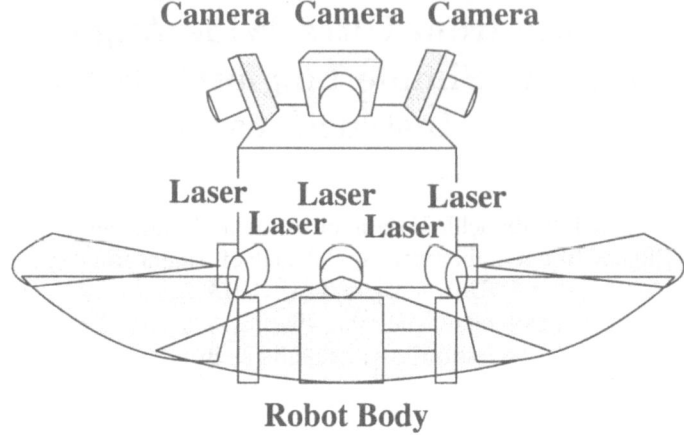

Robot Body

Figure 1. A schematic illustration of mobile robot with laser range sensor

to the floor. The three CCD cameras whose lens has wide view angle of 108 degrees detect the reflections of laser beam on the objects around the robot. The sensor can measure the ranges at every one degree from the front center to 130 degrees right and left of the robot. Hence, it can detect the shape of walls and obstacles in the corridor. This detected shapes are used in an algorithm which is developed by the authors [5]. It functions as a "navigator" to lead the robot along a corridor with wall following. The robot takes distance from the wall when the robot comes close to the wall during the wall following. If any obstacle appears in front of the robot, it must avoid the obstacle. Such behaviors arise from the "rich" sensory information, not from any detailed map. These behaviors are regarded as "sensory data based behaviors", where so many issues are reported [1, 2]. In this paper, the mobile robot navigation in corridors is realized by associating such sensory data based wall following function with the simple set of instructions, whose grammar is formulated in a "showing a person the way manner" by the authors.

In Section 2, formulations of the instructions are presented. In Section 3, the snapshots of the navigation experiment followed by the instructions are illustrated.

## 2. Action sequence instructions in "showing a person the way manner"

A principle to formulate the instructions is based on the intuitive manner of guidance, such as "Go along this corridor. You may find 'T' junction. It must be 2nd crossing. Then turn left and go straight 10 meters. You will finally find the lab of your destination." Inspired from such ideas, Shibata *et al.* [3] proposed the linguistic instructions of a navigation route associated with making correspondences to symbolic representations of corridors. In Shibata's approach, however, abstracted representation of the linguistic instructions are defined. Although the authors do not intend to study natural language under-

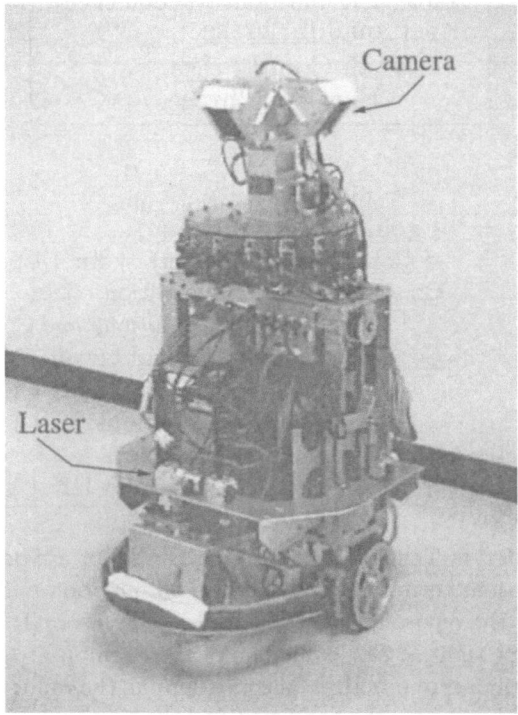

Figure 2. A mobile robot with laser range sensor

standing, instructions in a natural linguistic style has been preferred, because it may be good for human-robot interface.

### 2.1. Formulations

The authors define a unit of instruction clause which consists of three reserved words – *Keep_on, When, Exec* and three phrases – <u>JobPhrase</u>, <u>CondPhrase</u> and <u>ActPhrase</u>. Generation rules for the pherases are presented in Table 1 and Table 2.

The first line of Table 1 associated with the first line of Table 2 yields "*Keep_on* **going_along_corridor**", which commands the robot to begin keeping on going along the corridor relying on the wall following function [5].

The second line of Table 1 beginning from *When* associated with <u>CondPhrase</u> specifies the condition(s) to check whether walls disappear or the robot moves specified distance while the robot keeps on going along corridor. One must be selected from the entries in <u>Condition</u>. In <u>Condition</u> entries, **no_front_wall** means "there is no wall in front" and **right_wall** means "there is a wall on the right", for examples. Therefore, if someone specifies "*When* no_right_wall and no_left_wall, *Exec* turn_left", the robot will turn left when the both walls in left and right disappear at the same time, *i.e.* the robot will turn left at either crossroad or T-junction.

This unit as in Table 1 is described in sequence. One example of instruc-

Table 1. A unit of instruction clause

| |
|---|
| Keep_on JobPhrase. |
| When CondPhrase, |
| Exec ActPhrase. |

Table 2. Generation rules

| JobPhrase | = | { going_along_corridor } |
|---|---|---|
| CondPhrase | = | { Condtions [*at* Number] \| for Dist m} |
| | | Conditions = { Condition |
| | | \| Conditions *and* Condition} |
| | | Condition = { no_right_wall \| right_wall |
| | | \| no_left_wall \| left_wall |
| | | \| no_front_wall \| front_wall } |
| | | Number = {1st \| 2nd \| ...} |
| ActPhrase | = | { stop \| turn_right \| turn_left \| go_straight} |

tions are illustrated in Table 3. Assume that the robot are desired to pass the first (cross) junction, turn right at the second T-junction and turn left at the last crossroad in the environment illustrated in the left of Table 3. The set of instructions in the right of Table 3 is one of the examples to put the assumed behavior of the robot into practice. In this example, the robot will pass through the first crossroad, because the condition to take turn right action is specified by the case when there are no walls in the right and left but there is a wall in front, which is not a case of the crossroad junction.

Table 3. Instructions example

| Situation | Instructions |
|---|---|
| | Keep_on going_along_corridor. When no_right_wall and front_wall and no_left_wall, Exec turn_right. Keep_on going_along_corridor. When no_left_wall and no_front_wall and no_right_wall, Exec turn_left. |

## 2.2. Implementations

### 2.2.1. Compilation of the instructions

The basic programming language of the mobile robot of the authors is the Robol/0 language [4], which allows the programmer to describe the robot behavior in sensory data driven manner. The program in the Robol/0 is compiled to make executable object module and it is downloaded onto the robot. The robot runtime system will finally execute the program. The set of instructions in Section 2.1 is translated into the Robol/0 (Figure 3). A pre-processor for this translations is designed by using *yacc* and *lex*.

### 2.2.2. Junction detection

Figure 4 illustrates how a junction will be detected. Detection of the junction is, in other words, the detection of an incidence *When* CondPhrase conditions are fulfilled. For this detection, range data from the ultra-wide-angle laser range sensor is also utilized while the wall following functions. The rectangle areas are settled in front, left and right of the robot. If any long line segment which is fitted to the range data is detected inside of one of the rectangles, such a rectangle will be flagged to indicate that there is a wall. If not, there is no wall. For instance, if the long line segment is inside of the left rectangle, "*When* left_wall " condition fires, but "*When* no_left_wall" condition will not fires.

The size of the right and left rectangles has 3m width and 60cm depth, which are sufficient for wall detection because the corridors as experimental environment has at most 4m width and detectable range of the laser range sensor is at most within 3.5m. The depth of 60cm is because of the depth of the mobile robot. Similarly, the front rectangle has 60cm width and 3m depth.

## 3. Navigation experiments

Figure 5 illustrates the floor plan of the corridor environment of the buildings in the university of the authors. A navigation scheme based on the framework in this issue has been examined in this environment. The environment extends about 200 meters by 200 meters. Width of corridors are at most 4 meters.

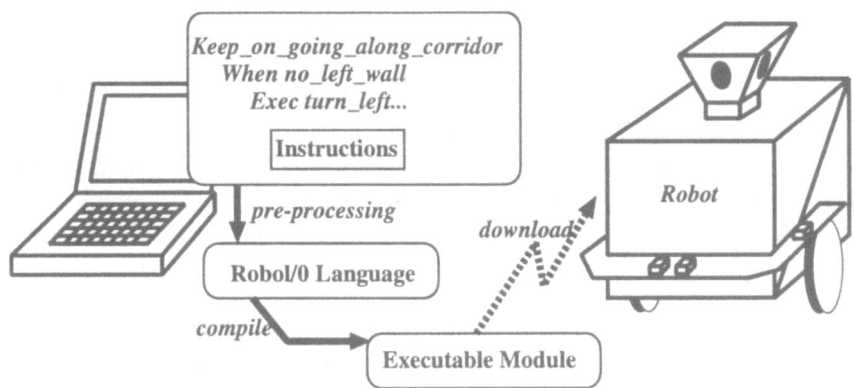

Figure 3. Implementation

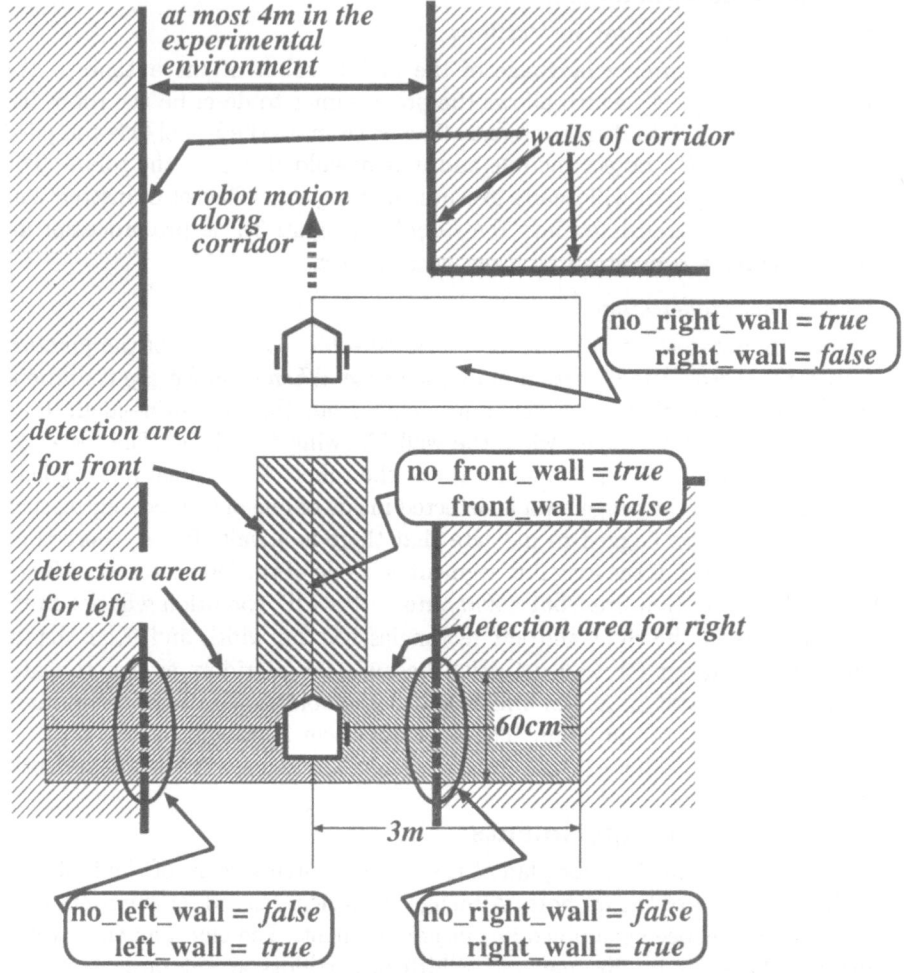

Figure 4. Detection of a junction

There are several crossroad junctions and T-junctions.

A mission of the robot is to move from the start in front of Laboratory A to the goal in front of Laboratory B in the corridor. Together with the floor plan, A path for the mission is illustrated in Figure 5. On the middle way of the path, the locations are numbered with $m1$ through $m8$. The robot must pass through the 1st crossroad near the location $m1$, turn right at the 2nd crossroad near $m4$, and turn left at the 3rd crossroad near $m6$ and $m7$. After the last turn, the robot must move more 5 meters and finally it stops. Set of instructions which is shown in Table 4 is an example to accomplish the mission.

The robot could come to the goal as desired. Figure 6 includes snapshots of the navigation experiment at every numbered location in Figure 5. This experiment will succeed in almost 7 trials within 10 trials.

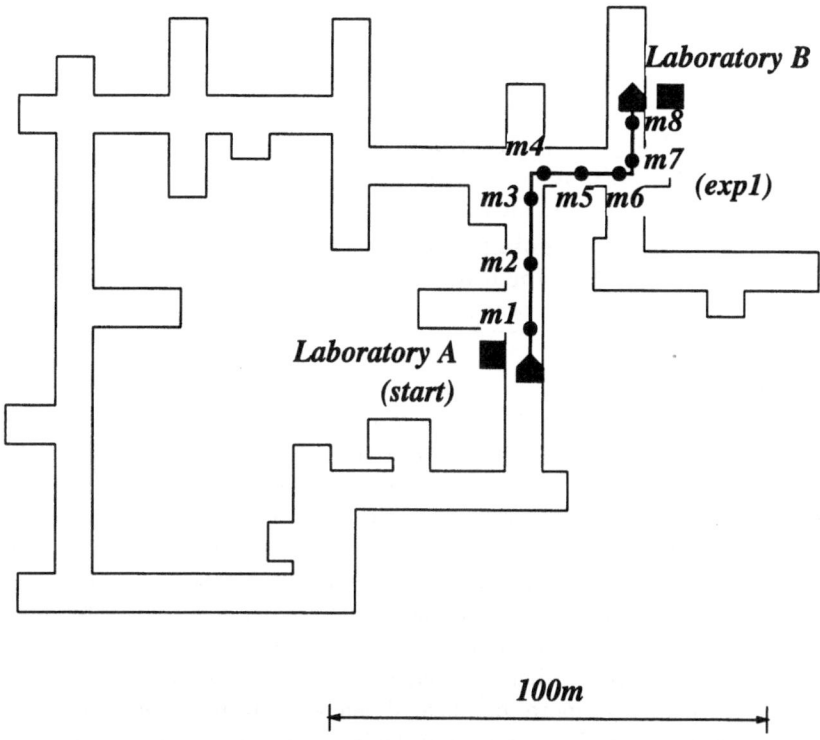

Figure 5. Environment for navigation experiment and path to the goal

Failing cases are mostly caused by collisions with unexpected obstacles in the corridor. The laser range sensor detects only the object which the laser beam comes across, and a laser beam extends only 20 cm above the floor. Hence, such an object with low hight like a hand cart could not be detected, nor the robot could not avoid such an invisible object. Unless we wave the laser fan beam vertically or we remove such sensor invisible objects, there is no help for it. In other case, the robot sometimes collides with a corner of the junction. This reason arises from weakness of implementations for **turn_right**, **turn_left**, and **go_straight** directives in the instructions. The robot is simply commanded to follow the directives and sensory data will not be referred while the robot passes through the junction because of the limit of processing speed of the laser range sensor. However, the wall following function which leads the robot to the junction can not guarantee that the robot heading is accurately parallel to the center line of the corridor. Thus, the robot will collide with a corner if the robot heading has some angle to the center line of the corridor. Even while the robot passes through the junction, the robot should watch sensory information and avoid collisions.

Table 4. An example of instructions for navigation experiment

> **Keep_on going_along_corridor.**
>   **When no_right_wall**
>     **Exec turn_right.**
> **Keep_on going_along_corridor.**
>   **When no_left_wall**
>     **Exec turn_left.**
> **Keep_on going_along_corridor.**
>   **When for 5 m,**
>     **Exec stop.**

## 4. Conclusions

A mobile robot navigation scheme in a corridor environment is presented, where the instructions to command the robot to move from start to goal are provided in a "showing a person the way manner". The grammar of the instructions are defined, which consists of three phrases. The one phrase to specify the robot to go along the corridor is associated with the wall following function based on the ultra-wide-angle laser range sensor which was proposed by the authors [5]. In the realistic environment, the scheme has been experimented. Examples of the experiment and evaluations are reported. It is realized that improvement of motions when the robot passes the junctions is necessary. The more experiments with more complicated behaviors in the corridor environment is necessary to assure the grammar of the instructions. These problems will be addressed in future.

## References

[1] Arkin R. C., 1998, Behavior-Based Robotics. *The MIT Press.*

[2] Brooks R. A., 1986, A Robust Layered Control System for a Mobile Robot. *J. Robotics and Automation* RA-2:1,14–23.

[3] Shibata F., Ashida M., Kakusho K., Kitahashi T., 1998, Correspondence between Symbolic Information and Pattern Information for Mobile Robot Navigation under Linguistic Instructions of a Route. *J. Japanese Soc. of Artificial Intelligence* 13:2,221–230 (in Japanese).

[4] Suzuki S. and Yuta S., 1991, Analysis and Description of Sensor Based Behavior Program of Autonomous Robot Using Action Mode Replesentation and Robol/0 Language. *Proc. IEEE/RSJ Int. Workshop on Intelligent Robots and Systems* 1497–1502.

[5] Ando Y., Tsubouchi T. and Yuta S., 1999, Development of Ultra Wide Angle Laser Range Sensor and Navigation of a Mobile Robot in a Corridor Environment. *J. Robotics and Mechatronics* 11:1,25–32.

[6] Ando Y., Tsubouchi T., and Yuta S., 1996, A Mobile Robot Navigation Using Both Wall Following Control and Position Based Control. *Video Proceeding of IECON'96.*

[7] Ando Y. and Yuta S., 1995, Following a Wall by an Autonomous Mobile Robot with a Sonar-Ring. *Proc. IEEE Int. Conf. Robotics and Automation* 3:2599–2606.

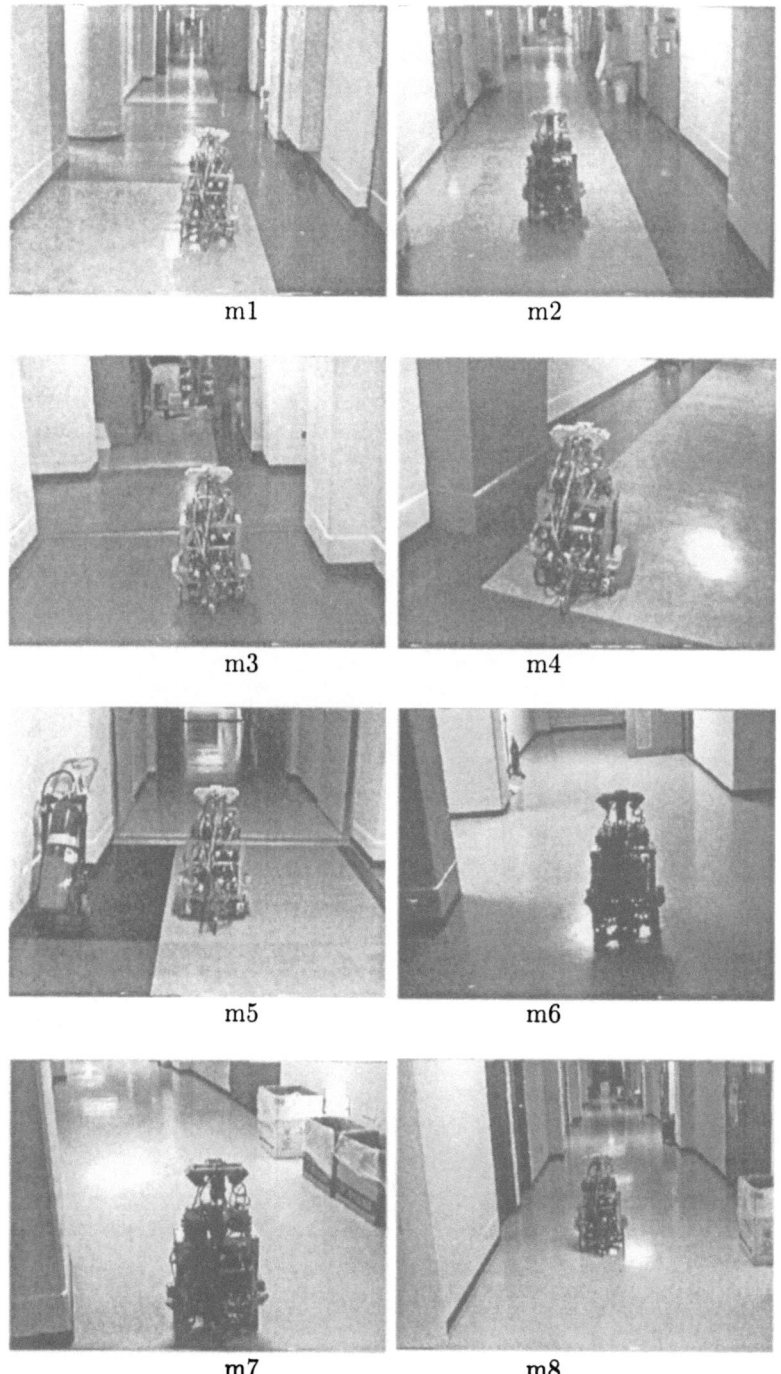

Figure 6. Snapshots of the navigation experiment (*m1–m8* correspond to the locations illustrated in Figure 5)

# Explicit incorporation of 2D constraints in vision based control of robot manipulators

**Guillaume Morel, Thomas Liebezeit**
University of Strasbourg, ENSPS
CNRS, LSIIT, Groupe de Recherche en Automatique et Vision Robotique
Boulevard Sébastien Brant F-67400 ILLKIRCH – France
Guillaume.Morel@ensps.u-strasbg.fr

**Jérôme Szewczyk, Sylvie Boudet and Jacques Pot**
Electricité de France, Direction des Etudes et Recherches
Groupe Téléopération et Robotique
6, Quai Watier F-78400 CHATOU – France

**Abstract.** This paper concerns eye-in-hand robotic applications that involve large relative displacements between the camera and the target object. A major concern for this type of task is to achieve simultaneously two different objectives. Firstly, the robot has to reach its desired final location relative to the target. Secondly, it must be guaranteed that the target stays in the camera's field of view during the whole motion. This paper introduces a new vision based controller that guarantees a convergence towards the final location while keeping the whole target in the camera's field of view.

## 1. Introduction

A growing number of advanced robotic applications requires the exploitation of an end effector mounted camera in the control of robot's motions. Our research particularly focuses on developing robust vision based control strategies for telemaintenance operations in hostile environments [1].

A characteristic of these applications is that, potentially, very large rotational and translational displacements have to be produced between the initial and final configurations. The final configuration is defined relative to a known object (e.g. a tool, a part to be grasped, ...). The image viewed by the camera in the final configuration is recorded in advance, during a teaching phase. Then, during the real task execution, the operator roughly moves the robot towards the object of interest, until the visual target fits in the camera's field of view. This defines the initial configuration for the vision based controller, which shall move the robot towards the learned final configuration. It is desirable to provide a system robust enough to converge towards the final configuration, even if it requires a large translational and/or rotational displacement. An important problem is to guarantee that during the vision based motion, the entire target stays inside the camera's field of view.

## 2. 3D and 2D trajectories

### 2.1 Position-based Control

Position-based visual servoing [3, 4] is a simple way of achieving relative positioning tasks using vision based control. In this approach, the position and orientation of the target with respect to the camera are first estimated using a pose reconstruction algorithm (see, e.g. [7]). This algorithm usually exploits image features and a geometrical model of the visual target. Without loss of generality, we consider that the vision based controller is provided at each sampling period, with the homogeneous transform from the camera frame $\mathcal{F}_c$ to the target frame $\mathcal{F}_t$, which will be noted $^c\mathbf{H}_t$. Also, the homogeneous transform $^{c*}\mathbf{H}_t$, from the frame $\mathcal{F}_{c*}$ attached to the camera in its final desired location and the target frame is supposed to be known.

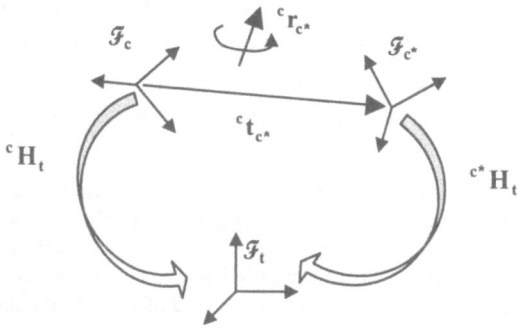

*Figure 1: Moving the camera in an absolute frame*

A widely used method (see e.g. [3]) consists in computing at each sampling period the positioning error between $\mathcal{F}_c$ and $\mathcal{F}_{c*}$. Then the camera displacement is controlled with respect to a fixed absolute frame (see Figure 1). The twist $^a\mathbf{T}_c$ of the current camera frame in the absolute frame $\mathcal{F}_a$ is computed by :

$$^a\mathbf{T}_c = \begin{bmatrix} ^a\mathbf{V}_c \\ ^a\mathbf{\Omega}_c \end{bmatrix} = \lambda \begin{bmatrix} ^c\mathbf{t}_{c*} \\ ^c\mathbf{r}_{c*} \end{bmatrix} \tag{1}$$

where $^a\mathbf{V}_c$ and $^a\mathbf{\Omega}_c$ are the linear and rotational velocities, respectively ; $^c\mathbf{t}_{c*}$ is the translation from $\mathcal{F}_c$ to $\mathcal{F}_{c*}$ ; $^c\mathbf{r}_{c*} = \theta\mathbf{u}$ is the finite rotation between from $\mathcal{F}_c$ and $\mathcal{F}_{c*}$ ; $\lambda$ is a proportional gain. This controller will be referred in the rest of the paper as « camera motion based » (CMB) 3D control.

In terms of trajectory, this control law tends to move the camera frame's origin towards its final location following a straight line. In the mean time, it rotates the camera following a geodesic, that is the shortest path between the initial and final locations. The camera's motion is independent form the target location. Consequently, there is no control of the 2D target image trajectory.

This is illustrated in the next simulation result. In this example, robot and camera dynamics were neglected to emphasize the influence of the control strategy on the trajectories. The camera is supposed to satisfy a perfectly calibrated pin hole model (see Figure 2). The pose reconstruction is then perfect. The simulations use a target model constituted by 7 coplanar points placed in order to form a π.

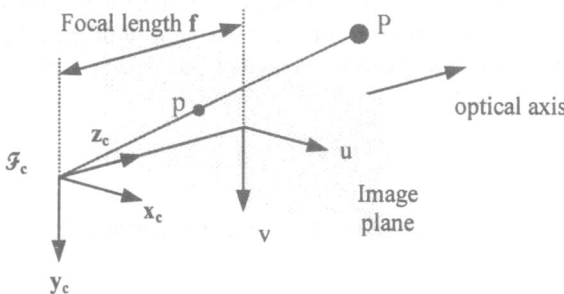

*Figure 2 : Camera model used for the simulation*

Figure 3 gives the simulation results for the CMB controller. Figure 3.a shows the straight line motion of the camera, which confirms analysis. In this figure, the sphere represents the target location ; the cones depict the optical axis, and the bold sticks indicate the rotation of the camera frame around its optical axis. Figure 3.b shows the motion of the target in the image plane. The initial position of the target is indicated by squares, while the final desired configuration is indicated by the circles. The gray rectangle represents the image range. One can see that due to the large rotational motion around the optical axis, all the seven points are leaving the camera's field of view. This is because the control law (1) does not allow any control of the 2D trajectory.

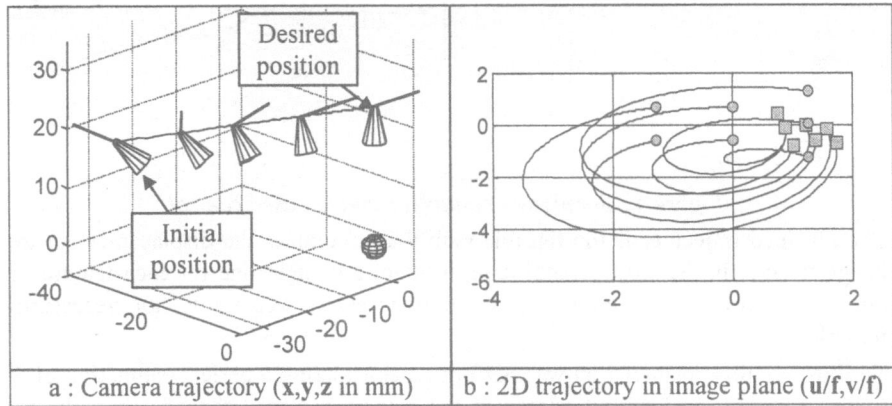

| a : Camera trajectory (x,y,z in mm) | b : 2D trajectory in image plane (u/f,v/f) |

*Figure 3: Simulation results for CMB 3D control*

## 2.2 Image-based control

Image-based control [5, 6], is a way of constraining the 2D trajectory of the target. With this approach, which is fundamentally different from position-based control, a set of features is extracted from the current image, and compared to its desired final value. An estimation of the *inverse image jacobian*, that relates the image feature velocity to the camera velocity is then used to compute the command velocity from the 2D error vector.

Here, we will consider that the image features are the image coordinates of a group of points. Although it is probably not optimal, this choice is the most popular over the literature. In terms of 2D trajectory, an ideal solution would be to have all the feature points following straight lines. However, in most cases, the dimension **m** of the image feature space is bigger than the dimension **n=6** of the robot motion space. The image jacobian **J** that maps robot velocities into image features velocities $\dot{s}$ is not square :

$$\dot{s} = J \begin{bmatrix} {}^aV_c \\ {}^a\Omega_c \end{bmatrix}, \qquad J : m \times n, \; n = 6 \qquad (2)$$

Instead of the inverse image jacobian $J^{-1}$, a pseudo inverse $J^+$ is often used [2]. The control law is then :

$$ {}^aT_c = \begin{bmatrix} {}^aV_c \\ {}^a\Omega_c \end{bmatrix} = -\lambda J^+ (s - s^*) \qquad (3)$$

where s is the vector grouping the current value of the image features, and s* its final desired value.

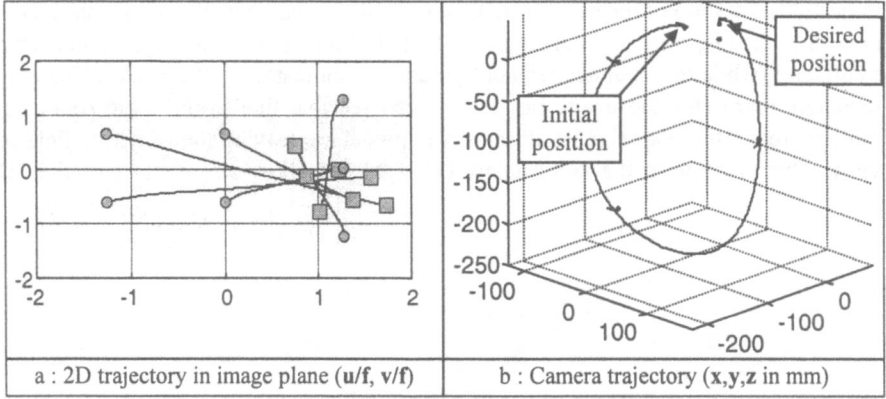

| a : 2D trajectory in image plane (u/f, v/f) | b : Camera trajectory (x,y,z in mm) |

*Figure 4 : Simulation results for image based control*

The obtained trajectory is the feasible path that minimizes the displacement of the image feature. In the most general case, it is not a straight line for each feature. In effect, the straight line motion between two feature vectors is not necessarily feasible.

To illustrate this, a simulation was performed considering a perfect estimation of $J^+$ (which supposes to know at each step the camera-to-target distance). Results are plotted in Figure 4. One can see that the target points motion is much more direct than the with CMB-3D control (Figure 4.a). However, the robot displacement is very large compared to what it should be. In fact, it is absolutely not compatible with realistic experimental conditions, since the robot moves behind the target (Figure 4.b ; note that the axis scale differs by a factor 10 from Figure 3.a).

Thus, the constraints imposed by image-based control on the 2D trajectory are clearly too tight.

Alternatively to position-based control and image-based control, a desirable solution for large range vision based positioning is to simultaneously provide a certain amount of control on both the 3D and 2D motions.

### 2.3 2D½ control

In order to combine the two approaches, 2D½ visual servoing has been recently proposed in [9]. In this approach, the rotational motion is controlled based on a 3D reconstruction algorithm, while the translation is controlled based on image features :

$$\begin{cases} {}^{a}V_{c} = -\lambda J_{v}^{-1}(s-s^{*}) + \lambda J_{v}^{-1}J_{\Omega}{}^{c}r_{c^{*}} \\ {}^{a}\Omega_{c} = \lambda^{c}r_{c^{*}} \end{cases} \qquad (4)$$

where **s** is a 3×1 vector constituted by the image coordinate of a point and a scaled estimation of the target depth ; $J_v$ and $J_\Omega$ are the 3×3 image jacobians for the linear and angular velocities, respectively.

In (5), the rotational velocity control is exactly the same as in (1), which implies that the rotational path will follow a geodesic. The translation velocity control is constituted by the sum of :

• an uncoupling term, that cancels the effect of the rotational velocity in the image ;

• a term aimed at moving a given point of the target image along a straight line.

The behavior of this approach is illustrated in Figure 5. Figure 5.a shows clearly that the target point incorporated in **s** (drawn in black) follows a straight line. One can also observe on Figure 5.b that the 3D trajectory is curved as compared to Figure 3.a. It shall be noted that a very similar behavior is obtained with another form of 3D servoing, that could be called "target motion based" (TMB) control. This approach considers the camera's view point and try to minimize the displacement of the target frame in the camera frame (see [4], [2]). It is then also possible to constrain a point of the target along a 3D straight line trajectory in the camera frame. As a result, its image, follows a 2D straight line trajectory in the image plane.

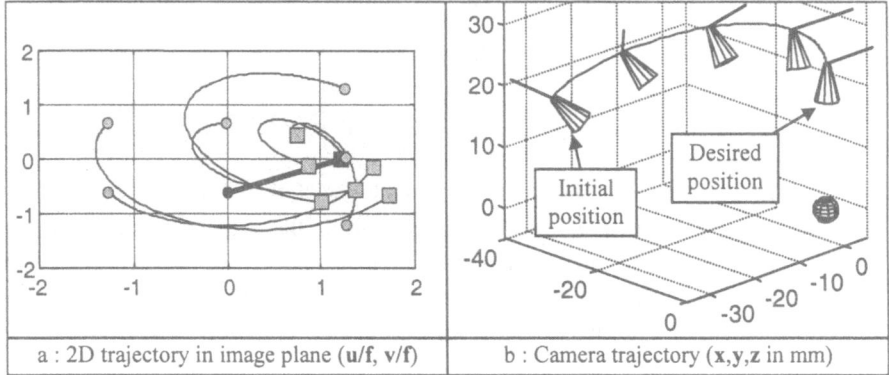

| a : 2D trajectory in image plane (u/f, v/f) | b : Camera trajectory (x,y,z in mm) |

*Figure 5 : Simulation results for 2D½ control*

While both 2D½ and TMB servoing allow to provide a certain amount of control on 2D and 3D trajectories, none of them guarantee that the entire target will stay in the image during large range motions. Actually, we were not able to find such a method in the literature. This had motivated the development of a novel approach presented in the next section.

## 3. Proposed approach

Our approach can be viewed as a 2D½ control method since it respects the task decomposition described by equation (4). However, it is different from the existing method because the 2D feature vector s was designed in order to be able of specifying constraint that will guarantee that the target entirely stays in the camera's field of view.

### 3.1 Design of the control law

We assume in the next that the target image is characterized by a set of $p$ points, which correspond to $p$ real points in the physical target.

Functionally, in order to fit the target in the image, we need to be able of controlling both its size and its center location in the image. Also, in order to keep the exact decoupling form of equation (4), we had to design a three components 2D feature vector. Naturally, this feature vector was designed to include two components for the image centering and one component for the target image sizing.

Among the numerous possible definitions of "image center", we have successively considered :

- the center of gravity of the $p$ points ;
- the image of a point of the physical target (e.g. its center, if the target is planar) ;
- the center of the smallest circle including all the $p$ points ;
- the center of the smallest rectangle including all the $p$ points, ...

Similarly, a number of definitions for the "image size" were considered :

- the radius of the smallest circle including all the $p$ points ;
- the diagonal length of the smallest rectangle including all the $p$ points ;
- the maximum distance between two of the $p$ points, ...

A major difficulty was to find a set of three parameters that guarantees a good conditioning of both $\mathbf{J}_v$ and $\mathbf{J}_\Omega$. This is achieved by the following choice :

$$s \triangleq \begin{pmatrix} u_c \\ v_c \\ \sigma = \sqrt{(u_c - u_d)^2 + (v_c - v_d)^2} \end{pmatrix} \tag{5}$$

where $(u_c, v_c)$ are the coordinates of the image center $p_c$, which is defined as the image of a given point of the physical target ; $\sigma$ is the distance between $p_c$ and $p_d = (u_d, v_d)$, the further from $p_c$ among the $p-1$ other points. It is clear that according to this definition, all the $p$ points lie in a disk centered at $p_c$ with a $\sigma$ radius. The image jacobian for the linear velocity is then given by :

$$\dot{s} = \mathbf{J}_v{}^a \mathbf{V}_c \Rightarrow \mathbf{J}_v = \begin{bmatrix} \frac{f}{z_c} & 0 & -\frac{u_c}{z_c} \\ 0 & \frac{f}{z_c} & -\frac{v_c}{z_c} \\ \frac{\sigma_u}{\sigma}\left(\frac{f}{z_d} - \frac{f}{z_c}\right) & \frac{\sigma_v}{\sigma}\left(\frac{f}{z_d} - \frac{f}{z_c}\right) & \left(\frac{\sigma_u}{\sigma}\left(\frac{u_c}{z_c} - \frac{u_d}{z_d}\right) + \frac{\sigma_v}{\sigma}\left(\frac{v_c}{z_c} - \frac{v_d}{z_d}\right)\right) \end{bmatrix} \tag{6}$$

where it was assumed that the coordinates of $^a\mathbf{V}_c$ are expressed in the $\mathcal{F}_c$.

Similarly, for the rotational motion, we get :

$$\dot{s} = J_\Omega{}^* \Omega_c \Rightarrow J_\Omega = \begin{bmatrix} -\dfrac{u_c v_c}{f} & \dfrac{u_c{}^2 + f^2}{f} & -v_c \\[2mm] -\dfrac{v_c{}^2 + f^2}{f} & \dfrac{u_c v_c}{z_c} & u_c \\[2mm] \dfrac{-\sigma_u(u_d v_d - u_c v_c) - \sigma_v(v_d{}^2 - v_c{}^2)}{f\sigma} & \dfrac{\sigma_v(u_d v_d - u_c v_c) + \sigma_u(u_d{}^2 - u_c{}^2)}{f\sigma} & 0 \end{bmatrix}$$

(7)

where $\sigma_u = u_d - u_c$ and $\sigma_v = v_d - v_c$.

The overall control law is the given by the combination of equations (4), (6) and (7). This control law uses the inverse of the linear velocity image jacobian. Thus the conditioning of $J_v$ is of particular importance in the controller behavior. Its determinant is given by :

$$\det(J_v) = \frac{f^2}{\sigma z_c{}^2} \left( -\frac{\sigma_u{}^2 + \sigma_v{}^2}{z_d} \right) = -\frac{f^2 \sigma}{z_c{}^2 z_d}$$

(8)

Thus $J_v$ is regular over the workspace except for $\sigma=0$, which corresponds to a case where all the *p* points reduced to one point.

### 3.2 Simulation results

According to the detailed stability analysis for 2D ½ control given in [11], this fundamental property of $J_v$ is a proof of stability if we consider ideal conditions (perfect calibration). Furthermore, the methodology developed in [11] can be used to investigate stability in the presence of calibration errors.

Also, the control law guarantees a perfect decoupling between the rotational motion and the 2D feature motion. This is illustrated in Figure 6, where a simulation of our control law was run for a trajectory that was chosen different from the previous ones.

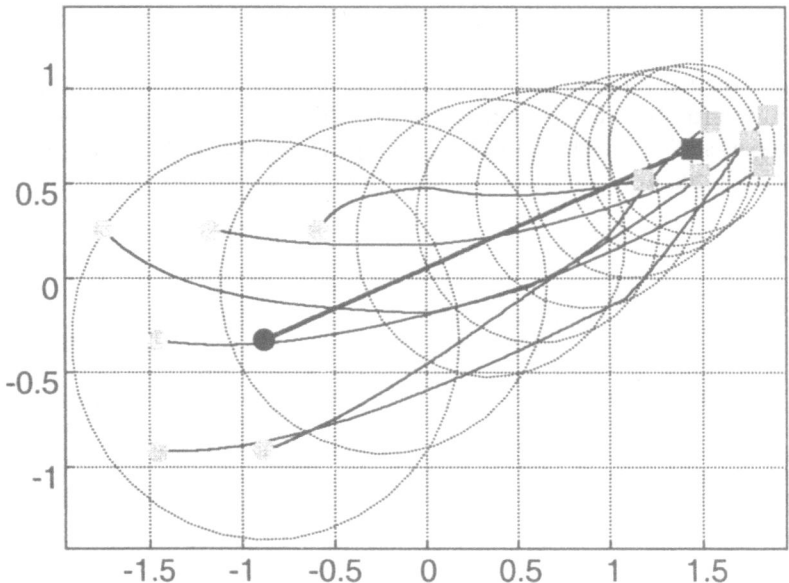

*Figure 6 : Controlling both the center trajectory and the target image size*

While the 3D trajectory of the camera relative to the target involves a large displacement along all the six degrees of freedom, one can see that :

• one point, arbitrarily selected as the target center, evolves along a linear path ;

• the target size, represented by a circle, varies linearly with the center displacement. This is due to the fact that the same exponential convergence rate $\lambda$ has been selected for the sizing and centering actions.

Thus, we are now able of controlling independently the rotational error, the centering and the size of the target. It is clear from Figure 6 that if the circle drawn from the image center, with a radius equal to the size, fits in the image in both the initial and final configurations, then all the circles will fit in the image during the motions. Consequently, in this case, none of the target point will ever leave the image.

However, when the target is large in the image, the circle may become bigger than the image limits. Thus, some points may leave the image. Figure 7 illustrates this for a simulation example similar to the one used in Figure 3 to 5.

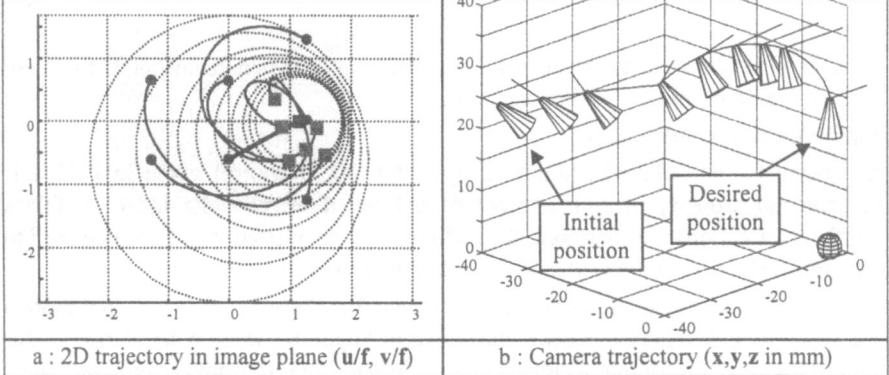

| a : 2D trajectory in image plane (u/f, v/f) | b : Camera trajectory (x,y,z in mm) |

*Figure 7 : Simulation results with the proposed approach*

### 3.3 Integrating constraints

Coping with this problem is possible, since we are able of controlling independently the rotational 3D camera motion and the 2D target size and position. Thus, constraints surfaces can be defined in a configuration space that include both 3D and 2D specifications.

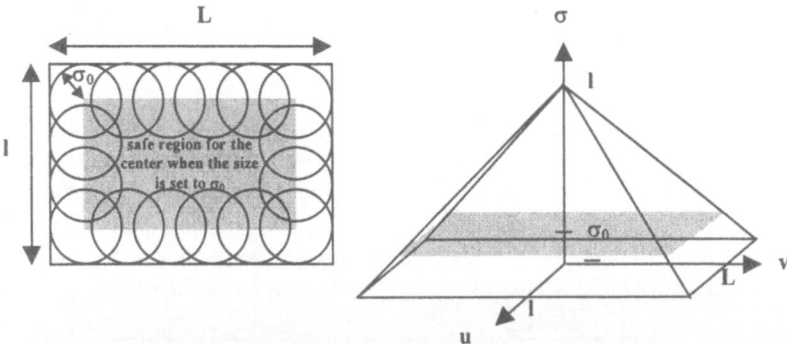

*Figure 8 : Expressing constraints on the 2D vector.*

For example, we can define the set of "safe configurations", that are configurations for s such that the target will fit in the image for all the possible rotations between the camera and the target. This correspond to a pyramid in the s configuration space (see Figure 8).

Additional constraints can be set in order to correctly condition the behavior of the vision signal process. For example, one can impose a minimum size of the target in the image to guarantee a minimum signal to noise ratio.

Constraint volumes can also be shrunk to increase the security margins in order deal with calibration errors and other undesirable dynamics in real experiments.

Figure 9 shows an example of a simple strategy that was used to perform the example simulation task :

1. The constraint volume is the pyramid depicted in Figure 8, truncated from its base (minimum size) and its summit (maximum size).

2. Initially, as the system violates the constraint, the rotational velocity command is null. Thus, the initial motion is a pure translation aimed at reaching the constraint volume.

3. Once the constraint volume has been reached, the reorientation begins, while the 2D error is being minimized within the constraint volume.

4. When the reorienting motion is finished (small rotational error) the constraint drops and a final translation is made to zero the remaining 2D error.

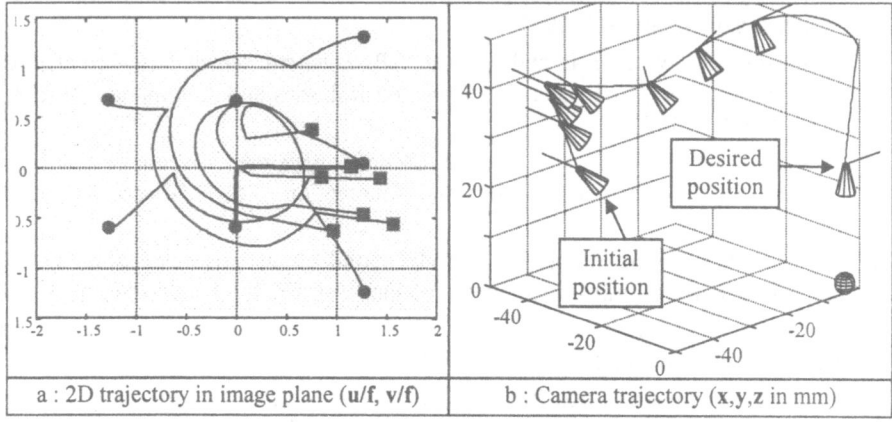

| a : 2D trajectory in image plane (u/f, v/f) | b : Camera trajectory (x,y,z in mm) |

*Figure 9 : Simulation results with constraints*

The trajectory of the center point is no longer a straight line. Rather, the center is constrained to stay approximately at the image center during the reorienting motion.

Also, notice that the 3D trajectory is not straightforward anymore. One can distinguish between the successive phases of the task. The transition between the different phases may result in sudden changes in the velocity commands. This is due to the fact that we are using sharp constraint surfaces in this example. There is no doubt that smoothening the constraint surfaces would be a way of smoothing the 3D trajectory of the camera.

## 4. Conclusion

Conserving the entire target in the image should be a major concern when studying vision based control of robot manipulators. However, the literature does not provide much data on this point.

This paper is an attempt of addressing this problem by means of control. An original controller has been proposed and experimented. It has been compared successfully with major existing controllers in simulation.

Further work will focus on experimental validation of this controller, and stability properties investigation in the presence of robot dynamics and calibration errors.

**References.**

[1] Morel G., Boudet S., Mauvilet G., Mezzadri R. and Pot J., *Integrating force and vision feedback control in teleoperated manipulation*, To appear in the proceedings of RoManSy 98, 1998.

[2] Hutchison S., Hager D. and Corke P.I., *A tutorial on visual servo control*, IEEE Trans. on Robotics and Automation, Vol. 12, 5, pp. 651-670, 1996.

[3] Wilson W.J., Williams Hulls C.C and Bell G.S., *Relative end-effector control using cartesian position based visual servoing*, IEEE Trans. on Robotics and Automation, Vol. 12, 5, pp. 684-696, 1996.

[4] Martinet P., Daucher N., Gallice J. and Dhome M., *Robot control using 3D monocular pose estimation*, IEEE/RSJ Int. Conf. on Intelligent Robots and Systems, Vol. 4, pp. 1-12, 1997.

[5] Sanderson A.C., Weiss L.E. and Neuman C.P., *Dynamic sensor-based control of robots with visual feedback*, IEEE Trans. on Robotics and Automation, Vol. 3, 4, pp. 404-417, 1987

[6] Espiau B., Chaumette F. and Rives P., *A new approach to visual servoing in robotics*, IEEE Trans. on Robotics and Automation, Vol. 8, 3, pp. 313-326, 1992

[7] DeMenthon D.F. and Davis, L.S., Model based object pose estimation in 25 lines of code, Int. Journal of Computer Vision, Vol. 15, 2, pp. 123-141, 1995

[8] Espiau B., *The effect of camera calibration errors on visual servoing in robotics*, Int. Symp. on Experimental Robotics, 1993.

[9] Malis E., Chaumette F. and Boudet S., *Positioning a coarse-calibrated camera with respect to an unknown target using 2D½ visual servoing*, IEEE Int. Conf. on Robotics and Automation, pp. 1352-1357, 1998

[10]Morel G., Malis E. and Boudet S., *Impedance based combination of visual and force control*, IEEE Int. Conf. on Robotics and Automation, pp. 1743-1748, 1998

[11]Malis E., *Contributions à la modélisation et à la commande en asservissement visuel*, PhD Thesis, University of Rennes, France, Nov. 1998.

# Chapter 4

# Control

Control and robotics have a long standing connection...

Bruyninckz, Lefebvre and De Shutter describe a force control system which takes into account the contact geometry which is identified from measured force and velocity information. They provide experimental results for various tasks such as vertex-face contact and placing a cube in a corner.

The paper by Zsewczyk and Bidaud describes how multiple robots can share a cooperative task without centralized control. Each robot implements a local impedance control scheme based on information from a force/torque sensor and the performance is demonstrated experimentally.

Fibreglass reinforced plastics (FRP) are important engineering materials but the molding process exposes workers who push the material into molds using rollers, to noxious fumes. Ohkami, Sunaoshi and Sugiura investigate the application of robot force control to the fast back and forth rolling of this visco-elastic material.

Control of nonholonomic mobile robots has been the subject of much recent research. Fourquet and Renaud describe an experimental wheeled mobile platform with a manipulator arm, and compare the performance of two different control strategies.

The final paper, by Quaid and Rizzi, concerns small planar robots moving within a minifactory. The robots are driven by planar stepping motors and the base platen acts as the motor's stator. Their paper describes how they are able to achieve micron level positioning accuracy using only onboard sensing.

PIC

# Experiments with Medium-Level Intelligent Force Control

Herman Bruyninckx,*Tine Lefebvre and Joris De Schutter†
Katholieke Universiteit Leuven, Dept. Mechanical Engineering
Celestijnenlaan 300B, B-3001 Heverlee, Belgium

**Abstract:** This text gives experimental results, comments and interpretations of a set of compliant motion tasks that form the basis for force-guided identification of the geometry of the most common contact situations. The experiments show the influence of (i) nonideal effects, (ii) nonlinearities, and (iii) different motion trajectories.

## 1. Introduction

Force-controlled compliant robot motion is an inherently more difficult control problem than the motion of mobile robots or cameras. Hence, most research focusses on the lowest servo control level (how to move the manipulated object ("MO") against the constraints), and much less on "intelligent" sensing (what are the geometry and/or the type of the contact situation?). *Vertex–face* is an example of a type of contact situation; the *position* of the contact point, and the *orientation* of the contact normal are examples of geometric parameters in the given contact type. Because of the higher complexity of the servo control level, force/torque sensing is much less used in robotics than ultrasound sensing and vision. In addition, state of the art research in force-controlled compliant motion is divided in two complementary areas:

1. *Repetitive tasks*, for example assembly: the goal of force sensing in these tasks is to detect when the transition between two contact situations takes place, and to predict what the new contact situation is. The most general and successful techniques are variations on Hidden Markov Models, [1, 2, 3]: the transition probabilities between contact situations, and the typical sensor images during each contact situation, are "learned" during repetitive execution of the task.

2. *One-off tasks*, for example detecting and localizing (partially) unknown objects: the robot should be able to execute a given task in an unstructured environment from the first time right. In general, this is still too ambitious for the current state of the art.

Previous work by the authors, [4, 5], presented a **general** theory, based on a kinematic contact model, to describe all possible contact situations, to specify any desired task

*Postdoctoral Fellow of the Fund for Scientific Research–Flanders (F.W.O.) in Belgium.
†This work was sponsored in part by the Belgian Programme on Inter-University Attraction Poles (IUAP).

(desired force and velocity setpoints), and to find a set of linear equations linking the measured force and velocity with the geometric uncertainty parameters in the contact models. This identification uses (non-linear) Kalman Filters. The experimental results in "complex" contact situations were promising; however, it was impossible to distinguish the influence of several factors: nonideal effects, nonlinearities, and the executed motion trajectory. Therefore, this text uses a **simple** contact situation (the *point contact*) to show to what extent the above-mentioned factors are important.

In a point contact, the MO has five degrees of freedom with respect to the environment. Three rotational degrees of freedom and two translational degrees of freedom. This contact situation occurs with many different kinds of geometric uncertainty: vertex–face with known vertex but unknown orientation of the face; vertex–face with known face but unknown position of the vertex; vertex–face where the face is part of a curved surface; edge–edge; face–vertex; etc. (The first "feature" is assumed to belong to the MO, the second feature to the environment.)

This text assumes that the "Task Frame" [6, 7] has its origin in the contact point, and its $Z$ axis along the normal direction. Hence, the bases for the possible velocities ("$J$") and ideal contact forces ("$G$") are:

$$J = \begin{pmatrix} 0 & 0 & 1 & 0 & 0 \\ 0 & 0 & 0 & 1 & 0 \\ 0 & 0 & 0 & 0 & 1 \\ 1 & 0 & 0 & 0 & 0 \\ 0 & 1 & 0 & 0 & 0 \\ 0 & 0 & 0 & 0 & 0 \end{pmatrix} \left. \begin{matrix} \\ \\ \end{matrix} \right\} \omega \atop \left. \begin{matrix} \\ \\ \end{matrix} \right\} v \qquad G = \begin{pmatrix} 0 \\ 0 \\ 1 \\ 0 \\ 0 \\ 0 \end{pmatrix} \left. \begin{matrix} \\ \\ \end{matrix} \right\} f \atop \left. \begin{matrix} \\ \\ \end{matrix} \right\} m . \qquad (1)$$

The first three components in $G$ represent the force, the last three components are the moment. The first three components in $J$ represent the angular velocity, the last three components are the translational velocity. The experiments are done with the hybrid force/position controller of [8]. In a point contact, it transforms any force error in the translational $Z$ direction into a translational velocity; this velocity is added to the velocity set-points in the other directions of the Task Frame.

## 2. Kalman Filter based identification

The reader is referred to standard references for a detailed description of the theory behind Kalman Filters (KF), e.g., [9, 10, 11]. This Section discusses only how KFs are used for the identification of contact situations.

**State.**  A KF is a state estimator. In this paper, the state $x$ are the *errors* on the estimates of the geometric parameters of the contact situation, given its type.

**Process.**  Moving the MO over the environment changes the geometric parameters of the contact situation. This change is *predicted* with the current contact type model and the current values of the estimated geometric parameters.

**Measurements.**  Force $(f, m)$ and velocity $(\omega, v)$.

**Measurement equation.**  This relation must predict the measurements given the current model and the current motion. As "measurement" (or "innovation") this text uses the *instantaneous power* of the (ideal) contact forces and the velocity of the MO that maintains the contact. The motion is called "compliant" (or "reciprocal") if this power vanishes, [12]: $f \cdot v + m \cdot \omega = 0$. This *reciprocity condition* is independent of both the chosen physical units and the reference frame.

**Identification equations.** The reciprocity condition links the state to the measurements. In the *real* contact point, the bases $J$ and $G$ have the values as in Eq. (1). However, one only knows the measured force and motion in the *nominal* Task Frame. Nominal and real Task Frame differ by a homogeneous transformation matrix that is a non-linear function of the state vector $x$, with a rotation matrix part $R$ and a translation vector $p$. The transformation of forces and velocities is done by means of a $6 \times 6$ matrix $S = \begin{pmatrix} R & 0 \\ RP & R \end{pmatrix}$, with $P$ the matrix corresponding to taking the vector product with $p$. Hence, the velocity-based and force-based reciprocity relations become:

$$\begin{pmatrix} v_m & \omega_m \end{pmatrix}^T S(x)\, G = 0, \tag{2}$$

$$\begin{pmatrix} m_m & f_m \end{pmatrix}^T S(x)\, J = 0. \tag{3}$$

The KF works with the *linearization* of this transformation, and the linearization is taken in the point corresponding to the last estimate of the parameters.

**Initialization.** Every KF has to be initialized with "noise" covariances on the state, the measurements, and the process. In the experiments described in the later Sections, *artificial process noise* is always added, because the linearity assumption underlying the classical KF does not hold: contact surfaces are curved, the process and measurement equations are non-linear, etc.

This text illustrates how we applied the different well-known interpretations of KFs to the problem of "intelligent" force-controlled compliant motion:

1. *Least-squares filter*, for underdetermined as well as overdetermined systems. The former case occurs when not enough information is available to *instantaneously* estimate all geometric parameters of the contact situation geometry. The latter is used to "average" estimates of the same parameter(s) over a window of several measurements.

2. *Fusion tool*: to combine information from different sources (i.c., force and velocity measurements) weighted by their respective "uncertainty" (i.e., covariances).

3. *Monitoring tool*: as long as the innovations are small, the type of the contact model is assumed to be valid. The magnitude of the innovation is calculated from the measurements by using the covariance as weighting matrices. The covariances are often interpreted as the "accuracy" of the estimates; but they are known to evolve *independently* of the measurements, such that this interpretation is to be used with care. The "belief" in the current contact model is represented by the *SNIS*: Sum of Normalized Innovations Squared, i.e., the sum of the above-mentioned innovation magnitudes over a certain window.

## 3. Vertex–face: unknown face

Figure 1 shows the vertex–face contact situation. The nominal Task Frame has its $Z$ axis normal to the face. If the orientation of the "face" is unknown, this uncertainty

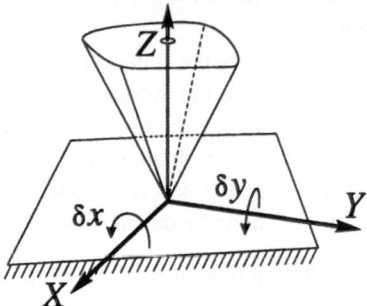

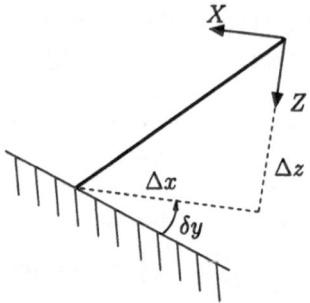

Figure 1. Vertex–face contact: $\delta x$ is the angular error in the orientation of the $Y$ axis; $\delta y$ gives the angular error of the $X$ axis.

Figure 2. Planar vertex–face contact, with uncertain vertex and face.

is modelled by a state vector $x = (\delta x \ \delta y)$. Thus, the real Task Frame is rotated with respect to the nominal Task frame over the yet unknown angles $\delta x$ and $\delta y$. The corresponding rotation matrix $R$ has the following form up to first order in the parameters:

$$R = \begin{pmatrix} 1 & 0 & \delta y \\ 0 & 1 & -\delta x \\ -\delta y & \delta x & 1 \end{pmatrix}. \tag{4}$$

The matrix $P$ is the identity. The reciprocity conditions Eqs. (2) and (3) yield, respectively:

$$v_x \, \delta y - v_y \, \delta x + v_z = 0, \tag{5}$$

and

$$f_x - f_z \, \delta y = 0, \tag{6}$$
$$f_y + f_z \, \delta x = 0, \tag{7}$$
$$m_x - m_z \, \delta y = 0, \tag{8}$$
$$m_y - m_z \, \delta x = 0, \tag{9}$$
$$m_x \, \delta y - m_y \, \delta x + m_z = 0. \tag{10}$$

The last three equations are identities, since the measured moments are nominally zero: the contact point is assumed to be known, such that the contact force (ideal or not) does not generate a moment about that point. However, the vanishing of these components could be checked, and used as monitoring information.

Active sensing is straightforward: velocity-based identification needs motion in both $X$ and $Y$ directions, since $\delta x$ and $\delta y$ are not instantaneously identifiable: Eq. (5) is only one equation in the two unknown parameters, and their coefficients are only non-zero if these velocities in $X$ and $Y$ are non-zero. The $v_z$ velocity is generated by the force control, as explained above. Force-based identification does give both parameters instantaneously (two equations (6)–(7) in two unknowns), but needs smoothing to cope with noise. Friction disturbs the force-based KF estimates.

## 4. Vertex–face: unknown vertex

The uncertainty in the vertex–face contact situation with unknown position of the "vertex" is modelled by a state vector $x = (\Delta x\ \Delta y\ \Delta z)$. The orientation of the face is known; hence, the transformation matrix $R$ is the identity matrix (if the orientation of the Task Frame is *chosen* normal to the face). The matrix $P$ is

$$P = \begin{pmatrix} 0 & \Delta z & -\Delta y \\ -\Delta z & 0 & \Delta x \\ \Delta y & -\Delta x & 0 \end{pmatrix}. \tag{11}$$

This yields the following velocity-based and force-based equations:

$$\omega_x\, \Delta y - \omega_y\, \Delta x + v_z = 0, \tag{12}$$

and

$$f_x = 0, \tag{13}$$
$$f_y = 0, \tag{14}$$
$$-f_y\, \Delta z + f_z\, \Delta y + m_x = 0, \tag{15}$$
$$f_x\, \Delta z - f_z\, \Delta x + m_y = 0, \tag{16}$$
$$-f_x\, \Delta y + f_y\, \Delta x + m_z = 0. \tag{17}$$

The $f_x$ and $f_y$ components are nominally zero. Hence: (i) only Eq. (12), and Eqs (15)-(16) (with $f_x = f_y = 0$) are used as identification equations, and (ii) the distance to the face (i.e, the parameter $\Delta z$) is not instantaneously observable. Active sensing consists of rotating the MO about the (nominal) contact point: this makes the components $\omega_x$ and $\omega_y$ large in the velocity-based identification. This allows to estimate $\Delta x$ and *Deltay, and* it changes the relative orientation between Task Frame and face, such that part of the instantaneously unidentifiable $\Delta z$ becomes identifiable.

Force-based identification boils down to finding the intersection of a number of lines, each being the line carrying the contact force. (Friction is *not* a disturbance, since all forces (ideal or not) go through the same contact point.) This line intersection is not well conditioned if the lines are close to each other; hence, the KF should work over a wide angular range.

## 5. Vertex–face: unknown vertex and face

Both the vertex and the face can be uncertain in a point contact. Figure 2 shows the planar case, with $\delta y \neq 0$. The identification equations are more complicated, but are derived by exactly the same procedure as before. For example, the (non-linearized and planar!) velocity-based and force-based identification equations are:

$$(v_z - \omega_y\, \Delta x)\cos(\delta y) + (v_x + \omega_y\, \Delta z)\sin(\delta y) = 0, \tag{18}$$
$$f_x \cos(\delta y) - f_z \sin(\delta y) = 0. \tag{19}$$

(These equations simplify to those in the previous Section when the face is parallel to the Task Frame, i.e., $\delta y = 0$.) The Kalman Filter works with the linearization of this equation in the *current best estimate* of the parameters. Hence, an inaccurate estimate of $\delta y$ has a big influence on the identification. Equation (18) identifies $\Delta x \cos(\delta y) - \Delta z \sin(\delta y)$, and not both parameters independently. However, measurements at different values of $\delta y$ allow to separate both parameters.

## 6. Experimental results

This Section gives a critical discussion of the experimental results for the tasks described above. The experiments show that the output of Kalman Filters can be very sensitive to (i) non-ideal effects, (ii) non-linearities, and (iii) the executed trajectory.

**Non-ideal effects.** Figure 3 shows the outcome of the velocity-based and force-based KFs, with a constant translational velocity as trajectory. We give only the simplified "2D" version of Section 3, because this is most suitable for interpretation. Velocity-based identification gives the correct result, $\delta y \approx 0.33$ rad. Force-based identification is disturbed by friction; the SNIS gives a much higher value than the SNIS of velocity-based identification.

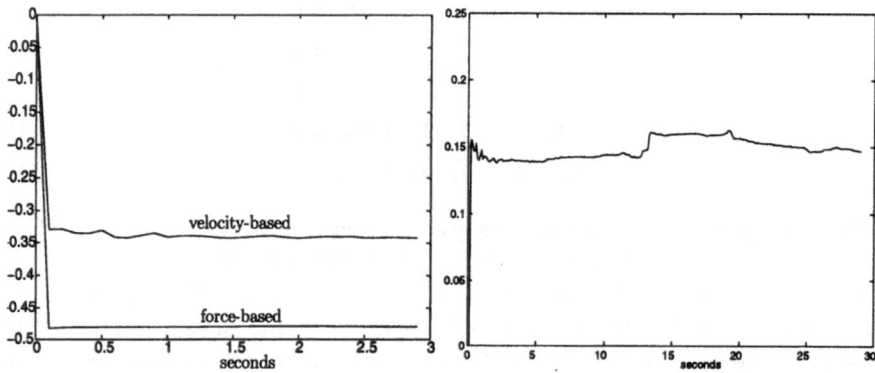

Figure 3. Vertex–face contact: estimation of orientation $\delta y$.

Figure 4. Vertex–face contact: estimation of friction coefficient.

Combining velocity-based and force-based identification gives an estimate of the friction coefficient $\mu$ in the contact: the friction coefficient is $\tan(\delta y_v - \delta y_f)$, with the subscripts denoting velocity-based and force-based identification, respectively. In the experiment, $\mu \approx 0.15$, as shown in Fig. 4.

Because force-based identification is instantaneously identifiable (i.e., the identification equation (19) is solvable at each instant in time) its results converge rapidly, Fig. 5. However, the numerical values of the estimates are very sensitive to the *calibration* of the force sensor and the *compensation* of the gravity in the force measurements: systematic errors in these transformations (i.e., first from raw sensor measurements into forces and moments, and then from total force to gravity-compensated force) result in systematic errors in the KF outcome. For example, in Fig. 5, the true value of $\Delta z$ is about 250 mm, and not the 270 mm given by the Kalman Filter; the major systematic error during this particular experiment was the moment component of the measured force. Systematic errors appear more in force sensors than in the robot encoders from which the velocity signals are derived. Hence, the velocity-based KF results are less susceptible to these errors.

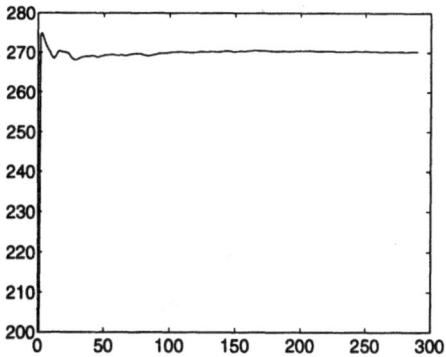

Figure 5. Vertex–face contact: force-based estimation of position error $\Delta z$.

**Nonlinearities.** Figures 6–9 show velocity-based identification of the situation in Sect. 5. In Figs 6–7, $\Delta x, \Delta z$ and $\delta y$ are identified at the same time, by an active sensing motion that combines translation in $X$ with rotation about $Y$. The influence of *linearization* is clearly seen if all three parameters are identified at the same time: a wrong estimate of $\delta y$ yields wrong estimates of $\Delta x$ and $\Delta z$; even after the estimate of $\delta y$ has improved, the estimates of $\Delta x$ and $\Delta z$ remain ppor because the filters have already "stiffened" too much. A classical solution for this well-known behavior is to use a "fading memory," or to add artificial process noise that keeps the filter alert for changes in the state.

In Fig. 7, $\delta y$ is identified first, by a pure translation; then, $\Delta x$ and $\Delta z$ are identified by a pure rotation.

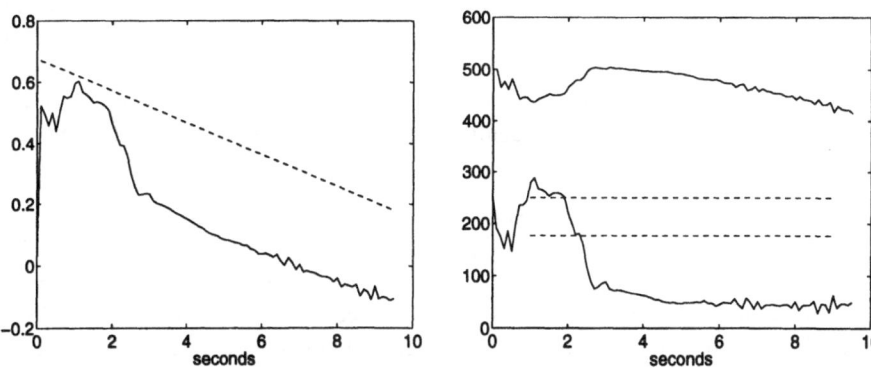

Figure 6. Vertex–face contact with unknown vertex and face: velocity-based identification results for the angle $\delta y$, with combined translation ($v_x = $ 10mm/sec) and rotation $\omega_y = $ 3deg/sec). The dashed line represents the real $\delta y$.

Figure 7. Same experiment as in Fig. 6, with the velocity-based identification results for $\Delta x$ (bottom) and $\Delta z$ (top). The dashed lines represent the real values.

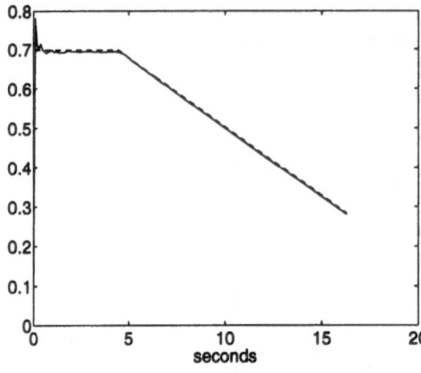
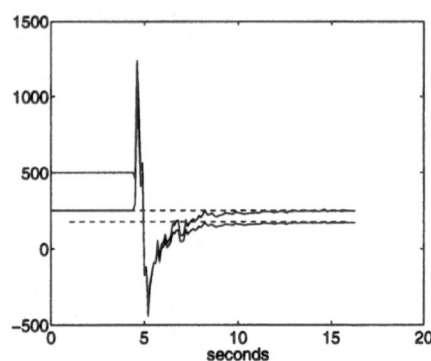

Figure 8. Vertex–face contact with un-known vertex and face: velocity-based identification of the orientation $\delta y$, with both $\Delta x$ and $\Delta z$ unknown. The exe-cuted motion is first a pure translation ($v_x = 10$mm/sec during the first 5 sec-onds), followed by a pure rotation ($\omega_y =$ 2deg/sec).

Figure 9. Velocity-based identification of $\Delta z$ (top) and $\Delta x$ (bottom), in the same experiment as in Fig. 8. The translation of the first 5 seconds doesn't "excite" the parameters, and hence the Kalman Fil-ter gets no new information and doesn't change its estimates.

**Motion trajectory.** The executed motion also has a big influence on the identifi-cation results. Figures 8 and 9 have already shown that a clever sequencing of the motions contributes to a better conditioning of the identification equations. These Figures show the best possible sequence (first pure translation, then pure rotation), as far as the conditioning for the nonlinearity caused by $\delta y$ is concerned. Figures 10 and 11 show another experiment, with simultaneous translation (20 mm/sec) and rotation ($0.2 \sin(2\pi t/10)$). The sinusoidal rotation trajectory results in (almost) "sufficient ex-citation": the variation in both $\delta y$ and $\omega_y$ yields a variation in the magnitudes of the coefficients of $\Delta x$ and $\Delta z$ in the identification equation (18), and hence the identifi-cation is better conditioned.

**Contact situation monitoring.** Some of the identification results above could also be obtained without *continuous* motion along the contact, for example by "haptic" touching of the environment. However, the continuous motion in combination with the SNIS value (Sect. 2) give a test to monitor the current contact model: if the SNIS is "low" the current model is very good in predicting the measurement; a jump in the SNIS is an indication that the contact situation has changed. This is illustrated by the "cube in corner" contact situation, Fig. 12: a cube is moved from free space until it makes contact, and is then guided through a sequence of contact situations until it is fixed in the corner. The figure shows the SNIS of four different contact hypotheses: one for each vertex–face contact between the vertices of the bottom of the cube and a face in the environment. This contact situation is the most probable one when the cube is moved from free space towards the environment. Three of the four hypotheses have a very high SNIS; hence, the fourth possibility is the most probable one. Figure 13 shows the evolution of the SNIS during the whole sequence of contact situations, from

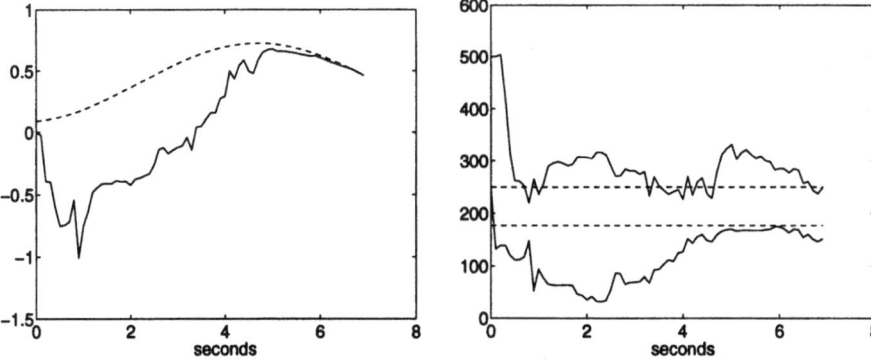

Figure 10. Same as in Fig. 8, but $\delta y$ is better identifiable with a sinusoidal rotation. The initial estimate is 0 radians, but after about five seconds, the estimate follows the sinusoidal trajectory of the applied angular velocity.

Figure 11. As in Fig. 10, but for $\Delta x$ (bottom) and $\Delta z$ (top). The convergence is not yet perfect, but nevertheless much closer to the real values than in Fig. 7.

free space to full constraint. The jumps are very pronounced.

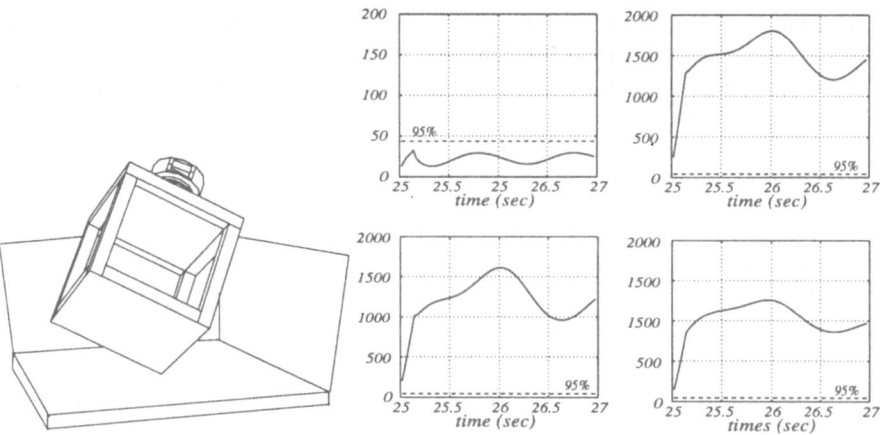

Figure 12. Cube in corner: monitoring of four possible vertex–face contacts when cube has first made contact when moving from free space. Only one of the four alternatives has a low SNIS value.

## 7. Conclusions

This paper presents experimental results for continuous force-based and velocity-based estimation of errors in a geometric contact model, using Kalman Filters. A blind application of the general theoretical solution procedure can lead to very suboptimal filters. The influence of non-ideal effects, nonlinearities, and the executed motion trajectory is studied. The problem of generating optimal active sensing motions is still

120

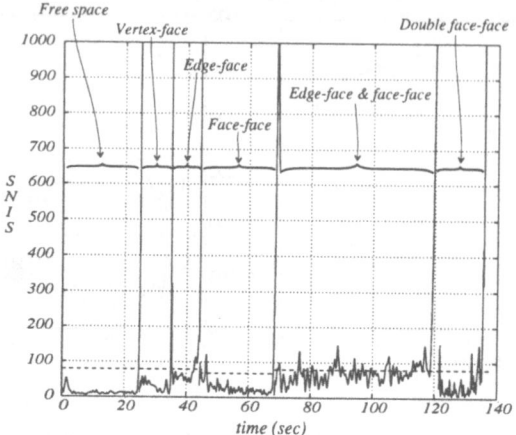

Figure 13. Cube in corner: monitoring results during sequence of contact situations.

largely unsolved.

## References

[1] Brian S. Eberman and J. Kenneth Salisbury, Jr. Application of change detection to dynamic contact sensing. *Int. J. Robotics Research*, 13(5):369–394, 1994.

[2] Brian Eberman. A model-based approach to Cartesian manipulation contact sensing. *Int. J. Robotics Research*, 16(4):508–528, 1997.

[3] Geir Hovland and Brenan J. McCarragher. Hidden Markov models as a process monitor in robotic assembly. *Int. J. Robotics Research*, 17(2):153–168, 1998.

[4] H. Bruyninckx, S. Demey, S. Dutré, and J. De Schutter. Kinematic models for model based compliant motion in the presence of uncertainty. *Int. J. Robotics Research*, 14(5):465–482, 1995.

[5] S. Dutré, H. Bruyninckx, S. Demey, and J. De Schutter. Solving contact and grasp uncertainties. In *Int. Conf. Intel. Robots and Systems*, pages 114–119, Grenoble, France, 1997.

[6] Herman Bruyninckx and Joris De Schutter. Specification of force-controlled actions in the "Task Frame Formalism": A survey. *IEEE Trans. Rob. Automation*, 12(5):581–589, 1996.

[7] M. T. Mason. Compliance and force control for computer controlled manipulators. *IEEE Trans. on Systems, Man, and Cybernetics*, SMC-11(6):418–432, 1981.

[8] J. De Schutter and H. Van Brussel. Compliant robot motion II. A control approach based on external control loops. *Int. J. Robotics Research*, 7(4):18–33, 1988.

[9] Yaakov Bar-Shalom and Xiao-Rong Li. *Estimation and Tracking, Principles, Techniques, and Software*. Artech House, 1993.

[10] Arthur (Ed.) Gelb. *Optimal Estimation*. MIT Press, Cambridge, MA, 3rd edition, 1978.

[11] Harold Wayne Sorenson. *Kalman filtering: theory and application*. IEEE Press, New York, NY, 1985.

[12] M. S. Ohwovoriole and B. Roth. An extension of screw theory. *Trans. ASME J. Mech. Design*, 103(4):725–735, 1981.

# Coordinated Manipulation under Distributed Impedance Control

Jérôme Szewczyk and Philippe Bidaud
Laboratoire de Robotique de Paris
France
sz@robot.uvsq.fr

**Abstract:** Distributed impedance is an efficient approach for controlling any kind of cooperating manipulation system. It combines the advantages of decentralized control solutions and the robustness of impedance control with force feedback. Distributed impedance was implemented on a dual-arm system composed of two industrial manipulators. Experimental results were obtained for free and constrained manipulation tasks.

## 1. The distributed impedance approach

Coordinated manipulation is a generic term used to designate the manipulation of a single body by multiple limbs. This is illustrated by two manipulators carrying an heavy, volumic and/or flexible object. Dextrous hands constitute an interesting sub-class of cooperative systems. They offer a high degree of flexibility compared to basic on-off grippers. They are particularly useful for manipulating fragile objects.

An analysis of the control problem of coordinated manipulation systems shows that besides the control of the trajectory and forces experienced on the manipulated object, it is necessary to provide a control of the internal forces. Control of internal forces indeed is critical with respect to grasp stability in that it permits to avoid slippage or lost of the contact on the surface of the object.

Most previous approaches for cooperative system control consist in global and centralized methods [1]. Object coordinates and internal efforts must be evaluated in line to be independently servoed. As a consequence, a specific hardware development is needed for data centralization.

Conversely, local approaches consider each cooperative manipulator as an autonomous, independently controlled system. Practically, each manipulator's controller is used without any hardwaremodification in the multi-robot system. This provides flexibility and modularity. Furthermore, undesirable phenomenons (actuators limitation, non-colocated modes,...) that limit the performances from a particular manipulator, can be locally treated, without major influence on the whole system.

A first possible approach consists in implementing pure joint control [2]. In this case, computation of the grasp and robot inverse kinematic models are required. This makes joint level approaches very sensitive to uncertainties in robots and grasp geometries.

A second alternative consists in controlling cooperative manipulation at the manipulator end-effector level. One focuses here on controlling positions and efforts at the interface between the manipulators and the manipulated object. In this way, important problems imparted to point contact manipulation (i.e.: control of impact, sliding on the object surface,...) can be addressed in a more direct and suitable manner.

End-effector control of cooperative systems ([3] among others...) has been investigated for many years but it did not give complete satisfaction up to now. In this case, two major difficulties inherent to coordinated manipulation have to be faced :

1) From one cooperating manipulator point of view, the environment, which consists of the manipulated object and the other manipulators, exhibits a complex dynamic behavior.

2) Internal load regulation is a fundamental component of coordinated manipulation, as well as trajectory control. Thus, at end effector level, both forces and motion have to be simultaneously and precisely controled.

To overcome these difficulties we proposed the distributed impedance control approach [4]. This approach is based on individual impedance control of all manipulators participating to the task.

Basically, impedance control establishes a linear relationship between the applied effort $F^i$ and the position error $(X^i - \tilde{X}^i)$ at the $i^{\text{th}}$ manipulator end-effector :

$$-F^i = M^i \ddot{X}^i + B^i \left( \dot{X}^i - \dot{\tilde{X}}^i \right) + K^i \left( X^i - \tilde{X}^i \right) \tag{1}$$

In case of point contact manipulation, $F^i$ is a pure linear force and the matrices $K^i$, $B^i$, and $M^i$ that compose the $i^{\text{th}}$ individual target impedance are $3 \times 3$ diagonal matrices.

Robustness of impedance control has been slightly evidenced in [5] among others. More generally, impedance control is well suited for controlling contact tasks involving a complex or unpredictable environment. Moreover, explicit force control can be introduced by simply activating an external force loop over the existing impedance scheme [6]. All these propertises make impedance control very attractive regarding the above mentioned difficulties encountered in coordinated manipulation control.

## 2. System description

We consider an object $S_p$ being grasped by $n$ manipulators through $n$ frictional point contacts (figure (1)). $\mathcal{R}_o$ is a frame fixed with respect to the object at point $O^o$ which can be chosen arbitrarily.

The absolute configuration of $\mathcal{R}_o$ with respect to a base frame $\mathcal{R}_b$ is represented by the $3 \times 1$ vector $X^o$ and the $3 \times 3$ rotation matrix $^oR_b$, respectively. The velocity of $\mathcal{R}_o$ with respect to $\mathcal{R}_b$ is represented by the $6 \times 1$ vector $V^o = \left( \dot{X}^{oT}, \Omega^{oT} \right)^T$, where $\dot{X}^o = \frac{dX^o}{dt}$ is a $3 \times 1$ linear velocity vector and $\Omega^o$ is a $3 \times 1$ angular velocity vector.

The $i^{\text{th}}$ effector contact point on the object is called $O^i$. Its location in $\mathcal{R}_o$ is given by the constant position vector $r_i$ while its location in $\mathcal{R}_b$ is given by the variable position vector $X^i$ (see figure (1)).

The object's equation of motion is:

$$F^{\text{ext}} + F^o \;\; = \;\; \Lambda^o \dot{V}^o + C^o \tag{2}$$

where $F^o$ is the resultant force/moment vector applied to $S_p$ at point $O^o$ by the set of commanded end-effectors (figure 1). $F^{\text{ext}}$ is the external force/moment vector at $O^o$, $\Lambda$ is the object's generalized inertia matrix and $C^o$ represents Coriolis and gravitational effects [4].

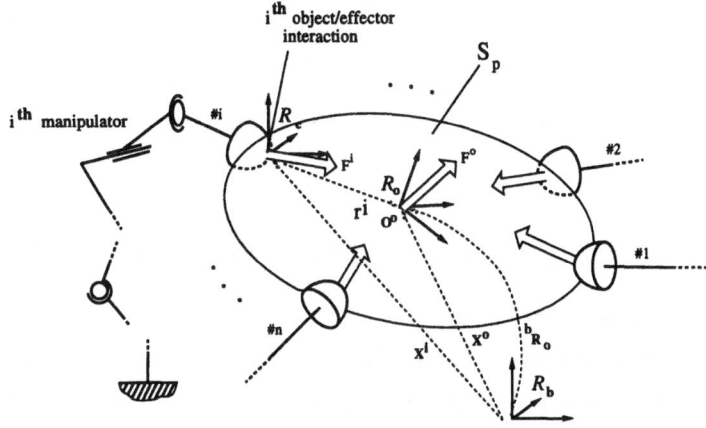

Figure 1. Grasp geometry

A pure force $F^i$ is applied to the object through the $i^{\text{th}}$ contact point. The $F^i$'s are grouped into a $3n \times 1$ vector $F = (F^{1T}, ..., F^{nT})^T$. $F$ is related to $F^o$ through the $6 \times 3n$ grasp matrix $W$:

$$F^o = W F \tag{3}$$

If $F$, $F^o$ and $r^i$, $(i = 1, ..., n)$ are all expressed with respect to the base frame $\mathcal{R}_b$, we have :

$$W = \begin{pmatrix} \mathbb{I}_3 & \cdots & \mathbb{I}_3 \\ -S(r^1) & \cdots & -S(r^n) \end{pmatrix} \tag{4}$$

Where, $\mathbb{I}_3$ is the $3 \times 3$ identity matrix and $S(.)$ represents the cross product operator.

The object's internal load comes from the component of $F$ that belongs to the null space of $W$. Assuming that $W$ is full rank and that the $(3n - 6)$ independent vectors forming a basis for its null space are grouped to form the columns of the matrix $N$, we have the relation :

$$F = W^\sharp F^o + N \eta^o \tag{5}$$

where $W^\sharp = W^T (W W^T)^{-1}$ is the generalized inverse of $W$. The first term in equation (5) represents the contribution of the contact forces to the object's motion. As $W N = 0_{6 \times 3n-6}$, the second term represents their contribution to the object's internal load. The $(3n - 6 \times 1)$ vector $\eta^o$ in (5) can be viewed as the intensity of the object internal load.

## 3. End effector control

The impedance behavior described by equation (1) can be realized considering the manipulator dynamic equation expressed in its operational space [7]:

$$J^{i-T} \tau^i_{com} - F^i = \Lambda^i \ddot{X}^i + C^i \tag{6}$$

Here, $J^i$ represents the Jacobian matrix of the $i^{th}$ manipulator which is assumed to be non-redundant. $\tau^i_{com}$ represents the commanded effort produced by its actuators. $\Lambda^i$ is the $i^{th}$ manipulator operational inertia matrix and $C^i$ represents the Coriolis, centrifugal and gravitational effects. According to (1) and (6), the commanded joint level effort to be realized is :

$$
\begin{aligned}
\tau^i_{com} = \quad & J^{iT} [\Lambda^i M^{i-1} \left( B^i \left( \dot{X}^i - \dot{\tilde{X}}^i \right) \right. \\
& \left. + K^i \left( X^i - \tilde{X}^i \right) \right) + C^i + \left( \mathbb{I}_3 - \Lambda^i M^{i-1} \right) F^i ]
\end{aligned}
\tag{7}
$$

## 4. Reference trajectories

The reference trajectory $\tilde{X}^i(t)$ in (7) is computed according to task specification, i.e. desired object trajectory $\tilde{X}^o(t)$, $^o\tilde{R}_b(t)$, $\tilde{V}^o(t)$, $\overset{\circ}{\tilde{V}}(t)$ and internal load $\tilde{\eta}^o(t)$. For that, each individual reference trajectory must be computed from an *accompanying* trajectory $\tilde{X}^i_a(t)$ and an *additional* trajectory $\tilde{X}^i_+(t)$.

- $\tilde{X}^i_a(t)$ is a local image of the system global motion seen from the $i^{th}$ contact point. It is derived according to the grasp geometry as :

$$\tilde{X}^i_a(t) = \tilde{X}^o(t) - r^i \tag{8}$$

- $\tilde{X}^i_+(t)$ is aimed to generate the $i^{th}$ desired end-effector/object interaction force $\tilde{F}^i(t)$. Namely, according to equations (2) and (5), we have :

$$\tilde{F}^i(t) = u^{iT} \tilde{F} \quad \text{and} \quad \tilde{F} = \left( W^\dagger \left( \Lambda \tilde{V}^o(t) + \tilde{C}^o - F^{ext} \right) + N \eta^o \right) \tag{9}$$

where $u^{iT}$ represents a suitable projection operator to extract $F^i$ from $\tilde{F}$. According to (1), $\tilde{X}^i_+(t)$ is finally deduced from $\tilde{F}^i(t)$ by the following low-pass filter application :

$$\tilde{X}^i_+(t) = \mathcal{L}^{-1} \left( \frac{\mathcal{L}(\tilde{F}^i(t))}{B^i p + K^i} \right) \tag{10}$$

where $\mathcal{L}(.)$ represents the Laplace transformation and $p$ represents the derivative operator.

# 5. Active force control

In distributed impedance control, so far, contact forces $F^i$ are not controlled explicitly (i.e.: existence of a feedback loop). Instead, they are specified off-line through a desired position terms $\tilde{X}^i_+$ (see equation (10)) and applied in a feed-forward manner. Thus, the system is unable to reject disturbances that make $F^i$ deviate from its desired value $\tilde{F}^i$.

However, introducing a complete active force control for all the components of the $F^i$'s would lead to a loss of control on the object's position. In this case, the system would be completely unable to reject any external force applied on the object.

A solution to this problem consists in selecting a subset among the $3\,n$ components of $F$ that will be actively force controlled while pure impedance control is kept for the other components. The two different subsets include respectively $n_\varphi$ and $n_\pi$ end-effector/object interaction components satisfying : $n_\pi + n_\varphi = 3n$. $n_\pi$ must be sufficient to guarantee a complete configuration control of the object. In other words, the $6 \times n_\pi$ grasp matrix $W_\pi$ extracted from $W$ and associated with the purely impedance controled interactions must be full rank. In case of free spatial manipulations, this implies to have $n_\pi \geq 6$.

To specify *which* components of $F$ should be force or purely impedance controlled, we can rely on the directions along which object internal forces are produced. This is illustrated on figures (2-a), (2-b) in the case of two simple two fingers grasps. When the object is in contact with the environment (figure (2-b)), internal forces also include the realization of the external force $F^{ext}$. Obviously, the active force control will be more efficient if it is implemented

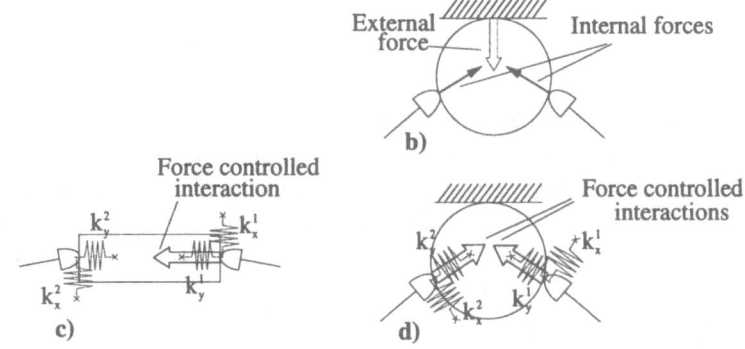

Figure 2. Example of force and pure impedance controllers repartition

along some of these particularly directions rather than along directions principally devoted to motion control. Possible repartitions for the grasps (2-a) and (2-b), are shown on figures (2-c), (2-d).

The resulting distributed impedance control scheme, including active force control, is represented in figure (3). Theoritical stability of this control scheme has been studied in [4].

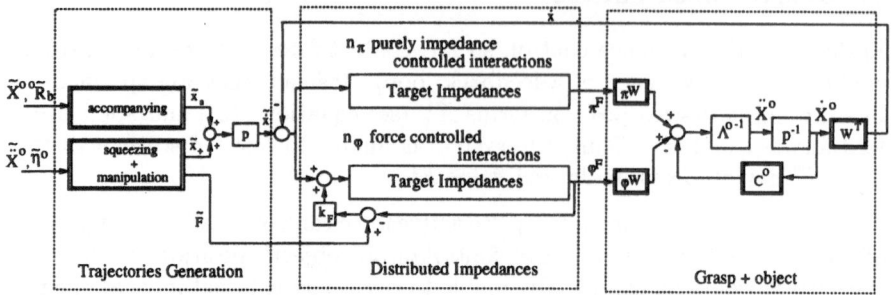

Figure 3. Distributed impedance scheme including active local force control

## 6. Experimental setup

Distributed impedance control has been applied to an experimental dual arm system, involving an IBM 7576 Scara manipulator and a Puma 560 manipulator (figure 4-a). Each manipulator is equipped with a 6 axis force/torque sensor. The two end-effectors are enhanced by punctual fingertips with friction.

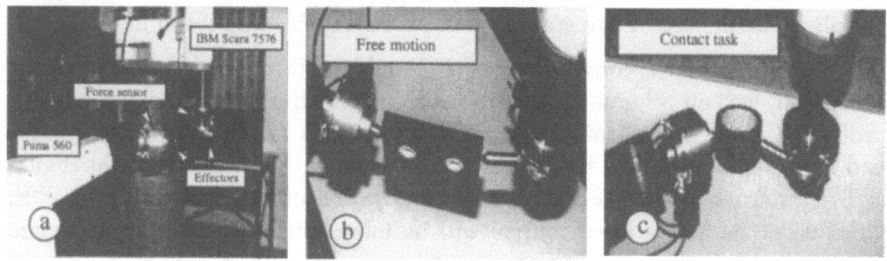

Figure 4. Experimental set-up in different task

## 7. Dynamic transporting task

The first experiment consists in transporting an object (15cm×5cm×3cm, 0.5kg) in the horizontal plane (figure 4-b). According to the repartition shown on figure 2-c, the y-component of the interaction with the SCARA robot $(k_y^1)$ is selected to be force controlled. We have then $F^o = W F$ with :

$$F^o = \begin{pmatrix} F_x^o \\ F_y^o \\ M^o \end{pmatrix}, F = \begin{pmatrix} F_x^1 \\ F_y^1 \\ F_x^2 \\ F_y^2 \end{pmatrix} \text{ and } W = \begin{pmatrix} c(\theta) & -s(\theta) & -c(\theta) & s(\theta) \\ s(\theta) & c(\theta) & -s(\theta) & -c(\theta) \\ r\,c(\delta\theta) & r\,s(\delta\theta) & r\,c(\delta\theta) & r\,s(\delta\theta) \end{pmatrix}$$

$$(11)$$

where $r$ is the half of the distance between the two contact points, $\theta$ is the angle characterizing the object orientation in the plane and $\delta\theta = (\tilde{\theta} - \theta)$ is the deviation with respect to its desired orientation $\tilde{\theta}$.

### 7.1. Stability considerations

As we can see, the resulting applied moment $M^o$, which depends on the last row of $W$, is affected by $\delta\theta$. Moreover, within the context a of decentralized control scheme, the actual object orientation error $\delta\theta$ can not be computed and must then be approximated to zero. This makes the system very sensitive with respect to object angular deviations.

However, according to the grasp repartition we propose, the sub-matrice $_\pi W$, extracted from $W$ :

$$_\pi W = \begin{pmatrix} c(\theta) & -c(\theta) & s(\theta) \\ s(\theta) & -s(\theta) & -c(\theta) \\ r\,c(\delta\theta) & r\,c(\delta\theta) & r\,s(\delta\theta) \end{pmatrix} \tag{12}$$

is non-singular as long as $\delta\theta < 90°$, which is always true in practice.

Moreover, we can show that the stability conditions [4] are verified as soon as :

$$\frac{k_x^1 k_x^2}{k_x^1 + k_x^2} > \frac{\eta^o}{2\,r} \quad \text{and} \quad k_y^2 > 0 \tag{13}$$

From a physical point of view, the first inequality in (13) can be interpreted as a condition on the serial combination of the two tangential stiffnesses $k_x^1$ and $k_x^2$. Moreover, we can show that (13) implies : $k_x^1 + k_x^2 > \frac{2\eta^o}{r}$, which is, in turn, a condition on the parallel combination of $k_x^1$ and $k_x^2$. Practically, this combination has to overcome the angular destabilizing effect of the object internal load $\eta^o$. As we can see, increasing the object's width $2\,r$ increases stability boundaries.

## 7.2. Tuning of the controllers

The controllers parameters were set according to the following reasoning :

1. Individual target impedances were initialized at their usual values as if the manipulators were performing independent tasks.
2. These values were then adjusted according to the type of closed kinematic chain constituted. Particular attention was paid to the reflected impedance seen from each manipulator at its contact point.
3. Preliminary tests were achieved along elementary trajectories leading to an empirical final adjustment of impedance parameters and force feedback gains.
4. We verified that stability conditions (13) were satisfied.

Applying this reasoning to the transporting task of figure (4-b), we obtained the following set of parameters :

| Components | $m_\beta^i \; (Nm/s^2)$ | $k_\beta^i \; (N/m)$ | $b_\beta^i \; (Ns/m)$ | $k_{F_\beta}^i \; (m/sN)$ |
|---|---|---|---|---|
| $f_x^1$ | 30-60 | 18900 | 1260 | 0 |
| $f_y^1$ | 30-60 | 6200 | 1930 | $7.10^{-3}$ |
| $f_x^2$ | 40-60 | 5400 | 85 | 0 |
| $f_y^2$ | 40-60 | 6100 | 170 | 0 |

Table 1 : parameters tuning for the transporting task

### 7.3. Results

In the present case, the commanded object trajectory involves both linear and angular object's large displacements within a wide range of velocities and accelerations (dotted lines of figures (5-b), (5-c) and (5-d)). The path of the object center of mass is an horizontal square with a five order polynomial chronological profile. Its orientation has a 20 degrees alternative movement. The desired internal load is set to $12N$.

Individual reference trajectories were computed according to task specification and equations (8)-(10). Kinematic redundancy of the system was solved by choosing a constant orientation of the two end-effectors during the task execution.

Figure (5) shows results in trajectory tracking and internal load regulation. With linear and angular errors less than 1mm and 0.03 radian respectively, the trajectory is accurately followed. Internal load (figure (5-a)) has been filtered for noise measurement attenuation using a low-pass filter with 30Hz bandwidth. Largest errors occur when the system starts or completely stops (i.e.: at the beginning, the middle, and the end of the trajectory). This is mainly due to important joint stiction effects. Anywhere else and particularly for phases at maximal velocity, internal load is correctly regulated.

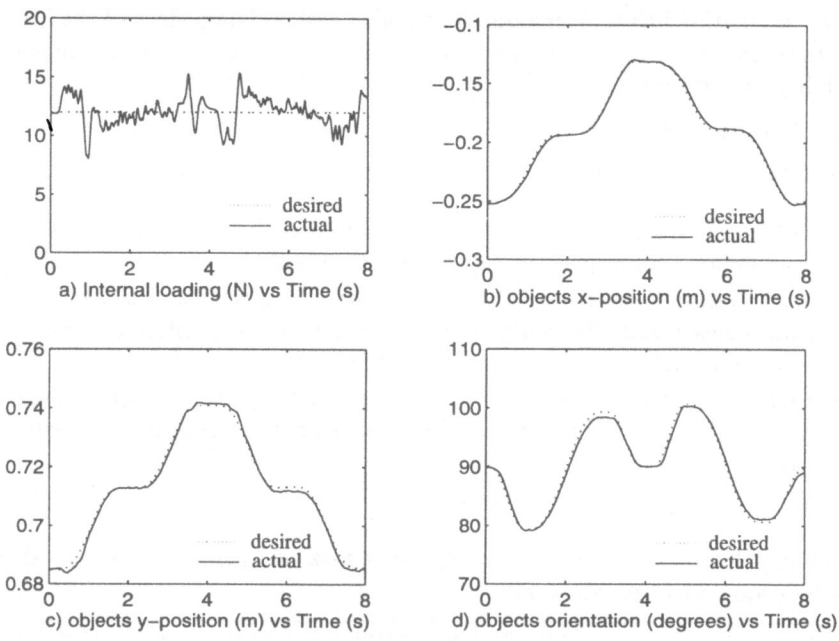

Figure 5. Results of the transporting task

## 8. Contact task

In order to illustrate the generality and the evolutivity of the distributed impedance approach, we also performed a contact task using the same dual-arm cooperative system without any modification in the controller structure.

This task is illustrated on figure (4-c). The manipulated object is a rigid cylinder with radius $0.05\,m$. It comes into contact with a planar, vertical, rigid surface. Because there is no friction between the object and the environment, the applied external force $F_{ext}$ is normal to the contact surface (figure 2-b).

As depicted on figure (2-d), two active force control loops can be implemented here. They are applied along the two y-directions of the two contact frames. They match the directions of interaction forces needed for producing both the object internal load (12N) and the desired contact force $F_{ext}$ (between 0.5N and 7.5N). Because $F_{ext}$ is changing during the task, contact frames orientations also evolute.

## 8.1. Tuning of the controllers

Controllers parameters were tuned according to the same reasoning as for the transporting task. For that, we just considered the contact between the object and the environment as an additional object-effector interaction involving an infinite target impedance in the direction normal to the contact and a null impedance in its tangential direction.

The following set of parameters was obtained :

| Components | $m_\beta^i\,(Nm/s^2)$ | $k_\beta^i\,(N/m)$ | $b_\beta^i\,(Ns/m)$ | $k_{F_\beta}^i\,(m/sN)$ |
|:---:|:---:|:---:|:---:|:---:|
| $F_x^1$ | 30-60 | 5800 | 630 | 0 |
| $F_y^1$ | 30-60 | 5800 | 630 | $7.10^{-3}$ |
| $F_x^2$ | 40-60 | 5400 | 85 | 0 |
| $F_y^2$ | 40-60 | 5400 | 85 | $6.10^{-3}$ |

Table 2 : parameters tuning for the contact task

## 8.2. Results

The object internal load and contact force that we obtained are represented on figure 6. As we can see, both of them are correctly regulated around their desired values with final deviation less than 1 N in both cases.

## 9. Conclusion

In this paper, we proposed the distributed impedance approach for controlling cooperative systems such as multi-arm manipulator systems or articulated hands.

Details of its implementation were discussed considering the example of an object being grasped through a set of punctual contacts with friction. Experimental results were presented. Different kind of manipulation tasks were performed using a Scara-Puma dual arm system without any modification in the controller structure. Moreover, impedance control is implemented at end-effector level. Thus, it also provides an efficient regulation of object-effector interactions. Particularly, it allows to introduce a local, active force control at the object-effector interface. This brings robustness in object internal loading realization while the object position is still controlled in a closed loop manner.

130

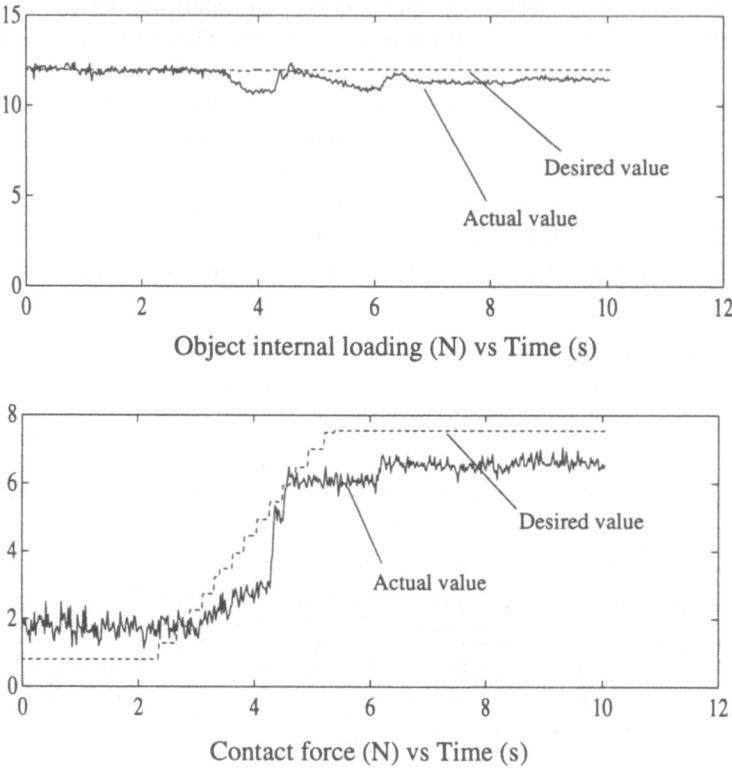

Object internal loading (N) vs Time (s)

Contact force (N) vs Time (s)

Figure 6. Results of the contact task

## References

[1] Chiacchio P, Chiaverini S, Sciavicco L, Siciliano B 1989 Task Space Dynamic Analysis of Multiarm System Configurations. *The international Journal of Robotics Research* Vol.10 No.6 : 708-715

[2] Choi H R, Chung W K, Youm Y 1995 Control of Grasp Stiffness using a Multi-fingered Hand with Redundant Joint. *Robotica* Vol.13 No.3 : 351-362

[3] Perdereau V, Drouin M 1996 Hybrid External Control for Two Robot Coordinated Motion. *Robotica* Vol.14 : 141-153

[4] Szewczyk J 1998 Planification et contrôle de la manipulation coordonnée par distribution d'impédance. PhD thesis (in french), University of Paris 6

[5] Kazerooni H, Waibel B J, Kim S 1995 On the Stability of Robot Compliant Motion Control : Theory and Experiments. *Journal of Dynamic Systems, Measurement and Control* Vol.112 : 417-426

[6] Morel G, Bidaud Ph 1996 A reactive external force loop approach to control manipulators in the presence of environmental disturbances. *Int. Conf. on Rob. and Automation* : 1229-1234

[7] Khatib O, Burdick J 1986 Motion and Force Control of Robot Manipulators. *Int. Conf. on Rob. and Automation* : 1381-1386

# Experiments and Factory Application of Force Control and Contour Following over Visco-Elastic Surface

Y. Ohkami,   T. Sunaoshi,   Y. Sugiura
Tokyo Institution of Technology
Tokyo, Japan
e-mail : ohkami@mes.titech.ac.jp

Abstract: This paper concerns force control and contour following over visco-elastic surfaces such as the case with FRP molding. Successful FRP molding requires that a bounded force be applied to the surface and that the pressing of the FRP results in an even surface thickness. To accomplish this task using a robot, we analyze the visco-elastic properties of FRP. The surface is expressed as a continuous model so that we can look the surface deformation caused by the roller motion. Numerical simulations are conducted with the model, and are verified with experiments. Results suggest that to get an even thickness of FRP, we should control not only the magnitude of the contact force but also the time that it acts on the surface.

## 1. Introduction

It is required that in the future industrial FRP molding, which is currently carried out manually, be automated. FRP is a useful material in various industries such as construction, ship-building, airplanes, aerospace, sports, and leisure. It is used to make tanks, unit baths, fishing boats, etc. The distinctive properties of FRP are that it is light, rigid and resistant to corrosion. An example of factory FRP molding is shown in Fig.1.1. Workers use a roller to mold the pre-hardened FRP surface and to make it even, while also pushing out the air pockets in the FRP mixture. The more complicated the surface is, the better the skill that is required. FRP burns easily and has a noxious smell. Furthermore there are few experienced workers who can perform the task. For these reasons, there is a strong desire within the industry to automate the task.

Generally force control is used to automate this task. Much research on force control has been presented in the past several years. It is noted approaches such as hybrid control and impedance control have not focused sufficiently on actual industry applications and their particular characteristics. For example, with FRP molding there is a unique high-speed backward and forward motion

Fig.1.1 FRP molding in a factory

that makes the force control task difficult. This paper addresses another important characteristic of this application: the visco-elastic nature of the contact surface, which is analyzed using a continuous-contact model.

## 2. Modeling

### 2.1 Visco-elastic surface model

First, we consider a model of the FRP surface at the point of contact. In this paper, FRP refers to the material in its pre-hardened state. The characteristics of FPR are:

1. It shows continuous displacement if pushed.

2. The displacement does not return to its original value even if the pushing force is taken away.

3. The displacement is a function of the magnitude of the contact force and the total time the force is applied.

4. The elastic properties of the FRP surface are dependent on environmental conditions such as temperature and humidity.

We consider only the first three of the above characteristics, and arrive at the model shown in Fig 2.1. In order to demonstrate that the model is appropriate, we compare numerical simulations and experiment results as shown in Fig.2.2. The characteristics mentioned above can be easily identified in these results.

Rolling on a visco-elastic layer is shown in Fig.2.3 and Fig.2.4. We can divide the surface deformation as resulting from two sources. One is created by pushing with the roller directly. The other is created indirectly by neighboring surface deformations.

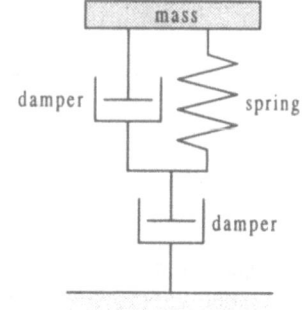

Fig.2.1   Visco-elastic point contact model

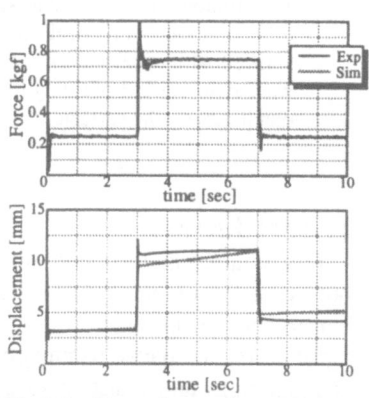

Fig.2.2   Numerical and experimental verification of above model.

Fig.2.3 Rolling over visco-elastic surface

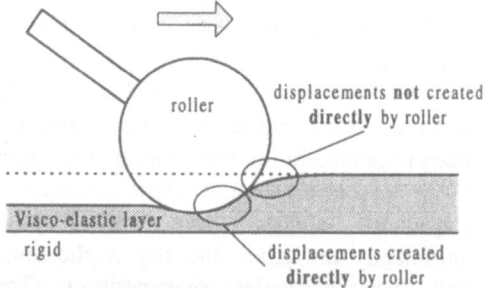

Fig.2.4 Two types of visco-elastic displacements

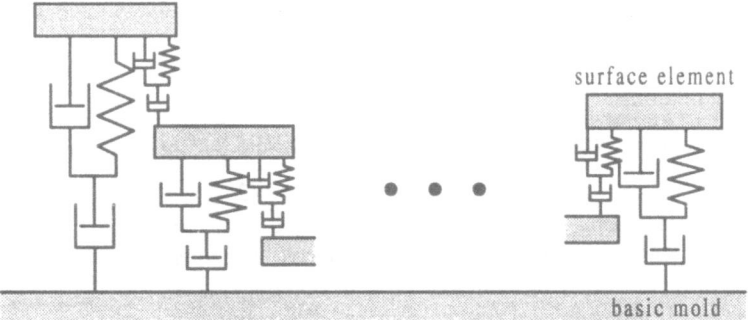

Fig.2.5 Visco-elastic surface model

The parameters we want to control are the magnitude of the normal force and the thickness of the surface. Taking into account these two parameters and the two sources of surface deformation, we consider the model shown in Fig.2.5 for the visco-elastic surface. We divide the visco-elastic layer into many sections. These sections and the distance between them are smaller than the roller radius. The sections are constrained to move only in the vertical direction. The roller contacts several sections at once while rolling across the surface.

## 2.2 Formulations

To arrive at a robotic contour following model for the FRP molding, we first consider a model of a contouring following robot and then consider the addition of the visco-elastic surface model.

### 2.2.1 Contouring following robot

We consider a force control mechanism mounted on the tip of kinematically controlled industrial manipulator. The force controlled mechanism is given velocity $v$ by the industrial manipulator. $\{a\}$ is the inertial frame of reference.

$$v = \{a\}^T [v_x \quad 0 \quad v_z]^T \tag{2.1}$$

$$f = \{g\}^T [f_t \quad 0 \quad f_n]^T \tag{2.2}$$

At the point where the roller contacts the surface, a contact force $f$ acts on the robot. The coordinate frame $\{g\}$ is defined such that one axis is tangent to the surface and another is normal to it. We can thus derive a formulation of the robot motion as follows,

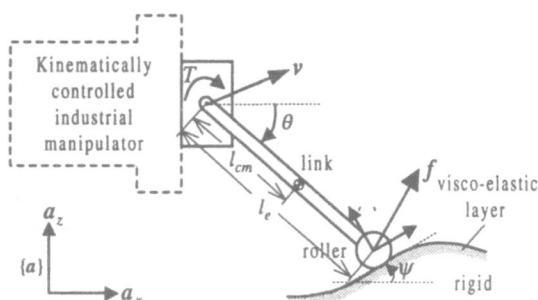

Fig.2.6 Model of contouring following robot

$$I'\ddot{\theta} - ml_{cm}\{\dot{v}_x \sin\theta + (\dot{v}_z + g)\cos\theta\} = T - f_n l_e \cos(\theta + \psi) - f_t l_e \sin(\theta + \psi) \tag{2.3}$$

134

### 2.2.2 Visco-elastic layer

First, we consider one surface section. The model parameters are defined as shown in Fig.2.7 and we get the following relationship.

$$m_e(c_1 + c_2)\ddot{d} + (m_e k + c_1 c_2)\dot{d} + c_2 k d = (c_1 + c_2)f + k\int f$$

(2.4)

$f$ is total force acting on a section. Next, we consider the relationship with neighboring sections. Adjacent sections connect with each other as shown in Fig.2.8. Force $f_{next}$ acts between the $i_{th}$ division and $(i+1)_{th}$ division. The following equation can be written:

$$(c_{n1} + c_{n2})\dot{f}_{next} + k_n f_{next} = c_{n1} c_{n2} \ddot{d}_n + k_n c_{n2} \dot{d}_n \qquad (2.5)$$

Therefore, the total force $f$ is expressed as follows, where $f_i$ is the force imparted by the roller.

$$f = f_i + f_{next(i,i+1)} - f_{next(i-1,i)} \qquad (2.6)$$

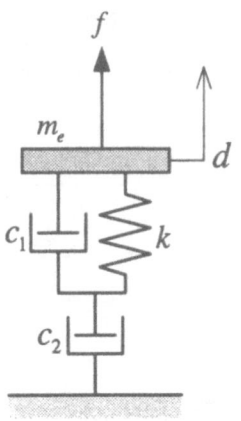

Fig.2.7 Definition of model parameters

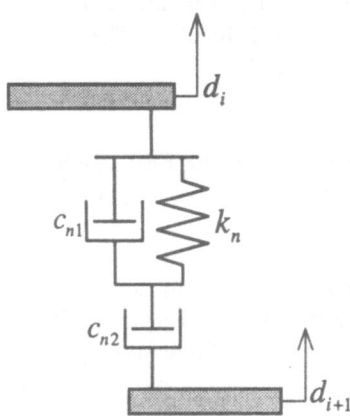

Fig.2.8 Connection between neighboring sections

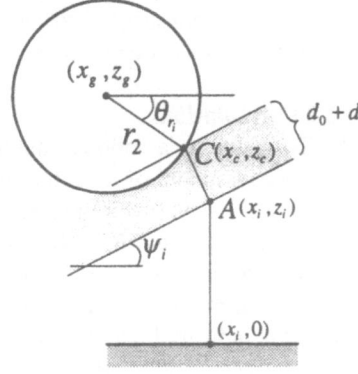

Fig.2.9 Contact condition

### 2.2.3 Contact conditions

Contact between the various sections and the roller is assured by the following constraint condition.

$$z_g - r_2 \sin\theta_{r_i} = z_i + (d_0 + d_i)\cos\psi_i$$

The above equation simply states that the contact point $C$ is the same whether seen from the robot or from the base plate is the same. We differentiate the above equation twice with respect to time and get the following,

$$\left(\cos\psi_i + \frac{\sin\psi_i \cos\theta_{r_i}}{\sin\theta_{r_i}}\right)\ddot{d}_i + l_e\left(\cos\theta + \frac{\sin\theta\cos\theta_{r_i}}{\sin\theta_{r_i}}\right)\ddot{\theta}$$

$$= \ddot{z} + l_e\dot{\theta}^2\sin\theta - (\ddot{x} - l_e\dot{\theta}^2\cos\theta - r_2\dot{\theta}_{r_i}^2\cos\theta_{r_i})\frac{\cos\theta_{r_i}}{\sin\theta_{r_i}} + r_2\dot{\theta}_{r_i}^2\sin\theta_{r_i} \qquad (2.7)$$

## 3. Numerical simulations

Numerical simulations were conducted using the equations derived in the previous section. A 4th order Runge-Kutta-Gill method is use on three types of surface shapes: flat, ramp, and wavy. The roller moves one cycle along these surfaces. Fig. 3.1 shows the result of contouring following on a flat surface. The torque is given as eq.(3.1).

Constant applied torque: $T = f_{des} l_e \cos(\theta + \psi) - mgl_{cm} \cos\theta$      (3.1)

PI control            : $T = f_c l_e \cos(\theta + \psi) - mgl_{cm} \cos\theta$      (3.2)

where $f_c$ is $f_c = f_{des} + K_p f_{err} + K_i \int f_{err}$ .

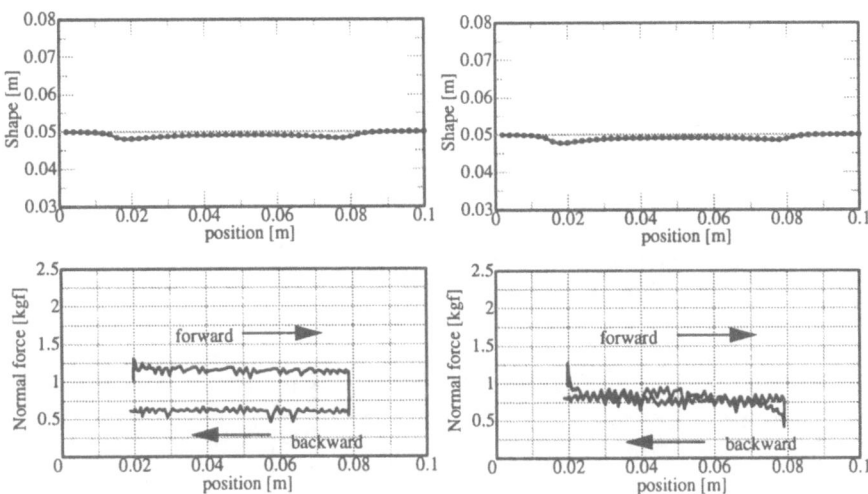

Fig. 3.1. Simulation results for flat surface. Constant applied torque (left) and PI control (right)

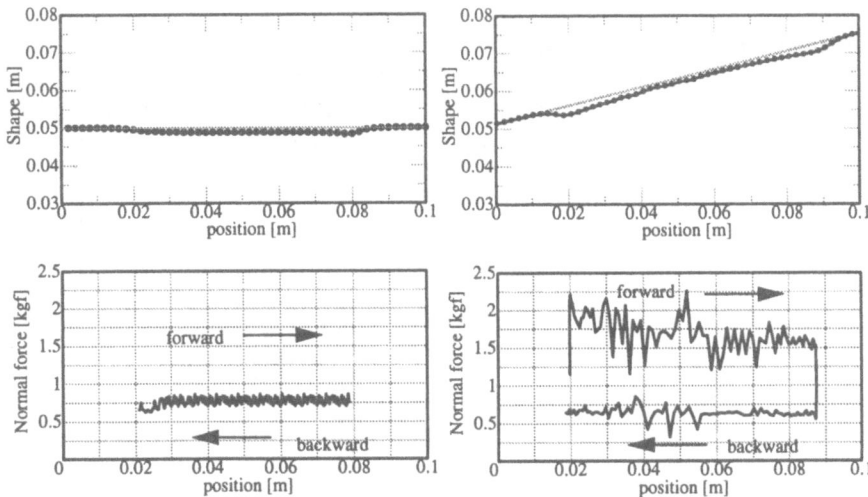

Fig. 3.2. Simulation results for flat surface with constant velocity (left), ramp surface (right)

Note that the level of the normal force is different between the forward motion and backward motion. This is a result of the friction between the surface and roller, as well as the fact that the contact point acts on only one side of the roller. At both ends of the work area, we notice a dip in the surface, even though the magnitude of the normal force is constant over the motion. If the motion speed were to be constant, the thickness of the surface would remain even. The reason for the dips is that acting time of the contact force is longer in those areas.

The proceeding figures show results for a ramp and wavy surface. For the wavy surface, at faster motions the contour following is not sufficiently good at the top of the wave. The faster the speed of the motion, the larger the inertia of the roller, so that the contact force at that point becomes very small.

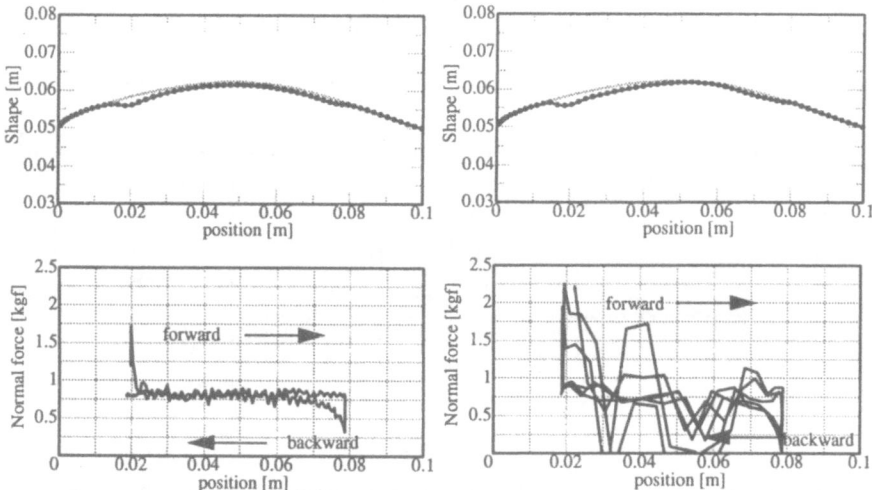

Fig. 3.3. Simulation results for wavy surface. Slow motion (left) and fast motion (right).

## 4. Laboratory Experiment

In order to validate the model, a laboratory testbed was constructed as shown in Fig.4.1. A 6-axis force/torque sensor is mounted on the outer link and is used for closed-loop force control. The work surface consists of a visco-elastic material (clay) spread out over a rigid base.

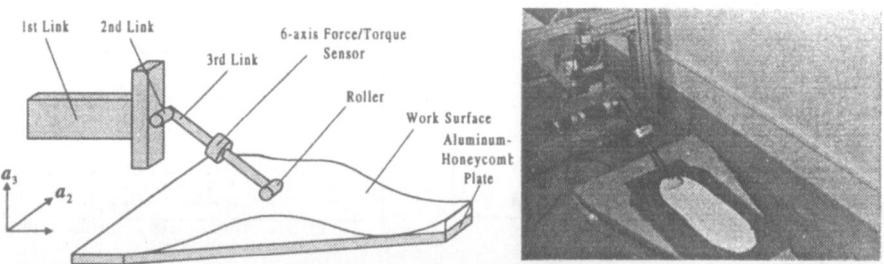

Fig.4.1 Laboratory testbed

Experimental results are shown in Fig.4.2. For the flat surface case, we see again the dips at both ends of the work area. These dips also appear for the ramp surface case.

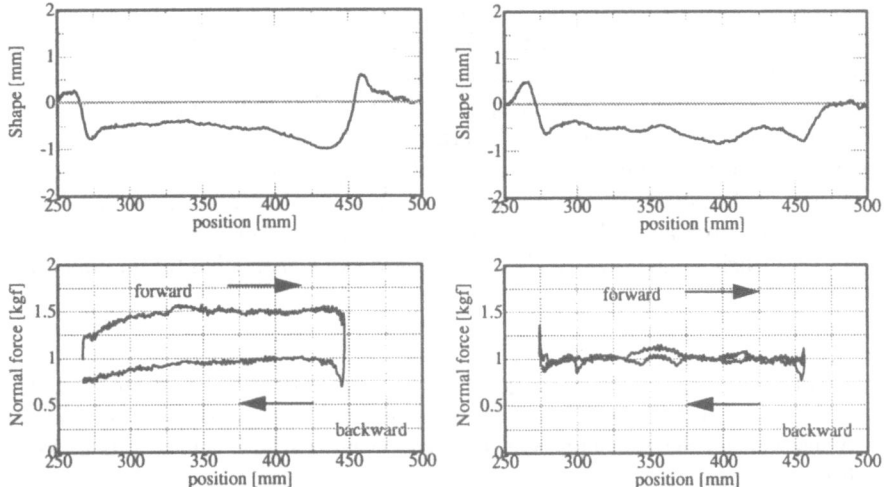

Fig. 4.2. Experimental results for flat surface. Constant applied torque (left) and PI control (right)

In the case of the wavy surface, again for faster sweeps of the contour not enough force is applied at the top of wave. This is consistent with the results of numerical simulations.

We next consider using the acting time of the force as a control variable to regulate the surface displacement. It is difficult to directly measure the acting time on a given section, so we use the velocity of the roller instead of the acting time. The desired force is given as follows:

$$f_{des} = f_0 + f_a v_g{}^2 \qquad (4.1)$$

A torque is applied to the outer link so as to achieve this force. With this approach, the displacement level of the surface is more even than that of the first result.

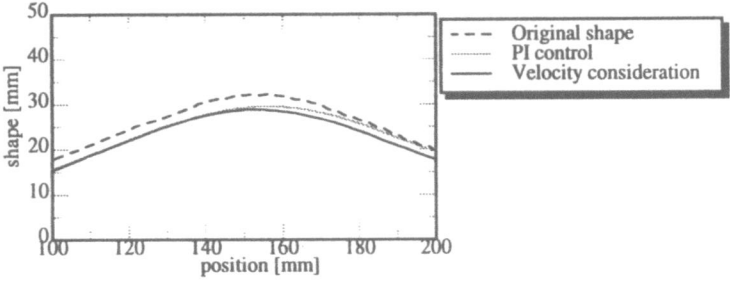

Fig.4.3 Experimental results for wavy surface

## 5. Conclusions

In this paper, a continuous-contact model for force control and contour following over visco-elastic surfaces is presented. Numerical and experimental results with the model show that the visco-elastic surface deformation is related not only to the magnitude of the applied contact force, but also the time that the force acts on a given point on the surface. In a factory application such as FRP molding, it is believed that the acting time of the applied force can be used as a control variable to achieve better quality FRP rolling. Efforts are currently under way at implementing the ideas expressed in this paper on an actual factory testbed.

A part of this study was sponsored by the R & D grant from Japan Science and Technology Corporation. The authors wish to express their thanks to the Corporation and to Mr. Nishida and Mr. Honda of Hirata Kyogyo Corporation for the supports at Tsuyama Works.

### References

[1]John J. Craig 1989 Introduction to ROBOTICS –mechanics and control. Addison-Wesley.

[2]A. Sharon, N. Hogan, D. E. Hardt 1993 The Macro/Micro Manipulator : An Improved Architecture for Robot Control. Robotics and Computer Integrated Manufacturing. Vol.10 No.3

[3]K. Nagai, T. Yoshida 1994 Impedance Control of Redundant Macro-Micro Manipulators. Proc. of IEEE/RSJ Int. Conf. on Intelligent Robots on Systems.

[4]G. R. Naghieh, H. Rahnetjat, ZM. Jim 1997 Contact mechanics of visco-elastic layered surface. Int. Conf. on Contact Mechanics Proc. pp59-66

[5]D. Gorinnevsky, A. Formalsky, A. Schneider 1997 Force Control of Robotics Systems. CRC

[6]W. Flugge 1975 Viscoelasticity. Springer-Verlag, New York

[7]Y. Ohkami, H. Lakhani, T. Sunaoshi 1998 Experimental Studies on Robotic Force Control and Contour Following. JSME Vol.64 No.621 pp1764-1771

# Coordinated Control of a Nonholonomic Mobile Manipulator

Jean-Yves Fourquet and Marc Renaud
LAAS-CNRS
7, Avenue du Colonel Roche, 31077 Toulouse Cedex 4
FRANCE
jeanyves@laas.fr renaud@laas.fr

**Abstract:** In this paper, we present a control algorithm for a nonholonomic mobile manipulator built from a $n$ joints robot manipulator mounted on a nonholonomic mobile platform. We want the manipulator end-effector *location* (position and orientation) to evolve between starting and final required locations and to follow a required *trajectory* (defined as a function of the time). We describe a *global control* which calculates the generalized coordinates for both the manipulator and the platform. Two types of control are presented; the first uses a set of additional tasks and the second minimizes a quadratic criterion. A comparison between these two types is made on an academic mobile manipulator, which consists of a double-pendulum mounted on the platform. The implementation of the first type is made on a real mobile manipulator which consists of the GT6A six revolute joints manipulator mounted on the HILARE H2bis platform.

Key words : Mobile manipulators, Nonholonomy, Additional tasks, Minimization of a quadratic criterion.

## 1. Introduction

When we write across or erase a blackboard, we locate our body so that our arm is in the most comfortable configuration to write. We place our body, in an instinctive manner, in the admissible workspace of our arm by moving our legs. So we consider our legs like a mobile platform which is able to locate and move our torso along a parallel to the board. We can describe two different ways to move and realize our goal: first, by moving the platform towards the board, keeping the robot manipulator stopped, and then by moving the robot manipulator; second, by moving both the robot manipulator and the mobile platform as a human does.

In this paper we consider the second way in order to develop a *global control* of a mobile manipulator built from a $n$ joints robot manipulator mounted on a nonholonomic mobile platform. More precisely we are looking for the combinated evolution of the platform and of the manipulator so that the *location* (position and orientation) of the manipulator end-effector follows a desired *trajectory*, defined as a function of the time.

In the litterature, the methods of coordinated control of mobile manipulators are rather limited. A typical characteristic of mobile manipulators is their high degree of kinematic redundancy created by the addition of the platform degrees of mobility to the manipulator ones. The coordinated control is then based upon the resolution of this redundancy. It is necessary to distinguish these methods in accordance with the problem to be solved; in a first class the end-effector motion is imposed whereas in a second class only the final end-effector location is imposed.

The most important contributions belonging to the first class must be themselves divided into two subclasses. In the first subclass the coordinated control is obtained by adding tasks in order to solve the redundancy problem. To our knowledge the first work in this area is due to H. Seraji [1] [2] which considers an instantaneous kinematic approach. Others works in the same area are due to C.-C. Wang et al. [3], U. M. Nassal [4] and F. G. Pin et al. [5]. Always in this class Y. Yamamoto [6] considers moreover a dynamic approach. In the second subclass the coordinated control is obtained by minimizing a quadratic criterion; this is for example the case in the works of C.-C. Wang et al. [3] and of U. M. Nassal [4]. The works concerning the second class are very limited; a few contributions are due to G. Foulon et al. [7] and C. Perrier et al. [8]. Finally, among a lot of interesting contributions in the field of mobile manipulators, let us mention the works of O. Khatib et al. [9], K. Nagatami et al. [10], T. Sugar et al. [11]...

In this paper we compare global controls based, respectively, on the definition of a set of additional tasks and on the minimization of a quadratic criterion. This control is implemented on a mobile manipulator made of the GT6A six revolute joints manipulator mounted on the HILARE H2bis platform. First we introduce the modelling of the mobile manipulator. Then, we define the methodology of the global control. Finally, we present the experimental system and the results.

## 2. Modelling of the mobile manipulator

### 2.1. Notations

We suppose that the mobile manipulator is built from a $n$ joints robot manipulator mounted on a nonholonomic mobile platform. In order to be general we assume that the position of the manipulator base, with respect to the platform, is defined by two parameters $a$ and $b$. Let $\mathcal{R} = (O, \vec{x}, \vec{y}, \vec{z})$ be the fixed frame in which the *trajectory* of the manipulator end-effector is given and $\mathcal{R}' = (O', \vec{x}', \vec{y}', \vec{z}')$ the moving frame attached to the platform. Furthermore let $\mathcal{R}_0 = (O_0, \vec{x}_0, \vec{y}_0, \vec{z}_0)$ be the moving frame attached to the manipulator base[1] and $\mathcal{R}_n = (O_n, \vec{x}_n, \vec{y}_n, \vec{z}_n)$ a moving frame attached to the manipulator end-effector and rotating around the point $O_n$. The point $O_{n+1}$, which is the center of this end-effector, is fixed relative to the frame $\mathcal{R}_n$.

### 2.2. Kinematics equations of the mobile manipulator

#### 2.2.1. Mobile platform subsystem

If we leave the wheels of the platform out of account we can identify the three *generalized coordinates* of the platform - defining its *configuration* - with the

---

[1]Remark that the moving frames $\mathcal{R}_0$ and $\mathcal{R}'$ are parallel.

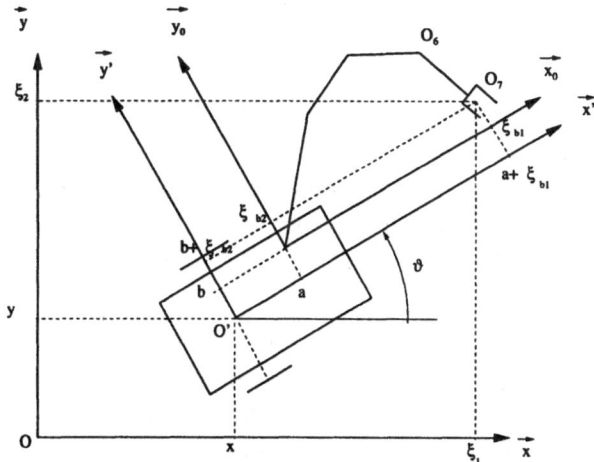

Figure 1. Mobile manipulator when $n = 6$

three *operational coordinates* of this platform - defining its *location*. If we choose
the three operational coordinates as: $\boldsymbol{\xi}_p = (\xi_{p1}\ \xi_{p2}\ \xi_{p3})^t = (x\ y\ \vartheta)^t$ (see Fig. 1),
the three generalized coordinates are defined by: $\mathbf{q}_p = (q_{p1}\ q_{p2}\ q_{p3})^t = (x\ y\ \vartheta)^t$.
With the previous abstraction it is necessary to suppose that the control of this
platform is given by: $\mathbf{u} = (u_1\ u_2)^t = (v\ \omega)^t$, in which $v$ represents the linear
velocity of the point $O'$, on its trajectory in the plane $(O,\ \vec{x},\ \vec{y})$, and $\omega$ the
angular velocity of the platform, around the vector $\vec{z}$.
$v$ is such that: $v = \dot{\sigma}$, where $\sigma$ is the curvilinear abscissa of the point $O'$ on
its path in the previous frame, and $\omega = \dot{\vartheta}$. The configuration kinematic model
[12] of this platform is then: $\dot{\mathbf{q}}_p = S(\mathbf{q}_p)\mathbf{u}$, with:

$$S(\mathbf{q}_p) = \begin{bmatrix} \cos q_{p3} & 0 \\ \sin q_{p3} & 0 \\ 0 & 1 \end{bmatrix} = \begin{bmatrix} \cos \vartheta & 0 \\ \sin \vartheta & 0 \\ 0 & 1 \end{bmatrix}.$$

From this model the nonholonomic constraint can be written in the form:

$$G_p(\mathbf{q}_p)\dot{\mathbf{q}}_p = 0 \qquad (1)$$

with: $G_p(\mathbf{q}_p) = \begin{bmatrix} \sin q_{p3} & -\cos q_{p3} & 0 \end{bmatrix} = \begin{bmatrix} \sin \vartheta & -\cos \vartheta & 0 \end{bmatrix}$.

### 2.2.2. Robot manipulator subsystem

The *configuration* of the manipulator is defined by the $n$ *generalized coordi-
nates*: $\mathbf{q}_b = (q_{b1}\ q_{b2}\ \ldots\ q_{bn})^t$, and the end-effector *location*, i.e. the position
of the point $O_{n+1}$ (defined here by its Cartesian coordinates) and the ori-
entation of the frame $\mathcal{R}_n$, relative to $\mathcal{R}_0$, by the $m$ *operational coordinates*:
$\boldsymbol{\xi}_b = (\xi_{b1}\ \xi_{b2}\ \ldots\ \xi_{bm})^t$. The direct kinematic model of the manipulator [13],
relative to $\mathcal{R}_0$, is: $\boldsymbol{\xi}_b = f_b(\mathbf{q}_b)$ and the direct instantaneous kinematic model,
with $J_b(\mathbf{q}_b) = \frac{\partial f_b}{\partial \mathbf{q}_b}(\mathbf{q}_b)$:

$$\dot{\boldsymbol{\xi}}_b = J_b(\mathbf{q}_b)\dot{\mathbf{q}}_b. \qquad (2)$$

As $\mathcal{R}'$ is obtained from $\mathcal{R}_0$ by a translation characterized by the two constant parameters $a$ and $b$ the end-effector location, relative to $\mathcal{R}'$, is defined by the $m$ operational coordinates: $\xi'_b = (\xi'_{b1} \ \xi'_{b2} \ \ldots \ \xi'_{bm})^t$, with: $\xi'_{b1} = a + \xi_{b1}$, $\xi'_{b2} = b + \xi_{b2}$, $\xi'_{b3} = \xi_{b3}$, $\ldots$, $\xi'_{bm} = \xi_{bm}$. And we can write: $\dot{\xi}'_b = \dot{\xi}_b = J_b(q_b)\dot{q}_b$.

### 2.2.3. Mobile manipulator

The *configuration* of the mobile manipulator is defined by the $\nu = n + 3$ generalized coordinates: $q = (q_1 \ q_2 \ q_3 \ q_4 \ q_5 \ \ldots \ q_\nu)^t = (x \ y \ \vartheta \ q_{b1} \ q_{b2} \ \ldots \ q_{bn})^t$. From (1) it is then possible to write:

$$G(q)\dot{q} = 0 \tag{3}$$

with: $G(q) = [\ G_p(q_p) \ \ 0 \ ]$. We can characterize the end-effector *location*, i.e. the position of the point $O_{n+1}$ and the orientation of the frame $\mathcal{R}_n$, relative to $\mathcal{R}$, by the $\mu$ *operational coordinates*: $\xi = (\xi_1 \ \xi_2 \ \ldots \ \xi_\mu)^t$. The direct kinematic model of the mobile manipulator [13], relative to $\mathcal{R}$, is: $\xi = f(q)$ and the direct instantaneous kinematic model, with $J(q) = \frac{\partial f}{\partial q}(q)$:

$$\dot{\xi} = J(q)\dot{q}. \tag{4}$$

From (3) and (4) we obtain:

$$\begin{bmatrix} G(q) \\ J(q) \end{bmatrix} \dot{q} = \begin{bmatrix} 0 \\ \dot{\xi} \end{bmatrix}. \tag{5}$$

But in fact it is more judicious to use the new set of coordinates of the mobile manipulator defined by [14]: $p = (\sigma \ \vartheta \ q_{b1} \ q_{b2} \ \ldots \ q_{bn})^t$. Then: $\dot{q} = M(q)\dot{p}$, with: $M(p) = \begin{bmatrix} S(q_p) & 0 \\ 0 & E \end{bmatrix}$ where $E$ is the $n$ order unit matrix. Taking into account that $G(q)M(q) = 0$, we obtain, with $\bar{J}(q) = J(q)M(q)$:

$$\bar{J}(q)\dot{p} = \dot{\xi}. \tag{6}$$

## 3. Methodology of the global control

We want to control the mobile manipulator in such a way to obtain, in the frame $\mathcal{R}$, a desired *trajectory* of the manipulator end-effector location. This trajectory is defined as a function of the time: $t \longrightarrow \xi_d(t)$. It is then necessary to calculate the column matrix $\dot{p}$ corresponding to the imposed column matrix $\dot{\xi}_d$, that is to invert the relation (6). This column matrix $\dot{p}$ gives the global control of the mobile manipulator (indeed it is always possible to calculate $\dot{q}$ from $\dot{p}$). The calculation can be done by imposing a set of additional tasks or by minimizing a quadratic criterion.

Note: in fact from a practical point of view the control we propose is closed loop; the imposed column matrix $\dot{\xi}_d(t)$ as to be replaced by $\frac{\xi_d(t) - \xi(t-dt)}{dt}$.

## 3.1. Additional tasks

We can impose a set of $\nu - \mu - 1$ additional tasks in the form: $\bar{J}_a(\mathbf{q})\dot{\mathbf{p}} = 0$. This leads to an augmented system of equations and (6) becomes:

$$\bar{K}(\mathbf{q})\dot{\mathbf{p}} = \dot{\Xi}_d \tag{7}$$

with: $\bar{K}(\mathbf{q}) = \left[ \begin{array}{c} \bar{J}(\mathbf{q}) \\ \bar{J}_a(\mathbf{q}) \end{array} \right]$, and: $\dot{\Xi}_d = \left[ \begin{array}{c} \dot{\xi}_d \\ 0 \end{array} \right]$. Then the global control is given by:

$$\dot{\mathbf{p}} = \bar{K}^{-1}(\mathbf{q})\dot{\Xi}_d, \tag{8}$$

out of the singularities $\mathbf{q}$ given by the equation: $\det \bar{K}(\mathbf{q}) = 0$.
Let us notice that the additional tasks can be chosen over a space of dimension $(\nu - \mu - 1)(\nu - 2)$, since the rows of the matrix $\bar{J}_a(\mathbf{q})$ can always be divided by a constant.

## 3.2. Minimization of a quadratic criterion

We can also minimize the quadratic criterion $\bar{I} = \dot{\mathbf{p}}^t \bar{Q} \dot{\mathbf{p}}$ where $\bar{Q}$ is a given symmetric definite positive matrix of order $\nu - 1$. The global control is given by:

$$\dot{\mathbf{p}} = \bar{Q}^{-1}\bar{J}^t(\mathbf{q})[\bar{J}(\mathbf{q})\bar{Q}^{-1}\bar{J}^t(\mathbf{q})]^{-1}\dot{\xi}_d, \tag{9}$$

out of the singularities $\mathbf{q}$ given by the equation: $\det(\bar{J}(\mathbf{q})\bar{Q}^{-1}\bar{J}^t(\mathbf{q})) = 0$.
Let us notice that the criteria can be chosen over a space of dimension $\frac{\nu(\nu-1)}{2} - 1$, since the elements of the matrix $\bar{Q}$ can always be divided by a constant. When the matrix $\bar{Q} = E$, where $E$ is the $\nu - 1$ order unit matrix the solution is given by:

$$\dot{\mathbf{p}} = \bar{J}^+ \dot{\xi}_d \tag{10}$$

where: $\bar{J}^+$ is the pseudo-inverse of the matrix $\bar{J}$ ($\bar{J}^+ = \bar{J}^t(\mathbf{q})[\bar{J}(\mathbf{q})\bar{J}^t(\mathbf{q})]^{-1}$ when $\det(\bar{J}(\mathbf{q})\bar{J}^t(\mathbf{q})) \neq 0$).

## 3.3. Comparison between additional tasks and minimization of a quadratic criterion

The same column matrix $\dot{\mathbf{p}}$ is obtained if the set of additional tasks or the given criterion are such that: $\bar{J}_a(\mathbf{q})\bar{Q}^{-1}\bar{J}^t(\mathbf{q}) = 0$.
In the general case the additional tasks, corresponding to a given criterion, belong to a space of dimension $(\nu - \mu - 1)(\nu - \mu - 2)$ and the criteria, corresponding to a given set of additional tasks, belong to a space of dimension $\frac{\nu(\nu-1)}{2} - 1 - (\nu - \mu - 1)\mu$. Thus, in general, the space of solutions generated by minimizing a quadratic criterion has greatest dimension than that generated by adding tasks.

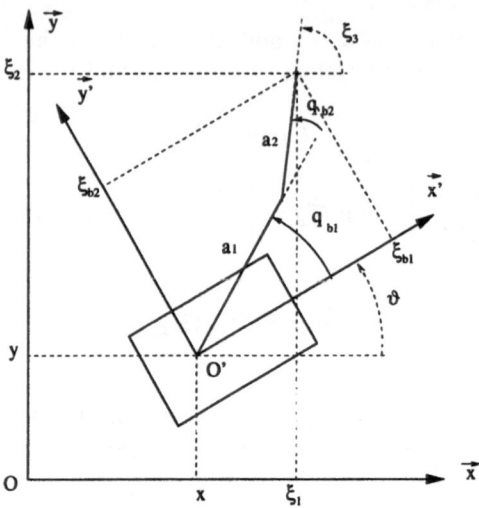

Figure 2. A planar system

## 3.4. Examples

### 3.4.1. First example

The first example consists of an academic horizontal double-pendulum mounted on the platform; then $\nu = 5$. In the frame $\mathcal{R}$ the position of the point $O_3$ is given by the Cartesian coordinates $\xi_1$ and $\xi_2$ and the orientation of the end-effector by the angle $\xi_3$; then $\mu = 3$. Furthermore we suppose that $a = b = 0$ (see Fig. 2). The matrix $\bar{J}(\mathbf{q})$ is:

$$\begin{bmatrix} \cos q_3 & -a_1 \sin q_{34} - a_2 \sin q_{345} & -a_1 \sin q_{34} - a_2 \sin q_{345} & -a_2 \sin q_{345} \\ \sin q_3 & a_1 \cos q_{34} + a_2 \cos q_{345} & a_1 \cos q_{34} + a_2 \cos q_{345} & a_2 \cos q_{345} \\ 0 & 1 & 1 & 1 \end{bmatrix}$$

with: $q_{34} = q_3 + q_4$ and $q_{345} = q_3 + q_4 + q_5$.

The control which corresponds to the minimization of the quadratic criterion such that $\bar{Q} = E$, where $E$ is the unit matrix of order 4, is given by (10) with:

$$\bar{J}^+ = \frac{1}{a_1 \cos q_4} \begin{bmatrix} a_1 \cos q_{34} & a_1 \sin q_{34} & a_1 a_2 \sin q_{34} \\ -\frac{1}{2} \sin q_3 & \frac{1}{2} \cos q_3 & -\frac{1}{2} a_2 \cos q_{45} \\ -\frac{1}{2} \sin q_3 & \frac{1}{2} \cos q_3 & -\frac{1}{2} a_2 \cos q_{45} \\ \sin q_3 & -\cos q_3 & a_1 \cos q_4 + a_2 \cos q_{45} \end{bmatrix}$$

with: $q_{45} = q_4 + q_5$, when $q_4 \neq \pm\frac{\pi}{2}$. Another expression is obtained for $\bar{J}^+$ when $q_4 = \pm\frac{\pi}{2}$.

This control leads to only one additional task defined by: $\dot{q}_3 - \dot{q}_4 = 0$.

For example when: $a_1 = a_2 = 1$, $\mathbf{q}(0) = 0$ and when the desired trajectory of the end-effector is such that: $\xi_{1d}(t) = 2, \xi_{2d}(t) = t, \xi_{3d}(t) = 0$, the evolution $q(t)$ of the mobile platform is given by: $q_1(t) = 1 - \cos t, q_2(t) = t - \sin t, q_3(t) = \frac{1}{2}t, q_4(t) = \frac{1}{2}t, q_5 = -t$, and the path of the mobile platform is a cycloïd.

### 3.5. Second example

This second example consists of a six joints manipulator mounted on the platform; then $\nu = 9$. In the frame $\mathcal{R}$ the position of the point $O_7$ is given by the Cartesian coordinates $\xi_1$, $\xi_2$, $\xi_3$ and the orientation of the manipulator end-effector by the angles $\xi_4$, $\xi_5$, $\xi_6$; then $\mu = 6$. Furthermore we suppose that $a$ and $b$ are arbitrary values (see Fig. 3). The matrix $\bar{J}(\mathbf{q})$ is:

$$
\begin{bmatrix}
\cos q_3 & -(a + f_{b1}(\mathbf{q}))\sin q_3 - (b + f_{b2}(\mathbf{q}))\cos q_3 & \cos q_3 J_{b1}(\mathbf{q}) - \sin q_3 J_{b2}(\mathbf{q}) \\
\sin q_3 & (a + f_{b1}(\mathbf{q}))\cos q_3 - (b + f_{b2}(\mathbf{q}))\sin q_3 & \sin q_3 J_{b1}(\mathbf{q}) + \cos q_3 J_{b2}(\mathbf{q}) \\
0 & 0 & J_{b3}(\mathbf{q}) \\
0 & 1 & J_{b4}(\mathbf{q}) \\
0 & 0 & J_{b5}(\mathbf{q}) \\
0 & 0 & J_{b6}(\mathbf{q})
\end{bmatrix}
$$

where $f_{bi}(\mathbf{q})$ are the components of $f_b(\mathbf{q}) = f_b(\mathbf{q}_b)$ and $J_{bi}(\mathbf{q})$ the rows of the matrix $J_b(\mathbf{q}) = J_b(\mathbf{q}_b)$.

We desire to control the mobile manipulator in such a way to avoid extension of the manipulator out of its limits. The simplest thing to do in this sense is to impose fixed values to the components $\xi_{b1}$ and $\xi_{b2}$. The corresponding set of additional tasks is such that: $\bar{J}_a(\mathbf{q}) = \begin{bmatrix} 0 & 0 & J_{b1}(\mathbf{q}) \\ 0 & 0 & J_{b2}(\mathbf{q}) \end{bmatrix}$, and the control is then given by (8), out of the singularities. For this set of additional tasks: $\det \bar{K}(\mathbf{q}) = (a + \xi_{b1})\det J_b(\mathbf{q}_b)$, and consequently the singularities are those of the manipulator (if we choose of course $a + \xi_{b1} \neq 0$); so they can be easily avoided. We have implemented this global control on the real system described in the next section.

## 4. Experimental System

Figure 3. Experimental mobile manipulator

## 4.1. Technical data

The experimental system, shown at Fig. 3, is composed with a HILARE type [15] nonholonomic mobile platform, named H2bis, and with a GT6A robot manipulator. This system has been realized in 1997 from the existing mobile platform. The arm has been chosen from the following criteria: low weight, minimum 1 kg load capacity, large workspace.

### 4.1.1. Mobile platform subsystem

The mobile platform is driven by two wheels mounted on the same axle. Each wheel is actuated independently by a DC motor powered by batteries. With this locomotion system, pure rotation of the platform is possible. This platform is equipped with different sensors (ultrasonic sensors, laser rangefinder, ...) and has been devised in order to be autonomous. For this purpose, the VxWorks Real Time OS [16] is installed on on-board processors together with the perception and locomotion softwares.

The general architecture and the technical data of the mobile platform are the following:

| Wheels | 2 driven wheels + 4 castors |
|---|---|
| Batteries | 48V |
| Bus | 1 Rack VME |
| Processors | 4 x Motorola 680x0 + 1 Motorola PPC 604 + 1 Datacube MV200 |
| Operating system | VxWorks |
| Communication | Ethernet radio modem (3 Mbit/s) |
| On-board Sensors | 32 US laser rangefinder 2D 1 camera B&W |
| Dimensions (LxWxH) | 130cm x 80cm x 80cm |
| Maximum accelerations | linear: $1 \text{ m}/s^2$ angular: $3\pi/2 \text{ rd}/s^2$ |
| Maximum velocity | linear: 0,9 m/s angular: 1 rd/s |
| Weight | 400 kgs |

### 4.1.2. Robot manipulator subsystem

The robot manipulator is a six revolute joints one built by GT Robotique. VxWorks has been chosen as operating system of this subsystem in order to obtain global coherence of the whole experimental mobile manipulator. The top of the mobile platform has been realized in order to have different places for attaching the robot manipulator. Thus, the global system can be adapted to different kinds of tasks.

Here are presented some technical data concerning the robot manipulator:

| Number of joints | 6 |
|---|---|
| End-effector | 2 digits without force sensor |
| Bus | 1 rack RSVME 808-1 |
| Processor | 1 Motorola 68020, 16 MHz |
| Operating system | VxWorks |
| Communication | Ethernet radio modem (3 Mbit/s) |
| Maximum angular displacement | 270 degres (joints 1,2,3,4,6) 180 degres (joint 5) |
| Maximum angular velocity | 1,4 rd/s (joints 1,2,3) 2,0 rd/s (joints 4,5,6) |
| Weight | 35 kg |
| Workspace | 0,78 m radius sphere |

### 4.2. Control Architecture and Simulation Tools

All the functionalities are implemented as libraries of C-functions. They are integrated in a more general structure, called GenoM [17], that has been devised to make easy the creation of modules or functions in a real time distributed systems framework and which is common to all the mobile robots of the HILARE family at LAAS.

The VxWorks architecture has allowed to work with a 2 milliseconds sampling time. The same structure is used for real experiments and for fine emulation. The only difference is on the racks that are addressed: in real experiment, they are the on-board racks on the mobile manipulator; in emulation, they are racks only devoted to development and test. In this latter case, one board on the rack is used by a graphical tool performing real time data collection and display, called Stethoscope (see Fig. 4).

Other simulation tools have been used for the test of the trajectories generated. The most useful has been the graphical 3D simulator, called GDHE [18], that allows to represent the mobile manipulator within its environment (see Fig. 4).

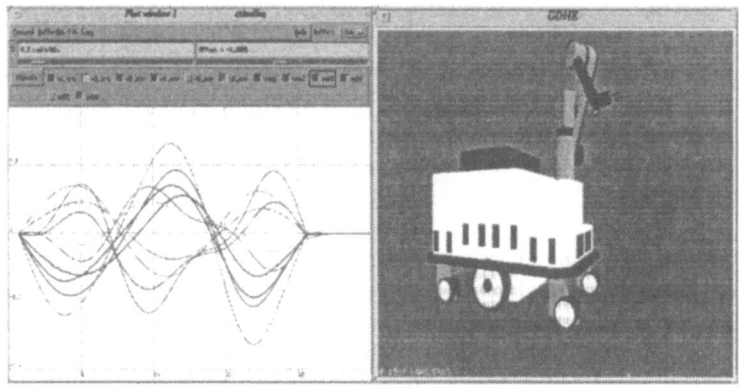

Figure 4. Stethoscope (left) and GDHE (right)

Figure 5. A writing task

## 5. Results

Various experiments have been conducted for differents choices of end-effector trajectories. The method based on additional tasks gives satisfactory results in all cases when the tasks consist in fixing $\xi_{b1}$ and $\xi_{b2}$. Others tasks may fail depending on the trajectories desired. In particular, as the scheme is an incremental one, the tasks may require an extension of the robot manipulator out of its workspace. A particularly interesting experiment is depicted at Fig. 5. In this experiment, the mobile manipulator has to write the name of the platform ('H2'), made of line segments, on a board. The previous choice for the additional tasks is shown to perform well on a video relating this experiment and many other.

**Acknowledgements**: We want to acknowledge G. Bauzil, S. Fleury, G. Foulon, M. Herrb and M. Khatib for their participation to the development of the experimental system.

## References

[1] H. Seraji, *An on-line approach to coordinated mobility and manipulation*, 1993 IEEE International Conference on Robotics and Automation, Atlanta, USA, pp. 28-34, May 1993.

[2] H. Seraji, *A unified approach to motion control of mobile manipulators*, The International Journal of Robotics Research, Vol. 17, No. 2, pp. 107-118, February 1998.

[3] C.-C. Wang, and V. Kumar. *Velocity control of mobile manipulators*, 1993 IEEE

International Conference on Robotics and Automation, Atlanta, USA, pp. 713-718, May 1993.

[4] U. M. Nassal, *An approach to motion planning for mobile manipulation*, 1994 IEEE/RSJ International Conference on Intelligent Robots and Systems, Munich, Germany, pp. 831-838, September 1994.

[5] F. G. Pin, C. J. Hacker, K. B. Gower, and K. A. Morgansen, *Including a non-holonomic constraint in the full space parametrization method for mobile manipulators motion planning*, 1997 International Conference on Robotics and Automation, Albuquerque, USA, pp. 2914-2919, April 1997.

[6] Y. Yamamoto, *Control and coordination of locomotion and manipulation of a wheeled mobile manipulator*, Ph. D. Dissertation, University of Pennsylvania, 1994.

[7] G. Foulon, J-Y. Fourquet, and M. Renaud, *Planning point to point paths for nonholonomic mobile manipulators*, 1998 IEEE/RSJ International Conference on Intelligent Robots and Systems, Victoria, Canada, pp. 374-379, October 1998.

[8] C. Perrier, P. Dauchez, and F. Pierrot, *A global approach for motion generation of non-holonomic mobile manipulators*, 1998 IEEE International Conference on Robotics and Automation, Leuven, Belgium, pp. 2971-2976, May 1998.

[9] O. Khatib, K. Yokoi, K. Chang, D. Ruspini, R. Holmberg, and A. Casal, *Coordination and decentralized cooperation of multiple mobile manipulators*, Journal of Robotic Systems, Vol 13, No. 11, pp. 755-764, 1996.

[10] K. Nagatami, and S. Yuta, *Door-opening behavior of an autonomous mobile manipulator by sequence of action primitives*, Journal of Robotic Systems, Vol. 13, No. 11, pp. 709-721, 1996.

[11] T. Sugar, and V. Kumar, *Decentralized control of cooperating mobile manipulators*, 1998 IEEE International Conference on Robotics and Automation, Leuven, Belgium, pp. 2916-2921, May 1998.

[12] G. Campion, G. Bastin, and B. D'Andréa-Novel, *Structural properties and classification of kinematic and dynamic models of wheeled mobile robots*, IEEE Transactions on Robotics and Automation, Vol. 12, No. 1, pp.47-62, February 1996.

[13] B. Gorla, and M. Renaud, *Modelling of robot manipulators. Control applications*, to be published 1999.

[14] G. Foulon, J.Y. Fourquet, and M. Renaud, *Coordinating mobility and manipulation using nonholonomic mobile manipulators*, Control Engineering Practice, to appear 1999.

[15] G. Giralt, R. Chatila, and M. Vaisset, *An integrated navigation and motion control system for autonomous multisensory mobile robots*, First International Symposium of Robotics Research, Bretton Woods, USA, pp.191-214, August 25-September 2, 1983.

[16] VxWorks, *Programmer's Guide 5.3.1*, Edition 1, Wind River Systems Inc., Alameda, USA, March 1997.

[17] S. Fleury, M. Herrb, and R. Chatila, *GenoM: A tool for the specification and the implementation of operating modules in a distributed robot architecture*, 1997 IEEE/RSJ International Conference on Intelligent Robots and Systems, Grenoble, France, pp. 842-848, September 1997.

[18] M. Herrb, *Graphic Display for Hilare Experiments*, personal communication, 1999.

# Exploiting Redundancy for Autonomous Calibration of a Planar Robot

Arthur E. Quaid and Alfred A. Rizzi
Carnegie Mellon University
Pittsburgh, Pennsylvania, USA
{aquaid,arizzi}@ri.cmu.edu

**Abstract:** Autonomous calibration techniques are applied to the calibration of two sensors of a planar robot: a platen sensor and a coordination sensor. The redundant angle measurements of the platen sensor are used to identify its parameters, improving its output accuracy by better than a factor of two. Redundant position measurements using both sensors are also used in an attempt to further improve the platen sensor accuracy and identify the relative position of the two sensors.

## 1. Introduction

In the Microdynamic Systems Laboratory[1] at Carnegie Mellon University, we are developing modular robotic components and tools to support the rapid deployment and programming of high-precision automated assembly systems [1]. The overall goal is to provide mechanically, computationally, and algorithmically modular factory *agents* and a collection of tools to support a user's interaction with the agents in an effort to expedite the process of designing, integrating, and deploying automated assembly systems. The work presented here focuses on the calibration of one of the most basic components of this larger system – the sensor systems associated with *courier* agents.

A *courier* (shown schematically in Fig. 1) is a planar robot which serves as both a product-carrier and local manipulation device in a minifactory assembly system [1]. The device is built upon existing planar linear (Sawyer) motor technology [2], in which four linear stepping motors are combined in a housing to provide large $x, y$ motions over a tabletop *platen* surface, as well as a rotation capability of a few degrees. Closed-loop control of these devices recently became possible with the development of a magnetic *platen sensor* [3]. This sensor detects the toothed structure of the platen surface and can interpolate between the nominal 1 mm tooth pitch to roughly 1 part in 5000, yielding an overall position sensing resolution of 0.2 $\mu$m. The complete sensor (see Fig. 2) consists of four quadrature-pairs, shown schematically in Fig. 3. Each pair measures motion along one of the two cardinal directions by monitoring changes in the magnetic coupling between the drive and sense coils. When integrated, information from the four pairs allows measurement of the full 3-DOF position of the robot. In addition, the planar robot is augmented with a *coordination sensor*, shown in Fig. 4, based on an upward-facing lateral

---

[1]See http://www.cs.cmu.edu/~msl

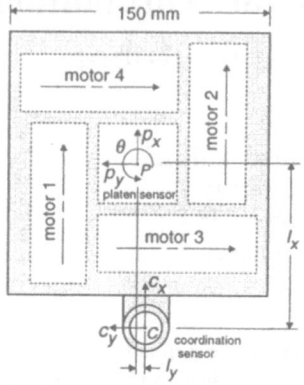

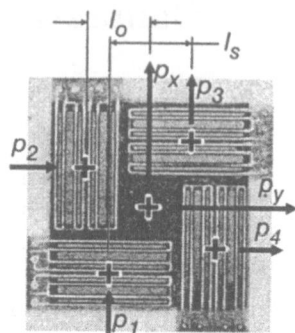

Figure 1. Courier top view          Figure 2. Platen sensor detail

effect position sensing diode. This device can precisely measure the relative displacement between the courier and overhead devices outfitted with LED beacons to sub-micron resolution [4].

Whereas basic models of planar linear motor sensors and actuators are sufficient for undemanding positioning applications, improved models are crucial for high-performance closed-loop operation. Poorly modeled effects including magnetic saturation, eddy-current damping, and non-linear position and angle dependencies limit positioning accuracies, controller robustness, and accurate force generation capabilities, all of which are vital for successful use in an environment such as the minifactory. Unfortunately, global effects such as platen tooth variations and thermal deformations are not easily captured by practical models, fundamentally limiting the accuracy of the models considered here.

Prior work in our laboratory has developed parameterized models for both the actuators and sensors by recourse to complicated and time-consuming processes which rely on high-precision independent sensors, specifically a laser interferometer and load cell [5, 3]. Realizing that model parameters will inevitably vary from device to device, and recognizing the need for stand-alone self-calibrating devices, a more practical procedure is necessary to support the autonomous calibration of individual devices in the field.

Fortunately, it has been shown that external measuring devices are not always necessary to accomplish such parameter identification tasks. Identification of the kinematic parameters of redundant manipulator systems can be performed by recording the joint angles during self-motions with the end-effector fixed in the workspace [6]. This technique has also been used for calibration of parallel mechanisms with redundant sensing [7]. Similarly, two 6-DOF force sensors pressing against each other in different configurations have been shown to provide sufficient information to enable their calibration [8]. The lesson to be taken from these works is that redundant measurements, even if individually inaccurate, can allow for autonomous calibration. In this paper, we focus on the application of these techniques to the calibration of the redundant sensors of the planar robot and experimentally demonstrate their effectiveness.

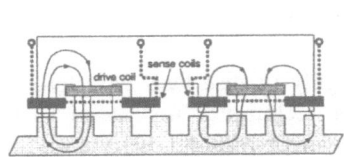

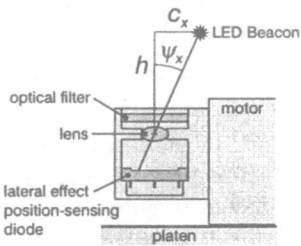

Figure 3. Schematic of a single platen sensor segment

Figure 4. Coordination sensor

## 2. Approach

In general, the autonomous calibration methods mentioned above presume the existence of a parametric model for each sensing component of the form

$$y_i = f_i(k_i, u), \qquad (1)$$

where $y_i$ is the scalar sensor output, $k_i \in I\!\!R^r$ is the vector of model parameters, and $u \in I\!\!R^n$ is the vector of input variables. Given a set of these models, $i \in \{1 \ldots N\}$ with $N > n$, the redundancy of the sensor system can be expressed through a constraint equation, $g(\cdots)$, of the form

$$g(y_1(k_1, u), y_2(k_2, u), ..., y_n(k_N, u)) = 0. \qquad (2)$$

The task is to recover the parameter vectors, $k_i$, given a series of measurements of the input vectors, $u_j$ $j \in \{1 \ldots M\}$. This is, in general, a non-linear optimization or root-finding problem and iterative methods are traditionally used to minimize $\sum_{j=1}^{M} g_j^2$ over the parameter vectors, $k_1 \ldots k_N$. However, if the models, $f_i(\cdots)$, are linear in the parameter vectors and the constraint equation, $g(\cdots)$, is linear in the outputs, $y_i$, it is well known that an analytic closed-form solution can be found.

The following section presents our application of these ideas to the autonomous calibration of both the platen and coordination sensor for a courier robot. To quantitatively evaluate and understand the benefit of including multiple sensing modalities, we have applied a number of different verification methods to these experiments. These include examining the condition number of the second derivatives of $\sum_j g_j^2$ (which provides evidence for the identifiability of the parameters), and the magnitude of the residual error both over the calibration data and independent test data. While these methods can detect a number of deficiencies, they are insensitive to certain model defects or violations of assumptions. These defects can often be detected through the use of an external sensor, and thus we also include comparisons against an independent sensor whenever possible.

## 3. Results

In this section, autonomous calibration of the platen sensor using a linear constraint equation is first presented. The platen sensor error model is then extended to include additional effects, and nonlinear constraint equations are derived to simultaneously calibrate the platen sensor parameters and the mounting parameters of the coordination sensor.

## 3.1. Platen sensor autonomous calibration

### 3.1.1. Formulation

Prior work [3] used a cumbersome but highly-precise laser interferometer to calibrate the platen sensor error. Fourier techniques were used to identify the significant frequency components of the error (difference from an idealized sensor model) as a function of the courier position. This demonstrated that the error of each sensor segment is well-modeled by an equation of the form

$$e_i := \begin{array}{l} k_{i,0} + k_{i,1}\sin(f\bar{p}_i) + k_{i,2}\cos(f\bar{p}_i) + k_{i,3}\sin(2f\bar{p}_i) + k_{i,4}\cos(2f\bar{p}_i) + \\ k_{i,5}\sin(4f\bar{p}_i) + k_{i,6}\cos(4f\bar{p}_i) + k_{i,7}\sin(6f\bar{p}_i) + k_{i,8}\cos(6f\bar{p}_i), \end{array} \tag{3}$$

where $\bar{p}_i$ is the unmodified output of the $i^{\text{th}}$ sensor, $f := 2\pi/\rho$, and $\rho$ is the pitch of the sensor teeth – 1.016 mm for our device. This model assumes the error is small enough that $\bar{p}_i$ can be used instead of the actual positions $p_i^*$ – i.e. the relation between the "measured position," $\bar{p}_i$, and the actual position, $p_i^*$, is monotonic. Effects of the courier angle, $\theta$, on the output are similarly neglected. As a result of these assumptions, the error, $e_i$, is linearly related to the parameter vector $k_i := [k_{i,0}, \dots, k_{i,8}]^T$ and can be written as $e_i = \Gamma_i^T k_i$, where $\Gamma_i$ is a vector of sine and cosine terms. Neglecting noise and unmodeled effects, the calibrated sensor output can be written as $\hat{p}_i := \bar{p}_i + e_i$. The *sensor segment pair* in either cardinal direction can then be used to independently compute the angle of the courier:

$$\hat{\theta}_x := \frac{\hat{p}_1 - \hat{p}_3}{l_s} \quad \text{and} \quad \hat{\theta}_y := \frac{\hat{p}_2 - \hat{p}_4}{l_s}, \tag{4}$$

where $l_s$ is the distance between sensors (9.7 mm), as shown in Fig. 3. In normal operation, the angles in (4) are averaged to reduce measurement noise, but here their redundancy is used to form a constraint equation,

$$\begin{aligned} g_p &= (\hat{\theta}_x - \hat{\theta}_y)l_s \\ &= (\hat{p}_1 - \hat{p}_3) - (\hat{p}_2 - \hat{p}_4). \end{aligned} \tag{5}$$

Given a set of $M$ observations, $\bar{p}_{i,j}$, indexed by $j \in \{1, \dots, M\}$, we construct the overall cost $\sum_{j=1}^{M}(g_{p,j})^2$ and find the set of parameters, $k_i$, which minimize it. Here, the constraint is linear in the parameters and a closed-form solution exists. Defining the constraint error for a single observation as

$$\tilde{g}_{p_j} := (\bar{p}_{1,j} - \bar{p}_{3,j}) - (\bar{p}_{2,j} - \bar{p}_{4,j}) \tag{6}$$

we can rewrite (5) as $g_{p_j} = \tilde{g}_{p_j} + (e_{1,j} - e_{3,j}) - (e_{2,j} - e_{4,j})$. Making explicit the linear dependence of (3) on $k_i$ yields

$$g_{p_j} = \tilde{g}_{p_j} + \sum_i \Gamma_{i,j}^T k_i = \tilde{g}_{p_j} + \Gamma_j k, \tag{7}$$

where $\Gamma_j$ and $k$ combine the $\Gamma_{i,j}$ and $k_i$ into single vectors. This form directly admits a linear least squares solution. Collecting the data over the index $j$

leads to $G_p = \tilde{G}_p + \Gamma k$, with $G_p := [g_{p_1}, \ldots, g_{p_M}]^T$, $\tilde{G}_p := [\tilde{g}_{p_1}, \ldots, \tilde{g}_{p_M}]^T$, and $\Gamma := [\Gamma_1, \ldots, \Gamma_M]^T$. The optimal parameter set is then given by

$$k = \Gamma^\dagger \tilde{G}_p, \tag{8}$$

where $\Gamma^\dagger$ is the pseudo-inverse of $\Gamma$.

### 3.1.2. Experiments

For this and later experiments, the quality of calibration is evaluated by examining the *residual error* vector, computed from the errors in the relevant constraint equation over the dataset, and also by examining the *position error* and *angle error* vectors, computed by comparing the calibrated platen sensor outputs to laser interferometer ground truth measurements for each datapoint. These errors are reported for the calibrated and uncalibrated sensor for comparison purposes, and error magnitudes are reported in terms of the standard deviation of the error vector elements.

Using an open-loop controller, the courier was moved to a series of uniformly distributed random $(x, y, \theta)$ positions over a range of $0 < x, y < 10\rho$ and $-0.5° < \theta < 0.5°$. The output of the data collection routine consisted of a vector of platen sensor outputs for each of 1000 courier positions.

Using this dataset, $\Gamma$ and $\tilde{G}_p$ were constructed from $\bar{p}$ using the method described in Sec. 3.1.1. To confirm the identifiability of $k$, the condition number of $\Gamma$ was computed and found to be small[2] (about 8) suggesting that the parameters were sufficiently excited by the data. The impact of the remaining unmodeled effects was computed by evaluating the residual error, $r := \Gamma k - \tilde{G}_p$. Using the identified parameters, the residual was 4.1 $\mu$m, significantly smaller than the 29.3 $\mu$m residual using a zero vector for $k$, but still higher than the sub-micron resolution of the sensor.

To examine the generality of $k$, the courier was commanded to move a distance of 40 mm passing through the calibration area. For this *fly-by* test, the residual decreased from 12.1 $\mu$m to 5.5 $\mu$m with calibration, indicating that $k$ was also valid over this dataset. To provide an independent verification and better characterize the performance of the autonomous calibration, retroreflectors were mounted on the courier, allowing a laser interferometer to precisely measure two axes. The 40 mm fly-by test was repeated, and the translation along the direction of motion and rotation in the plane were sampled by both the interferometer and platen sensor. Using the interferometer measurements as ground-truth, the sensor translation errors are shown in Fig. 5 and the errors for each of the sensor angle measurements are shown in Fig. 6. The position error decreased from 8.9 $\mu$m to 6.5 $\mu$m with calibration. The error of $\theta_x$ stayed about the same at 0.4 mrad, while the error of $\theta_y$ decreased from 1.2 mrad to 0.4 mrad with calibration.

Although the autonomous calibration does yield a nominal improvement in position error and an improvement in the $\theta_y$ angle error by a factor of 3, there is a significant systematic error in $\theta_x$ that does not decrease. For this test, the courier moves in the $y$ direction, nominally leaving fixed the outputs

---

[2]As the arbitrary offset parameters $k_{i,0}$ cannot be individually identified, the condition number computation considered only their sum.

156

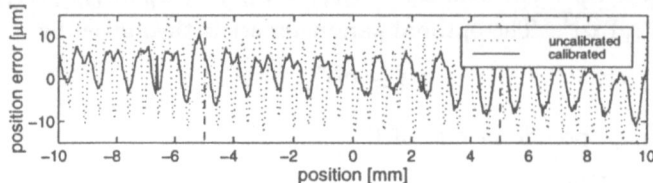

Figure 5. Position error in direction of motion during motion through platen sensor calibration region (indicated by dashed lines).

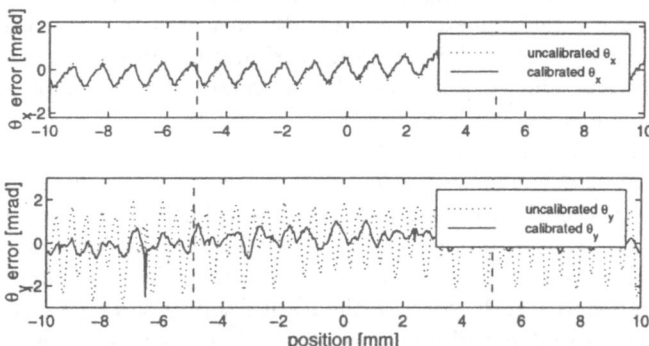

Figure 6. Angle error during motion through platen sensor calibration region.

of the sensors that measure motion in the $x$ direction, which are also used to compute $\theta_x$. Errors in $\theta_x$ are therefore best explained as an unmodeled dependence of the output on lateral position. As these *lateral dependencies* are roughly the same size as the residual errors and are unmodeled by (3), they are likely to be limiting further error reduction. However, the structure of the constraint equation (5) makes it difficult to distinguish lateral dependencies of one sensor segment pair from translational effects of the perpendicular sensor segment pair.

### 3.2. Dual sensor autonomous calibration

In hopes of identifying these lateral effects, the coordination sensor can be used to provide additional constraints. However, the precise mounting position and other parameters of the coordination sensor are not known. In this section, the autonomous calibration method is applied to the calibration of both the coordination sensor and an extended model of the platen sensor.

To model the lateral motion effects, the platen sensor models are augmented with additional parameters based on their lateral motion

$$
\begin{aligned}
e_1' &= e_1 + k_{1,9} \sin(f(\bar{p}_y - l_o\bar{\theta})) + k_{1,10} \cos(f(\bar{p}_y - l_o\bar{\theta})) \\
e_2' &= e_2 + k_{2,9} \sin(f(\bar{p}_x + l_o\bar{\theta})) + k_{2,10} \cos(f(\bar{p}_x + l_o\bar{\theta})) \\
e_3' &= e_3 + k_{3,9} \sin(f(\bar{p}_y + l_o\bar{\theta})) + k_{3,10} \cos(f(\bar{p}_y + l_o\bar{\theta})) \\
e_4' &= e_4 + k_{4,9} \sin(f(\bar{p}_x - l_o\bar{\theta})) + k_{4,10} \cos(f(\bar{p}_x - l_o\bar{\theta})),
\end{aligned}
\tag{9}
$$

where $\bar{p}_x = (\bar{p}_1 + \bar{p}_3)/2$ and $\bar{p}_y = (\bar{p}_2 + \bar{p}_4)/2$ are the average of the uncalibrated sensor readings along each axis, and $\bar{\theta}$ is the average of the uncalibrated angle measurements computed as in (4). The error of each sensor segment is now dependent on the measurements of the other segments.

Although the coordination sensor requires internal calibration for electronic gains, mounting inaccuracies, optical distortions, and PSD nonlinearities, we assume for now that these effects have been calibrated[3] so that it outputs a perfect measurement of the angles to the LED beacon ($\psi_x$ and $\psi_y$). Assuming a stationary beacon, the motion of the sensor is given by

$$c_x = h\tan(\psi_x) \quad \text{and} \quad c_y = h\tan(\psi_y). \tag{10}$$

As the coordination sensor supplies two additional measurements, two more constraint equations can be derived. A physically meaningful constraint is to equate the displacements of coordination sensor ($c_x$, $c_y$) with that of the platen sensor ($p_x$, $p_y$)

$$\begin{bmatrix} g_x \\ g_y \end{bmatrix} = \begin{bmatrix} -\hat{p}_x + \cos(\hat{\theta})c_x - \sin(\hat{\theta})c_y - l_y\sin(\hat{\theta}) + l_x\cos(\hat{\theta}) + o'_x \\ -\hat{p}_y + \sin(\hat{\theta})c_x + \cos(\hat{\theta})c_y + l_y\cos(\hat{\theta}) + l_x\sin(\hat{\theta}) + o'_y \end{bmatrix}, \tag{11}$$

where $c_x, c_y$ are relative to coordinate frame $C$ shown in Fig. 1, which has a stationary origin under the LED beacon but rotates with $\theta$. The platen sensor positions $\hat{p}_{x,y}$ are relative to stationary coordinate frame $P$, and are simply the average of the two platen sensor measurements in each direction. Parameters $l_x$ and $l_y$ are the physical offsets between the centers of the two sensors as measured on the body of the courier, while offsets $o_x$ and $o_y$ account for the arbitrary zero position of the platen sensor. The angle of the courier, $\hat{\theta}$, is the average of $\hat{\theta}_x$ and $\hat{\theta}_y$ in (4).

These constraint equations contain five new parameters to be identified ($h, l_x, l_y, o_x, o_y$), in addition to the platen sensor parameters ($k_1, k_2, k_3, k_4$). The three constraint equations $g_x, g_y$, and $g_p$ are combined by taking the sum of their squares to provide a scalar valued function for minimization.

### 3.2.1. Experiments

The courier was positioned so that the coordination sensor was directly under an LED beacon and was moved to a series of uniformly distributed random ($x, y, \theta$) positions over a range of $0 < x, y < 2\rho$ and $-1.5° < \theta < 1.5°$. In this case, courier rotation was about the center of the coordination sensor in order to minimize the effects of coordination sensor nonlinearities. The output of this data collection process consisted of a vector of platen sensor outputs, $\bar{p}$, and coordination sensor outputs, ($\psi_x, \psi_y$), for each of 1000 courier positions.

Using this dataset, a cost function $J = \sum_j (g_{x,j}^2 + g_{y,j}^2 + g_{p,j}^2)$ was computed using the procedure described above. As the constraint equations are non-linear in the parameters, definitive testing for identifiability of parameters is difficult. The condition number of a numerically computed linear approximation of the second derivative of $J$ was examined at both the initial parameter and the calibrated parameter results, and was very large at both ($8.11 \times 10^8$ and $3.38 \times 10^8$ respectively). The largest singular value was also large (at least $5 \times 10^4$

---

[3]In practice these parameters would be identified prior to integration with the courier.

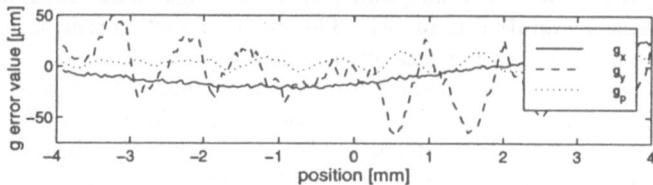

Figure 7. Constraint equation errors during motion through dual sensor calibration region.

for both cases), suggesting a stiff system rather than completely unidentifiable parameters. To investigate further, a synthetic data set was generated using nominal parameter values and Gaussian noise added to the sensor outputs. The cost function, $J$, was used with a BFGS Quasi-Newton method with a mixed quadratic and cubic line search procedure[4] and identified the correct parameters to within 20 $\mu$m for $h, l_x$, and $l_y$, and within a micron for the remaining parameters, even with initial errors of 5% for the coordination sensor parameters, $3\,\mu$m random errors in the platen sensor parameters, and all sensor readings corrupted by 1.0 $\mu$m standard deviation Gaussian noise.

Encouraged by the simulation results, the same cost function and minimization routine were applied to the experimental dataset. To provide a good initial parameter vector, the platen sensor autonomous calibration results were used to initialize the basic set of platen sensor parameters, while the coordination sensor parameters $(o_x, o_y, l_x, l_y, h)$ were initialized by minimizing a partial cost function $J' = \sum_j (g_{x,j}^2 + g_{y,j}^2)$ with the platen sensor parameters fixed. With these initial parameter estimates, the complete cost, $J$, was minimized over the full set of parameters, including the lateral effect terms $(k_{i,9}, k_{i,10})$. The residuals of the three constraints $(g_x, g_y, g_p)$ were computed at the initial and final parameter vectors, with values (8.4, 31.6, 4.2) $\mu$m and (8.06, 25.2, 5.19) $\mu$m, respectively. The main change is a decrease in $g_y$, which includes the large lever-arm term $l_x \sin(\theta)$ (with $l_x \approx 90$ mm) and is very sensitive to the angle error of the platen sensor. To test the calibration over a different dataset, residuals for the three constraints were measured using the calibrated sensor outputs for an 8 mm flyby through the calibration area, shown in Fig. 7. The errors in $g_y$ and, to a lesser extent, $g_p$ vary systematically with tooth pitch, indicating unmodeled platen sensor errors. However, the $g_x$ error varies gradually over multiple pitches, suggesting unmodeled coordination sensor effects.

The laser interferometer was again sampled to provide an independent verification for a fly-by of 8 mm, while simultaneously sampling the platen sensor and coordination sensor outputs. The platen sensor position error (using the interferometer as ground truth) decreased from 7.3 $\mu$m uncalibrated to 4.4 $\mu$m using the initial parameters from the separate calibrations to 2.9 $\mu$m for the final calibration, a significant improvement. The error of $\theta_y$ (using the moving sensor segment pair) decreased from 1.4 mrad uncalibrated to 0.40 mrad for

---

[4]The fminu function in Mathworks Inc.'s Optimization Toolbox for Matlab.

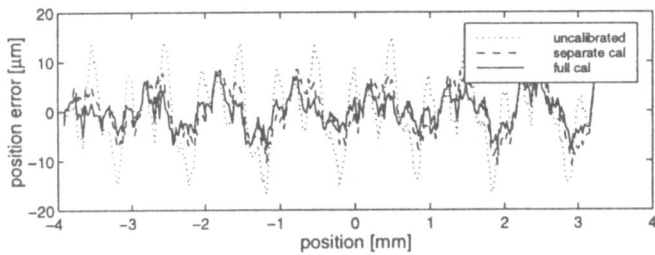

Figure 8. Position error in direction of motion during motion through dual sensor calibration region.

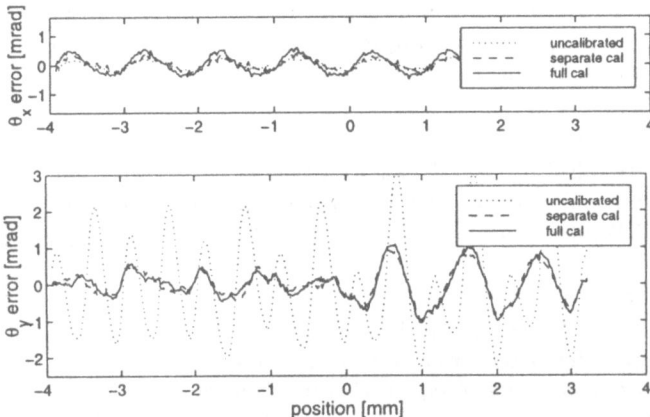

Figure 9. Angle error during motion through dual sensor calibration region.

the separate calibration, but increased slightly to 0.43 mrad for the combined calibration. The error of $\theta_x$ (using the nominally stationary sensor segment pair), which was expected to decrease when adding the lateral platen sensor parameters, remained small but increased from 0.13 mrad uncalibrated to 0.17 mrad for the separate calibration to 0.30 mrad for the combined calibration. In addition, the identified values of the $k_{i,9}$, $k_{i,10}$ parameters were all under $2\,\mu$m, which were smaller than the lateral effects seen in previous data, strongly suggesting that these less significant parameters are not being correctly identified.

## 4. Conclusions

Experiments showing the autonomous calibration of the two planar robot sensors demonstrated a marked improvement in accuracy by a factor of three, to a level of $3\,\mu$m in translation and 0.4 mrad in rotation, a level of accuracy comparable to results of an interferometer-based calibration [3]. However, these values are significantly higher than the $0.2\,\mu$m sensor resolutions. Two effects are limiting further accuracy improvements. First, the platen teeth are assumed to be uniform, but actually have significant manufacturing deviations. Using interferometer position measurement errors of the platen sensor over a series of seven teeth to compute an average sensor error waveform gave an error from platen tooth variations of 1.9 $\mu$m. Further accuracy improvements would require a global model of the platen or more uniform platens – which are currently under development in our laboratory. The second limitation is from unmodeled nonlinearities in the coordination sensor. Although the experiments were designed to keep the coordination sensor angles small, it was necessary to move several pitches for identifiability reasons, and even this small motion seems to have produced errors of several microns, as shown in the $g_x$ error in Fig. 7.

We intend to calibrate the internal coordination sensor parameters and re-attempt identification of the lateral effect parameters of the platen sensor. In addition, we are currently applying the autonomous calibration technique to calibrate the force output of the redundant planar linear motor actuators. Preliminary results indicate improvements in both force ripple and linearity. Finally, we are integrating the improved models into the closed-loop controllers, enabling improved high-performance operations.

## Acknowledgements

We wish to thank Ralph Hollis, Zack Butler, Michael Chen, and Jimmy Ma for their contributions. This work is supported in part by NSF grants DMI-9523156 and DMI-9527190. Quaid is supported by a Lucent Technologies fellowship.

## References

[1] A. A. Rizzi, J. Gowdy, and R. L. Hollis, "Agile Assembly Architecture: an agent based approach to modular precision assembly systems," in *Proc. IEEE Int'l Conf. on Robotics and Automation*, pp. 1511–1516, April 1997.

[2] E. R. Pelta, "Two-axis Sawyer motor for motion systems," *IEEE Control Systems magazine*, pp. 20–24, October 1987.

[3] Z. J. Butler, A. A. Rizzi, and R. L. Hollis, "Integrated precision 3-DOF position sensor for planar linear motors," in *Proc. IEEE Int'l Conf. on Robotics and Automation*, 1998.

[4] W.-C. J. Ma, "Precision optical coordination sensor for cooperative 2-DOF robots," Master's thesis, Carnegie Mellon University, Pittsburgh, PA, 1998.

[5] A. E. Quaid, Y. Xu, and R. L. Hollis, "Force characterization and commutation of planar linear motors," in *Proc. IEEE Int'l Conf. on Robotics and Automation*, pp. 1202–1207, April 1997.

[6] D. J. Bennett and J. M. Hollerbach, "Autonomous calibration of single-loop closed kinematic chains formed by manipulators with passive endpoint constraints," *IEEE Transactions on Robotics and Automation*, vol. 7, no. 5, pp. 597–606, 1991.

[7] J. M. Hollerbach and D. M. Lokhurst, "Closed-loop kinematic calibration of the RSI 6-DOF hand controller," in *Proc. IEEE Int'l Conf. on Robotics and Automation*, (Atlanta, GA), pp. 142–148, May 1993.

[8] R. M. Voyles and P. K. Khosla, "Collaborative calibration: Extending shape from motion calibration," in *Proc. IEEE Int'l Conf. on Robotics and Automation*, (Albuquerque, NM), pp. 2795–2800, April 1997.

# Chapter 5

# Applications

To me, the most exciting part of robotics research is applications, and researchers in this field often complain it is hard to find them. In fact, as has already been remarked in the introduction to this volume, applications are commonplace. The reason why researchers have difficulty finding them is that they look for really interesting ones, ones which are worth devoting whole sessions to at conferences. This chapter presents three of them in the one place!

The first two discuss earth moving which is very apposite in an Australian setting where most of the wealth is gained by moving earth (to expose valuable minerals) or farming it (to grow food and animals). I was relieved to find that Howard Cannon and Sanjiv Singh have confirmed, on a large scale and large budget, what some of my students have attempted (on a tiny scale with a tiny budget). Their remarkable demonstration of an entirely automatic backhoe digging the ground made a big impact at the conference. The backhoe even knows when it is about to undermine itself and sensibly backs away from the edge before digging some more dirt. In their paper, they focus on the model of the digging process: readers who want more details on the machine itself will have to buy one from their nearest Caterpillar agency!

My co-chair, Peter Corke, with his colleagues Jonathan Roberts and Graeme Winstanley have been working on an even bigger machine — a walking dragline with a bucket almost big enough to swallow Cannon and Singh's Cat in one bite. They show how they have set about controlling this monster with the knowledge that tiny efficiency gains provide huge financial rewards. They report that two laser scanners work much better than vision for tracking the movements of the haulage cables in the pouring rain which commonly occurs over the coalfields of Queensland. In their case, the human operator controls the digging and the automation is reserved for the tedious task of swinging the bucket around for dumping the load, and back to the working face for the next bite.

David Tristano, John Hollerbach and Robert Christensen take us into virtual worlds with a treadmill, but this is no machine for the everyday masochist.

The treadmill is the means by which we walk into the virtual world. In his video presentation, John showed how a mechanical tether can pull or push the user's waist to simulate sudden changes in slope which no ordinary treadmill can. Makers of motion simulators have long known some clever techniques which can dramatically enhance the experience of motion. An aircraft simulator tilts backwards, the sound of roaring engines fills the cockpit and the trainee pilot's mind immediately accepts the sensation of accelerating along a completely flat runway. Here, the trainee walker is pulled backwards by the tether and he leans forward, imagining he is climbing a steep slope. John showed video footage of soldiers learning to fight virtual terrorists who pop out from behind virtual doors and stairways. The only development left is a silent treadmill: the noise seems to conflict with the deadly quiet steps I thought would be needed to surprise them!

JPT

# Models for Automated Earthmoving

Howard Cannon
Caterpillar Inc.
Technical Services Division
Peoria, IL 61656
cannohn@cat.com


Sanjiv Singh
Robotics Institute
Carnegie Mellon University
Pittsburgh, PA 15213
ssingh@cmu.edu

**Abstract:** We present a composite forward model of the mechanics of an excavator backhoe digging in soil. This model is used to predict the trajectories developed by a closed-loop force based control scheme given initial conditions, some of which can be controlled (excavator control parameters), some directly measured (shape of the terrain), and some estimated (soil properties). Since soil conditions can vary significantly, it is necessary that soil properties be estimated online. Our models are used to both estimate soil properties and predict contact forces between the excavator and the terrain. In a large set of experiments we have conducted, we find that these models are accurate to within approximately 20% and run about 100 times faster than real-time. In this paper we motivate the development of these models and discuss experimental data from our testbed.

## 1. Introduction

We have developed a robotic earthmoving system that completely automates the task of mass excavation and truck loading. Our excavator uses scanning laser rangefinders to recognize and localize trucks, measure the soil face and detect obstacles (Figure 1). The excavator's software decides where to dig in the soil and where to dump in the truck and how to quickly move between these points. This system was fully implemented and was shown to load trucks as fast as human operators [13].

Previously we have proposed a method to automatically plan digging motions for excavators [11],[12]. Digging actions are described by a compact set of action parameters and the space spanned by these parameters is searched for an action that both satisfies constraints and optimizes a cost function. Much of the success of such a constraint optimization system hinges on the ability to accurately model the effect of actions— there is little use in comparing candidate actions unless there is some way of scoring each action. Here we report on models used to predict digging trajectories before they are executed. These models capture machine dynamics, soil properties and the soil-tool interaction.

*Figure 1    Our testbed excavator is a commercial grade 25 ton machine. It is equipped with onboard computing, joint sensing, and range sensing that allows mapping of the surrounding terrain. The joints are powered by hydraulic actuators.*

## 2. Related Work

For purposes of automatic planning we require models that are both accurate and computationally fast. Speed is essential since many candidate actions must be examined in only a few seconds. The literature in soil-tool interaction tends to the extremes. On one end are finite element models such as [4] that are accurate but computationally taxing—these models are typically three orders of magnitude slower than real-time. At the other end of the spectrum are models that represent gross behavior of machines [1],[14]. The level of detail closest to what we require is found in research that seeks to estimate cutting resistance for agricultural tilling implements [7],[8]. While somewhat simplistic, these models provide the basic mechanics of earthmoving. We have extended these models to suit our application and use them to estimate soil properties, as well as to predict contact forces between the excavator and the terrain. In a somewhat similar vein we have found that the methods described in the literature on the modeling of hydraulic actuators (such as [2],[6]) do not provide the requisite computational speed. We have sacrificed some accuracy in favor of prediction speed by using empirical models of hydraulic actuators.

## 3. Forward Model of Excavation

Digging efficiently requires solution of two qualitatively different problems— *where to dig* and *how to dig*. We solve the first problem by examining candidate actions and choosing the best one based on a number of criteria. The second problem, which is to guide the bucket through the soil once a digging location has been selected, is solved by the use of a force based control law. The two problems are coupled, however. To select digs it is necessary to know the effect of the closed-loop system, that is, the resultant trajectory for a given action. In our case the action corresponds to the starting conditions of a dig. Note that this trajectory will not only vary with the control parameters of an action, but also with the shape of the terrain as well as soil properties. Terrain shape can be measured directly, but soil properties must be estimated.

We have developed a feedforward model of the excavation process as shown in Figure 2. A model of the machine's actuators is used to predict the motion of the bucket in response to the actuator commands and reaction forces. The resultant motion is used by a soil-tool model to predict reaction forces on the bucket. Reaction forces and actuator positions are used by a control law to dictate actuator commands. This cycle is started with initial conditions (starting location and orientation of the bucket) and continued until the bucket is out of the ground. Predicted trajectories are scored by the use of a utility function composed of factors such as the time spent during digging, the energy expended, and the volume of soil captured.

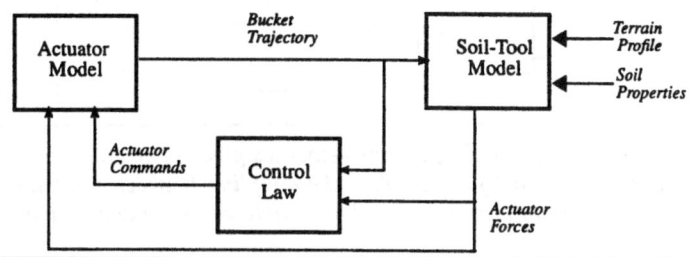

*Figure 2    Composite model of the digging process.*

### 3.1. Actuator Model

A hydraulic excavator has four revolute joints: swing, boom, stick, and bucket (Figure 1). Since digging requires a planar motion, we model only the last three degrees of freedom. The boom, stick, and bucket joints are controlled by extending and retracting prismatic hydraulic actuators attached across each joint.

Several factors contribute to the difficulty of modeling this system. Open-center valves are used to control the hydraulic actuators. Use of these valves requires that some of the flow is leaked back to a reservoir while the remainder is used to move the actuator. The leakage is partly dependent on the load on the actuator. The implements are also coupled in motion because a single hydraulic pump is used to power two joints and is limited in the amount of flow it can produce. There are also many transient effects, such as the time required for a pump to stroke up to meet demand, and the compressibility of the hydraulic fluid. Finally, unlike many typical robot systems, the contact forces at the tool are very high, and significantly contribute to system non-linearity. For instance, when the pressure in an actuator exceeds a threshold, a pressure relief valve opens stopping the actuator.

Figure 3 shows an overview of the model required to predict motion of the hydraulic actuators. Models of such complex mechanisms tend to be computationally expensive. Since we need both accuracy and speed to model the system we use an empirical model composed of neural nets. Neural nets are particularly suited to this application because system performance does not change significantly over time and a large set of training data (operation in various geometric configurations as well as

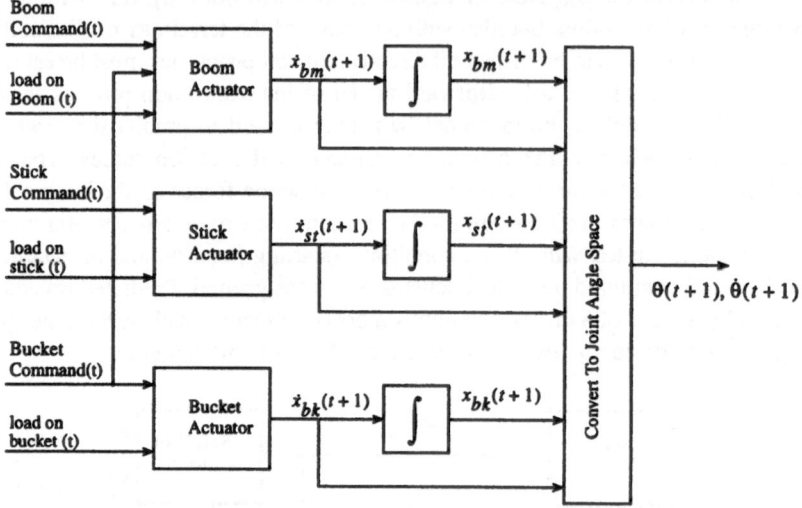

*Figure 3    The actuator model predicts velocities given commands and loads (due to gravitational and contact forces). Note the coupling between the bucket and boom due to the fact that both are powered by a single source. Each actuator is modeled with a neural net.*

soil conditions) is available prior to prediction. The advantage of neural nets is that they can capture the non-linearity not possible with globally linear methods and are much faster than locally linear methods. The disadvantage is that good performance requires trial and error experimentation with various configurations of neural nets. We model each actuator with a separate two layer network consisting of three input nodes, five hidden nodes, and one output node. The networks are *recurrent*— the predicted velocity is fed back to the network as one of the inputs for the next time step. Inputs are delayed by a small amount to account for system latency [3].

### 3.1.1. Actuator Model Performance

Figure 4 shows a comparison between the actuator velocities and their predictions. The networks were trained using 30 digs (1710 data points) from various terrain profiles and soil hardnesses. The plots show the predicted velocities for a separate set of 10 dig cycles not in the training set.

### 3.2. Soil-Tool Model

Forces acting against the bucket are a function of the geometry of the bucket-terrain intersection ($\Gamma$), and the inherent properties of the soil ($\Psi$).

$$F = f(\Gamma, \Psi)$$

where $F$ represents the generalized force acting at the bucket tip.

$\Gamma$ is found by intersecting the bucket pose with a topological map of the area to be excavated. Terrain topology is obtained prior to every dig cycle using onboard laser range sensors [13]. $\Psi$ is obtained by solving for a set of parameters that minimize the

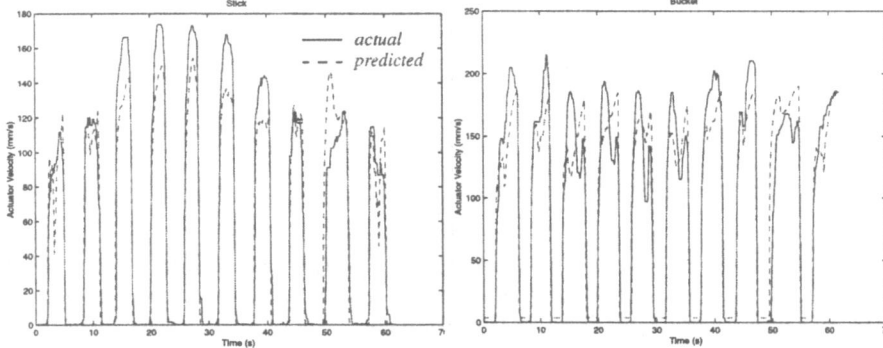

*Figure 4    Comparison of predicted versus measured actuator velocities for the stick (left)
and the bucket (right) and stick over 10 digging cycles. Mean absolute velocity errors for the
boom, stick and bucket were 4.3 mm/s, 11.8mm/s and 18.6 mm/s respectively. Peak velocities
for these actuators were 87 mm/s, 243 mm/s and 250 mm/s.*

difference between modeled and measured resistive forces.

What remains is a means, $f()$, to predict contact forces at the bucket due to resistance from the soil. We have implemented two methods for estimating reaction forces on the bucket. The first method uses an analytical model based on the well known *Fundamental Earthmoving Equation* (FEE) in soil mechanics [9]. The second method uses a empirical model with terms from the FEE as a basis set.

Reaction forces can vary dramatically depending on terrain geometry, soil compaction, and even climate. For instance at our test site, the soil varied from hard and frozen in the winter, wet and soupy in the rainy seasons, and dry and powdery during the summers. Therefore it is necessary that our soil reaction force models reflect the changing soil characteristics encountered. For each of the models mentioned above we have implemented a method for extracting soil properties based on the forces encountered during digging.

In our system, reaction forces are estimated based on the measurement of pressure in the hydraulic actuators. These pressures can be transformed into joint torques (see [5] for details), which can subsequently be used to estimate the reaction forces:

$$\hat{F} = J^{-T}(\tau - G)$$

where $J$ is the manipulator Jacobian of the mechanism (from the boom joint to the bucket tip), $\tau$ represents the joint torques, and $G$ the torques on the mechanism due to the weights of the implements. The bucket is intersected with a terrain map resulting in an estimate of the intersection geometry $\hat{\Gamma}$. $\hat{\Gamma}$ and $\hat{F}$ can be combined to find the best estimate of soil properties $\Psi$ (Figure 5). Given $\Psi$, we find $\Gamma$ for each step in the simulated dig, and predict reaction forces at the bucket. The model shown in Figure 2 is used to predict motion of the bucket tip.

### 3.2.1. Analytical Model

The FEE which was originally developed to estimate the forces on tilling equipment in agricultural applications, predicts resistive forces acting against a flat blade moving horizontally through a level surface at constant depth. Resistive force is computed

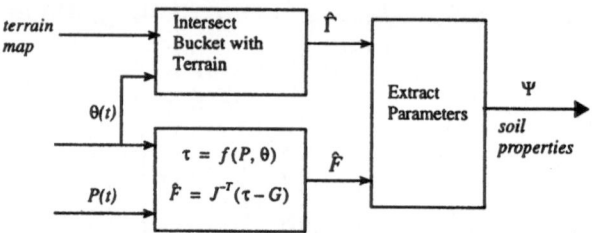

*Figure 5    Actuator pressures, P(t), measured during digging are used to estimate contact forces, $\hat{F}$. These forces and the intersection geometry are used to estimate soil properties. This process may occur as frequently as once after each dig.*

as the force required to shear a wedge of soil along a series of failure planes. The FEE uses first principle physics and the parameters are identifiable physical properties.

We have modified the FEE to be more applicable to our application (Figure 6).

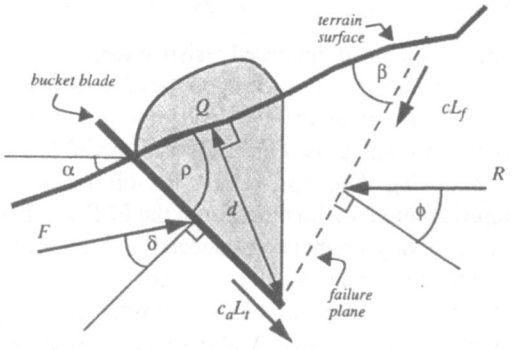

*Figure 6    Pseudostatic model of earthmoving at critical point just before failure. This model extends the FEE by accounting for the material being retained in the bucket and for the slope of the terrain (α). $L_t$ is length of the tool, $L_f$ is the length of the surface along which the wedge slides, Q is the surcharge, or the displaced soil that rests on the wedge, φ is the soil-soil friction angle, c is the cohesiveness of the soil, $c_a$ is the adhesion between the soil and blade, δ is the soil-tool friction angle, R is the force of the soil resisting the movement of the wedge, and F is the force exerted by the tool to cause failure. The material in the shaded region constitutes all of the material that has passed over the top of the bucket tip during the bucket's motion, and is referred to as the swept volume, $V_s$.*

Chiefly, the model was modified to account for a variably sloped terrain and the calculation of depth is based on the local shape of the terrain. Ignoring the small adhesion forces and eliminating R via substitution, we obtain:

$$F = d^2 w \gamma g N_w + c w d N_c + V_s \gamma g N_q$$

$$N_w = \frac{(\cot\beta - \tan\alpha)(\cos\alpha + \sin\alpha\cot(\beta + \phi))}{2[\cos(\rho + \delta) + \sin(\rho + \delta)\cot(\beta + \phi)]}$$

$$N_c = \frac{1 + \cot\beta\cot(\beta + \phi)}{\cos(\rho + \delta) + \sin(\rho + \delta)\cot(\beta + \phi)}$$

$$N_q = \frac{\cos\alpha + \sin\alpha\cot(\beta + \phi)}{\cos(\rho + \delta) + \sin(\rho + \delta)\cot(\beta + \phi)}$$

The terms representing the intersection geometry correspond to the depth of the bucket tip, the terrain slope, rake angle, and swept volume: $\Gamma=(d, \alpha, \rho, V_s)$. These values can be calculated once bucket pose and terrain shape are known. The inherent soil properties correspond to the soil-tool friction angle, the soil-soil friction angle, the failure surface angle, the density of the material, and the cohesiveness of the material: $\Psi=(\delta, \phi, \beta, \gamma, c)$.

Note that each of $N_w$, $N_c$ and $N_q$ is composed of highly non-linear terms. Since these equations are not invertible, we have implemented a numerical method that searches for the set $\Psi$ that minimizes $|f(\hat{\Gamma}, \Psi) - \hat{F}|$. This method uses a combination of a stochastic search and efficient gradient descent [5].

### 3.2.2. Empirical Model

Since the FEE requires a potentially expensive computational process, we have also investigated the use of a linear, lumped-parameter model. The independent variables of this model are terms from the FEE that can be measured directly or can be controlled. We also used a statistical analysis to ensure that the terms used were good predictors of the contact force. We assume that the model for each of the planar components ($F_x$ and $F_y$) and the moment at the bucket tip ($M$) takes the form:

$$f = \sum_i \Psi_i \Gamma_i = \Psi_1 \Gamma_1 + \Psi_2 \Gamma_2 + \Psi_3 \Gamma_3 + \dots$$

We selected the basis set $\Gamma = (d^2, \cos(\rho), \alpha, V_s)$. An intelligent choice of $\Gamma$ can help remove some of the non-linearity. Linear regression is used to find the soil properties, $\Psi_i$. Note that the soil properties estimated with this method are not identifiable physical quantities. Their values however are intrinsically dependent on the overall soil characteristics. Thus we cannot use this model to extrapolate soil forces for harder or stickier soils, whereas with the analytical model, we have a physical understanding of how the soil properties should change.

### 3.2.3. Soil-Tool Model Performance

We performed a cross validation test using data from 23 separate digs (approximately 1100 data points) and used a sliding window of approximately 300 data points to estimate the forces for the next dig.

Referring to Figure 7, somewhat surprisingly, the difference in error between the two models is small. However, while the time required to predict reaction forces using the empirical model is only slightly faster than the analytical model (13 ms vs. 16 ms), extraction of the soil properties with the empirical model is significantly faster (8 ms vs. 3400 ms). (Tests were conducted with an SGI R10000 processor). Soil properties estimated via the analytical model appear to make physical sense, and are relatively consistent during an extended sequence of operations. The empirical model parameters however vary a great deal since they are not tied to any physical relationship [3].

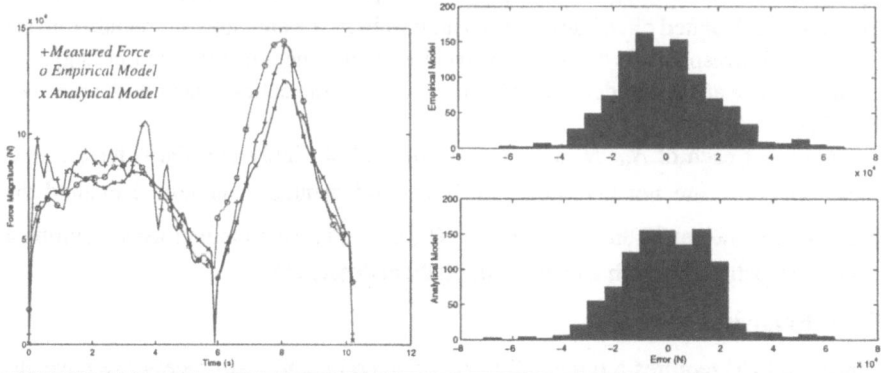

*Figure 7    Cross-validation test for resistive force using the analytical and empirical models. The left plot shows a comparison of the force magnitudes for two digs.The right plots show histograms of errors. Using the analytical model, the mean absolute error in force magnitude is 13,980 N vs. 14,790 N when the empirical model is used. Standard deviations are 10,970 N and 11,880 N respectively. Peak forces are approximately 142,000N.*

### 3.3. Control Law

Excessive force on any one implement during digging can result in an inefficient and time consuming digging operation. In order to control the forces in our machine, we used a control law (AutoDig) designed to mimic the operation of a human operator [10]. AutoDig uses the position of the implements and cylinder pressures to directly control cylinder velocities. The behavior of AutoDig can be modified by changing a small set of parameters some of which have to do with implement position, while others are used to map cylinder pressures to actuator velocities. The latter can be thought of as an adjustment of end-effector stiffness or conversely a means of indicating soil hardness. Experiments show that the control law is effective in most cases where it is possible to capture a full bucket. When a stiff inclusion such as a boulder is encountered, AutoDig guides the bucket along the boundary of the object without stalling the excavator. AutoDig requires prior specification of certain joint angles at which the digging should end. In some cases, this means that digging is terminated early, while in others, the bucket sweeps a volume of soil greater than bucket capacity. More details about AutoDig can be found in [3],[10].

# 4. Excavation Model Results

We have conducted approximately 1900 digging experiments. We find that most significant discrepancies between the model and measured trajectories are either due to non-homogeneous material effects such as boulders, or, errors in the range data required to estimate terrain shape. Figure 8 shows a qualitative comparison between the predicted and measured trajectories for four digs.

Figure 9 shows a comparison between the digging statistics predicted by the forward model and the statistics calculated by monitoring the actual digging cycle. These quantities (time, energy, and swept volume) are important for evaluating the overall utility of a dig. The time calculation corresponds to the length of time that the

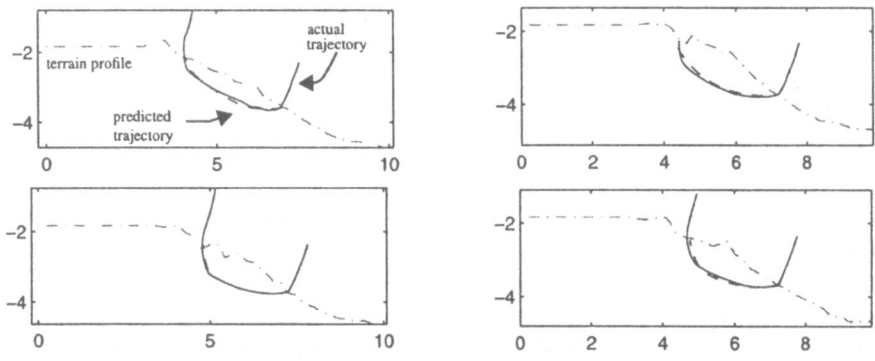

*Figure 8   Comparison of modeled versus actual dig trajectories for 4 separate dig cycles.
The model predicts the bucket trajectory reasonably well for a number of different terrain
profiles and starting conditions.*

bucket tip is below the surface of the terrain. The energy is calculated by integrating
the product of the actuator forces and displacements. The swept volume is computed
by intersecting the bucket trajectory with the terrain map at each time step.

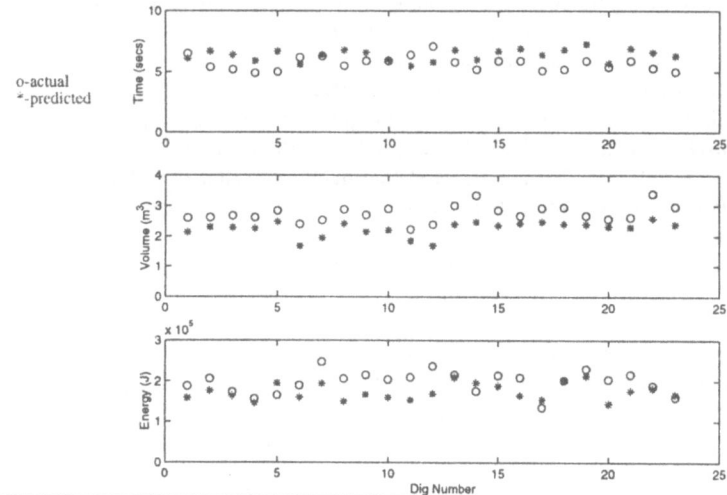

*Figure 9   Comparison of digging statistics produced by the feed forward model to the
statistics computed from actual test data. Mean errors in estimating time, volume and energy
are 13.7%, 14.7% and 12.5%. We note a small systematic bias: captured volume is consistently
underestimated, while  time is generally overestimated.*

As previously indicated, the purpose of our model is to allow an evaluation of a
number of candidate trajectories as specified by a starting location and orientation of
the bucket. A utility function is used to select the best candidate and this action is
then executed. See [3], [12] for more details about this approach.

# 5. Conclusions

We have created a faster than real-time model which is able to estimate the trajectory of the bucket during a typical excavation cycle. The model combines a force based control law, a model of the vehicle's hydraulic actuators, and a soil reaction force model. Two methods were investigated for modeling the soil reaction forces, and methods were developed for estimating the characteristics of the soil. The models are comparable in accuracy, but differ in the ability to extrapolate and in the speed in which the soil characteristics can be estimated. We have implemented a planner that uses the forward model of the excavation process to examine the utility of numerous candidate digs and chooses the one that meets some objective criteria. The planner is able to examine approximately 100 candidate digs in a matter of just a few seconds. Our experiments have shown that the combination of the model with this planning methodology produces results comparable to operation by an expert human operator.

# 6. References

[1]     Alekseeva, T. V., Artem'ev, K.A. & Bromberg, A., *Machines for Earthmoving Work, Theory and Calculations*, A. A. Balkema, Rotterdam, 1992

[2]     Bilodeau, G. and Papadopoulos, E., Modeling, Identification and Experimental Validation of a Hydraulic Manipulator Joint for Control, In *Proc. International Conf. on Intelligent Robots and Systems*, Grenoble, France, September 1997.

[3]     Cannon, H., Extended Earthmoving with an Autonomous Excavator, Masters Thesis, Robotics Institute, Carnegie Mellon University, May 1999.

[4]     Cundall, P., Board, M. A Microcomputer Program for Modeling Large Strain Plasticity Problems, *Numerical Methods in Geomechanics*, Balkema, Rotterdam,1988.

[5]     Luengo, O., Singh, S., Cannon, H., Modeling and Identification of Soil-Tool Interaction in Automated Excavation, In *Proc. International Conf. on Intelligent Robots and Systems*, Victoria, B.C, Canada. October 1998.

[6]     Krishna, M., and Bares. J., Hydraulic System Modeling through Memory-based learning, In *Proc. International Conf. on Intelligent Robots and Systems*, Victoria, B.C, Canada. October 1998.

[7]     Mckyes, E. *Soil Cutting and Tillage*, Elsevier, 1985.

[8]     Hettiaratchi, D. R. P., Theoretical Soil Mechanics and Implement Design, *Soil and Tillage Research*, Vol 11, 1988, pp 325-347. Elsevier Science Publishers.

[9]     Reece, A. R., The Fundamental Equation of Earthmoving Mechanics, In *Proc. of the Institution of Mechanical Engineers*, 1964.

[10]     Rocke, D. J., Control system for automatically controlling a work implement of an earthmoving machine to capture material, US Patent 5528843, 1994.

[11]     Singh, S., Synthesis of Tactical Plans for Robotic Excavation, Ph.D Thesis, January, 1995, Robotics Institute, Carnegie Mellon University, Pittsburgh, PA 15213.

[12]     Singh, S., Cannon, H., Multi-Resolution Planning for Earthmoving, In *Proc. International Conference on Robotics and Automation*, Leuven, May 1998.

[13]     Stentz, A., Bares, J., Singh, S. & Rowe, P., A Robotic Excavator for Autonomous Truck Loading, In *Proc. International Conf. on Intelligent Robots and Systems*, Victoria, B.C, Canada. October 1998.

[14]     Zelenin, A. N., Balovnev, V. I. & Kerov L.P., *Machines for moving the earth*, A. A. Balkema, Rotterdam, 1992.

# Robot Control in Maxillofacial Surgery

Andreas Hein
Department for Navigation and Robotics
Clinic for Maxillofacial Surgery
Charité – Berlin – Germany
e-mail: a.hein@ieee.org


Tim C. Lueth
Department for Navigation and Robotics
Clinic for Maxillofacial Surgery
Charité – Berlin – Germany,
e-mail: t.lueth@ieee.org

**Abstract** : This paper presents the architecture and the implementation of a robot controller for a Delta-3 manipulator used for the implantation of bone fixtures for Anaplastology. The main feature of this robotics system for maxillofacial surgery is its interactive tool handling in conjunction with an image based control. The controller allows the use of the system as an intelligent tool which carries the drilling machine, allows the handling of the drilling machine as usual during the whole intervention, and extents the potential position and orientation accuracy of the surgeon. A control algorithm is described which allows a stable guidance of the manipulator in contact and non-contact situations. This algorithm works without the use of two force/torque sensors for distinguishing contact and guiding forces. Currently, the system undergoes a technical approval and first clinical experiments with cadavers.

## 1. Introduction

Surgery is a relatively new and a rapidly growing field of application for robotics technology. Especially in fields of surgery where the human hand is the limiting factor for further optimization of the surgical techniques –like in neurosurgery, orthopedic and maxillofacial surgery– robotics technology can be applied.

In the Surgical Robotics Lab (SRL) at the Virchow Hospital (*Fig. 1*) a robotics system is under development which allows the use of the system for different tasks within an intervention and which is interactively controllable by the surgeons [1]. The robotics system is used to enhance the accuracy and the dexterity of a surgeon and to decrease the tremble of the human hand. An open controller architecture has been developed which allows the easy extension of the system with new skills while retaining the safety of the operation. In a first step, the system will be clinically used for the implantation of bone fixtures for epitheses.

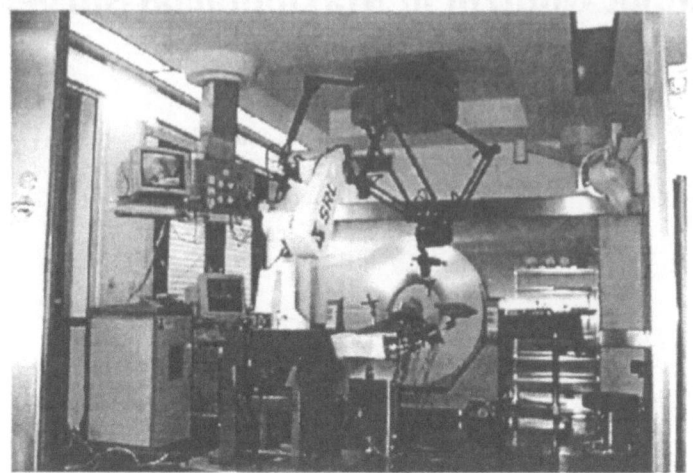

*Fig. 1: Experimental OR at the Virchow Hospital*

## 2. State of the art

During the last 10 years robotics techniques have been introduced to operating theaters for only a few surgical applications. The first systems were used in neurosurgery for the guidance of surgical instruments like biopsy needles, catheters, or microscopes [2][3]. During the treatment the manipulators are either switched off or the brakes are locked.

Active robotics systems are used in orthopedic surgery and for laparoscopic applications. The Robodoc system has been developed for milling of holes exactly fitting for a hip prosthesis [4]. This system works fully automatic during the operation. That means, the robot executes exactly the preoperative planned movements and can react to environmental changes only with an interruption of the program.

The only interactively controllable robotics system for invasive application has been designed for knee surgery [5]. This experimental 2-link arm is force controlled and can be guided by the surgeon manually to the cutting position at the knee bone. The workspace of the cutter is restricted by decreasing controller gains in the near of the boundaries. Commercially available systems for non-invasive procedures are the SurgiScope by Elekta and the MKM by Zeiss. In contrary to these systems, fully tele-operated systems like AESOP and Zeus by Computer Motion, EndoAssist by Armstrong-Healthcare, and Intuitive by Intuitive Surgical are not equipped with any degree of autonomy. Interactive control of robots for non-surgical applications can be found in [6][7][8].

In addition to the author's system [1], for maxillofacial surgery only two systems are under development [9][10]. One system consists of a passive manipulator used for positioning instruments and the other system is an automatic system for drilling of holes. For both systems the interaction with the surgeon is not considered in detail.

# 3. Description of the application

Due to the high social impact of the face and the great number of vital structures (vessels, nerves, brain, eyes) in this area accuracy is of paramount importance during an intervention in maxillofacial surgery. In contrary to the orthopedic applications (milling for hip implants or cutting at the knee for surface implants), in maxillofacial surgery no intervention exists which is carried out in a great number of cases. This is due to the great individual differences between the patients and the highly personalized planning of an intervention. Additionally, the access to the bony structures is restricted and the swelling of tissue leads to changes of the access path.

This difficulties especially on the field of maxillofacial surgery require a system design of the robot controller, which allows a cooperation between the surgeon and the robotics system. In this shared control task the surgeon remains the full control over the robotics system, but the system helps him/her to find positions and orientations and restricts the working area.

In the first step the robotics system is designed to position a drilling machine relative to the patient in a predefined position and orientation. During the drilling operation the robot only allows movements in the z-axes of the TCP. For an easy step by step extension of the functionality of the robotics system a special architecture of the robot controller has been developed.

In this paper the architecture of the robot controller (chapter 4) and a task-level control scheme (chapter 5) is described. In chapter 6 experimental results are presented.

# 4. Interactive controller architecture

The robotics system has been modeled as a discrete event system similar to the hybrid dynamic system with human integration in [11]. This architecture reflects the fact that an interactive robotics system can derive the parameters of a movement from two distinct sources – from a world model and from correction commands of a human supervisor. For the task described in [11] – the guiding of a mobile robot around obstacles – the world model consists of a map modeled by potential fields or by constraints. In surgical applications of robotics systems it is not possible to completely model the environment. Only a part of the patient and the near surroundings can be modeled using image data of the patient. Due to the uncertainty of the environment in surgical tasks safe paths and regions have to be defined by the surgeon or a nurse according to the real situation for the ongoing intervention.

From the perspective of a robotics system which has two distinct information sources this system can be divided into a Human Discrete Event System (HDES) with the user interface, a Plan Discrete Event System (PDES) with the plan interface, a Discrete Control System (DCS), and a Continuous Time System (CTS). *Fig. 2* shows an overview of the system.

The PDES is a finite state machine, which encodes the general command sequence for a type of intervention, specific patient data as target positions $^l t$ and registration points $^I p$ in respect to the image data coordinate system, and paths between working points outside the patient model. Depending on the current state $s^P$ of the PDCS an elementary operation $o(k)$ in the DCS will be activated by the command

$$u^P(k) = (o(k), {}^I t, {}^I p, path) . \qquad (1)$$

Each elementary operation terminates by setting one condition out of the vector $y^P(k)$ true and then the PDCS switches to the next state, which corresponds to the true condition.

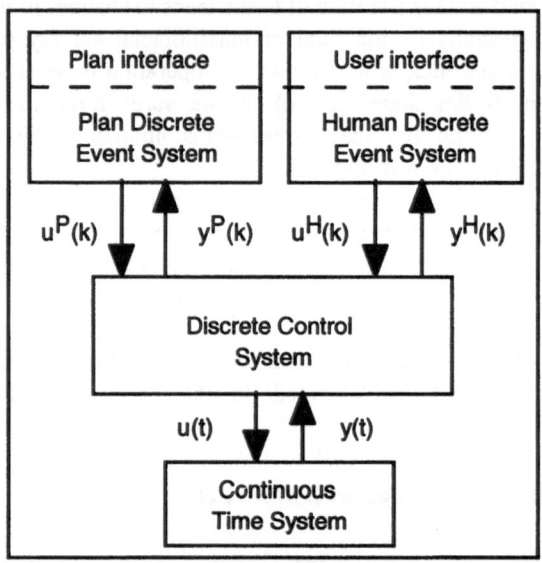

*Fig. 2: Architecture of the interactive robot controller*

Depending on the active elementary operation $o(k)$ and the current position and velocity of the TCP in respect to the image data coordinate system

$$y^H(k) = (o(k), {}^I p_{TCP}, {}^I \dot{p}_{TCP}) \qquad (2)$$

the HDES generates a command

$$u^H(k) = ({}^I \dot{p}_{TCP}) \qquad (3)$$

which will be delivered to the DCS.

In addition to $y^H(k)$ the HDCS gets a force/torque vector through the user interface, which consists of a force-torque sensor. It should be noticed that the HDES can not distinguish exactly between a users intervention and an interaction of the manipulator with the environment.

In the DCS the manipulator (CTS) will be controlled by a position, velocity and/or force scheme. The position of the manipulator is given by the vector $y(t)$ that consists of the six encoder values of the joints. The output of the DCS is the vector $u(t)$ that consists of the six values for the DAC for each joint motor.

## 5.    Control scheme

The following sensor data are used during the intervention:

- a vector of the encoder values of the manipulator $enc = (e_0,...,e_6)$ and the resulting position of the TCP, $^{BKS}T_{TCP}$ as a homogeneous 4×4 Matrix calculated by the inverse kinematic,

- the position of the TCP measured by the navigation system $^{CAM}T_{TCP}$, and

- the vector of forces $^{FTS}F = (f_x, f_y, f_z)$ and moments $^{FTS}M = (M_x, M_y, M_z)$ in the coordinate system of the force/torque sensor.

*Fig. 3* shows the coordinate systems and the applied forces/moments during the application of the robotics system.

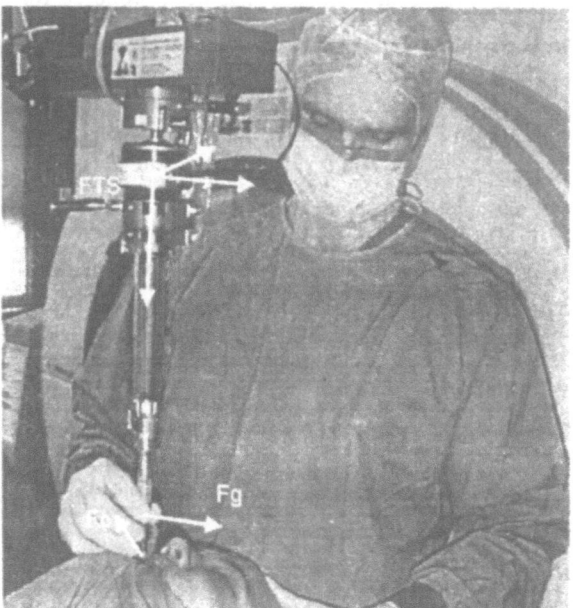

*Fig. 3: Interactive tool handling during an intervention*

An important design assumption for the interactive control of the robot is that the surgeon guides the tool as usual by grasping it and applying forces in the desired direction. Due to this assumption, two implications for the sensor data should be noticed. Firstly, due to the passive compliance of the manipulator in cases of environmental forces, the calculated position $^{BKS}T_{TCP}$ can differ from the position $^{Cam}T_{TCP}$ (assuming that the transformation $^{Cam}T_{BKS}$ is constant). It is assumed that the transformation between the locator position and the TCP is constant, because the tool system itself is stiff. Secondly, a distinction between contact forces $^{FTS}F_c$ with the patient or other objects and guiding forces $^{FTS}F_g$ applied by the user is not possible.

During an intervention two different modes are distinguished. In the guiding-mode, the tool has no contact with the environment. In this case the measured force $^{FTS}F_r$ is equal to the guiding force $^{FTS}F_g$. In the contact-mode the force $^{FTS}F_r$ is the sum of the contact force $^{FTS}F_c$ and the guiding force $^{FTS}F_g$.

The moving-mass control scheme has been proposed in [12]. In the following, a slightly different control scheme and a feedforward control extension in case of a restricted move along the z-axes of the TCP will be described.

In [13] and [7] the damping control scheme has been proposed for the guiding of a surgical robot. In this control scheme the velocity of the TCP is proportional to the force applied to it:

$$\dot{X}_d = K_f \cdot F_r \tag{4}$$

where $\dot{X}_d$ is the desired velocity, $K_f$ a scalar force feedback gain, and $F_r$ the force applied to the TCP.

Due to instability problems, a control scheme has been chosen, which is based on the model of a moving mass with friction. The virtual mass can be accelerated by external forces and will be slowed down by the virtual friction. The physical model is described by the friction energy $E_f$:

$$E_f = -\mu \cdot m \cdot g \cdot |X(k-2) - X(k-1)| \tag{5}$$

the linearized kinetic energy of the system (the term $|\dot{X}(k-1)|^2$ has been transformed to $|\dot{X}(k-1)| = |(X(k-2) - X(k-1)) / \Delta k|$ to damp the systems dynamic behavior):

$$E_{kin} = \frac{1}{2} \cdot m \cdot |\dot{X}(k-1)| \tag{6}$$

and the velocity caused by the acceleration of the system by external forces:

$$\dot{X}_a(k) = F_r(k-1) \cdot \frac{\Delta k}{m} \tag{7}$$

where $X(k)$ is the next delivered position of the TCP, $\dot{X}(k)$ is the velocity of the TCP, $F_r$ is the measured external force, and $\dot{X}_a(k)$ is the desired velocity of the TCP caused by the applied force $F_r(k-1)$. In every controller cycle (every 30 msec) the velocity of the system is decelerated by the virtual friction:

$$\dot{X}_f(k) = \dot{X}(k-1) \cdot \frac{E_{kin} + E_f}{E_{kin}} \tag{8}$$

if $E_{kin} + E_f \geq 0$, otherwise the velocity $\dot{X}_f(k)$ is set to zero. The equation (8) can be rewritten using equation (5) and (6):

$$\dot{X}_f(k) = \dot{X}(k-1) \cdot (1 - \frac{1}{2} \cdot \mu \cdot g \cdot \Delta k) \tag{9}$$

The resulting control law of the TCP position delivered to the underlying position controller of the robot is the sum of the velocities from equation (7) and (9):

$$X(k) = K_p.(K_f \cdot \dot{X}(k-1) + K_F \cdot F_r(k-1)) \cdot \Delta k \tag{10}$$

where $K_f$ is the control gain depending on the chosen virtual friction, $K_F$ is the control gain depending on the virtual mass and $K_p$ is the proportional gain for the position controller. This controller scheme works as a low pass filter which smoothes the movement of the robot. Distortions of a movement can arise from sensor noise and from scattering of the human arm during the guiding of the robot.

For restricted moves during the drilling operations the controller scheme (*Fig. 4*) has been extended by a feedforward controller based on the forces applied to the tool orthogonal to the z-axes of the TCP. In a first approach the disturbance force results over the proportional gain $K_d$ into a position offset. A direct measurement of the position deviation of the TCP is not possible due to the limited resolution of the encoder and the great distance between the last joint and the TCP.

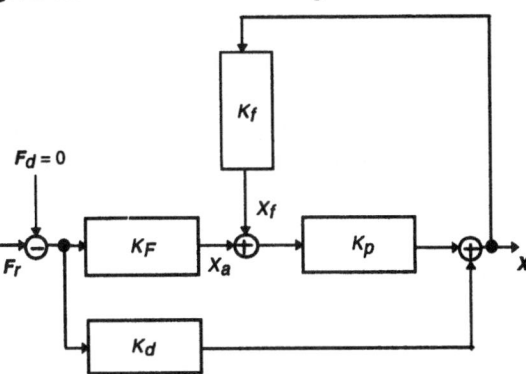

*Fig. 4: Structure of the controller with feedforward control for restricted movements*

Using the feedforward controller gain $K_d$ the control law (10) is extended to:

$$X(k) = K_p \cdot (K_f \cdot \dot{X}(k-1) + K_F \cdot F_r(k-1)) \cdot \Delta k + K_d \cdot F_r \tag{10}$$

## 6.    Results and Conclusions

*Fig. 5* shows the behavior of the robot before and during a drilling operation. In the first phase (0-5 sec) the robot has been guided along the z-axes without contact to the environment. It can be seen, that the noise of the force measurement is filtered and the velocity of the TCP remains stable. In the second phase (5-20 sec) the drilling is carried out and in the third phase (20-25 sec) the drill is pulled out of the hole. Although the force changes rapidly the contact to the skull is stable.

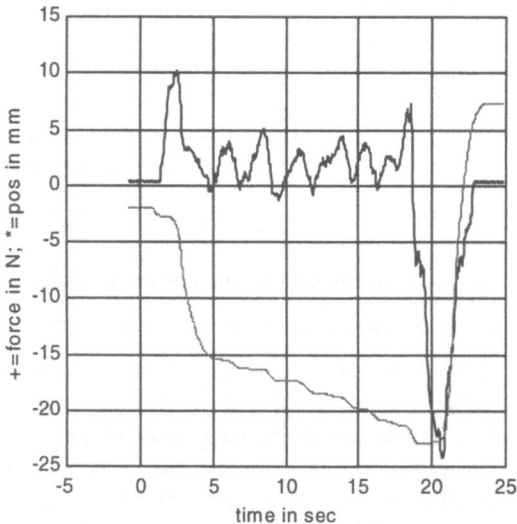

*Fig. 5:*    *Force and position of the TCP during a drilling operation. The black line shows the applied force in N. The lower line is the z-position of the TCP in mm. At approx. 5 sec. the contact point is reached and the drilling begins.*

The feedforward controller gain $K_d$ has been determined by applying forces to the TCP of the robot and while measuring the position deviation of the TCP by the optical measurement system. Due to the mechanical construction of the manipulator the deviations are small in the z-direction of the TCP. Therefore, only position deviations within the x-y-plane are taken into account. The deviation in the x- and y-direction (*Fig. 6*, left) is approximated by polynomials of second degree:

$$dx = 0.00005 \cdot f_x^2 + 0.062 \cdot f_x - 0.734 \tag{11}$$

and:

$$dy = 0.00001 \cdot f_y^2 + 0.07 \cdot f_y - 0.553 \tag{12}$$

Hence, neglecting the small parameters of the quadratic term the controller gain $K_d$ is simplified to:

$$K_d = \begin{bmatrix} 0.062 \\ 0.07 \end{bmatrix} \qquad (13)$$

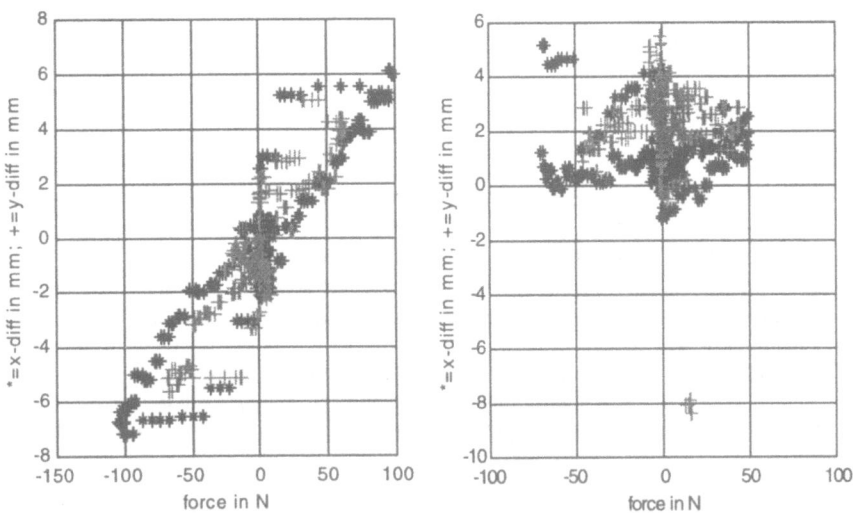

*Fig. 6: Position deviations of the TCP caused by applied forces. The deviation in the x-direction caused by the x-component of the force is represented by \* and the deviation in the y-direction is represented by +. Left hand side the measurement without the feedforward controller and right hand side with the controller extension.*

The extended controller has been tested by applying forces to the manipulator (*Fig. 6*, right). It can be seen that the errors have been decreased significantly. Great position errors only occur if the force is changing rapidly. Therefore, the extension of the feedforward controller to a PD or PID controller is necessary.

In the paper new concepts for an interactively usable robotics system have been described. Advantages of this concept are the better acceptance by the surgeons, the improved safety of the procedure, and the easy adaptation of the treatment to the actual requirements. Especially the behavior of the robot as a normal tool is of great importance for the surgeons. The surgeons can adapt the dynamic of the system by choosing the appropriate values of mass and friction.

## 7. Acknowledgment

This research work has been performed at the Department for Maxillofacial Surgery, Prof. Dr. Dr. Juergen Bier, within the Surgical Robotics Lab, Prof. Dr. Tim C. Lueth, Medical Faculty Charité, Humboldt-University Berlin. The work has been supported by the Deutsche Forschungsgemeinschaft with the Graduiertenkolleg 331 „Temperaturabhängige Effekte für Therapie und Diagnostik" (granted to Prof. Dr. Dr. h.c. R. Felix, PD Dr. N. Hosten) and by the Real-Time Control Group, Prof. Dr.-

182

Ing. Guenter Hommel, of the Technical University Berlin. Parts of the research have been supported financially by the Deutsche Krebshilfe (granted to Prof. Dr. Dr. J. Bier, PD Dr. P. Wust) and the Berliner Sparkassenstiftung Medizin (granted to Prof. Dr. T. Lueth, Dr. Dr. Ernst Heissler, Prof. Dr. Dr. Berthold Hell). Special thanks to the companies Elekta, Metalor and Philips for their support of the project. We would like also to thank Thomas Hölper, Edgar Schüle, Dr. h.c. Hervé Druais, Dr.-Ing. Armin Freybott, and W. Scholz. Their personal inspiration was the basis for this challenging research.

## 8.    References

[1]    Lueth T C, Hein A, Albrecht J, *et al.* 1998 A Surgical Robot System for Maxillofacial Surgery. IEEE Int. Conf. on Industrial Electronics, Control, and Instrumentation (IECON), Aachen, Germany, Aug. 31-Sep. 4, pp. 2470-2475.

[2]    Kwoh Y S, Hou J, Jonckheere E, Hayati S 1988 A robot with improved absolute positioning accuracy for CT guided stereotactic surgery. *Transactions on Biomedical Engineering*, Vol. 35, No. 2, pp. 153-161.

[3]    Lavallée S 1989 A new system for computer assisted neurosurgery. Proc. 11th IEEE Engineering in Medicin and Biology Conf., Seatle, November, pp. 926-927.

[4]    Taylor R H, Mittelstadt B D, Paul H A, *et al.* 1994 An Image-Directed Robotic System for Precise Orthopaedic Surgery. *IEEE Trans. on Robotics and Automation*, Vol.10, No.3, pp. 261-275.

[5]    Ho S C, Hibberd R D, Cobb J, Davies B L 1995 Force control for robotic surgery. ICAR IEEE Int'l. Conf. on Advanced Robotics, pp. 21-31.

[6]    Hirzinger G, Landzettel K 1985 Sensory feedback structures for robots with supervised learning. ICRA IEEE Int. Conf. on Robotics and Automation, pp. 627-635.

[7]    Kosuge K, Fujisawa Y, Fukuda T 1993 Mechanical System Control with Man-Machine-Environment Interactions. Proc. of the 1993 IEEE Int. Conf. of Robotics and Automation, pp. 239-244.

[8]    Colgate J E, Wannasuphoprasit W, Peshkin M A: Cobots 1996 Robots for collaboration with human operators. Proc. of the Int. Mechanical Engineering Congress and Exhibition, Atlanta, pp. 433-439.

[9]    Cutting C, Bookstein F, Taylor R H 1996 Applications of simulation, morphometrics and robotics in craniofacial surgery. In Taylor R H, Lavallee S, Burdea G C, Mösges R (eds) 1996 *Computer-integrated surgery: technology and clinical applications*, MIT Press, pp. 641-662.

[10]   Bohner P, Haßfeld S, Holler C, Damm M, Schloen J, Raczkowsky J 1996 Operation planning in cranio-maxillo-facial surgery. Medicine Meets Virtual Reality 4 (MMVR4'96), San Diego, California.

[11]   Aigner P, McCarragher B 1997 Contrasting Potential Fields and Constraints in a Shared Control Task. IROS IEEE/RSJ Int. Conf. on Intelligent Robots and Systems, pp. 140-146.

[12]   Hein A, Lueth T C 1999 Sensing and Control in Interactive Surgical Robot Systems. IFAC World Automation Congress, Bejing, China, in print.

[13]   Kazanzides P, Zuhars J, Mittelstadt B, Taylor R H 1992 Force sensing and control for a surgical robot. ICRA IEEE Int. Conf. on Robotics and Automation, Nice, France, May, pp. 612-617.

# Experiments and Experiences in Developing a Large Robot Mining System

Peter Corke, Jonathan Roberts and Graeme Winstanley
CSIRO Manufacturing Science and Technology
CRC for Mining Technology and Equipment
www.cat.csiro.au/cmst/automation

**Abstract:** We discuss our experiences in automating a very large mining machine, an electric walking dragline. Initially this appeared to be a very simple robotics problem and that it was "just a matter of scaling up" in terms of machine size. In actuality the robotics part of the problem was the easiest to solve. The more difficult problems were to do with issues like machine environment (dust, rain, insects), complex dynamic effects that are only manifested in the full scale machine, understanding the skills of human operators, and people (machine operators and management).

This paper briefly describes the major technical elements of this project, but more importantly tells how our thinking about the problem, and the solution, has changed over time.

## 1. Introduction

Draglines are large electrically powered machines used to remove overburden (unwanted material) from above coal seams in open-cut mines. They are large (weight 3000-4000 tonnes) and expensive ($50-100M) and generally in production 24 hours per day, every day of the year. The machines, see Figure 1, fill their bucket by dragging it through the ground; it is then lifted and swung to the spoil pile and dumped. Each swing moves upto 100 tonnes of overburden, and a typical swing takes 60 seconds. The work is cyclic and very repetitive, yet requires considerable operator skill. The *only* performance metric is tonnes of overburden moved per hour (without breaking the machine).

From a robotics point of view the machine can be considered as a 3DOF robot with a flexible, or un-actuated, final link, and the task can be thought of as comprising:

- constrained motion (digging, interaction between the bucket and ground),

- unconstrained position control (moving the bucket in free space).

In our project we chose to automate the second task since it is simpler and also comprises the bulk (80%) of the cycle time. However even a 1% improvement in machine productivity is valued at around $1M per machine per year (and repay the cost of the entire research program).

The remainder of this paper describes significant technical aspects of the project, but also describes how our thinking changed over the duration of the

Figure 1. A Bucyrus-Erie 1370 dragline.

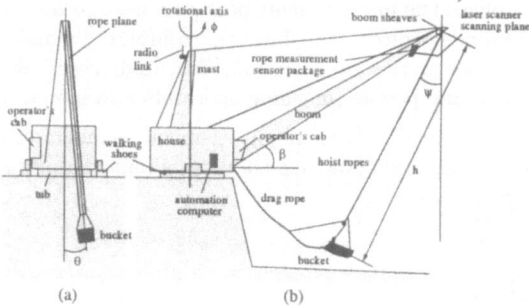

Figure 2. A schematic of a typical dragline showing the main elements of the machine. (a) front view, (b) side view.

project as we became aware of other, generally non-technological, issues. A demonstration based on a visually servoed Puma robot was made in 1993. This was scaled up by an order of magnitude in 1995 to operate on a one-tenth scale model dragline, and then by another order of magnitude in 1998. At each stage new problems are introduced which involve significant engineering to overcome.

Section 2 discusses load position sensing, and describes our attempts to use machine vision, and our final solution based on a laser scanner. Section 3 discusses issues related to path planning and axis control, and Sections 4 describes the operator interface. Section 5 summarises our achievements to date, and Section 6 lists conclusions and areas for future work.

## 1.1. Prior work

Control of swinging loads is not a new area: open-loop control strategies are used in commercially available anti-swing controllers for cranes, and closed-loop systems are used for helicopter sling loads and some cranes. However open-loop control is critically dependent upon knowledge of the load oscillation period which is a function of rope length and is unable to compensate for external disturbances or model errors.

An early industry attempt to automate a dragline using a PLC system to simply "replay" the sequence of operator generated reference signals was unsuccessful[1]. That approach was not able to robustly handle cycle by cycle variation in starting point, variations in payload or external disturbances. There was no sensory feedback of the state of the machine, and in particular the angle of the swinging load, and we believed that this was the core technical problem to be solved.

## 1.2. The dragline

The dragline has three driven degrees of freedom, the hoist and drag rope lengths, and rotation of the whole 'house' and boom assembly on the base (tub). 'Our' dragline has a boom lengths of 100 m, 2000 hp DC motor sets on each axis, and a 60 tonne bucket that holds around 100 tonnes of dirt.

The human operator visually senses the state of the bucket, its rigging and the local terrain and actuates the drag, hoist (joystick) and slew (pedal) controls in a coordinated fashion. A good deal of skill is required to control, and exploit, the bucket's natural tendency to swing. This skill is not easily learnt and a novice operator typically requires 6 months training to become proficient. Performance can vary by up to 20% across the operator population, and for an individual will vary over a shift (up to 12 hours).

Our swing automation system is activated by the operator once the bucket is filled. It automatically hoists, swings and dumps the bucket (at an operator defined reference point) before returning the bucket to a predefined digging point.

# 2. Robust load position sensing

The operator uses his eyesight and perceptual skills to infer instantaneous bucket position and velocity, and this functionality must be replicated by the automation system. The extremely harsh nature of the bucket's interaction with the ground precludes instrumentation on the bucket itself, and requires a remote, or non-contact, sensing approach.

Our first approach to load position sensing was to use a TV camera and build a visual servo system. This was prototyped in our laboratory in 1993. The laboratory example was contrived to have high contrast thus simplifying the task of image segmentation. Clearly the system would have to work in an outdoor environment, 24 hours per day in all weather conditions (in particular heavy rain and dust). We continued the machine vision approach[2] (because we knew how to do it) for some time before we abandoned it. The main problems were robustly segmenting the load from the background in an environment where both camera and load are moving, background is unknown, lighting is variable, and target scale varies by two orders of magnitude. Another significant problem was to find a camera location that will not be 'blinded' by the sun at certain times of the day.

Our final solution was based on the insight that it is far easier to find the ropes that support the load than the load itself[3]. We also changed the sensor to a scanning laser rangefinder which produces a much more consise output

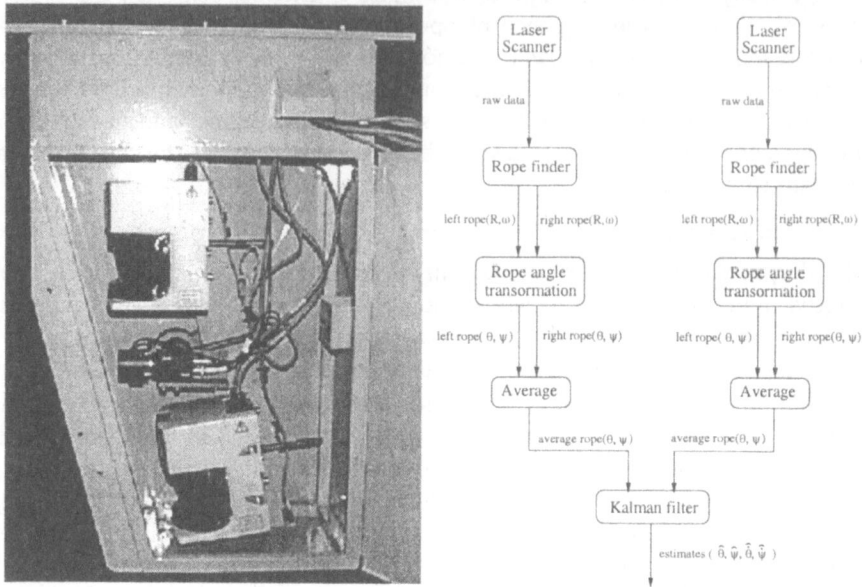

Figure 3. Left, the boom-tip mounted laserrangefinder package. Right, schematic of the hoist rope angle estimation system.

that can be more easily segmented. Considerable work was still required to robustly find the ropes in the presence of spurious targets such as raindrops and insects (which are attracted to the dragline lights at night). This retains the key advantage of non-contact position sensing, the devices are rugged and low in cost, and the difficult problem of robust scene segmentation is side stepped.

## 2.1. Details

Our final rope finding system provides estimates of the swing angle, $\theta$, and the hoist angle, $\psi$ at a rate of 5 Hz. The system uses two of the ubiquitous PLS sensors manufactured by Sick Opto-electronics, Germany. The PLS returns range, $R$, and bearing, $\omega$, data across a 180° field-of-view. Figure 2 shows the location of the rope sensing system on the dragline. The scanners are mounted in a weatherproof unit (Figure 3) that is located near the boom tip. Two sensors are used for redundancy to hardware failure and sun dazzle.

A schematic of the hoist rope angle measurement system is shown in Figure 3. The main processing steps are:

1. Rope finding, which uses *a priori* knowledge to robustly discriminate rope targets. It comprises:

   (a) Range gating, where only targets in a specified range are accepted. We can do this because we know that the ropes are constrained by the physical setup.

   (b) Adjacent target merging, where targets located at adjacent bearing angles, and with a similar range, are considered to belong to the same object.

(c) Rope discrimination (pairing data). We know that there are two ropes and that they are constrained in the way they move relative to one another. This fact allows us to *pair up* data points into possible rope pairs (an $n^2$ search task). Experiments have shown that the hoist ropes are always within 0.5m of one another in the $x$-direction and are between 0.5 and 1.0m of one another in the $y$-direction.

(d) Tracking and data association, where we match pairs between successive scans, and hence identify the hoist ropes and reject the false target pairs.

2. These positions are transformed to positions of the ropes in the scanning plane, and then to swing and hoist angles of the hoist rope.

3. The swing and hoist angles for each rope are averaged.

4. A Kalman filter 'fuses' the measurements from the separate and asynchronous scanners (running at approximately 3 Hz), and provides estimates at a fixed frequency (5 Hz). The filter was implemented as a discrete-time Kalman filter with a variable time step. The filter covariance estimates are used to reject incompatible measurements, or to diagnose sensor failure.

## 2.2. False targets: rain, bugs etc.

In practice we will observe false targets due to rain (most common) and insects, birds (less common). To investigate this we undertook experiments with the scanner in the rain. The important measure, for the rope tracking system to work reliably, is the actual probability of 'seeing' a rain drop in a given scan segment, which for the two experiments conducted were:

| Rain rate | Probability (%) |
|-----------|-----------------|
| light     | 0.02            |
| medium    | 0.28            |

The algorithm described above was implemented in MATLAB and run on a twenty minute long data file captured from the dragline. Simulated rain was added to test the robustness of the algorithm to false targets and the results are shown in Figure 4. One line is the success rate for a single scanner, which shows that the success rate drops to about 90% when there is a probability of 20%. The other line shows the success rate when two scanners are used. This line shows that the success rate increases to just over 99% when two scanners are used with a rain rate of 20% (over two orders of magnitude worse than the medium rain probability).

## 3. Control

The final control system is vastly more complex than the small-scale implementations, and comprises three main components. The first is a finite-state-machine which sequences through various phases of the task, with state transitions based on time and measured quantities such as rope tension. It enables and sets parameters for the remaining components.

188

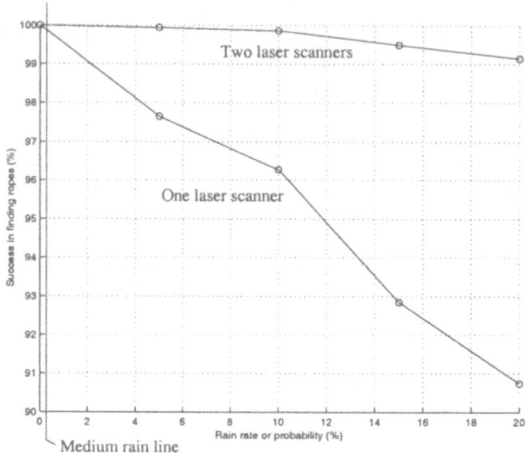

Figure 4. Success in finding the ropes versus rain rate.

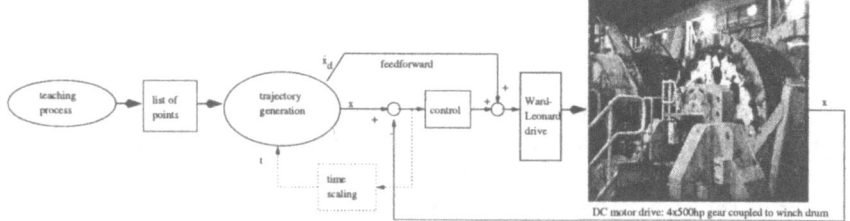

Figure 5. Axis control loop schematic.

Second are the trajectory generators which create a reference demand for the position controlled axes. Standard robotic techniques are used to achieve coordinated multi-axis motion. The bucket is required to follow a trajectory through space joining points marked by the operator during a 'training' operation. Each segment of the motion is of constant velocity with cubic polynomial blends whose duration is set online to meet the acceleration limits of the particular drive.

Thirdly are the axis control loops. The two winch drives are voltage controlled, and using encoder feedback and reference velocity feedforward result in reasonable performance position controlled devices. See Figure 5. The loop gains are direction dependent due to the motor field weakening strategy, and typical following error performance is shown in Figure 6.

The most significant difficulty encountered is that for minimum time paths at least one drive must be in saturation. The motor regulator enforces a strict volt-amp limit, see lower left quadrant of Figure 7, so that drive speed falls off with increasing load. The dependence on load is critical and the slope of the line shows that speed falls off at around 20 mm/s/tonne. Load varies not only as a function of the amount of dirt in the bucket but also as the rope configuration changes since much of the rope force is antagonistic. Our first

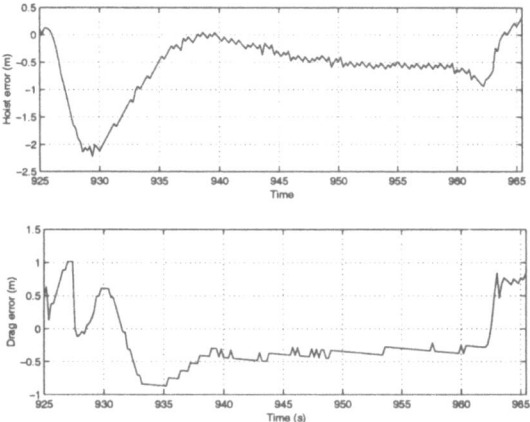

Figure 6. Measured axis error performance.

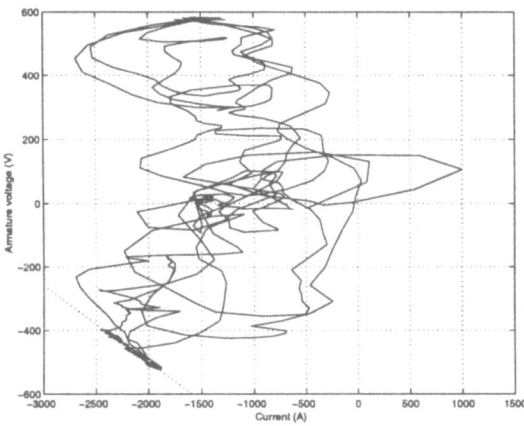

Figure 7. Volt-amp plot for hoist drive.

approach was to set conservative limits on demanded speed. However if the limits are too low this has a negative effect on cycle time, and if too high the drives will lag significantly behind the reference and cause a collision between the bucket and a spoil pile. Techniques for simple torque limited drives have been proposed by Dahl[4]. Another approach to increasing speed is to reduce the rope antagonistic tension by feedback of motor current. Figure 8 shows the reduction in each rope tension as the amount of tension feedback is increased.

The basic principles of pendulum stabilization are covered elsewhere[5, 6]. The slew drive has a nested loop structure with feedback of axis rotation angle and also bucket swing angle rate. The result of a swing motion with and without swing stabilization are compared in Figure 9. Due to inter-cycle variability it is more difficult to compare actual performance with and without swing stabilization.

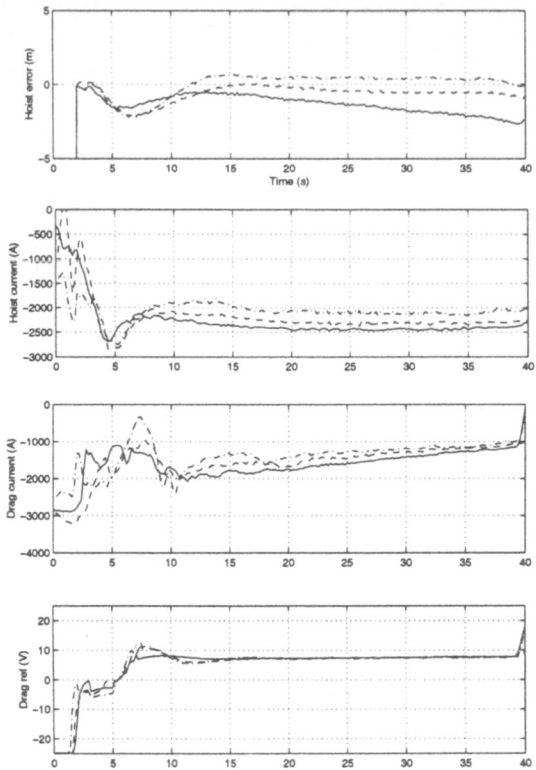

Figure 8. Effect of drag tension control on drag rope following error.

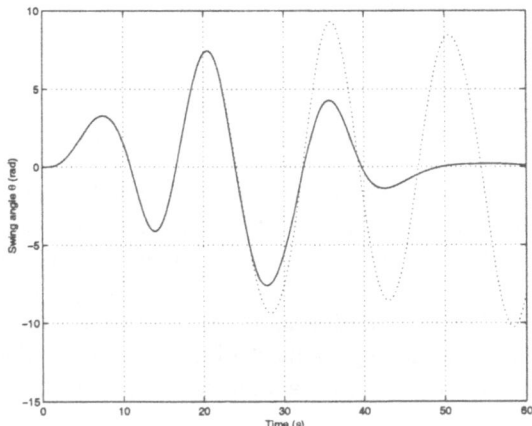

Figure 9. Swing stabilization starts at $t = 21\,\mathrm{s}$.

## 4. User interface

The automation system 'drives' the dragline by physically moving the control joysticks and pedals; like the cruise control in a car. To achieve this we fitted

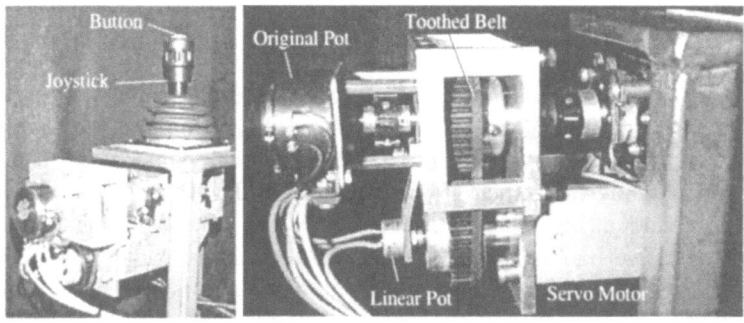

Figure 10. A modified operator joystick.

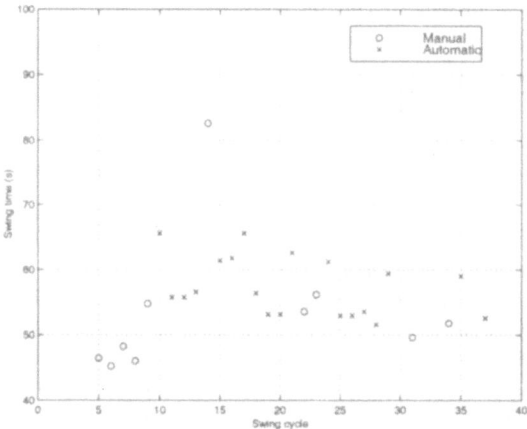

Figure 11. Comparison of manual and automatic swing cycle times.

brushless servo motors to each of the control devices, see Figure 10.

Servoing these controls facilitates the smooth transfer of system set points in the transitions between automatic and manual control modes. Sensing on the drag and hoist joystick servo motors allows the automation system to sense if the operator is opposing the motion of the joysticks — in which case it smoothly transfers control back to the operator. Thus the servoed operator controls impose on the control computer the same safety interlocks and overrides that apply to an operator.

The servoed joystick could also be used to provide a kinesthetic interface to the operator. Draglines are very poorly instrumented and the driver has no direct indication of payload weight or drag force. Feedback of force through the joystick may provide this data in an intuitive manner.

## 5. Results

As stated in the introduction the only performance measure of interest to the client is swing cycle time. Figure 11 compares swing cycle times for manual and automatic modes of operation. The automation system is competitive with the operator but the statistical base for comparison is still very small.

## 6. Conclusions

At the outset this seemed like a straightforward robotics problem, albeit a large one. The basic principle of vision-based swing stabilization was demonstrated in 1993 and we knew then that the final control problem would be easier because the larger machine has much slower dynamics. By the end of the project it was clear that swing stabilization was only one of a significant number of problems that we had to solve. These included time-variant and non-linear drive characteristics, robust sensing, user interface and people management — our client is interested only in production metrics, not the technology.

While our early results are encouraging we still have much more to do. A significant problem is that the automation system has a lack of knowledge of its environment which makes it incapable of planning its own paths. The human operator takes the bucket *very close* to the spoil piles, whereas the automation system must plan conservative paths that have considerable clearance from potential objects. Vertical clearance in particular requires extra bucket hoisting which adds seconds to the cycle time, and time is money. In order to plan better paths we need to measure, or map, the environment online.

## Acknowledgements

The authors gratefully acknowledge the help of their colleagues Stuart Wolfe and Frederic Pennerath. The present work is funded by a consortium consisting of the Australian Coal Association Research Programme as project C5003, Pacific Coal Pty Ltd, BHP Australia Coal Pty Ltd and the Cooperative Research Centre for Mining Technology and Equipment (CMTE). Valuable inkind support has been provided by Bucyrus Erie Australia, Tritronics Pty Ltd, and Tarong Coal for allowing the automated swing system to be installed on their BE 1370.

## References

[1] L. Allison, "Creative engineering — a key to the future of the australian coal industry," in *Proc. Conf. Inst. Mining Electrical and Mining Mechanical Engineering*, (Newcastle), Oct. 1992.

[2] J. Roberts, P. Corke, and G. Winstanley, "Applications of closed-loop visual control to large mining machines," in *Video Proceedings of IEEE Int. Conf. on Robotics and Automation*, 1996.

[3] P. I. Corke, J. M. Roberts, and G. J. Winstanley, "Modelling and control of a 3500 tonne mining machine," in *Experimental Robotics V* (A. Casals and A. de Almeida, eds.), no. 232 in Lecture Notes in Control and Information Sciences, pp. 262–274, Barcelona: Springer Verlag, June 1998.

[4] O. Dahl and L. Nielsen, "Torque-limited path following by on-line trajectory time scaling," *IEEE Trans. Robot. Autom.*, vol. 6, pp. 554–561, Oct. 1990.

[5] P. Corke, J. Roberts, and G. Winstanley, "Vision-based control for mining automation," *IEEE Robotics and Automation magazine*, Dec. 1998.

[6] P. I. Corke, *Visual Control of Robots: High-Performance visual servoing*. Mechatronics, Research Studies Press (John Wiley), 1996.

# Slope Display on a Locomotion Interface

David Tristano, John Hollerbach, and Robert Christensen
University of Utah
Depts. Mechanical Eng. and Computer Science
Salt Lake City, Utah, USA
davidt@gse.upenn.edu,jmh@cs.utah.edu,robchris@es.com

**Abstract:** Smooth slopes can be displayed on treadmills through tilt mechanisms, but these mechanisms are typically too slow to present sudden slope changes. The Sarcos Treadport utilizes a mechanical tether that can apply horizontal forces to the user to simulate gravity forces. Psychophysical results are presented to show that slope display via tether forces is a reasonable facsimile for treadmill slope. Results are also presented on discrimination of slope while walking and of tether pulling direction.

## 1. Introduction

Locomotion interfaces are devices that require repetitive limb motion to propel ourselves through virtual worlds. In defining locomotion interfaces, one usually thinks of locomotion in terms of the repetitive motions of legs in weight-bearing situations (walking) or non-weight-bearing situations (bicycling). Arms may also be involved with the legs (Nordic skiing, rowing), or may be involved exclusively (walking on one's hands, turning a crank).

The key issue is that the distance covered by the range of limb motion is small compared to the desired distance to be traveled. Therefore the limbs must be cycled in a *gait* pattern, which requires energy expenditure. The energy expenditure may also result from acceleration of the body. The environment itself may impose loads: gravity in going uphill, viscosity when moving through a fluid or air.

This energy extraction is a key feature of locomotion interfaces. A contrasting approach to virtual world motion is the use of rate-controlled devices, such as joysticks. The velocity of moving through the virtual world is proportional to displacement of the joystick. The workspace of the joystick fits well within the workspace of our arms, and consequently a negligible amount of energy is employed in this navigation mode. Another example is a driving simulator, where the vehicle velocity is proportional to the deflection of a gas pedal [1].

It is hypothesized that locomotion interfaces increase the sense of realism and immersion in the virtual world, not only through energy extraction which affects decision-making processes but also through sensorimotor integration of proprioception and vision. It has been argued that locomotion calibrates

Figure 1. The Treadport comprises a large treadmill in a CAVE-like environment, and a 6-DOF mechanical tether for body motion measurement and application of axial forces.

vision, and that by itself vision is not a very accurate absolute distance sensor [2]. Therefore our situational awareness in virtual worlds may be enhanced with locomotion interfaces over the use of rate-controlled navigation interfaces.

With respect to leg-oriented locomotion interfaces, a variety of devices have been created: pedaling devices [3], programmable foot platforms [4], sliding surfaces [5], and treadmills. This paper is concerned with one form of the latter, the Sarcos Treadport.

### 1.1. Treadmill Locomotion

Work on treadmill-style interfaces initially employed passive treadmills, where the belt motion is entirely due to the user pushing with his or her legs [6]. The naturalness of locomotion is compromised for concerns of safety. More recently, active treadmills have been employed where the belt speed is controlled by measurements of user position. The Sarcos Treadport [7] comprises a large 4-by-8 feet treadmill and a 6-axis mechanical tether that acts as a position tracker (Figure 1). The mechanical tether is comprised of a Hookean shoulder joint, a linear axis, and a spherical joint attached to a harness worn by the user. The belt is servoed to roughly center the user based upon the tether measurements of user position. A mechanical tether is also employed in the Omni-Directional Treadmill [8], while in the ATLAS treadmill system, optical tracking of leg motion is employed [9].

Locomotion rendering may be defined similar to haptic rendering, as the simulation of natural locomotion through analogous mechanical control of the locomotion interface. Four aspects of locomotion rendering are linear motion, turning, unilateral constraints, and uneven terrain.

The display of linear motion is the major strength of treadmills. Constant-velocity locomotion on a treadmill is indistinguishable from constant-velocity motion on a stationary surface of similar characteristics. However, when a user accelerates to a run, the physics of treadmill locomotion is different, because the belt is servoed to center the user and applies an inertial force to the user [10]. This inertial force assists the runner and causes considerably less energy expenditure. In the Sarcos Treadport the linear axis of the mechanical tether is actuated and can push or pull on the user. This capability is employed to exert a force that counteracts the inertial force from the belt, resulting in more natural running [7]. Without a mechanical system like the active tether to provide such inertial force feedback, a treadmill system cannot realistically simulate running.

Turning control is less natural for one degree-of-freedom belt systems. In the case of the Sarcos Treadport, rate control is employed based upon the amount of user lateral displacement when walking or of user twist when standing. More natural are two-dimensional treadmills. The Omni-Directional Treadmill employs a belt comprised of rollers, which roll in a direction orthogonal to the belt by the action of a second belt. The Infinite Floor [11] comprises twelve small treadmills connected side-by-side and moved perpendicularly as one big belt. The ATLAS treadmill is mounted on a spherical joint. As the user turns, the treadmill is swiveled about a vertical axis in the direction that the user is moving.

The locomotion interface should somehow simulate unilateral constraints, to prevent the user walking through virtual walls or objects. One can simply stop the treadmill belt, although the user would still be able to move forward. For the Sarcos Treadport, the active mechanical tether is employed to apply a penalty force which restrains the user at the same time the belt is stopped.

## 1.2. Slope Display

Walking up a smooth slope can be readily displayed on treadmills that can tilt, whereas the display of uneven terrain such as steps is not feasible. Commercial treadmills are usually employed as exercise machines, and have a very slow tilt that limits the rate of slope change that can be displayed. The first-generation Sarcos Treadport employs a commercial Trackmaster treadmill with a tilt of 1 degree per second, which is too slow to be generally useful.

Instead of treadmill tilt, tether force has been employed in the Sarcos Treadport as a gravity display to represent slope. For uphill motion, the tether pulls on the user, who must then exert more effort. For downhill motion, the tether pushes the user, as if gravity were assisting.

It is a question as to how realistically tether force is interpreted by the user as a gravity force that represents a particular slope. This paper reports on three psychophysical experiments to answer this question. The first two experiments determine subjects' discrimination thresholds with regard to treadmill tilt and tether force vector or angle. The third experiment then determines to what extent tether pulling force is equivalent to treadmill tilt.

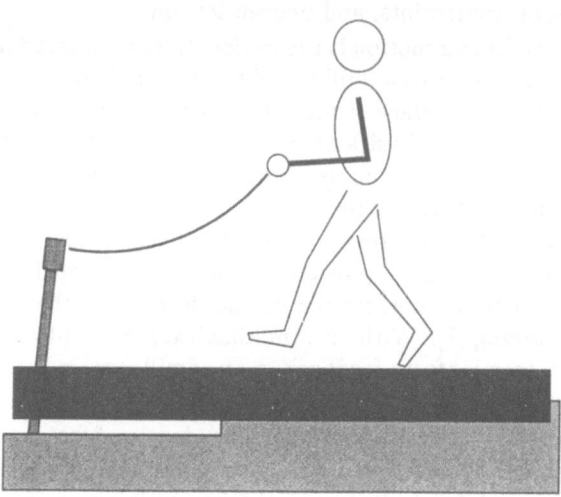

Figure 2. A string guide for vision-less walking.

## 2. Discrimination of Slope during Walking

To set a baseline for the interpretation of results with tether force simulation of slope, we wanted to know how well humans can discriminate slope differences while walking. Interestingly, the literature did not have this information, perhaps because there was no reason to ask this question before. The following experiment was devised to answer this question.

### 2.1. Method

Subjects were asked to walk on the treadmill at a constant pace at different slopes. Subjects kept their eyes closed. To provide kinesthetic cues to help them stay centered while their eyes were closed, a string attached to two posts at the front of the treadmill was held by the subjects (Figure 2). The string attachment points at the posts were through springs, to approximate a constant force source when pulling on the string. This hopefully avoids the use of any force cues in pulling on the string in judging slope. The spring constants were low, so that the subject could not pull as an aid in walking up slopes.

From a particular reference slope, the slope of the treadmill was adjusted either up or down, and subjects were asked to indicate the direction of the change. Five different reference slopes were used: 0 degrees, ±2 degrees, and ±4 degrees. For each reference slope there were six different adjustments: ±0.5 degrees, ±0.75 degrees, and ±1.0 degrees. Subjects wore headphones to help mask the sound of the tilt motor, but it appeared that the motor made the same noise whether the platform went up or down. Preliminary tests were done where the up-down motion was randomized before settling at the final slope to avoid kinesthetic cues about the adjustment. This random excursion stayed within 0.5 degrees and lasted approximately 10-12 seconds. The results were similar to just placing the treadmill at the final slope, which takes 2 seconds, and for the results reported below the treadmill motions were not randomized.

The order of the trials was randomized, then fixed for all subjects. There were 13 subjects, 10 male and 3 female, ranging in age from 20 to 50. Subjects were given as much time as they wished. No knowledge of results was provided. The entire experiment lasted approximately 15 minutes.

## 2.2. Slope Discrimination Results

Table 1 shows the percent of correct responses for all subjects for slope discrimination combining all reference slopes. The results are not strictly correlated with the magnitude of slope change. Given a criterion of 75% as the threshold of discrimination, then subjects were able to perceive a 0.5 degree slope change.

| Angle change (deg) | -1.0 | -0.75 | -0.5 | 0.5 | 0.75 | 1.0 |
|---|---|---|---|---|---|---|
| Success rate | 85% | 83% | 77% | 82% | 89% | 77% |

Table 1. Success rate for all subjects for slope discrimination.

Table 2 shows the success rate as a function of the reference slope for all subjects, combining all adjustments. People were most accurate when starting from level. There is not a significant difference between the 2 and 4 degree reference slopes.

| Reference angle (deg) | -4 | -2 | 0 | 2 | 4 |
|---|---|---|---|---|---|
| Success rate | 73% | 83% | 94% | 77% | 85% |

Table 2. Success rate versus reference slope.

Table 3 shows that people were far more accurate for upward adjustments than for downward adjustments. These results are for a non-zero reference angle. The figure of 93% for discriminating upward adjustments are about as good as judging all adjustments relative to a zero reference angle, 94% (Table 2).

| New angle greater than reference | 93% |
|---|---|
| New angle less than reference | 65% |

Table 3. Success rate versus direction of adjustment.

There was evidence that learning occurred. The success rate for the first 15 trials was 77%, while the rate for the last 15 trials was 87%. Subjects by and large felt they were guessing the whole time, and were surprised as to the extent of their correct responses.

# 3. Perception of Tether Force Angle

The main linear linkage of the tether stays approximately horizontal, as it is a straight-line path from the base of the tether mechanism to the subject's waist. It is possible that a different tether pulling angle should be employed when simulating uphill versus downhill walking. When going uphill, it might be better for the tether to be pulling at a downward angle, while when going downhill, a pushing force from an upward angle might be appropriate.

### 3.1. Methods

It was not possible to change the height of the tether base, in order to change the tether pulling angle. Instead, a manual setup was employed, where a string attached to a user's back was pulled at different angles by an experimenter. A spring gage was employed to measure the pulling force. Pulling angles of +10 and +20 degrees (up from horizontal) and -10 degrees were employed; for the upwards angles, the experimenter stood on a ladder. The subjects walked on a reference slope of the treadmill for 1 minute; only upward slopes were employed because one could only pull with the string. Then the treadmill was leveled, and the tether force for a given pulling angle was adjusted until subjects felt the tether force most resembled walking up the reference slope.

In the first part of the experiment, subjects were asked their preference for the different pulling angles to simulate the reference slope. The order of pairwise comparisons was randomized.

In the second part of the experiment, the ability of subjects to discriminate different pulling angles was teseted. Again a pairwise force-choice paradigm was used.

### 3.2. Results on Tether Force Angle

Results for preference on pulling angle were mixed, and the horizontal pulling direction in general was preferred. This may be explained by the results on discrimination of tether pulling angle, which were poor, on the order of 10-15 degrees. It is possible that backlash in the harness worn by the user masks the perception of pulling direction. In any case, the results seem to justify proceeding with the Treadport's horizontal pulling direction for the last experiment.

## 4. Correlation of Pulling Force with Perceived Slope

The correlation of pulling force with perceived slope was tested, by asking subjects to adjust tether pulling force to match a reference slope. The hypothetical result would be that the tether force is adjusted to $f = mg\sin\theta$, where $m$ is the person's mass, $g$ is gravity, and $\theta$ is the angle of the reference slope.

### 4.1. Methods

The experimental method was similar to the previous one. Subjects walked on a reference slope for a minute, then the slope was returned to horizontal. They then indicated what magnitude of pulling force was most appropriate to duplicate the sensation of the slope. Pulling was horizontal only.

### 4.2. Results on Pulling Force versus Perceived Slope

The results are plotted in Figure 3 as reference slope (horizontal axis) versus force represented as fraction of body weight. The hypothetical fit $f = mg\sin\theta$ is the steeper dotted line. Instead, a fit of $f = 0.65mg\sin\theta$, i.e., approximately 2/3rds of the expected force, best represents the results for all of the subjects. The relationship is linear between the reference slope and the equivalent tether force, and in that sense the tether force is a reasonable means for slope display. However, the preferred proportion of tether force is 2/3rds of the hypothetical.

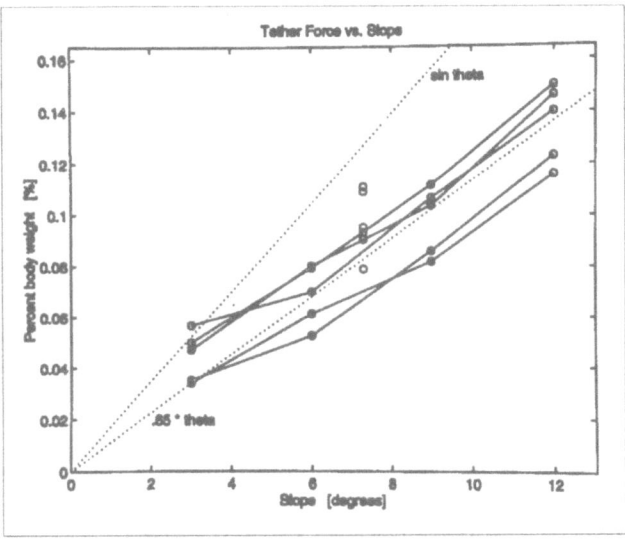

Figure 3. Reference slope versus tether force expressed as fraction of body weight. Data is from 5 subjects. The lower dotted line is a straight-line fit to the results. The higher dotted line is the hypothetical fit.

## 5. Discussion

The main result of this paper is to demonstrate that tether force is a reasonable means for displaying gravity force. The fit of tether force to equivalent treadmill tilt approximately satisfies the relation $f = 0.65mg \sin\theta$, or 2/3rds of the hypothetical. An analogous result was observed in subjects' preference for inertial force display [7]. The hypothetical inertial force feedback $f = ma$, where $a$ is the subject's acceleration, but the preferred relation was $f = 0.8ma$, i.e., 80% of the hypotheticals. One explanation is that gravity loads and inertial loads should be distributed over the whole body, but the tether just applies force to one point at the back of the body and through the harness.

The equivalence does not necessarily mean that the experience is the same; subjective notions of the naturalness of the slope presentation were not tested. Some subjects reported that they employed a sense of effort to relate tether force to treadmill tilt, rather than some sense of sameness. An objective way to quantify sameness might be a biomechanical study. Goswami and Cordier [12] studied the knee-hip cyclogram (plot of knee versus hip angle) during steady walking on different slopes. The cyclogram changed in characteristic ways as the slope changed. It is possible that the tether force changes the biomechanics of walking also, because a subject has to lean forward more. Thus a comparison of the knee-hip cyclograms for tether force versus slope walking would be of interest.

The results on slope perception of 0.5 degrees while walking also point to a high sensitivity to slope while walking. While tether force can substitute for treadmill tilt, it would be desirable also to have a fast tilt mechanism.

200

Figure 4. The new Sarcos Treadport.

Consequently a second-generation Treadport has been designed, with one of its key features a fast tilt mechanism (Figure 4). The specifications on tilt are ±20 degrees in 1 second. In the first-generation Treadport, the rotation occurs at the back of the platform for upward tilt from horizontal, whereas for downward tilt from horizontal the rotation occurs in the middle of the platform. The unfortunate result for the upward tilt is that the user is lifted as well as tilted. In the new Treadport, the tilt axis is fixed in the middle of the platform. The platform has also been made larger (6x10 feet) to allow more freedom of motion. The mechanical tether has been strengthened; the horizontal pulling direction has been retained due to the results on user preference and discrimination. The treadmill belt has been made white to allow for floor projection.

Although the tilt mechanism is fast, the tether is faster yet. It is possible that tilt and tether force can be combined and blended to display sudden slope changes which then persist.

### Acknowledgements

This research was supported by ONR Grant N00014-97-1-0355, by NSERC NCE IRIS Project AMD-2, and by an NSF Graduate Research Fellowship. We thank Herb Pick and John Rieser for suggestions concerning the experiments.

## References

[1] Cremer, J., and Kearney, J. 1995. Improvisation and opportunism in scenario control for driving simulation. *First Workshop on Simulation and Interaction in Virtual Environments*. Iowa City: pp. 114-119.

[2] Rieser, J.J., Pick Jr., H.L., Ashmead, D.H., and Garing, A.E. 1995. The calibration of human locomotion and models of perceptual-motor organization. *J. Experimental Psychology: Human Perception and Performance*. 21: 480-497.

[3] Brogan, D.C., Metoyer, R.A., and Hodgins, J.K. 1997. Dynamically simulated characters in virtual environments. *SIGGRAPH'97 Visual Proc.*. pp. 216.

[4] Roston, G.P., and Peurach, T. 1997. A whole body kinesthetic display device for virtual reality applications. *Proc. IEEE Intl. Conf. Robotics and Automation*. pp. 3006-3011.

[5] Iwata, H., and Fujii, T. 1996. VIRTUAL PERAMBULATOR: A Novel Interface Device for Locomotion in Virtual Environment. *Proc. IEEE 1996 Virtual Reality Annual Intl. Symposium*. pp. 60-65.

[6] Brooks, Jr., F.P. 1992. Walkthrough Project: Final Technical Report to National Science Foundation Computer and Information Science and Engineering. Technical Report TR92-026. Univ. North Carolina-Chapel Hill, Dept. Computer Science.

[7] Christensen, R., Hollerbach, J.M., Xu, Y., and Meek, S. 1998. Inertial force feedback for a locomotion interface. *Proc. ASME Dynamic Systems and Control Division*. pp. 119-126.

[8] Darken, R.P., and Cockayne, W.R. 1997. The Omni-Directional Treadmill: a locomotion device for virtual worlds. *Proc. UIST*. pp. 213-221.

[9] Noma, H., and Miyasato, T. 1998. Design for locomotion interface in a large scale virtual environment. ATLAS: ATR Locomotion INterface for Active Self Motion. *Proc. ASME Dynamic Systems and Control Division*. pp. 111-118.

[10] Moghaddam, M., and Buehler, M. 1993 (July 26-30). Control of virtual motion systems. *Proc. IEEE/RSJ Intl. Conf. on Intelligent Robots and Systems*. Yokohama: pp. 63-67.

[11] Iwata, H. 1999. Walking about virtual environments on an infinite floor. *Proc. IEEE Virtual Reality '99.*.

[12] Goswami, A., and Cordier, E. 1997. Moment-based parmaeterization of evolving cyclograms on gradually changing slopes. *Intl. Symp. of Computer Methods in Biomech. and Biomed. Eng.*. Barcelona:.

# Chapter 6

# Locomotion

Really interesting robots move themselves. And really interesting moving robots use difficult ways to move around like walking, or driving on one, five or six wheels instead of the usual three or four. That's why these papers challenge us.

Kenneth Waldron and Christopher Hubert lead off with their careful kinematic analysis of a Mars rover design in which they try to allocate contact forces to all six wheels. Thanks to Ken's inspiring work, this seems easy for six legs. They show that it is much more difficult for six wheels because there are fewer degrees of freedom and more uncertainty due to slip, both along and across the wheel direction. The problem, they report, is "still just beginning to be explored".

From six wheels to one. Yangsheng Xu with his colleagues K W Au and W K Yu have entranced audiences at many conferences with their gyro-stabilised wheel. The wheel is the robot! Made from a bicycle tyre with two plastic domes, the wheel climbs steps, can pick itself up after a fall and even drive (slowly) over water. Xu discusses models for the wheel dynamics and ways of implementing teleoperator control.

Samuel Setiawan and his colleagues are developing the perfect partner for any man learning to dance the Tango, whose wife will not willingly become a perfect follower! They have implemented a special form of walking gait on a humanoid robot (of which we shall have more to say later in chapter 11) which makes the robot follow a human partner, as in a dance. Like the previous paper, this work can only be really appreciated if one has been fortunate enough to see Setiawan's videotape. However, the team will need to understand that the male Tango partner directs his female partner only partly by hand pressure: the main cues come from pressure on her back between the shoulder blades and the distance between their chests! Only a few extra sensors. Also, fortunately, Tango permits long pauses (useful for computing new gaits) and slow movements. But the seductive female kicks will no doubt keep the team occupied for several years yet.

The last paper is definitely not about dancing. It is aimed at realising cheap and effective prosthetic legs for amputees or braces for people with partial paralysis. Most robots have actuators at every joint but recently, as in this paper, researchers are looking at passive joints as a way of reducing cost and complexity. Bernard Espiau, Isabelle Guigues and Roger Passard-Gibollet describe the dynamics of a leg with a passive knee joint and confirm their analysis with some elegant experiments. Hopefully we will see humans wearing these legs before long.

JPT

# Control of Contact Forces in Wheeled and Legged Off-Road Vehicles

Kenneth J. Waldron          Christopher J. Hubert
The Ohio State University
Columbus, OH, USA
waldron.3@osu.edu          hubert.11@osu.edu

**Abstract:** The use of variable configuration vehicles creates the possibility of direct control of the contact forces between the foot or wheel of a vehicle and the ground. That control is an important element of the coordination of practical walking machines. Control of the contact forces at each wheel is also attractive for wheeled vehicles. For example, the power consumed in driving such a vehicle is quite sensitive to wheel-ground conditions. In conditions of strict power budgets, as in planetary roving vehicles, it becomes attractive to attempt to minimize power consumption by optimizing the wheel-ground contact forces. Unfortunately, control of contact forces in a wheeled vehicle is very complex because of the non-holonomic nature of the system.

In this paper we will review the allocation of contact forces in legged systems and examine the attempts that have been made to apply similar principles to the guidance and propulsion of wheeled systems on uneven terrain. The Adaptive Suspension Vehicle (ASV) and Wheeled Actively Articulated Vehicle (WAAV) from the authors' laboratory will be used as examples, and as sources of experimental validation of theoretical and numerical results. The former is a six-legged walking machine with each leg having three actively controlled degrees of freedom relative to the vehicle body. The latter is a six-wheeled variable configuration vehicle with three-axis, actively controlled, spherical articulations between successive axles.

## 1. Introduction

Actively configurable vehicles offer the possibility of continuous optimization of the load conditions at the interfaces between a vehicle and the ground. This can yield benefits in terms of mobility, and in minimizing energy consumption. In a robotic vehicle operating on unstructured terrain, the power consumed at the interface with the ground is sensitive to slip. Management of the contact conditions to minimize slip is a productive approach to minimization of power consumption, a critical issue in applications such as planetary rovers.

Determination of the ground interface load conditions to minimize the tendency of each foot to slip was first demonstrated by the Adaptive Suspension Vehicle [1]. Active control of the interface load conditions was also implemented on the Ambler [2]. In the absence of means of determining local soil properties, the best that can be done is to minimize the ratio of the contact force component tangent to the ground at the contact to the normal component. This becomes a non-linear programming problem that can be converted to a linear problem by a simple, geometric substitution with very little reduction of the solution space [3]. The linear programming solution has not been used in real-time control both because of the iterative nature of the algorithm, and because the solution produces discontinuous force commands that "chatter" and are unattractive for use as control inputs. Nevertheless, the linear programming solution is useful as a baseline against which to compare other techniques [4].

A slightly different approach that focuses on the interactions between pairs of legs and requires that they not work against each other yields the criterion that the interaction forces in the contact force system should be zero. This leads to an elegant, closed-form solution that requires minimal computation in each coordination cycle [4]. This solution also turns out to be the minimum norm solution, equivalent to the result produced by the much more computationally intensive Moore-Penrose pseudoinverse. This solution is sub-optimal, with the deviation from the linear programming solution increasing with the severity of the terrain. Conceptually, the absence of any interaction forces prevents the vehicle from "grasping" the terrain to improve traction. This solution actually has the character of a canonical solution with the interaction force field forming a null set with the equilibrating solution.

It appears that there is no correspondingly elegant formulation and simple solution to the problem if the contact elements are wheels, rather than legs. The primary reason is the non-holonomic character of the wheel-ground interaction. However, active control of the wheel-ground interface is, perhaps, even more important in this case because of the slip inherent in the mechanics of the wheel ground interaction. Once again, power consumption is very sensitive to the interface load conditions, and is crucial in remote applications, such as planetary rovers. For this reason, we have extensively studied the problem of continuous control of the interface conditions for the configuration of our *Wheeled Actively Articulated Vehicle* (WAAV) [5] [6].

These studies have employed several different approaches to the problem of determining the contact forces to be commanded at the interfaces. The first was an attempt to extend the known solution for legged systems to the wheeled configuration [7]. A second approach employed computation of the instantaneous power consumption of the vehicle and employed system optimization methods to determine the commanded contact forces [8]. More recently we have been attempting to develop a better understanding of the kinematic geometry of the system with the objective of developing a more sophisticated, and efficient solution.

## 2. Legged Locomotion

In this paper we do not consider "dynamically stable" systems. Thus, the legged locomotion systems discussed are those that are loosely referred to as statically stable.

The contact between a foot and the ground is much simpler to model than a wheel contact because, to a good approximation, the contact force system can be modeled as being rotationally symmetric about the contact normal. This, in turn, leads to a much simpler system model for the overall vehicle. Also, in most circumstances, the foot-ground interface can be adequately modeled as being static.

Use of position/rate control of the legs, while natural to those who work with serial manipulators, does not adapt well to legged locomotion. The reason can be simply described by considering a four-legged table placed on an uneven floor. In general, only three legs will contact the floor, and there is no way to control the distribution of load among those legs. Even if the leg lengths are made variable so the fourth can be brought into contact, there is no way to control the load distribution and, unless compliance is introduced, contact of all four legs cannot be continuously maintained.

There are actually inconsistencies even in the case of three-legged contact when the system is in motion. When the machine is in motion, servo errors will result in changes in the distances between the foot contact points, which must, ideally, be constant. In practice, the result of all these effects is uncontrolled variations in the loads in the legs. That is not acceptable in a large, heavy machine.

It is much more effective, and conceptually simpler, to compute the force that each foot must generate, and command the leg servos to produce that force. The leg will continue to extend until it either generates the commanded force or approaches the boundary of its working envelope. Further, in this mode of operation the force allocation to each foot can be controlled in a physically meaningful way to optimize vehicle performance.

In the ASV [4] force allocation was achieved by means of a control structure consisting of two loops, an outer position-rate loop, with which the operator interacted, and an inner force loop. These two main loops were connected by a dynamic model of the machine. In brief, the way in which this worked was that the operator's commands with regard to the direction and rate of motion were continuously entered via a six-axis joystick system. The six axis rate commands were compared with six axis vehicle motion data measured by a simple, inertial motion measurement unit to generate a six axis rate error system. Division by the computational cycle time converted this into an acceleration system, which was multiplied into the vehicle inertia matrix to generate a commanded inertia force system. This system was combined with the vehicle weight to generate the force system to be generated. A force allocation algorithm adapted from that described in [4] was used to generate the commanded contact forces. These were then decomposed into actuator forces using a static/kinematic model of each leg using the familiar, serial chain actuator force computation. Dynamic models of the legs were researched, but were found to be unnecessary. Drift was corrected in the inertial system and a vertical reference was calculated by "weighing" the machine using the force sensor data from the leg actuators at deliberately low bandwidth.

The demonstrated performance of the machine was very tolerant of foot sinkage, and slip. Slippage of feet off rocks, curbs and obstacles occurred on numerous occasions and was automatically compensated for by the force allocation system. The machine was operated, on several occasions, in soft, slippery mud. Indeed, in one trial, the feet were sinking to a depth of at least 40 cm in soft mud. The machine was able to operate normally in these conditions without adjustment, demonstrating the robustness of this mode of control.

A further point of reference is that, in more congenial conditions, the feet caused minimal deformation of the soil. This demonstrated the effectiveness of contact force allocation based on elimination of interaction forces in eliminating foot scuffing and minimizing any tendency toward foot slip. Minimization of foot slip and sinkage is also effective in minimizing the soil work done by the system, which is a very important component of the power consumed by the vehicle. Figure 1 shows the ASV walking through a grassy field. Notice the minimal amount by which the environment is disturbed.

Figure 1. The ASV Operating with Minimal Terrain Disturbance

## 3. Wheeled Locomotion

### 3.1. Design and Configuration Issues

"Typical" wheeled vehicle designs differ from "typical" legged vehicle design. Most walking machines, such as the ASV [1], the AMBLER [2], and the Hexapod [9], have *a single* body module, from which several multi-DOF legs emanate, resulting in a "star" topological structure. Typically, the legs have three degrees of freedom; thus each leg can generate a force in any direction and the net combination of all legs have *complete* controllability over the force distribution system. Wheeled rover vehicles, by contrast, generally tend to have *rigid* axles. As shall be discussed shortly, the presence of these rigid axles limits both the force and motion control capabilities of a wheeled vehicle. Further, these vehicles frequently have several different body "modules," with a pair (or several pairs) of coaxial wheels attached to each module, resulting in a "tree" topology— a much more complex parallel-serial hybrid than the "star" topology.

Use of wheels affords designers of rover, or off-road, vehicles a choice that is not typically used in legged off-road locomotion: passive suspension. Examples of passive wheeled rover vehicles include the "Rocky" series [10], the JPL's CARD concept [11], and the elastic frame articulated vehicles (SLRV), discussed by Bekker [12]. In contrast are the *actively articulated* vehicles, such as the WAAV [5], the FMC Attached Scout [13], and Martin Marietta's actively controlled articulated wheeled vehicle [14]. Note that these vehicles exhibit not only an *active* suspension, but also a significant ability to adapt to an unstructured terrain [7].

Within the context of a discussion of force control, passive vehicles might perhaps seem somewhat out of place, as the passively actuated joints minimize any active control that can be exerted over the contact forces. Here, they have been included both for completeness and, as an example of why force control is, indeed, an important issue. The passive vehicles are unable to control and distribute the various tangential and normal contact forces. This can result in a significant waste of power, as the optimal terrain contacts are not exploited. Wheels rolling on an unstructured terrain will inevitably slip and sink. Some effective means of redistributing contact forces can significantly reduce losses due to "plowing through" the soil or slipping wheels [8]. Further, lateral slip may also be prevented or minimized by increasing normal forces where needed to provide adequate frictional resistance. Additionally, these passive vehicles lack the ability to adapt their terrain to overcome obstacles and resume operation after a fall that the actively articulated vehicles possess.

Another significant design issue in a wheeled (or legged, for that matter) vehicle is the size of the actuators to be used. Obviously, in the case of actively articulated vehicles, one has several more actuators to size and thus the choice becomes more significant. Power and weight budgets are frequently at a premium, as these classes of vehicles are generally intended for exploration of remote areas, to which minimal hardware can be carried and little regenerating power can be supplied. This dictates that smaller actuators should be used, with standard operation using a much larger percentage of full actuator capacity than typically seen in conventional road vehicles. In order for these smaller actuators to provide the large torques and forces that are needed to negotiate an unstructured terrain, significant speed reduction is most likely required. Unlike legged vehicles, the rolling of the compliant tire minimizes "shock" loading on the wheel and joint actuators, and allows gearing to be used. However, losses in the gearing or other transmission can deplete large amounts of the available power, and seriously impede efficient operation.

### 3.2. Wheel Slip

Wheeled vehicle systems are much less amenable to static modeling since slip between the wheel and the ground is necessarily present when the system is in motion, and may be substantial. Kinematically, the typical structure of several interconnected modules, each comprised of two (or sometimes more) wheels mounted upon rigid axles, demands some slippage. On *rough* terrain a pair of *non-slipping* wheels mounted on a common axle results in a 2 DOF mechanism. Excluding "exceptional" cases (such as planar terrains), parallel chains of these modules, interconnected by joints of 3 or less DOF quickly erodes any motion capability without slip. The Kutzbach constraint equation (1) leads to a very simple demonstration of the necessity of slip in an actively articulated vehicle.

$$M = 6\left(N_m - n_j - 1\right) + \sum_{i=1}^{n_j} F_i = 6\left[3n + 1 - (5n - 1) - 1\right] + 11n - 3 \quad (1)$$

Thus, (1) results in an $n$ module vehicle having a Kutzbach mobility of $3 - n$, *if pure rolling occurs at the wheel terrain contact point*. Of course, a number of mechanisms exhibit singularities with respect to the Kutzbach constraint equation, possessing an actual mobility that is greater than their Kutzbach mobility. For wheeled vehicles, however, the Kutzbach constraint is generally valid, as singular cases are essentially limited to the situation where the axle and the line connecting the contact points are parallel (as on planar terrain) [15] [16].

Moreover, when a wheel operates on an unstructured surface it will sink into the soil. As a result, the soil must be deformed before any tangential force can be measured [8]. Thus, wheel *slip must* occur if *any* tractive force is to be generated. Figure 2, which shows a plot of the normalized tangential force versus wheel slip using data collected on lunar soil simulants [17], confirms the necessity of slip for

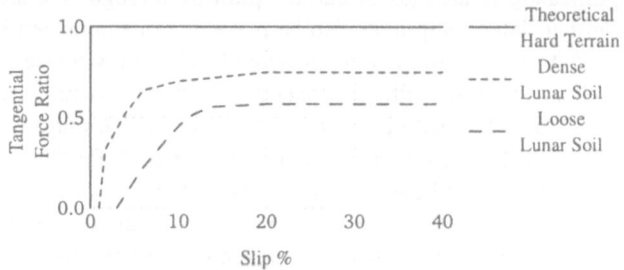

Figure 2. Normalized Tangential Force versus Slip

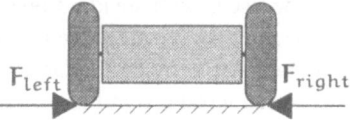

Figure 3. Illustration Infeasible Interaction Force

motion. Adequate models of the wheel-ground interface are, consequently, relatively complex, and this carries through to more complicated system models. However, even while wheel sink and slip makes modeling and controlling forces more difficult, it also makes it more important. Variations in terrain geometry and property are apt to lead to radically different contact conditions even across the limited span of the vehicle. Properly exploiting the "optimal" contact can significantly reduce power consumption [8]. Unfortunately, this also tends to suggest that the sub-optimal solution based on eliminating interaction forces is not reasonable for wheeled vehicles.

### 3.3. Incomplete Controllability

The wheel-ground interface has the directionality of the wheel motion that breaks the symmetry of the foot contact. In and of itself, this can limit the motion capabilities and complicate motion planning (a rolling wheel is perhaps the canonical example of a non-holonomic constraint). This asymmetry also limits the force control capabilities of wheeled vehicles. Further, as already discussed, wheels are typically mounted in pairs along a rigid axle.[1] The inherent difficulty in this design is the inability of the tires to "squeeze" the terrain along the wheel axle. Such "squeezing" pairs of forces are *interaction forces* [4]. Figure 3 illustrates an example of an *infeasible* wheel force pair. Mathematically, the inability of wheels to "squeeze" the terrain along their axle (with direction $a$) can be expressed as follows.

$$(F_{left} - F_{right}) \cdot a = 0 \tag{2}$$

The fundamental problem of allocating contact forces thus becomes more complex as one must now separate out the infeasible interaction force and assign forces only along feasible directions.

Force control of a robotic vehicle essentially amounts to optimizing some function (typically either net power consumed, power lost due to slip, or, in extreme ma-

---

[1] Even if some extra DOF are added at the axle wheel axle interface, motion along the axle is almost always prohibited.

neuvers, the contact force ratio), subject to the force balance.

$$GF = q \tag{3}$$

In (3) F is a $3n$ vector, denoting each of the three components of the contact forces at the $n$ contact points, G is the $6 \times 3n$ matrix denoting the screw axes along which the various contact forces act, and q represents the sum of all forces (including inertial forces) acting upon the body. Noting the lack of controllability over *lateral* forces, leads to partitioning G and F into "wheel plane" ($w$) and "lateral" ($l$) components [7] [8].

$$\left[ \; G^w | G^l \; \right] \left[ \frac{F^w}{F^l} \right] = q \tag{4}$$

Given that the vehicle can not squeeze the terrain along the axles, the lateral contact forces must correspond with the equilibrium solution, $F^l = F^{*\,l}$. The equilibrating force, $F^*$, can be found either using a pseudo-inverse or using the concept of an equilibrating force field [4]. Choosing $F^w$ such that it consists of the the wheel plane components of the equilibrating force, $F^{*\,w}$, plus components of the *null* space of $G^w$, $f_i^{w\,N}$, ensures that the wheel plane partition of the force balance is satisfied.

$$G^w \left[ F^{*w} + \sum_{i=1}^{n} \alpha_i f_i^{w\,N} \right] = q - G^l F^{*\,l} \tag{5}$$

Optimization of the contact forces (for whatever purposes) amounts to assigning the various scalar coefficients, $\alpha_i$, so as to extremize some objective function.

Conceptually, the wheel contact force allocation may not appear much more difficult than that for a legged vehicle. While it may result in a less optimal solution, partitioning is straightforward and does not appear to introduce any numerical complexities. The "catch" is that finding the null space of the wheel plane partition becomes much more complex. To begin with, while the general G matrix for the $n$ point contact will have a $3n - 6$ dimensional null space,[2] the partitioned $G^w$ matrix could possibly have a $2n - 5$ dimensional null space, in addition to the $2n - 6$ dimensional null space it typically possesses. More importantly, a very simple and elegant basis for the null space of the non-partitioned contact matrix, G, can be determined purely through geometric reasoning. By contrast, the basis vectors for the null space of $G^w$ are usually more complicated, and cannot all be found analytically [7].

### 3.4. The WAAV

To aid in the investigation of the functionality and control of actively articulated wheeled vehicles, the *WAAV*, Wheeled Actively Articulated Vehicle has been constructed at the authors' labs. It is a typical example of an actively articulated vehicle, consisting of three modules, each with two coaxial wheels, that are interconnected by three concurrent mutually orthogonal revolutes, which are effectively modeled as a spherical joint. Past work in the authors' labs has included a study of the force control characteristics, in an effort to extend many of the principles successfully applied to

---

[2] Technically, the G matrix does loose full rank if contact points are all co-linear. However, this is not a physically realistic situation.

Figure 4. The Wheeled Actively Articulated Vehicle

walking machines such as the ASV [7]. Additionally, acknowledging the importance of wheel slip and sinkage, a detailed wheel-terrain contact model has been developed and incorporated into an optimization scheme based on the minimization of power consumed at *all* actuator joints. Simulation of the optimization scheme versus a passively articulated vehicle has indeed shown that significant power reductions can be achieved. Similarly, this more advanced optimization scheme was compared with the minimum norm solution, and significant reductions in required power were again observed [8]. This further suggests that the stratgey that worked well for the ASV is most likely not an appropriate choice for vehicles like the WAAV. Unfortunately, the numerical complexity involved in first finding, then using the rather cumbersome null space vectors, as well as in the detail of the terrain model needed for the optimizer, results in an extremely computationally intensive scheme.

Recently, the basic rate kinematics of the WAAV have been reviewed in an attempt to develop a simpler and more elegant control algorithm. One possible approach involves what can be viewed as an inversion of the ASV control strategy. As discussed, with walking machines, the mechanics and dynamics of the system suggest force control as the preferred strategy. Desired rates are incorporated into this strategy by numerically differentiating and then processing through the inverse dynamics. For wheeled vehicles, force and position kinematics are much more complicated, owing to the non-holonomic nature of the system. Thus, rate control might seem to suggest itself. Force allocation must now be handled by somehow "converting" forces into the rate domain (as the ASV control scheme "converted" rates into the force domain). One possibility would be to literally invert the ASV scheme, by finding the desired accelerations through the forward dynamics and then integrating to get rates. However, this does not seem to free one from the many problems inherent in force allocation (or from the more involved dynamics associated with "tree" topologies). Another approach might be to let the difference between actual and desired forces be modeled as spring forces. From the spring forces, one could figure the deflection of the spring, which would represent additional desired displacements. These displacements could be differentiated to "fit" into a rate scheme. Conceptually, this could be seen to allow the vehicle to adjust articulation velocities to "press" wheels into the terrain (for greater normal forces) or perhaps to retard the rotating of a slipping wheel.

## 4. Conclusion

Walking machines and variable configuration wheeled robots both possess the ability to influence their terrain contact conditions. However, the nature of the force control problem differs radically from one type of machine to the other. Consequently, it appears the approaches to the force control problem must also differ. Human history and experience with wheeled machines is *much* longer and broader than with legged machines. Curiously though, within the limited realm of variably configured vehicles much more experience exists with legged machines. Wheeled locomotion is so strongly coupled to prepared surfaces, that the dominance of the wheel in society has not been extended to variable configuration vehicles. The exact nature of the issues associated with control of variably configured wheeled vehicles on rough terrain is still just beginning to be explored.

## References

[1] R. D. Pugh, E. A. Ribble, V. J. Vohnout, T. E. Bihari, M. R. Patterson, and K. J. Waldron. Technical description of the adapative suspension vehicle. *Int. J. of Robot. Res.*, 9(2):24–42, 1990.

[2] J. E. Bares, M. Hebert, T. Kanade, et al. Ambler: An autonomous rover for planetary exploration. *IEEE Computer*, pages 18–26, June 1989.

[3] C. A. Klein and S. Kittivatcharapong. Optimal force distribution for the legs of a walking machine with friction cone constraints. *IEEE Trans. on Robot. and Autom.*, 6(1):73–85, 1990.

[4] V. Kumar and K. J. Waldron. Force distribution in closed kinematic chains. *IEEE J. of Robot. and Autom.*, 4(6):657–664, December 1988.

[5] S. V. Sreenivasan, P. K. Dutta, and K. J. Waldron. The wheeled actively articulated vehicle (waav): an advanced off-road mobility concept. In *Proc. of 4th Int. Workshop on Advances in Robot Kinematics*, Ljubljana, Slovenia, 1994.

[6] J. C. Yu and K. J. Waldron. Design of wheeled actively articulated vehicle. In *Proc. of the Second National Applied Mechanisms Robot. Conf.*, Cincinnati, OH, 1991.

[7] S. V. Sreenivasan. *Actively Coordinated Wheeled Vehicle Systems*. Phd dissertation, The Ohio State University, 1994.

[8] S. C. Venkataraman. *Active Coordination of Wheeled Vehicle Systems for Improved Dynamic Performance*. Phd dissertation, The Ohio State University, 1997.

[9] R. B. McGhee and G. I. Iswandhi. Adaptive locomotion of a multi-legged robot over rough terrain. *IEEE Trans. on Sys., Man, and Cybernet.*, 9(4):176–182, 1979.

[10] J. E. Chottiner. Simulation of a six wheeled martian rover called the rocker bogie. Master's thesis, The Ohio State University, 1992.

[11] G. Klein, K. J. Waldron, and B. Cooper. Current status of mission/system desing for a mars rover. *Unmanned Systems*, 5(1):28–39, Summer 1986.

[12] M. G. Bekker. *Introduction to Terrain Vehicle Systems*. The University of Michigan Press, Ann Arbor, 1968.

[13] L. S. McTamaney et al. Final report for mars rover/sample return (mrsr)– studies of rover

mobility and surface rendezvous. Technical report, Federal Mining Corp., Santa Clara, CA, December 1989.

[14] A. J. Spiessbach et al. Final report for mars rover/sample return (mrsr)– rover mobility and surface rendezvous studies. Technical Report 958073, Martin Marietta Corp., Denver, CO, October 1988.

[15] C. J. Hubert. Coordination of variably configured vehicles. Master's thesis, The Ohio State University, 1998.

[16] S. V. Sreenivasan and P. Nanua. Kinematic geometry of wheeled vehicle systems. In *Proc. of the ASME Design Technical Conference*, 1996.

[17] K. J. Melzer and G. D. Swanson. Performance evaluation of a second generation elastic loop mobility system. Technical Report M-74-7, U.S. Army Engineer Waterways Experiment Station, Mobility and Environmental Laboratory, Vicksburg, June 1974.

# Control of a Gyroscopically Stabilized Robot

Yangsheng Xu, K.W. Au, and W.K. Yu
Dept. of Mechanical and Automation Engineering
The Chinese University of Hong Kong
ysxu@mae.cuhk.edu.hk

**Abstract:** We have been developing a single wheel, gyroscopically stabilized robot, initially at Carnegie Mellon University. It is a basically single wheel which connected to a spinning flywheel through a two-link manipulator at the wheel bearing. The nature of the system is nonholonomic, nonlinear and underactuated. In this paper, we first develop a dynamic model and decouple the model with respect to the control inputs. We then study the effect of the flywheel dynamics on stabilizing the single wheel robot through simulation and experimental study. Based on understanding of the robot dynamics, we design a linear state feedback controller for the lean angle of the robot. Finally, we discuss the possibility of transferring operator's manual control directly via learning the mapping between data acquired from the manual control interface and the sensor data from the robot on-board sensors. The experimental results supported the method. The work is significant for understanding the highly coupled dynamics system, and is valuable for developing the automatic control for such a dynamically stabilized robot.

## 1. Introduction

We have been developing a single-wheeled, gyroscopically stabilized robot called Gyrover. Gyrover is a sharp-edged wheel, with an actuation mechanism fitted inside the wheel. The actuation mechanism consists of three separate actuators: (1) a spin motor which spins a suspended flywheel at a high rate, imparting dynamic stability to the robot, (2) a title motor which controls the steering of Gyrover, and (3) a drive motor which produces forward and/or backward acceleration by driving the single wheel directly.

The behavior of Gyrover is based on the principle of gyroscopic precession as exhibited in the stability of a rolling wheel. Because of its angular momentum, a spinning wheel tends to precess at right angles to an applied torque. When a rolling wheel leans to one side, rather than just falling over, the gravitationally induced torque causes the wheel to precess so that it turns in the direction that it is leaning. Gyrover supplements this basic concept with the addition of an internal gyroscope – the spinning flywheel – nominally aligned with the wheel and spinning in the direction of forward motion. The flywheel's angular momentum produces lateral stability when the wheel is stopped or moving slowly.

This robot concept conveys significant advantages over multi-wheel, statically stable vehicles, including good dynamic stability and insensitivity to attitude disturbances; high maneuverability; low rolling resistance; ability to recover from falls; and amphibious capability. Presently envisioned applications include amphibious vehicles, surveillance robots, and lunar/planetary rovers. Because it can travel on both land and water, it may find amphibious use on beaches or swampy areas, for general transport, exploration, mine detecting, or recreation. Gyrover could use its slim profile to pass through doorways and narrow passages, and its ability to turn in place to maneuver in tight quarters. Another potential application is as a high-speed lunar vehicle, where the absence of aerodynamic disturbances and low gravity would permit efficient, high-speed mobility.

We have developed three prototypes of the robot and Figure 1 shows the third version built recently. Thus far it has been controlled only manually, using two joysticks to control the drive and tilt motors through a radio link. It has been our goal to develop an automatic control scheme for the robot so that the robot can be really used for many tasks. In this paper, we will first present a dynamic model of the system and simplify it by considering the center of flywheel coincident with the center of the robot. We decouple the tilting variable $\beta_a$ from dynamic equation and consider $\dot{\beta}_a$ as a new input of the system, such that the number of the generalized coordinates and the dimension of the inertial matrix are reduced. Then, we study the effect of the high spinning flywheel to the stability of the robot, through the simulations and experiments. Based on linearization, the motion control is decomposed into three separate parts: (1) controlling the rolling speed $\dot{\gamma}$, (2) controlling the tilting variable $\beta_a(t)$ to the desired trajectory, (3) designing a linear state feedback controller for the lean angle $\beta$, so as to the steering velocity. At last, we will discuss the implementation of transferring human strategy in controlling such a robot for automatic control of the system. The method provides an alternative to control such a dynamical system without modeling, and experimental results validated the approach.

## 2. Dynamics

In our previous work [4], we developed a dynamic model using the constrained generalized Lagrangian formulation, with five independent generalized coordinates $\alpha, \beta, \gamma, \beta_a, \theta$ and two nonholonomic velocity constraints. We assumed that the wheel is a rigid, homogeneous disk which rolls over a perfectly flat surface without slipping. We modeled the actuation mechanism suspended from the wheel bearing as a two-link manipulator, with a spinning disk attached at the end of the second link (Figure 2). For simplicity, $l_1$ and $l_2$ are assumed to be zero, so the mass center of flywheel will be coincident with the center of the robot. For steady motion of the robot, the pendulum motion of the internal mechanism is sufficiently small to be neglected, such that $\vartheta$ is set to zero. The spinning rate of the flywheel $\gamma_a$ is set to be constant.

Let $S_x := sin(x)$, $C_x := cos(x)$, $S_{\beta,\beta_a} := sin(\beta + \beta_a)$, $C_{\beta,\beta_a} := cos(\beta + \beta_a)$, and $S_{2\beta\beta_a} := sin[2(\beta + \beta_a)]$. The derivation in [4] showed the normal form of the dynamics model below.

$$M(q)\ddot{q} = F(q,\dot{q}) + Bu \tag{1}$$

$$\dot{X} = R(\dot{\gamma}C_\alpha + \dot{\alpha}C_\alpha C_\beta - \dot{\beta}S_\alpha S_\beta) \tag{2}$$

$$\dot{Y} = R(\dot{\gamma}S_\alpha + \dot{\alpha}C_\beta S_\alpha + \dot{\beta}C_\alpha S_\beta) \tag{3}$$

where $q = [\alpha, \beta, \gamma, \beta_a]^T$,

$$M = \begin{bmatrix} M_{11} & 0 & M_{13} & 0 \\ 0 & I_{xf} + I_{xw} + mR^2 & 0 & I_{xf} \\ M_{13} & 0 & 2I_{xw} + mR^2 & 0 \\ 0 & I_{xf} & 0 & I_{xf} \end{bmatrix},$$

$$F = [F_1, F_2, F_3, F_4]^T,$$

$$B = \begin{bmatrix} 0 & 0 & 1 & 0 \\ 0 & 0 & 0 & 1 \end{bmatrix}^T, u = \begin{bmatrix} u1 \\ u2 \end{bmatrix}$$

$$
\begin{aligned}
M_{11} &= I_{xf} + I_{xw} + I_{xw}C_\beta^2 + mR^2C_\beta^2 + I_{xf}C_{\beta,\beta_a}^2 \\
M_{13} &= 2I_{xw}C_\beta + mR^2C_\beta \\
F_1 &= (I_{xw} + mR^2)S_{2\beta}\dot{\alpha}\dot{\beta} + I_{xf}S_{2\beta\beta_a}\dot{\alpha}\dot{\beta} + I_{xf}S_{2\beta\beta_a}\dot{\alpha}\dot{\beta}_a \\
&\quad + 2I_{xw}S_\beta\dot{\beta}\dot{\gamma} + 2I_{xf}S_{\beta,\beta_a}\dot{\beta}\dot{\gamma}_a + 2I_{xf}S_{\beta,\beta_a}\dot{\beta}_a\dot{\gamma}_a \\
F_2 &= -gmRC_\beta - (I_{xw} + mR^2)C_\beta S_\beta\dot{\alpha}^2 - I_{xf}C_{\beta,\beta_a}S_{\beta,\beta_a}\dot{\alpha}^2 \\
&\quad - (2I_{xw} + mR^2)S_\beta\dot{\alpha}\dot{\gamma} - 2I_{xf}S_{\beta,\beta_a}\dot{\alpha}\dot{\gamma}_a \\
F_3 &= 2(I_{xw} + mR^2)S_\beta\dot{\alpha}\dot{\beta} \\
F_4 &= -I_{xf}C_{\beta,\beta_a}S_{\beta,\beta_a}\dot{\alpha}^2 - 2I_{xf}S_{\beta,\beta_a}\dot{\alpha}\dot{\gamma}_a
\end{aligned}
$$

where $(X, Y, Z)$ is the coordinates of center of mass of the robot with respect to the inertial frame as shown in Figure 2. $M(q) \in R^{4\times4}$ and $N(q,\dot{q}) \in R^{4\times1}$ are the inertial matrix and nonlinear terms, respectively. Eq.(1) and Eq.(2),(3) form the dynamics model and nonholonomic velocity constraints of the robot.

The model can be further reduced by decoupling the tilting variable $\beta_a$ from (1). Practically, $\beta_a$ is directly controlled by the tilt motor (position control), assuming that the tilt actuator has an adequate torque to track the desired $\beta_a(t)$ trajectory exactly.

When we consider $\dot{\beta}_a$ as a new input $u_{\beta_a}$, the dynamics model (1) becomes

$$\dot{\beta}_a = u_{\beta_a}$$

$$\tilde{M}(\tilde{q})\ddot{\tilde{q}} = \tilde{F}(\tilde{q}, \dot{\tilde{q}}) + \tilde{B}\tilde{u}. \tag{4}$$

where $\tilde{q} = [\alpha, \beta, \gamma]^T$,

$$\tilde{M} = \begin{bmatrix} M_{11} & 0 & M_{13} \\ 0 & I_{xf} + I_{xw} + mR^2 & 0 \\ M_{13} & 0 & 2I_{xw} + mR^2 \end{bmatrix}$$

$$\tilde{F} = \left[\tilde{F}_1, \tilde{F}_2, \tilde{F}_3\right]^T$$

$$\tilde{B} = \left[\begin{array}{ccc} 0 & 0 & 1 \\ \tilde{B}_{12} & 0 & 0 \end{array}\right]^T, \tilde{u} = \left[\begin{array}{c} u_1 \\ u_{\beta_a} \end{array}\right]$$

$$
\begin{aligned}
\tilde{F}_1 &= (I_{xw} + mR^2)S_{2\beta}\dot{\alpha}\dot{\beta} + I_{xf}S_{2\beta\beta_a}\dot{\alpha}\dot{\beta} \\
&\quad -2I_{xw}S_\beta\dot{\beta}\dot{\gamma} + 2I_{xf}S_{\beta,\beta_a}\dot{\beta}\dot{\gamma}_a \\
\tilde{F}_2 &= F_2 \\
\tilde{F}_3 &= F_3 \\
\tilde{B}_{12} &= I_{xf}S_{2\beta\beta_a}\dot{\alpha} + 2I_{xf}S_{\beta,\beta_a}\dot{\gamma}_a
\end{aligned}
$$

Let's now study the characteristics of the robot dynamics. Because its main component is a rolling wheel, it preserves typical characteristics of a rolling disk. For a rolling disk, it never fall when it is rolling, because of the gyroscopic torque resulting from the coupling motion between the roll and yaw motions, in balancing the gravitational torque. However, its rolling rate must be high enough to provide a sufficient gyroscopic torque for balancing the disk. For the robot as a whole, because of the high spinning flywheel, its gyroscopic torque is greater, thus it less depends on its rolling speed $\dot{\gamma}$.

Figures 3 and 4 show the simulation results of a rolling disk without the flywheel and the single wheel robot respectively, under the same initial conditions

$$
\left\{
\begin{array}{l}
\beta = 90^\circ, \beta_a = 0^\circ, \\
\dot{\beta} = \dot{\alpha} = \dot{\beta}_a = 0\ rad/s, \dot{\gamma} = 15\ rad/s, \\
\alpha = 0^\circ.
\end{array}
\right.
$$

Note that the lean angle $\beta$ of a rolling disk without flywheel decreases much rapidly than that of the single wheel robot as shown in Figures 3(b) and 4(b). It verifies the effect of flywheel for stabilization of the single wheel robot. From Figure 3(a), under the influence of friction in the yaw direction, the steering rate $\dot{\alpha}$ converges to a steady state, so does the leaning rate $\dot{\beta}$ which is not shown in figure.

Up to now, we consider only the case when the flywheel's orientation is fixed with respect to the robot, i.e., the main wheel. Now let's focus on the tilting effect of the flywheel to the robot. Based on the conservation of angular momentum, when the tilt angle of the flywheel $\beta_a$ changes, the whole robot rotates in the opposite direction in order to maintain a constant angular momentum. It implies that the lean angle of the robot can be controlled for steering. Simulation and experiment results are shown in Figures 5 and 6, respectively, under the same initial condition given above. Both Figures 5 and 6 show that if the tilt angle $\beta_a$ rotates in 73 deg/sec anti-clockwise at t = 2.4 second, the lean angle $\beta$ rotates in clockwise direction. In the experiment, the transient response of $\beta$ is more critical than the simulations. In 2.7 second, the tilt angle remains unchanged and then $\beta$ and steering rate $\dot{\alpha}$ converge to a steady position in both simulations and experiments.

## 3. Model-based Control

We derive a linear model of the robot dynamics based on the assumption that the spinning rate $\dot{\gamma}_a$ is sufficiently high so that the terms $\dot{\gamma}_a\dot{\beta}, \dot{\gamma}_a\dot{\alpha}$ are much greater than the terms $\dot{\beta}\dot{\gamma}, \dot{\beta}\dot{\alpha}, \dot{\alpha}^2$. We then linearize the system (4) around the vertical position where $\beta = 90^\circ + \delta_\beta, \dot{\gamma} = \Omega_o + \Omega, \beta_a = \delta_{\beta_a}$.

$$(I_{xf} + I_{xw})\ddot{\alpha} = 2(I_{xw}\Omega_o + I_{xf}\dot{\gamma}_a)\dot{\delta}_\beta - \mu_s$$
$$+2I_{xf}\dot{\gamma}_a u_{\beta_a} \tag{5}$$
$$(I_{xw} + mR^2)\ddot{\delta}_\beta = gmR\delta_\beta - (2I_{xw} + mR^2)\Omega_o\dot{\alpha}$$
$$-2Ixf\dot{\gamma}_a\dot{\alpha} \tag{6}$$
$$(2I_{xw} + mR^2)\dot{\Omega} = -\mu_g\Omega + u_1 \tag{7}$$

Because $\Omega$ is independent of the roll and yaw dynamics (5) and (6), we can decompose the pitch dynamics (7) and setup a close loop for controlling the angular velocity of the single wheel robot $\Omega$. The remaining yaw and roll dynamics form a state equation shown below.

$$\dot{x} = Ax + Gu, \tag{8}$$

where $x = \begin{bmatrix} \delta_\beta, \dot{\alpha}, \dot{\delta}_\beta \end{bmatrix}^T$, $A = \begin{bmatrix} 0 & 0 & 1 \\ 0 & a_{22} & a_{34} \\ a_{42} & a_{43} & 0 \end{bmatrix}$, $G = \begin{bmatrix} 0 \\ b_2 \\ 0 \end{bmatrix}$

where $a_{22}, ..., a_{43}$ and $b_2$ are derived from (5) and (6). Based on the controllability matrix, the system is controllable if $\dot{\gamma}_a \neq 0$ and $\dot{\gamma} \neq 0$. It is because if the single wheel robot is not rolling, it falls immediately. The system can be stabilized by using a linear state feedback

$$u_{\beta_a} = -k_1(\delta_\beta - \delta_{\beta ref}) - k_2\dot{\delta}_\beta - k_3(\dot{\alpha} - \dot{\alpha}_{ref}). \tag{9}$$

where $k_1, k_2, k_3$ are feedback gains, $\delta_{\beta ref}$ are the desired lean angle, and

$$\dot{\alpha}_{ref} = \frac{gmR\delta_{\beta ref}}{(2I_{xw} + mR^2)\Omega_o + 2I_{xf}\dot{\gamma}_a} \tag{10}$$

Let $\rho$ be the radius of curvature. If

$$X = -\rho S_\alpha, Y = \rho C_\alpha, \tag{11}$$

then, solving the constraints (2) and (3) yields

$$\Omega_o = -\dot{\alpha}_{ref}\left[\frac{\rho}{R} + C_{\beta ref}\right]. \tag{12}$$

By combining (10) and (12), we can calculate the desired $\Omega_o$ and $\dot{\alpha}_{ref}$ for a given $\delta_{\beta ref}$, in order to track a circle defined in (11).

In the first simulation, we stabilized the robot at the vertical position ($\beta = 90^o, \delta_{\beta ref} = 0^o$), such that the resulting trajectory is a straight line. Figure 7 shows that the lean angle $\beta$ exponentially converge to $90^o$, so that the steering rate will exponentially converge to zero. Therefore the trajectory of the center of the robot was slight curved at the beginning and then finally converged to a straight line. Then we set $\delta_{\beta ref} = 20^o$ and the simulation results are shown in Figure 8. Note that as $\delta_{\beta ref}$ was not zero, there was a steady steering rate $\dot{\alpha}_{ref}$ according to the Eq. (10). As a result, its trajectory was a circle.

## 4. Human-based Control

Because the nature of dynamically stabilized system is usually nonlinear, non-holonomic, and underactuated, it is potentially difficult to model the system accurately. Moreover, some of unmodeled parameters, such as friction and coriolis force, are relatively more important to such a system than a static or quasi-static system. The complexity in modeling such a system makes the design of model-based control scheme extremely difficult. Therefore, we have been looking into a possibility of generating control input by modeling the human operator's manual control.

We installed a CPU in the robot for dealing with on-board sensing and control. It communicates with a stationary PC via a pair of wireless modems. Based on this communication system, we can down-load the sensor data file from the on-board CPU and send supervising commands to the robot. We have also installed two pulse encoders to measure the spinning rate of flywheel and the single wheel. Furthermore, we have two gyros and accelerometer to detect the angular velocity of yaw, pitch, roll, and acceleration respectively. A 2-axis tilt sensor is developed and installed for direct measuring the leaning and pitch angle of the robot. These sensors and on-board computing is necessary for automatic control.

Taking advantage of our group's work on modeling human control strategy [5] [6], we used the flexible cascade neural network architecture with node-decoupled extended Kalman filtering (NDEFK) for the modeling the control decision pattern, i.e., relationship between the human operator's control command given through a radio control interface and the state variables of the robot measured from on-broad sensors in the robot. For the control commands, we will use data from two variables (drive torque and tilt torque). For the state variables, we must select a set of variables out of all available sensing parameters. We defined a sensitivity measure to compare the relative importance of the state variables with respect to the control decision that operators make, so as to select a set of minimum parameters in training.

In the experimental study, we picked up the tilt-up motion of the robot as a task to implement the method, because such a motion is typical, and also is observable for the operator and on-board sensors. The purpose here is to make the robot stand up, i.e., initially the lean angle $\beta$ is set to be approximately $18^o$, and at the final state we make it to reach $90^o$. First, data were recorded in both the radio-controlled interface and the on-board sensors simultaneously, and the sensor data of the selected set of state variables mentioned above were transmitted to computer through wireless communication. Figure 9 shows an example of the operator control commands (the right figure) and the state variables measured (the left and central figures). The recorded data was used for training human control strategy (HCS) model. The learned HCS model was then used as the control input for the robot. Figure 10 shows the corresponding variables of the learned neural network control inputs, and the robot motion under such control commands. Here, for simplicity, we listed three variables $(\beta, \beta_a,$ and $u_0)$ only where $\beta$ is the lean angle representing the target mission, $\beta_a$ is the lean angle of flywheel representing the major part of the tilt torque for stabilization, and $u_0$ is the control command. In every experiment, the target was always accomplished: $\beta \approx 90^o$ even if the initial angles we set and the execution time varied. It demonstrated that the learned control input is capable of controlling the robot with the similar pattern of the control command generated from the operator.

## 5. Conclusion

In this paper, we presented and simplified the dynamic model of the single wheel, gyroscopically stabilized robot by decoupling the tilting variable from the robot dynamics. The model was verified through the simulations and experiments. The lean angle of the robot can be directly controlled by tilting the flywheel for steering. We designed a linear state feedback to control the robot to the desired lean angle, so as to control the steering velocity. Finally, we discussed the method of transferring human strategy in controlling such a robot through learning based on the data acquired from the manual control interface and the robot on-board sensors. The work is significant in understanding the dynamics of the highly coupled and nonholonomic system, and is valuable in developing automatic control for such a dynamically stable robot.

## References

[1] A. V. Beznos, et al, "Control of autonomous motion of two-wheel bicycle with gyroscopic stabilization" *Proc. IEEE Int. Conf. on Robotic and Automation,* Vol.3, pp.2670-2675, 1998.

[2] N. H. Getz, "Control of balance for a nonlinear nonholonomic non-minimum phase model of a bicycle" *Proc. America Control Conference,* Baltimore, pp.148-151, 1994.

[3] H. B. Brown and Y. Xu, "A single wheel gyroscopically stabilized robot." *Proc. IEEE Int. Conf. on Robotic and Automation,* Vol.4, pp.3658-63, 1996.

[4] Y. Xu, K. W. Au, G. C. Nandy and H. B. Ben, "Analysis of actuation and the dynamic balancing for a single wheel robot" *Proc. IEEE/RSJ Int. Conf. on Intelligent Robots and Systems,* Vol.4, pp.3658-63, 1998.

[5] M.C. Nechyba and Y. Xu, "Cascade neural network with node-decoupled extended Kalman filtering", *Proc IEEE Int. Symp. on Computational Intelligence in Robotics and Automation,* Vol.1, pp.214-9, 1997.

[6] M.C. Nechyba and Y. Xu, "Human control strategy: abstraction, verification and replication", *IEEE Control Systems Magazine,* Vol.17, no.5, pp.48-61, 1997.

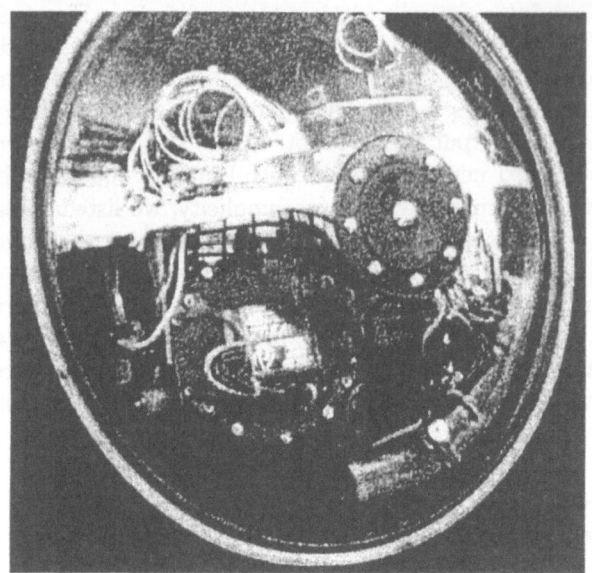

Figure 1. Photograph of Gyrover III.

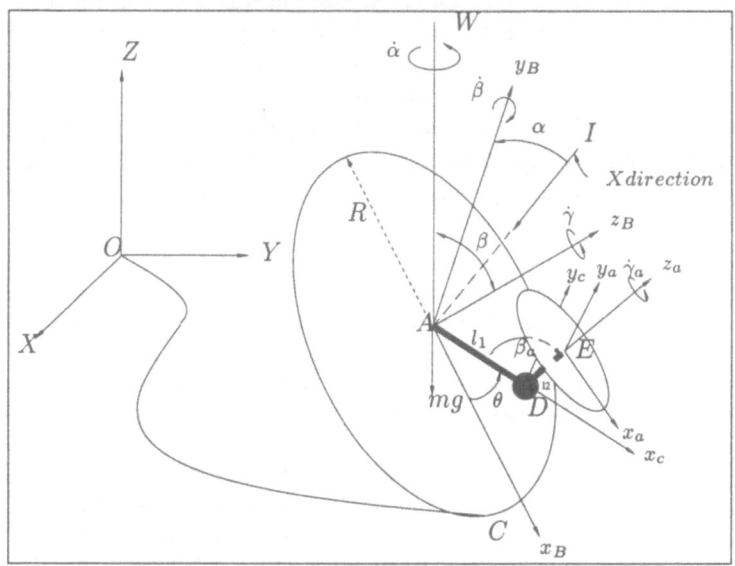

Figure 2. Coordinates assignment and parametes defined in Gyrover

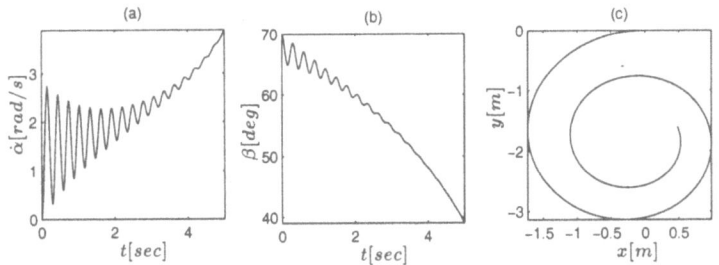

Figure 3. The simulation results of a rolling disk without flywheel.

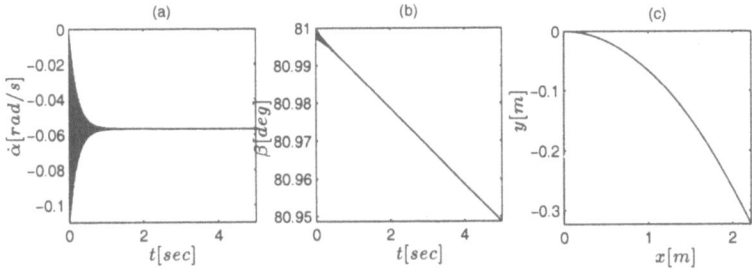

Figure 4. The simulation results of the single wheel robot.

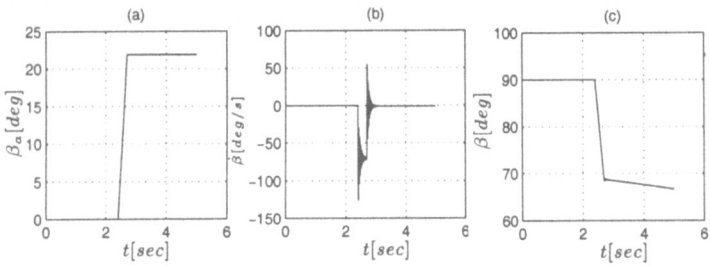

Figure 5. The simulation results of tilting the flywheel of the robot with $\dot{\beta}_a = 73 \ deg/s$

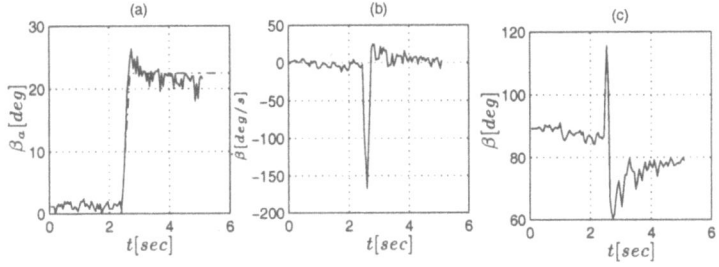

Figure 6. The experiment results of tilting the flywheel of the robot with $\dot{\beta}_a = 73 \ deg/s$

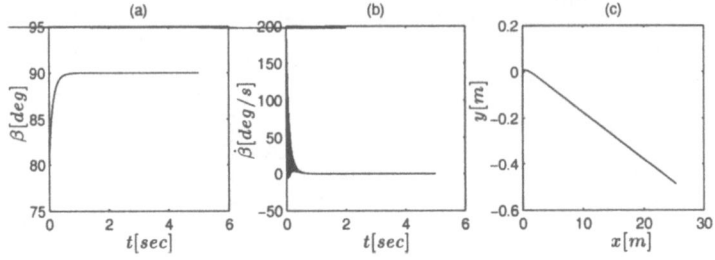

Figure 7. The simulation results of the single wheel robot stabilized to vertical position.

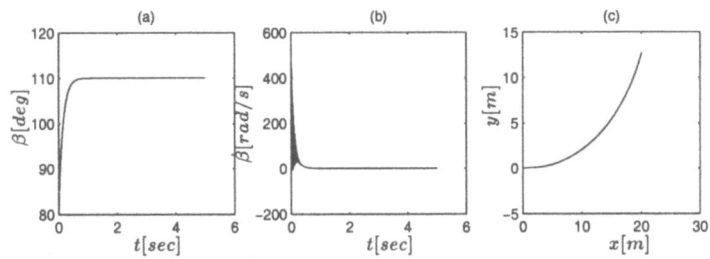

Figure 8. The simulation results of the single wheel robot moving in a circle.

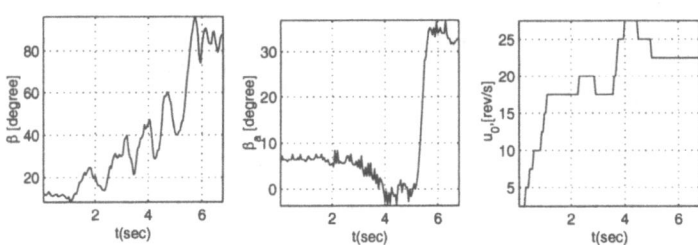

Figure 9. Human input-output skill data (a)$\beta$, (b)$\beta_a$,(c)$u_0$

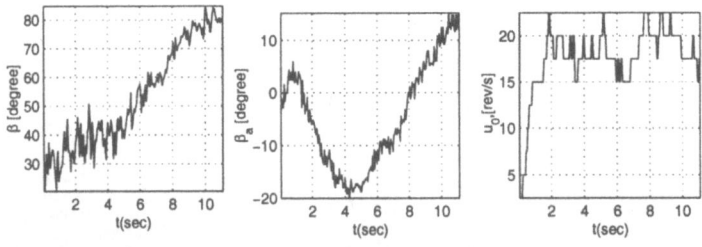

Figure 10. Experimental result using learned controller (a)$\beta$, (b)$\beta_a$,(c)$u_0$

# Localization and map building for a mobile robot

Patrick Rives
INRIA-centre de Sophia Antipolis
Valbonne, France
Patrick.Rives@sophia.inria.fr

José-Luís Sequeira , Pedro Lourtie
GCAR - Instituto Superior Técnico
Lisboa, Portugal
{jluis, lourtie}@gcar.ist.utl.pt

**Abstract**: In many applications of mobile robotics, it is significant to be able to locate the robot with respect to its environment. We present an approach based on the use of telemetric and odometry data for the mapping and the localization of a mobile robot moving in a structured but unknown environment. The algorithm of fusion between odometry data and US data with the aim of localization implement an Extended Kalman Filter using the odometry data in the equation of evolution and the measurement provided by the US sensors in the equation of observation. A second filter, permits to update, after each localization, an incremental representation of the scene defined in the initial reference frame, in a integrated strategy of perception.

## 1. Introduction

In many applications of mobile robotics, it is significant to be able to locate the robot with respect to its environment. This localization can be carried out by integrating the odometry data but it is not very reliable because of the slips of the wheels and the geometrical defects of the robot. It is then interesting to be able to compensate this drift while being readjusted with respect to the environment.

Successful autonomous navigation by a mobile robot depends on its ability to accurately determine its position within an environment mapped *a priori*. For real industrial environments seldom static, navigation systems have been developed which update a stored description of their environment or even model the environment without any *a priori* information while moving in it. Building the robot's map however requires that the robot's position be known accurately so that map consistency is ensured. Because of the present dilemma, many of the original work has addressed exclusively either on localization [1-3] or map building [4, 5], reducing the effectiveness of such systems (or claiming the knowledge of either a priori map or the accurate robot position).

To overtake this restriction, methods have been developed which simultaneously produce a map of the environment and achieve localization. Smith at al. [6] used an Extended Kalman Filter to estimate the system state and assume full knowledge of all the covariance and cross-covariance matrices as well as perfect data-to-target association. However the process can diverge quickly due to the linearisation errors introduced by the Extended Kalman Filter and the deficient data-to-target matching in real systems. Moutarlier and Chatila [7], which approach also relies on computing the cross variances between the robot and the targets, and between the targets themselves, proposed a three stage process, where the system state update is split into two stages. Here the data associated troubleshooting is directed by the implementation of a nearest neighbor validation technique, which produces unavoidable 'false alarms'. Furthermore, the resolution of the correlation between each feature at every update imposes expensive computation, preventing real time operation. Based on information obtained from Polaroid sonar sensors and a Bayesian Multiple Hypothesis algorithm to achieve localization and environmental modeling, Leonard and Cox [8] proposed a technique where the cross variances are eliminated. Nevertheless a good operation speed and small computational demand has not yet been achieved.

Borthwick and Durrant-Whyte [9] developed a multitrack, multi-target Extended Kalman Filter navigation system, while simultaneously constructing an environment map, updated with observations from the infra-red range scanner. Real time localization is achieved by one hand due to data quality of infrared scanning and increased scan rate and by another by reducing computational demand.

Our approach is based on the use of telemetric and odometry data for the mapping and the localization of a mobile robot moving in a structured but unknown environment. We study in this research orientation, techniques allowing at the same time the localization and the incremental construction of the map of the environment. The difficulty of the problem lies on the fact that the localization at current time k must be carried out starting from elements of environment discovered and rebuilt in the N previous perceptions and which remain visible at instant k. Of course, the precision of the localization is larger when the stage of rebuilding is reliable. In a dual way, the new elements of the environment discovered at the instant k are readjusted in the initial reference frame with k=0 by using the localization estimated at the instant k. A simple analysis of the method shows that it is all the more precise since the elements of the environment (natural landmarks) remain in the field of perception a long time. A trade-off is thus to be made between the function of discovering (i.e. discover the maximum of unknown elements of environment) and the function localization (i.e. keep the maximum of known landmarks during the motion). It will be the role of the strategy of perception.

In the case of relatively structured indoor scenes, the ultrasonic sensors (US) provide, at a low cost, measurements of distances being able to be used for the localization. On the other hand, these measurements are often of bad quality and contain many artifacts which must be filtered. Moreover, the geometrical model of the environment is often badly known. Our work is aimed at various aspects. First of all the robustness of the measurements provided by the ultrasonic sensors, as described in the first part of section 2.

A second aspect relates to the definition of an algorithm of fusion between odometry data and US data with the aim of localization. This algorithm implements an Extended Kalman Filter using the odometry data in the equation of evolution and the measurement provided by the US sensors in the equation of observation. A second filter, permits to update, after each localization, an incremental representation of the scene defined in the initial reference frame. A test on the covariance matrix of the rebuilt data, allows that at instant (k+1) to use in the process of localization only the data sufficient consistent, explained in section 2.

In order to validate these algorithms, a complete simulator was written in MATLAB, and is presented in section 3. This tool has as a characteristic to allow at the same time the validation of the algorithms starting from simulated data and the exploitation on the real data provided by our experimental robot. From the point of view of simulation, it makes possible to describe and handle polygonal environments, models of sensors and models of robots. In the case of a real experimentation, it permits to acquire and treat the ultrasonic or vision data coming from the robot and to compute the motion which is, afterwards, carried out by the robot.

Results of simulation and real experiment, will be discussed in section 4, and conclusions and directions for further work can be found in section 5.

## 2. Methodology

A good localization and knowledge of the environment is a crucial problem in the autonomous navigation systems research. On one hand, the problem of locating an autonomous robot in a given environment is based on the model it has of the environment. On the other hand, the mapping of a previously unspecified space area is dependent on a good knowledge of the position of the robot. Hence the duality of this process leads us to the problem of finding an equilibrium between the need to follow the landmarks found in the environment, so that the robot can securely locate itself, and the need to find new landmarks to identify and add to the map of the environment. This paper presents one approach to this task of building a map of an structured but unknown environment and simultaneously perform the localization of the autonomous robot, using ultrasonic sensors.

### 2.1. Sensor Model

In the case of relatively structured indoor scenes, the ultrasonic sensors (US) are largely used in robotics due to their low cost and simplicity of implementation. However, the data coming from pure time of flight measurements, give only a distance information to a target in one emission/reception cone.

From a purely geometrical point of view, an ultrasonic sensor is a telemeter providing the shortest distance to an obstacle located in its field of view. This target behaves, in general, like a mirror taking into account the wavelengths used. We have used this property to establish from two US elementary sensors, operating one in emission/reception and the other only in reception, a telemetric system based on triangulation, providing as output, at the same time, the distance and the orientation of a target, supposed locally planar. Taking into account the following figure, where the obstacle behaves like a mirror, we can deduce, from the distance received by sensor $S_1$, on emission/reception mode, and by sensor $S_2$ only on reception mode, what is the target orientation.

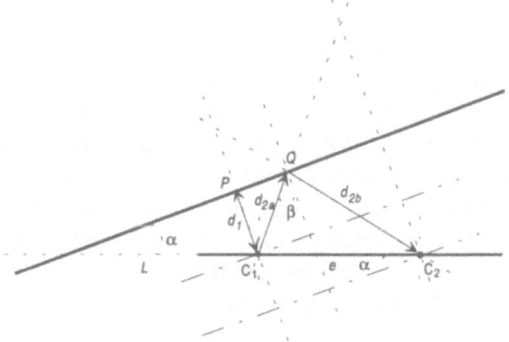

Then,

$$d_1 = L \sin \alpha$$
$$d_{2a} \sin \beta + d_{2b} \sin \beta = e \cos \alpha$$
$$d_{2a} \cos \beta = d_1 = L \sin \alpha$$
$$d_{2b} \cos \beta = d_1 + e \sin \alpha$$

from where,

$$\alpha = \arcsin\left(\frac{4d_2{}^2 - 4d_1{}^2 - e^2}{4d_1 e}\right)$$

Reversing the role of the sensors, $S_2$ on emission/reception mode and $S_1$ on reception mode, the reliability of measurement is reinforced by introducing redundancy. To verify this technique it was validated in simulation and on real data. The final precision is around the centimeter and around one degree for a slope of a target plane in a range of ±30 degrees.

The target plane is represented through a line segment, described by line parameters $(d, \alpha)$ and the points of *impact P, Q*.

Based on this algorithm, we may improve the information given by ultrasound sensors. In order to smooth noisy data, we use the sensor in a multiple shots mode (typically, each sensor measurement results from a series of 20 consecutive shots. Considering as example a pair of sensors S4 and S5, placed in front of the robot, and a target ahead of it, the 20 real US observation data are illustrated in figure below

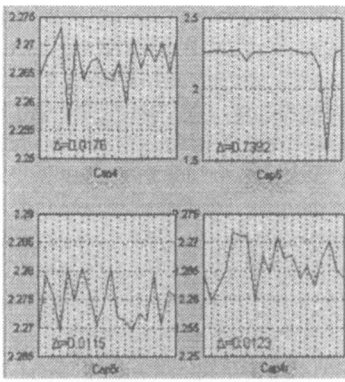
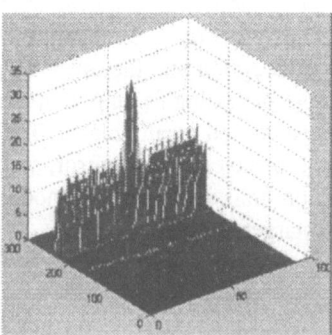

In this figure, the top plots correspond to data obtained by the sensor running in emission/reception mode and the bottom plots to those running in reception only. Using the proposed technique we obtained the distances and orientations, being respectively, dS4=226,7cm with a cap the 2,5degres and dS5=226,2cm with a cap the 2,7degres. From this it becomes clear the good performance even employing a simple median filter in order to eliminate aberrant US data. In spite of it, this approach does not make it possible to completely eliminate false measurements corresponding, for example, with phenomena of multiple reflections. We supplemented this work on the improvement of the quality of the ultrasonic data by the addition of a method of rejection using an algorithm of vote implemented by means of a Hough transform technique. From the right side of the figure above, we can see the plot of the accumulator with the maximum corresponding to the estimated value of the parameters of the line. The results are appreciably improved with the detriment, however, of a significant increase in the computing time.

## 2.2. Robot System

We denote $(x, y, \theta)$, the position and the orientation of the robot frame with respect to a global fixed frame.

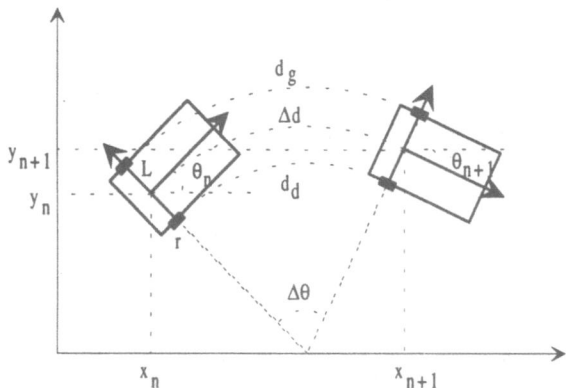

Suppose we know at a time $n$, the robot location $X_n=[x_n, y_n, \theta_n]$ and the wheels displacement $d_g$ and $d_d$ between $n$ and $(n +1)$, then we can compute $X_{n+1}$ by approximating the trajectory as a sequence of constant curvature elements of length $\Delta d$:

$$x_{n+1} = x_n + \Delta d \cos\left(\theta_n + \frac{\Delta\theta}{2}\right)$$

$$y_{n+1} = y_n + \Delta d \sin\left(\theta_n + \frac{\Delta\theta}{2}\right)$$

$$\theta_{n+1} = \theta_n + \Delta\theta$$

where, $\Delta d = \dfrac{d_g + d_d}{2}$ and $\Delta\theta = \dfrac{d_g - d_d}{L}$, being $d_g$ and $d_d$, the displacements which are expressed as a function of the rotations of the wheels given by the encoders. However it could be not very reliable because of the slips of the wheels, and the geometrical defects on radii of the wheels as well as in the length of the wheels axis of the robot. It is then interesting to be able to compensate this drift while readjusting with respect to the environment.

## 2.3. Localization and Map building approach

The localization process was carried out by a method of data fusion of ultrasonic observations and odometry data. This procedure was performed using an Extended Kalman Filter that provides at each instant the optimal state estimation in the sense of minimum mean square error. The state vector is the robot configuration $X_k=[x_k, y_k, \theta_k]$ and the state transition equation takes the form:

$$X_{k+1} = F(X_k, U_k, \nu_k)$$

being F the function witch describe the robot state, with $U_k$ the control vector and $\nu_k$ the state noise considered uncorrelated, gaussian with zero mean and variance $Q_k$

The covariance of the state prediction is given by,

$$P_{k+1|k} = \nabla F_{X_k} \times P_{k|k} \times \nabla F_{X_k}{}^T + \nabla F_{U_k} \times Q_k \times \nabla F_{U_k}{}^T$$

where $\nabla F_{X_k}$ is the state function linearised around $X_{k|k}$, and $\nabla F_{U_k}$ the state function linearised around $U_k$.

$$\nabla F_{X_k} = \frac{\partial F}{\partial X_k} = \begin{bmatrix} 1 & 0 & -l_k \sin\left(\theta_k + \frac{\Delta\theta_k}{2}\right) \\ 0 & 1 & l_k \cos\left(\theta_k + \frac{\Delta\theta_k}{2}\right) \\ 0 & 0 & 1 \end{bmatrix}$$

$$\nabla F_{U_k} = \frac{\partial F}{\partial U_k} = \begin{bmatrix} \frac{1}{2}\cos\left(\theta_k + \frac{\Delta\theta_k}{2}\right) - \frac{l_k}{2L}\sin\left(\theta_k + \frac{\Delta\theta_k}{2}\right) & \frac{1}{2}\cos\left(\theta_k + \frac{\Delta\theta_k}{2}\right) + \frac{l_k}{2L}\sin\left(\theta_k + \frac{\Delta\theta_k}{2}\right) \\ \frac{1}{2}\sin\left(\theta_k + \frac{\Delta\theta_k}{2}\right) + \frac{l_k}{2L}\cos\left(\theta_k + \frac{\Delta\theta_k}{2}\right) & \frac{1}{2}\sin\left(\theta_k + \frac{\Delta\theta_k}{2}\right) - \frac{l_k}{2L}\cos\left(\theta_k + \frac{\Delta\theta_k}{2}\right) \\ \frac{1}{L} & -\frac{1}{L} \end{bmatrix}$$

In order to compensate possible drifts due to the odometry process, we use US data in the observation process. The observation equation $Y_k = H(X_k + w_k)$, describes the current US measurements $(d_k, \alpha_k)$ with a gaussian noise $w_k$, associated, $H$ being given by:

$$H_{X_{k+1|k}} = \begin{bmatrix} d_0 - x_{k+1|k}\cos(\alpha_0) - y_{k+1|k}\sin(\alpha_0) \\ \alpha_0 - \theta_{k+1|k} \end{bmatrix}$$

where $(d_0, \alpha_0)$ are the values of the segments in the initial reference frame constituting the current map and which has been estimated and updated during the previous steps. The prediction of the observation, $Y_{k+1|k} = H_{X_{k+1|k}}$, is function of the state predicted and consists to express all the segments of the map in the predicted robot's frame at time $(k + 1)$. After executing the displacement, a new US measurement acquisition is achieved providing the set of observations $Y_{k+1|k+1}$. This set of observation is matched with the prediction using a is classic Mahalanobis distance,

$$DM = \left[Y_k - Y_{k+1|k}\right]^T S_{k+1|k}^{-1} \left[Y_k - Y_{k+1|k}\right]$$

where $S_{k+1|k}$ is given by

$$S_{k+1|k} = \nabla H_{X_k} \times P_{k+1|k} \times \nabla H_{X_k}^T + R_k$$

With the results obtained from $DM$ we make two tables of observations; those that passes the predefined threshold value will improve the map and others the will composing new map elements or will be "false alarms".

The observations that minimize the validation criterion are used to update the robot state estimation, given by:

$$X_{k+1|k+1} = X_{k+1|k} + K_{k+1}\left(Y_k - Y_{k+1|k}\right)$$

The Kalman gain, $K_{k+1}$, is given by

$$K_{k+1} = P_{k+1|k} \times \nabla H_{X_k}^T \times S_{k+1|k}^{-1}$$

and the covariance associated with the estimated robot state $X_{k+1|k+1}$ is:

$$P_{k+1|k+1} = \left(I - K_{k+1} \times \nabla H_{X_k}\right) \times P_{k+1|k}$$

This new estimated of the position is used to retroproject the measurement $Y_{k+1|k+1}$ in the reference frame. Then, we update the map by merging these measurements with the segments estimated during the previous step. To do that, we use a least square method working on the coordinates of the extremities of the segments built at each measurement step.

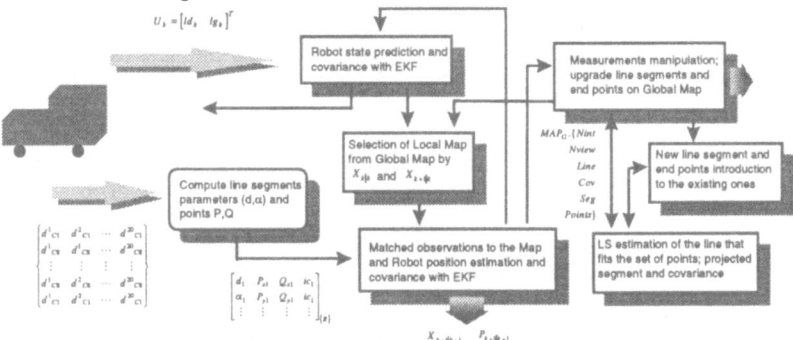

The strategy of building the environment map, either the local map which serves to locate the robot, either the global map that represents the whole observation, has to take into account the uncertainties associated with the estimated position of the robot and also with the confidence that we have on the segment that has been followed for a certain time and also on the segments which are seen more than once and recognized as so.

## 3. Experimental environment

A tool for simulation the mobile robot behavior and for the analysis and visualization of the experimental data was developed in general proposed framework, Matlab, in order to evaluate the methodologies approach. This simulator allow to define a polyhedral environment and to draw a linear trajectory, adjusted by lines and arcs, that the robot must follow. All low level functions are implemented in both, the robot and the simulator. We can manipulate simulated and real data.

A series of tests were carried out on simulated data. On the figure below, we can see the simulation of a simple environment, with one reference trajectory, with white noise on the movement and observations, we observe that the robot is away from the reference trajectory, but the position estimated is correct with the real robot pose, and therefore a well agreement has been observed between the obtained map and the environment.

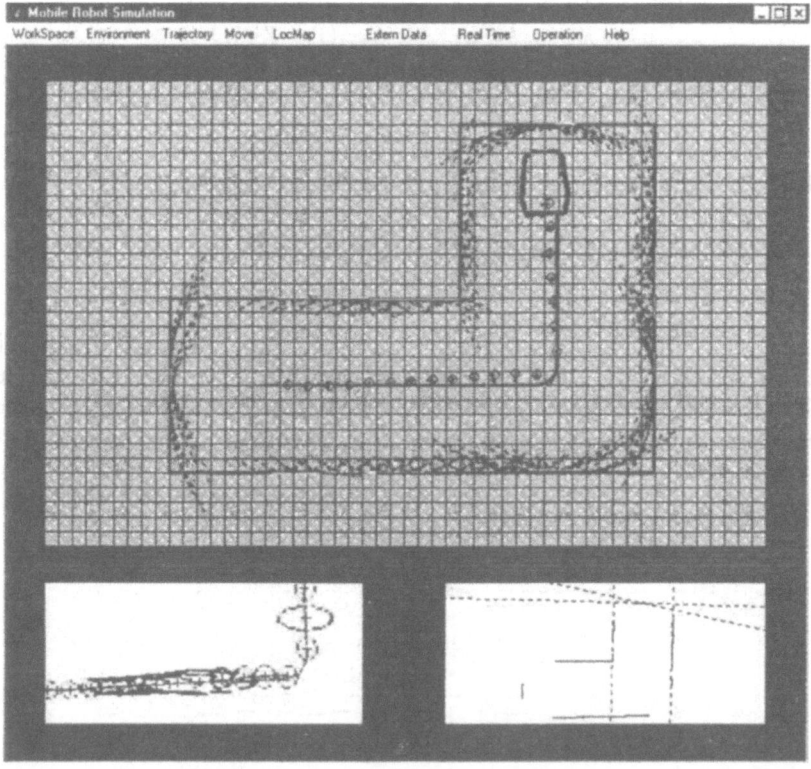

We also established a connection between Matlab and the experimental robot, Anis from Inria-Sophia, and with on board operating system, Vxworks, allowing to carry out real experiments previously developed and validated with off line data. This link are allowed to test and to analyze efficiently all the techniques and methods developed.

# 4. Results

The localization and map building algorithms have been validated on the mobile robot Anis at Inria Robotics Laboratory. Anis is equipped with a belt of wide angle ultrasounds sensors. During its mission, the mobile robot must follow a given trajectory, update its real position, since it may deflect from the reference trajectory, and create a map of its environment. At the start of the trajectory the robot follows it reasonably well, as shown in the figure bellow, but, after a while, as observed by the ellipses of incertitude, the robot trajectory drives away, illustrated by the cross marks, from the reference line. Therefore, the estimated position, represented by little circles, shows a good agreement with the real position. The obtained map, represented as a white line over the line that's limit the environment, shows that the errors inherent to the uncertainty on the robot position is emphasized on curves and vanished of landmarks pursued.

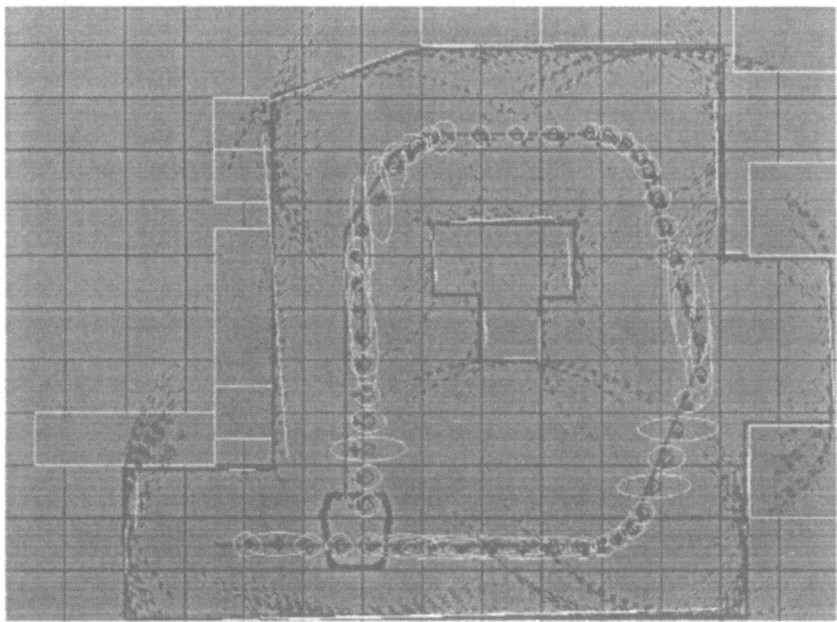

# 5. Perspectives and conclusions

This paper presents an approach for dealing with the mobile robot localization and map building processes. We introduce a system based on the fusion of odometry and ultrasound data, complemented with a perception strategy. The simulations show that the proposed algorithm is able to identify the robot position as well the environment reconstruction.

In a more prospective view, some of the remaining problems, like misclassification of the wall segments, from where inadequate localization could be overtaken using one filter with an inverse structure of the EKF. Or either, knowing the position and the map estimating and doing a backpropagation of the covariance matrices and the Kalman gain, applying the incertitude ellipsoids, one can perform a convergent more robust. On the other way, an increase accurate sensor, like a laser range finder, could improve mainly on the definition of the walls segment crossing.

Despite some of the remaining problems, the results obtained are very encouraging and auspicious for a fascinating future work.

# 6. References

[1] Holenstein A A, Muller M A, Badreddin E 1992 Mobile Robot Localization in a Structured Environment Cluttered with Obstacles. *Proceedings of the 1992 IEEE Conference on Robotics and Automation.* pp 2576-2581

[2] Chevanier F, Crowley J L 1992 Position estimation for a Mobile Robot Using Vision and Odometry. *Proceedings of the IEEE Conference and Robotics and Automation.* pp 2588-2593

[3] Kleeman L 1992 Optimal Estimation of position and Heading for Mobile Robots Using Ultrasonic Beacons and Dead-Reckoning. *Proceedings of the IEEE Conference and Robotics and Automation.* pp 2582-2587

[4] Zelinsky A 1991 Mobile Robot Map Making using Sonar. *J. Robot. Syst..* 8 (5): 557-577

[5] Gilbreath G, Everett H R 1989 Path planning and collision avoidance for an indoor security robot. *Proceedings SPIE-Inst. Soc. Opt. Eng. Mobile Robots III.* 1007: 19-27

[6] Smith R, Self M, Cheeseman P 1990 Estimating uncertain spacial relationships in robotics. In I. Cox and G. Wilfong. eds *Autonomous Robot Vehicles.* Springer Verlag

[7] Moutarlier P, Chatila R 1989 Stochastic multisensory data fusion for mobile robot location and environment modeling. *5th Int. Symposium on Robotics Research.* pp 85-94

[8] Leonard J J, Cox I J 1993 Modelling a Dynamic Environment Using a Bayesian Multiple Hypothesis Approach. *AAAI Journal of Artificial Intelligence.*

[9] Borthwick S, Durrant-Whyte H 1994 Simultaneous Localisation and Map Building for Autonomous Guided Vehicles. *Proceedings of the IEEE/RSJ Int. Conf. Intell. Robot. Syst..* pp 761-768

# Construction and Modelling of a Carangiform Robotic Fish

Richard Mason
California Institute of Technology
Pasadena, California, USA.
mason@robotics.caltech.edu

Joel Burdick
California Institute of Technology
Pasadena, California, USA.
jwb@robotics.caltech.edu

**Abstract:** We describe an experimental testbed which we have constructed to study carangiform fish-like swimming. We mimic a carangiform fish with a three-link robot, the tailfin link of which acts as a flat-plate hydrofoil. We experiment with sinusoidal signals to the joint actuators. The experimental results agree well with a simulation assuming quasi-steady flow around the tailfin.

## 1. Motivation

Many fish and marine mammals are very impressive swimmers: so impressive, in fact, that some have argued that fishlike swimmers are superior, in one way or another, to conventional man-made water vehicles. For decades researchers have tried to better understand how fish swim so effectively, [1, 2, 3] and a few have contemplated the mechanical imitation of fish, with the idea of building vehicles which are faster, more efficient, more stealthy, and/or more maneuverable than propeller-driven craft. [4]

Perhaps the best-known robotic fish was built by Triantafyllou et al. [5] Their highly sophisticated robot had many actuated degrees of freedom and represented a fairly faithful reproduction of tuna swimming. They found that by undulating its tail and body the tuna was able to reduce the drag it experienced while being moved through the water. This evidence tends to bolster the notion that fish swimming is particularly efficient. It may still be an open question whether robot fish can really outperform efficient propeller designs, which themselves have received much research and labor. The prospect of increased efficiency for robot submersibles remains tantalizing, however, since these submersibles typically run on electric batteries of limited duration, and anything which enabled them to operate for a longer period would be of significant value.

Increased stealth is another potential motive for building robot fish. Fish typically do not cause the noisy cavitation sometimes experienced by propellers. Indeed from a biological standpoint there is good reason to suppose that fish have evolved to swim quietly and stealthily. Ahlborn et al. [6] built an artificial fishtail to mimic fishlike swimming; they observed that the alternating creation and destruction of vortices in the wake behind the fish was not only an efficient way to swim, but also helped guard against detection by predators.

Furthermore, many fish are highly maneuverable. Some fish can perform a 180 degree turn within a fraction of their own body length. This is not generally possible for boats or ships, which typically have large turning radii. We suspect that improved agility may be the biggest advantage that robot fish enjoy over their propeller-driven cousins.

Finally, we have an abstract theoretical interest in understanding aquatic locomotion and hopefully unifying it in a single mathematical framework with other, terrestrial forms of locomotion in which an animal or robot uses quasiperiodic changes of internal shape variables to generate gross body motion. Past work at Caltech has brought the tools of differential geometry to bear on the control of other forms of undulatory locomotion, [7, 8] and we hope that these tools can be extended to fishlike swimming. But before we can proceed with this program, we must develop a model of fish locomotion with some experimental validation, which is the topic of this paper.

## 2. Description of the Model

There are a wide variety of fish morphologies and at least a few different types of fish locomotion. We focus on attempting to mimic the swimming of the *carangiform* fishes, fast-swimming fishes which resemble tuna and mackerel. Carangiform fishes typically have large, high-aspect-ratio tails, and they swim using only motions of the rear and tail, while the forward part of the body remains relatively immobile. For our model we consider an idealized carangiform fish that consists of only three links: a rigid body in front, a large wing-like tail at the rear, and a slender stem or *peduncle* which connects the two. The three rigid links are connected by rotational joints with joint angles $\theta_1$ and $\theta_2$. See Figure 1.

We continue to idealize the model by supposing that we can neglect three-dimensional effects and regard the problem as essentially planar. In particular, we assume that the large tail can be considered as a rectangular flat plate (although the tails of real carangiform fish are often lunate in shape.) We will presume the tail experiences a hydrodynamic lift force derived from quasi-steady two-dimensional wing theory. The peduncle we will regard as hydrodynamically negligible. The forward rigid body will experience a drag force quadratic in and opposed to its velocity.

There is a distance $l_b$ between the body's center of mass and the peduncle. The peduncle has length $l_p$. The tailfin has chord $l_t$ and area $A$. Let $\vec{le}$ be a unit vector pointing in the direction of the leading edge of the tailfin hydrofoil. In a coordinate frame aligned with the principal axes of the fish's body, then:

$$\vec{le} = -(\cos(\theta_2), \sin(\theta_2), 0) \tag{1}$$

The body of the fish has instantaneous translational velocity $\dot{x}$ and $\dot{y}$ along its longitudinal and lateral axes; it also has instantaneous rotational velocity $\dot{\phi}$.

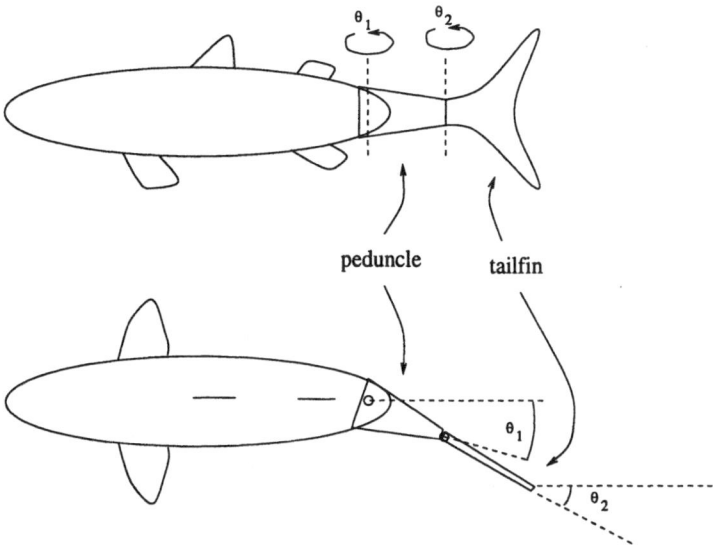

Figure 1. *A simplified carangiform fish with a two-jointed tail.*

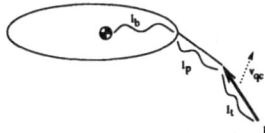

Figure 2. *The idealized model.*

Let $\vec{v}_{qc}$ be the velocity of the quarter-chord point of the hydrofoil.

$$\vec{v}_{qc} = \begin{pmatrix} \dot{x} - \dot{\phi}(l_p\sin(\theta_1) + (l_f/4)\sin(\theta_2)) - \dot{\theta}_1 l_p \sin(\theta_1) - \dot{\theta}_2(l_f/4)\sin(\theta_2) \\ \dot{y} + \dot{\phi}(l_p\cos(\theta_1) + (l_f/4)\cos(\theta_2)) + \dot{\theta}_1 l_p \cos(\theta_1) + \dot{\theta}_2(l_f/4)\cos(\theta_2) \\ 0 \end{pmatrix}$$

$$(2)$$

See Figure 2. Using the Kutta-Joukowski theorem and assuming that the tail hydrofoil is in a quasi-steady uniform flow with the velocity implied by the instantaneous velocity of the foil's quarter-chord point, we arrive at the following lift force on the hydrofoil:

$$L = \text{sign}(\cos(\text{arg}\vec{le} - \text{arg}\vec{v}_{qc}))\pi\rho A(\vec{v}_{qc} \times \vec{le}) \times \vec{v}_{qc} \qquad (3)$$

The purpose of the first factor in the expression for $L$ is to handle cases when the hydrofoil's angle of attack is between $\pi/2$ and $3\pi/2$ radians, when the trailing edge of the hydrofoil has in fact become the leading edge. The density of the fluid is $\rho$. For our current purposes we assume that the fish's body is constrained to move only in the longitudinal direction and $\dot{y}$ and $\dot{\phi}$ are identically zero. This will be the case for the experimental setup discussed in this paper. Then the only relevant equation of motion is the one in the $x$-direction. The rigid body has a translational inertia $I_x$ (this includes added-mass effects from the fluid

Figure 3. *A "cartoon" of carangiform swimming. The fish body is invisible in this picture–only snapshots of the tail are pictured as the fish swims from left to right and the tail moves up and down. The arrows indicate lift forces acting on the tail.*

surrounding the body) and a coefficient of drag $C_{D_{\text{body}}}$.

$$I_x \ddot{x} = L_x - \frac{1}{2} C_{D_{\text{body}}} \dot{x} \|\dot{x}\| \tag{4}$$

It is worth stressing how extremely simplified this model is. There are many things of which we have taken no account here: added-mass forces acting on the tailfin; the effect of vortices shed by both the body and the tailfin; the fundamentally unsteady nature of the flow around the tailfin; the effect of nearby walls and surfaces on the fluid flow (for our experimental fish will swim, not in an unbounded fluid, but in a confined tank.) We have little doubt that any and all of these effects are important in some regimes of the fish's operation, and we anticipate refining our model to include them. However, it is interesting that in the experimental results obtained so far, the simplest model does a good job of matching the experimental data.

We conclude this section with a rough sketch of how carangiform swimming is generally understood to work. In Figure 3, the tail of a fish is shown pitching and heaving up and down as the fish moves with some velocity from right to left. The basic idea is that the tail maintains a negative angle of attack on the upstroke, and a positive angle of attack on the downstroke, with the result that the lift force on the hydrofoil/tail is always oriented so as to propel the fish forward and maintain its velocity. One detail of the sketch as it is drawn here is that the pitch of the tailfin is zero when the angular excursion of the peduncle is at its maximum, and the pitch of the tailfin is at its maximum when the peduncle is horizontal. In other words, if the tailfin and the peduncle angles are considered to be sinusoids, they are ninety degrees out of phase. This picture makes some sense from an intuitive standpoint–but interestingly, we did not find it to be quite borne out by experiment.

## 3. Description of the Experiment

We built an experimental "fish" to test the validity of our models. See Figure 4. As in the model, the tailfin is a large flat plate–with a chord of 12.5 cm and a typical depth in the water of 40 cm. The peduncle is a thin supporting arm, 12.5 cm in length, which we believe experiences only negligible hydrodynamic forces. A support keel plays the role of a rigid forward body. In order to simplify the experiment, the motors which drive the joints of the fish are kept out of the water, on a carriage which rides on rails along the length of the water tank. Timing belts and two coaxial shafts transmit power from the motors to the submerged joints of the fish. The servomotors are quite small, with a maximum output torque of 7.7 kg·cm.

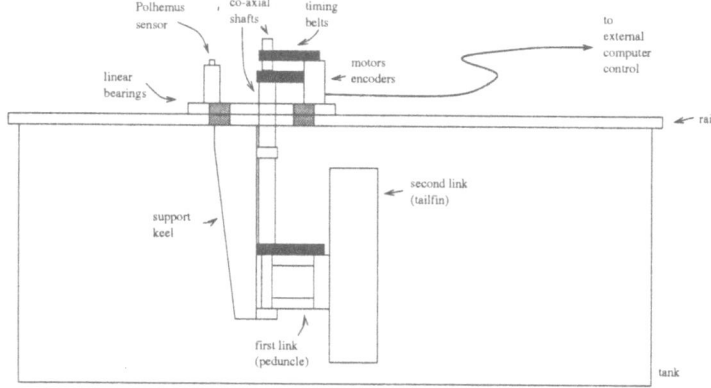

Figure 4. *Side view schematic of experiment.*

Also mounted on the carriage are optical shaft encoders, which record the fish's joint angles at each instant. The encoders are Hewlett-Packard HEDS-5500s, and are accurate to within about forty minutes of arc. Finally, the carriage carries a Polhemus receiver. Elsewhere in the lab we have a Polhemus transmitter which generates near-field, low-frequency magnetic field vectors, and in this way the relative position and orientation of the Polhemus receiver, and therefore the position of the carriage on the tank, can be determined.

In the current configuration of the experiment, the carriage is constrained to move only longitudinally down the tank. See Figure 5. On the right side of Figure 5 we see an enhanced carriage which can translate longitudinally or laterally, and also rotate, thus allowing the fish full freedom to swim about in the plane. These extra degrees of freedom have actually been built, but we do not present experimental data from the enhanced carriage in this paper. The support keel by itself is an inadequate "body" for full planar swimming–for example, it has inadequate rotational inertia–and it requires enhancement.

## 4. Experimental Results

Some experimental results are plotted in Figures 6-9 and compared to results of a computer simulation based on the quasi-steady flow model. The computer simulation received as input the actual joint angle trajectories for each gait as reported by the optical shaft encoders. For several sinusoidal gaits at several frequencies, distance of motion down the tank is plotted as a function of time. In addition, next to each figure is a small sketch of the joint positions in that gait at the moment of zero peduncle angle $\theta_1$; at the moment of maximum peduncle angle $\theta_1$; and at the moment of maximum tailfin angle $\theta_2$. In this way the reader can see the maximum angular excursion of each gait.

We make the following general observations:

- The computer simulation generally does a very good job of matching the experimental data, especially over the first meter or so of travel.
- Over greater distances, the simulation tends to deviate from what is found

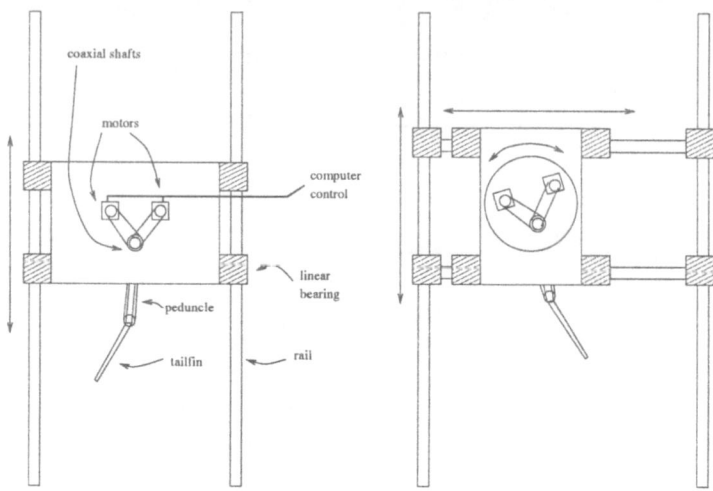

Figure 5. *Top view schematic of experiment.*

experimentally. We attribute this to the fact that the experimental carriage experiences a frictional resistance associated with the rails. The magnitude of this resistance varies at different points along the tank and it is not modelled in the simulation.

- Contrary to what we might have expected, the fastest sinusoidal gait found did not have a phase difference of ninety degrees between $\theta_1$ and $\theta_2$. Gaits in which there was a forty-five degree phase difference actually generated faster velocities at an equivalent frequency. It is not clear if this is a fair comparison since, for example, the maximum angular excursion of the tailfin was not the same between the two sets of gaits. Nevertheless, it is interesting that the experimental results so far were surprising and somewhat contrary to our original intuition.

## 5. Future Work

We have demonstrated that our experimental carangiform swimmer works and that it is reasonably well modelled under an assumption of quasi-steady flow. In future experimental work, we will add a more realistic body to the fish and allow the carriage to move with three degrees of freedom. This will allow us to test the lateral and yaw stability of the existing gaits and also experiment with turning and maneuvering gaits. If necessary, we will expand the model to incorporate more complex hydrodynamic effects.

We will use the experimentally validated model to search for gaits which give optimal performance in both forward swimming and in turning. Finally, we will move beyond open-loop joint angle trajectories (e.g. sinusoidal joint angle signals) to feedback control laws that take the velocity of the fish as an input. It seems likely that truly optimal fish locomotion will involve tuning the angle of attack of the tailfin at each instant, which in turn requires incorporating feedback about the fluid flow around the fish into the control scheme.

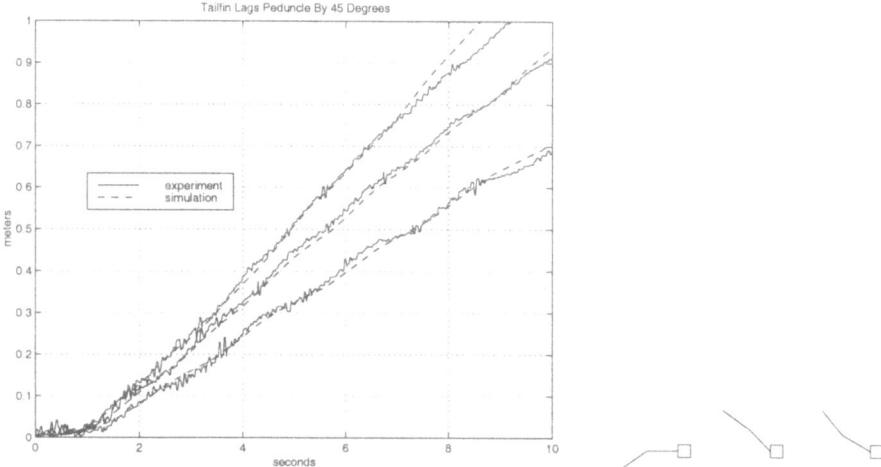

Figure 6. *Distance traveled by the fish in a given time, for gaits of the form* $\theta_1 = 0.7\cos(\omega t)$ *and* $\theta_2 = 0.9\cos(\omega t)$.

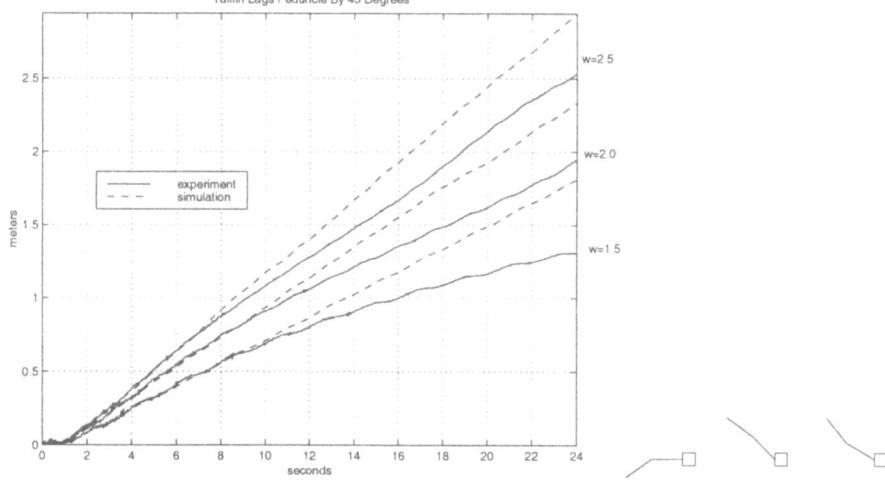

Figure 7. *Over a long distance, the experiment deviates from the simulation. This is because the carriage experiences a resistance associated with the rails which varies along the length of the tank, and is not modelled in the simulation.*

## References

[1] Childress S 1981 *Mechanics of Swimming and Flying.* Cambridge University Press, Cambridge

[2] Newman J N, Wu T Y 1974 Hydromechanical aspects of fish swimming. In: Wu T, Brokaw C J, Brennan C(eds) 1974 *Swimming and Flying in Nature, Vol 2.* Plenum Press, New York, pp 615-634

[3] Lighthill J 1975 *Mathematical Biofluiddynamics.* SIAM

242

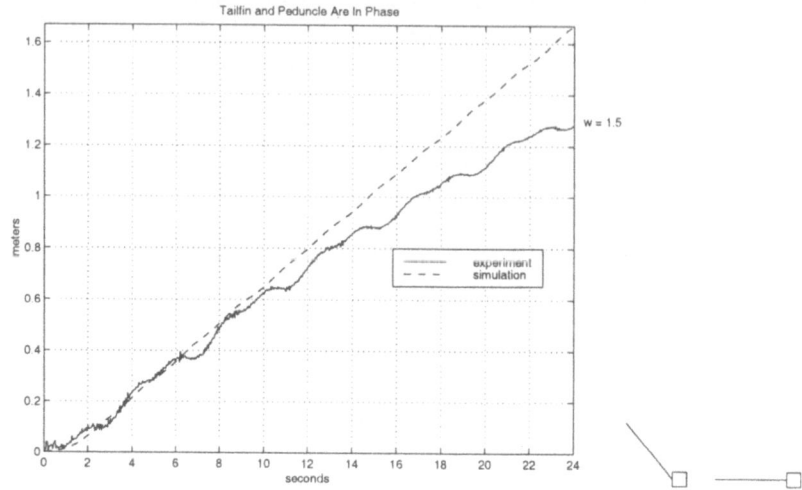

Figure 8. *Distance traveled in a given time, for a gait of the form* $\theta_1 = \theta_2 = \cos(\omega t)$.

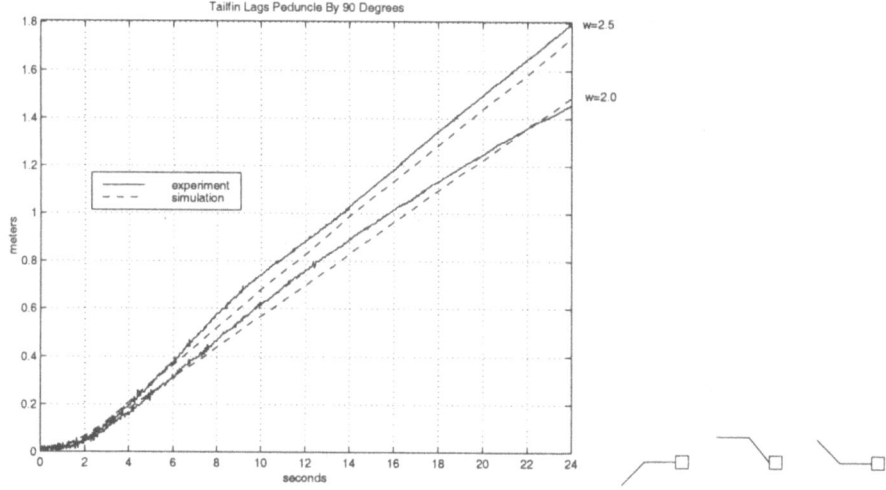

Figure 9. *Distance traveled in a given time, for gaits of the form* $\theta_1 = 0.8\cos(\omega t)$ *and* $\theta_2 = 0.9\cos(\omega t - \pi/2)$.

[4] Harper K A, Berkemeier M D, Grace S 1998 Modeling the dynamics of spring-driven oscillating-foil propulsion. *IEEE J Oceanic Eng.* 23:(3) 285-296

[5] Anderson J M, Streitlien K, Barrett D S, Triantafyllou M S 1998 Oscillating foils of high propulsive efficiency. *J Fluid Mech.* 360:41-72

[6] Ahlborn B, Harper D G, Blake R W, Ahlborn D, Cam M 1991 Fish without footprints. *J Theor Biol.* 148:521-533

[7] Ostrowski J P 1995 The mechanics and control of undulatory robotic locomotion. PhD thesis, California Institute of Technology

[8] Kelly S D, Murray R M 1995 Geometric phases and robotic locomotion. *J Robotic Syst.* 12:(6) 417-431

# Can an Underactuated Leg with a Passive Spring at the Knee Achieve a Ballistic Step?

Bernard Espiau, Isabelle Guigues, Roger Pissard-Gibollet
INRIA, 655 Avenue de l'Europe
ZIRST, 38330 Montbonnot, FRANCE
Corresponding author: Bernard.Espiau@inrialpes.fr

**Abstract:** In this paper, we present a control scheme aimed at moving an underactuated double pendulum simulating a leg. The system is actuated at the hip and includes a spring at the knee. It interacts with the ground, which is in fact a treadmill, through a telescopic foot made of a linear spring and a damper. The control scheme is different in the 2 phases, swing and stance, and a repetitive step is achieved by switching between these two controls. The paper describes the system, the used models and the control algorithms. It presents also experimental results of the approach.

## 1. Introduction

The origin of this work is a medical application. The problem we addressed was to design a reconfigurable knee orthosis with tunable passive springs and dampers. This orthosis was aimed at improving rehabilitation programs following an operation and helping patients having weak flexion or extension capabilities. An effect of the system is that, when walking, a patient who wears such a knee-pad may have to use dynamical effects in addition to his remaining actuation capabilities in order to counter the spring force and reach the leg extension which is required for starting the stance phase.

In complement to classical experimental protocols to be conducted on patients, we found interesting to experimentally investigate the purely mechanical and control issues of the problem on a dedicated robotic device. We thus designed and realized a double pendulum, with size and mass repartition close to human's ones and having a single actuator at the hip and a spring at the knee in order to study the synthesis of gaits for this system. Compared to a human or to a classical biped robot, further difficulty with a vertical planar double pendulum arises, because the hip position is fixed with respect to the ground. In order to simulate the motion, we therefore added a treadmill to the setup and mounted a telescopic foot using a second linear spring.

The purpose of this paper is to present the control scheme we propose for this system and to report the obtained results. Section 2 gives a brief review of existing works; section 3 presents the general approach we used; in section 4, we describe the experimental setup and the related models. Section 5 presents the control algorithms and section 6 the experimental results. More details on the approach can be found in [18].

## 2. A Brief Tour of the Literature

The present paper adresses simultaneously questions of ballistic walking, underactuation and effects of springs on the gait. One of the early references in human ballistic walking is [5]. The use of springs in robot joints have soon been recognized as a way of realizing energy storage or impact absorption and many works have been reported in the literature. Although running robots be obviously the first concerned by this approach (see [1, 11] and, more generally, the work done in the MIT LegLab[1]), the case of walking machines is also considered [2, 3, 10].

In the domain of automatic control, underactuated systems have been rather largely investigated. As an interesting class of nonlinear systems the tangent linear approximation of which is not controllable, underactuated robots, extending to flexible or non-holonomic ones, have been soon targeted by researchers from the automatic control area [4, 6, 8, 7, 9]. Finally, let us specially mention interesting works around Acrobot-like systems (these are 2 dof robots, with a single actuator either at the hip - the Pendubot - or at the knee , and submitted to the gravity): [12] proposes a decoupling and partial feedback linearization scheme with control switching; [13] uses a different control scheme for every phase of a hopping Acrobot. A last point to be mentionned is the fact that the control of a compass using the hip actuator as done in [15] can be viewed as a special case of control of an underactuated double pendulum.

In fact, and to our knowledge, the problem addressed in this paper has not been considered under the same form in the literature. Thus, although some parts of the control we propose are not really new, the global solution we bring to the overall problem of achieving a ballistic step with an underactuated double pendulum having a knee spring seems to be original.

## 3. Background in Modelling and Control

### 3.1. Unconstrained Case

Let us consider a robot with rigid links, $n$ joints and a fixed basis. Let us suppose that the (constant) dimension of the actuation space is $m < n$. By performing a change of variables if needed, its dynamical equation may always be written under the form

$$M(q)\ddot{q} + N(q, \dot{q}) = B\Gamma = \begin{pmatrix} \Gamma \\ 0 \end{pmatrix} \qquad (1)$$

where $q$ is here taken as the set of joint variables, $M(q)$ the s.d.p. matrix of kinetics energy and $N(q, \dot{q}) = N'(q, \dot{q}) + G(q)$. $N'$ is the vector of Coriolis/centrifugal forces and $G$ the vector of potential-based forces. $\Gamma$ is the $m-$dimensional actuation vector.

---

[1]http://www.ai.mit.edu/projects/leglab

The set of configuration points $q^*$ that we can try to stabilize is given by the equilibrium manifold of (1): $G_2(q^*) = 0$ where $G_2$ is the lower part of $G$ corresponding to the zero actuation. We already see that the most favourable situation is the one of a planar robot without joint elasticities, where the equilibrium manifold is the whole configuration space. This is not our case, since the double pendulum is submitted to gravity and has a spring at the knee joint. Let us now define a $m-$dimensional output function $e(q, t)$ to be regulated to zero as an user specification. Differentiating $e$ twice gives

$$\ddot{e} = J\ddot{q} + f(q, \dot{q}, t) \ , \quad J = \frac{\partial e}{\partial q} \tag{2}$$

Combining (1) and (2) gives the dynamics in the output space:

$$\ddot{e} = JM^{-1} \begin{pmatrix} \Gamma \\ 0 \end{pmatrix} - JM^{-1}N + f \tag{3}$$

Let us now partition $M$ as $\begin{pmatrix} M_{11} & M_{12} \\ M_{12}^T & M_{22} \end{pmatrix}$ with dim $(M_{11}) = m \times m$. It is easy to see that (3) may be written as

$$\ddot{e} = RK^{-1}\Gamma - JM^{-1}N + f \tag{4}$$

where

$$R = J_1 - J_2 M_{22}^{-1} M_{12}^T \ \text{and} \ K = M_{11} - M_{12} M_{22}^{-1} M_{12}^T \tag{5}$$

with $J = (J_1(m \times m) \ J_2)$. Assuming that $M_{22}$, $R$ and $K$ are nonsingular matrices, we can therefore finally write

$$\Gamma = E\ddot{e} + F \tag{6}$$

where $E = KR^{-1}$ and $F = KR^{-1}J_2 M_{22}^{-1} N_2 f + N_1 - M_{12} M_{22}^{-1} N_2$, with $N = \begin{pmatrix} N_1 \\ N_2 \end{pmatrix}$. Finally, and like in [12], a natural control scheme is

$$\Gamma = \hat{E}(-K_p e - K_v(J\dot{q} + \frac{\partial e}{\partial t})) + \hat{F} \tag{7}$$

This control achieves an approximate partial linearization and decoupling: *approximate*, because the hats on the terms in (7) indicate that the models used in the control may differ from the theoretical expressions; *partial* since the linearization/decoupling concerns only the $m-$ dimensional controlled manifold while the complementary evolves freely. We will use a control belonging to this class in the swing phase of our double pendulum (see section 5).

## 3.2. Constrained Case

It is generally possible to model a biped robot in the swing phase as an open articulated chain, satisfying therefore the previous equations. During the stance phase, new constraints appear, that we will model as bilateral. Like in [16], let us complete (1) with the constraints equation

$$h(q) = 0 \; ; \; \dim(h) = p < n \; ; \; \text{rank}\,(C = \frac{\partial h}{\partial q}) = p \qquad (8)$$

We further assume that the solution of (8) is not an empty set. The constrained dynamic equation can now be written as

$$M(q)\ddot{q} + N(q, \dot{q}) = B\Gamma - C^T(q)\lambda \qquad (9)$$

where $\lambda$ are the Lagrange multipliers. Differentiating twice (8) gives

$$C\ddot{q} + s(q, \dot{q}) = 0 \qquad (10)$$

which, with (9) leads to:

$$\lambda = (CM^{-1}C^T)^{-1}(CM^{-1}(B\Gamma - N) + s) \qquad (11)$$

Replacing in (9 gives:

$$\ddot{q} = PM^{-1}(B\Gamma - N) - C^\dagger s \qquad (12)$$

where

$$C^\dagger = M^{-1}C^T(CM^{-1}C^T)^{-1} \qquad (13)$$

It can now be shown [18] that the constrained dynamics writes as:

$$P(\ddot{q} - M^{-1}(B\Gamma - N)) = 0 \; ; \; h(q) = 0 \qquad (14)$$

where

$$P = I_n - C^\dagger C \qquad (15)$$

$P$, $n \times n$, is a projection operator onto $NS$, the null space of $C$ with respect to the kinetics metrics $M$ (this means that we have $P^T M P = M P$ instead of $P^T P = P$ as involved by the euclidean metrics). Its rank is $n - p = n_1$, while the LHS of equation (14) is of dimension $n$. We can therefore replace (14) by a set of $n_1$ equations in the following way: (14) means that the projection is $M$-orthogonal to $NS$. Therefore, the projected vector is $M$-orthogonal to any set of basis vectors of $NS$. Let us partition $C$ as

$$C = (C_1 \; C_2) \qquad (16)$$

where $C_1$ is $p \times p$, assumed to be nonsingular. We can therefore choose as a basis of $NS$ the columns of the $n_1 \times n$ matrix:

$$\bar{R} = \left( \begin{array}{c} -C_1^{-1}C_2 \\ I_{n_1} \end{array} \right)^T \qquad (17)$$

Finally, the $n$ equations describing the constrained dynamics can be defined as:

$$\bar{R}(M\ddot{q} - B\Gamma + N) = 0 \; ; \; h(q) = 0 \qquad (18)$$

The number of degrees of freedom left available for moving the system (18) is $n_1$. We can specify again our control goal as the regulation at zero of a desired output function of dimension $n_1$, $e(q,t)$. Combining (2) with (10) gives

$$\ddot{q} = \bar{K} \begin{pmatrix} s \\ \ddot{e} - f \end{pmatrix} \qquad (19)$$

where

$$\bar{K} = \begin{pmatrix} C \\ J \end{pmatrix}^{-1} \qquad (20)$$

For $\bar{K}$ to be nonsingular, the independency of $e$ and $h$ is required. Let us now partition $\bar{K}$ as

$$\bar{K} = (\bar{K}_1 \ \bar{K}_2) \qquad (21)$$

where $\bar{K}_1$ is $n \times p$ and $\bar{K}_2$ is $n \times n_1$, both assumed to be of full rank. Using (21) and (19) in (18) finally leads to the dynamics equation in the output space:

$$M'(q)\ddot{e} + f'(q, \dot{q}, t) = u \qquad (22)$$

where

$$M' = \bar{R}M\bar{K}_2, \qquad (23)$$

assumed to be nonsingular,

$$f' = \bar{R}M(\bar{K}_1 s - \bar{K}_2 f) + \bar{R}N \qquad (24)$$

and

$$u = RB\Gamma \qquad (25)$$

Now, a decoupling and feedback linearizing control is given by

$$u = M'(q)(-K_p e - K_v(J\dot{q} + \frac{\partial e}{\partial t})) + f'(q, \dot{q}, t) \qquad (26)$$

where $K_p$ and $K_v$ are diagonal positive matrices. This ideal control ensures a linear second-order decoupled behavior for $e$. Given $u$, it remains to compute the actuator torques, $\Gamma$, the dimension of which can be greater than $n_1$. This can be done by selecting, for example, the torques which minimize some energy-based criterion, or the ones which ensure that unilateral constraints will not be violated [17]. We will use a control of type (25,26) simplified in the stance phase of our double pendulum.

## 4. The Experimental Setup

The double pendulum (see pictures in section 6) includes a thigh (length 0.45 m, adjustable mass up to 6 kg), a leg (length 0.45 m, adjustable mass up to 4kg) and a telescopic foot made of a spring/damper system with a range of 4 cm. The effective contact area (when the leg is vertical) is at most of 2 $cm^2$. The system is equipped with a spring at the knee (see figure 1 and can touch a treadmill, the maximum speed of which is 25 cm/s.

The actuation is realized by a Parvex Motor (nominal torque 2 N.m) with a gear providing with a reduction ratio of 25. Encoders measure angular positions of the two joints.

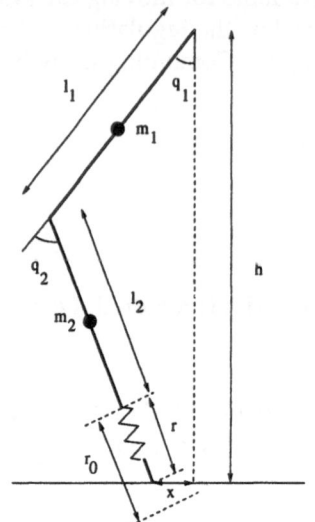
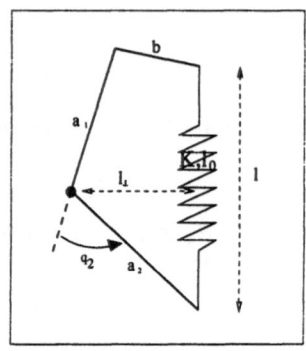

a- Pendulum in the Stance Phase     b- A spring at the knee

Figure 1. Pendulum Model

### 4.1. Model in the Swing Phase

The notation is given in figure 1. We have here $q = (q_1, q_2)$. In the following, $c_i$ and $s_{ij}$ stand for $cos(q_i)$ and $sin(q_i + q_j)$ respectively. The expressions of the inertia and centrifugal terms are given in [18]. The gravity vector is

$$G^1 = -g \left( \begin{array}{c} m_1 l_1 s_1 - m_2(\frac{1}{2} l_2 s_{12} + l_1 s_1) \\ m_2 l_2 s_{12} \end{array} \right) \qquad (27)$$

The torque exerted by the spring at the knee is

$$G^2 = \left( \begin{array}{c} 0 \\ K(l - l_0) l_\perp \end{array} \right) , \qquad (28)$$

$$l = \sqrt{(a_2 c_2 + a_1)^2 + (a_2 s_2 - b)^2} \qquad (29)$$

$$l_\perp = \sqrt{a_1^2 + b^2 - \frac{1}{4l^2}(l_2 + a_1^2 + b^2 - a_2^2)^2} \qquad (30)$$

and the term $G$ in (1) is given by $G = G^1 + G^2$. Finally, let us note that $\Gamma$ is here the 1D hip actuator torque.

### 4.2. Model in the Stance Phase

We assume at this stage that the foot spring is in the leg direction and that its mass and inertia can be neglected. We also suppose that the normal contact force and the friction coefficient are large enough to avoid sliding and lifting of the foot. Note that, clearly, these assumptions are not true at the beginning and the end of the stance phase. However, we will use them in order to derive the control, then we will experimentally verify that we nevertheless can cope with the cases where they are not satisfied.

Now, the configuration space of the system is of dimension 3: $\{q_1, q_2, r\}$, since it should include the spring length variable. The 2-dimensional constraint, modelled as bilateral, writes:

$$\begin{cases} l_1 c_1 + (l_2 + r)c_{12} = h \\ l_1 s_1 + (l_2 + r)s_{12} = x(t) \end{cases} \tag{31}$$

Therefore, the system is no more underactuated in this phase. Equations (31) can also be written as

$$\begin{cases} r = \frac{h - l_1 c_1}{c_{12}} & (a) \\ l_1 s_1 + h - l_1 c_1 t_{12} - x(t) = 0 & (b) \end{cases} \tag{32}$$

The equation (32) (a) can be taken as the constraint (8) and the related dynamics, (9), is now the 2-dimensional system:

$$M(q)\ddot{q} + N'(q, \dot{q}) + G^1(q) + G^2(q) + K'(q) = \begin{pmatrix} \Gamma \\ 0 \end{pmatrix} - C^T(q)\lambda \tag{33}$$

where

$$K'(q) = \begin{cases} k(r - r_0)l_1 s_2 & for \ r < r_0 \\ 0 & for \ r \geq r_0 \end{cases} \tag{34}$$

$r$ given in (32) (a), and $C^T$ is the jacobian matrix of the constraint (32) (b):

$$C^T = \begin{pmatrix} C^1 \\ C^2 \end{pmatrix} = \begin{pmatrix} l_1 c_1 + l_1 s_1 t_{12} + \frac{h - l_1 c_1}{c_{12}^2} \\ \frac{h - l_1 c_1}{c_{12}^2} \end{pmatrix} \tag{35}$$

## 5. Control

### 5.1. Control in the Swing Phase

In that phase, our objective is to have the leg almost fully extended (without further knee angle increase), despite gravity, spring and inertia effects, at a desired value of the hip angle, and after a given time, $T$. The initial conditions are the ones of the free system in equilibrium. We therefore simply define as the task function to regulate:

$$e_1(q, t) = q_1(t) - q_1^d(t) \ , \ \ t \in [0, T] \tag{36}$$

with $q_1(0) = q_1^d(0)$ and $\dot{q}_1^d(0) = \dot{q}_1^d(T) = 0$. We also need to ensure that $\dot{q}_2(T) = q_2(T) = 0$. It remains now to find an adequate trajectory, $q_1^d(t)$, satisfying the expressed goals. We will use the following result: *for the system modelled in section 4.1, if the joint velocities are bounded, there exists $\Theta$, $\frac{\pi}{2} \leq \Theta < \pi$ such that $\forall |q_2| < \Theta$, there exists $\ddot{q}_1^m > 0$ such that $\forall |\ddot{q}_1| > \ddot{q}_1^m$, then sign $(\ddot{q}_2) = -sign(\ddot{q}_1)$. The proof is extremely simple and given in [18].*

In fact, using this property will allow us to search for the final acceleration of $q_1$ able to place $q_2$ at the right value by inertia coupling effects only. The easiest way of finding the desired trajectory $q_1^d(t)$ is to compute a polynomial satisfying the constraints:

$q_1^d(0) = q_1(0)$ (measured); $\dot{q}_1^d(0) = 0$; $q_1^d(T) = q_1^*$; $\dot{q}_1^d(T) = 0$; $\ddot{q}_1^d(T) = \omega$. Given $T$ and $q_1^*$, $\omega$ is our single tuning parameter, and the polynomial is $q_1^d(t) = q_1(0) + \sum_{i=1}^{4} a_i t^i$, where $\delta = q_1^* - q_1(0)$ and:

$$a_1 = 0 \; ; \; a_2 = (\omega + 6\frac{\delta}{T^2})/T^2 \tag{37}$$

$$a_3 = -(\omega + 8\frac{\delta}{T^2})/T \; ; \; a_4 = (\frac{\omega}{2} + 8\frac{\delta}{T^2})/T^2 \tag{38}$$

To complete the control equations, it remains to choose the models to be used in (7). In order to ensure a better tracking of the reference trajectory, we chose to compensate for the gravity term $G^1(q)$ and for the acceleration of the reference trajectory. Knowing that, here, the matrix $R$ of (5) is equal to 1, the final control scheme is:

$$\Gamma_1 = \hat{K}(-K_p e - K_v(\dot{q}_1 - \dot{q}_1^d)) + G^1(q) + \hat{M}_{11}\ddot{q}_1^d(t) \tag{39}$$

In this expression, the inertia matrix (therefore $\hat{K}$ and $\hat{M}_{11}$) is taken constant; the spring torque, assumed unknown, and the coriolis/centrifugal term are not compensated for.

### 5.2. Control in the Stance Phase

We simply take here $e_2 = q_2$ and apply the control of section 3.2 using (33) and (35). The matrix $\bar{R}$ of (18) is then

$$\bar{R} = (-C^1/C^2 \quad 1) \tag{40}$$

and $\bar{K}$ (21) is

$$\bar{K} = (K_1 \quad K_2) = \frac{1}{C^1}\begin{pmatrix} 1 & -C^2 \\ 0 & 1 \end{pmatrix} \tag{41}$$

Note that, here, we have $\bar{R} = K_2^T$. The expression (25) writes as: $u = -\frac{C^2}{C^1}\Gamma$ and the final control (eqs. (22 to 26)) is:

$$\Gamma_2 = \hat{A}(-K_p q_2 - K_v \dot{q}_2) + \hat{B} \tag{42}$$

where $\hat{A}$ and $\hat{B}$ are suitable approximations for $A = -C^1/C^2 K_2^T M K_2$ and $B = K_2^T N$ respectively. Let us now consider the case where $q_2$ is small, which is required for having the leg extended and thus a stance phase long enough. Then, $M(q)$ is approximately constant and also:

$$r \approx (\frac{h}{c_1} - l_1) - l_2 \; ; \; C^1 \approx l_1 c_1 + C^2 \; ; \; C^2 \approx \frac{1}{c_1}(r + l_2) \tag{43}$$

Now, if the range of variation of $q_1$ is small (in the stance phase, for maintaining the contact, $|q_1|$ cannot be greater than 7 deg.), we can set $c_1 \approx 1$. We can therefore choose $\hat{A} = A_{(q_1=q_2=0)}$ in (42). Since the gravity effects are small when the leg is almost vertical and that the velocities are small because they are only due to the treadmill motion, we can simply take $\hat{B} = 0$ in (42). We have constated experimentally that reasonnable value of gains allow to compensate for the influence of these modelling approximations on the error.

## 6. Results and Conclusion

The control laws have been implemented with a hip extension angle $q_1^*$ of 20 deg and a swing time $T$ of 1.2 sec. The treadmill velocity varied from 1 cm/sec to 25 cm/sec. The system then undergone successive regular cycles, achieving therefore successfully the goal of ballistic walking. For illustration, we present, in the figures 2, six snapshots of a full step, showing the efficiency of the control scheme.

In this paper we have thus presented the analysis of a problem of ballistic "walking" and we have proposed a control scheme which has been validated experimentally. We intend now to use the experimental testbed for studying improvments of the approach, mainly in two directions: a first idea would be to specify the desired behavior as a limit cycle (as done in [15]) and to try to design a control allowing to stabilize it; a second possibility would be to specify only the structure of the control, its parameters having to be adapted from one step to another, through an iterative learning control approach. This last method would allow to cope with unknown physical parameters, like masses or stiffnesses, that we cannot presently do.

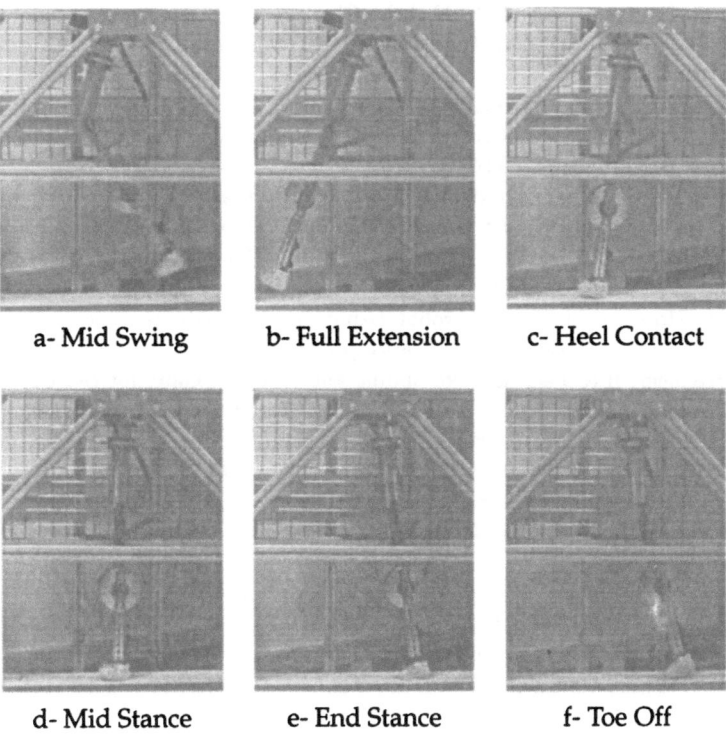

| a- Mid Swing | b- Full Extension | c- Heel Contact |

| d- Mid Stance | e- End Stance | f- Toe Off |

Figure 2. Snapshots of a full step

# References

[1] R. McN. Alexander 1990 Three Uses for Springs in Legged Locomotion, *The Int. J. of Robotics Research*, 9(2), pp 53-61

[2] G.T. Yamaguchi, A. Takanishi 1997 Development of a Biped Walking Robot Having Antagonistic Driven Joints Using Nonlinear Spring Mechanisms, *Proc. IEEE Int. Conf. on Robotics and Automation*, Albuquerque, pp 185-192

[3] R.Q. van der Linde 1998 Active Leg Compliance for Passive Walking, *Proc. IEEE Int. Conf. on Robotics and Automation*, Leuwen

[4] M. Bergerman, C. Lee, Y. Xu 1995 Experimental Study of an Underactuated Manipulator, *Proc. TEEE/RSJ Int. Conf. on Intelligent Robots and Systems*, Pittsburgh, pp 317-322

[5] S. Mochon, T.A. McMahon 1980 Ballistic Walking: An Improved Model, *Mathematical Biosciences* 52:241-260

[6] M. C. Laiou, A. Astolfi 1998 Underactuated Nonlinear Systems with Non-stabilizable Linear Approximation: A Case Study, *IFAC Int. Workshop on Motion Control*, Grenoble, pp 23-238

[7] A. De Luca, R. Mattone, G. Oriolo 1996 Control of Underactuated Mechanical Systems: Application to the planar 2R Robot, *Conf. on Decision and Control*, Kobe, pp 1455-1460

[8] M. Reyhanoglu, A. van der Shaft, N.H. McClamroch, I. Kolmanovsky: Nonlinear Control of a class of Underactuated Systems, *Conf. on Decision and Control*, Kobe, pp 1682-1687

[9] G. Oriolo, Y. Nakamura 1991 Control of Mechanical Systems with Second-order Nonholonomic Constraints: Underactuated Manipulators, *Conf. on Decision and Control*, Brighton, pp 2398-2403

[10] G.A. Pratt, M.M. Williamson 1995 Series Elastic Actuators, *Proc. IEEE/RSJ Int. Conf. on Intelligent Robots and Systems*, Pittsburgh, pp 399-406

[11] Pratt, J., Dilworth, P., Pratt, G. 1997 Virtual Model Control of a Bipedal Walking Robot, *Proc. IEEE Int. Conf. on Robotics and Automation*, Albuquerque

[12] 1994 Swing Up Control of the Acrobot Using Partial Feedback Linearization, *IFAC Symposium on Robot Control*, Capri, pp 833-838

[13] M.D. Berkemeier, R.S. Fearing 1998 Sliding and Hopping Gaits for the Underactuated Acrobot, *IEEE Trans. on Robotics and Automation*, 14-4:629-634

[14] F. Saito, T. Fukuda, F. Arai 1994 Swing and Locomotion Control for a Two-link Brachiation Robot, *IEEE Control Syst. Magazine*, 14:5-12

[15] A. Goswami, B. Espiau, A. Keramane 1997 Limit Cycles in a Passive Bipedal Gait and Passivity-mimicking Control Laws, *J. of Autonomous Robots*, 4:273-286, 1997

[16] B. Espiau, R. Boulic 1998 On the Computation and Control of the Mass Center of Articulated Chains, *INRIA Research Report* no 3479 (http://www.inria.fr/RRRT/publications-eng.html)

[17] F. Génot, B. Espiau 1998 On the Control of the Mass Center of Legged Robots Under Unilateral Constraints, *CLAWAR Symposium*, Bruxelles

[18] B. Espiau, I. Guigues, R. Pissard-Gibollet 1998 Can an Underactuated Leg with a Passive Spring at the Knee Achieve a Ballistic Step? *INRIA Research Report* no 3544 (http://www.inria.fr/RRRT/publications-eng.html)

# Chapter 7

# Localization and Map Building

As mentioned in Chapter 6, *really interesting* robots are mobile. But *useful* robots must be able to determine their location, build maps, plan paths and execute them. These topics were discussed in the double-session which incorporated papers from this Chapter and the next.

In this context maps are important. If I have a map then from observed features I can (perhaps not uniquely) infer my position. Conversely if I know where I am I can build a map based on what I observe. The 'holy grail' of many researchers is to be able to localize *and* build a map when started in an unknown environment, the so-called simultaneous localization and map building problem.

Dissanayake and colleagues describe a new theoretical approach to the simultaneous localization and map building (SLAM) problem which has been applied to navigation in an outdoor environment. Rives, Sequeira and Lourtie present experimental results for an indoor mobile robot using odometry and ultrasonic range sensor data which is fused by an extended Kalman filter.

A method by which multiple autonomous mobile robots, equipped with different sensors, can work together is described by Tercero, Paredis and Khosla. They model the environment as an occupancy grid and provide experimental results obtained using a computationally cheap Bayesian update technique.

The quality of sensor data will affect the performance of simultaneous localization and map building process. Castellanos et al. compare results for a mobile vehicle equipped with either a trinocular stereo system or a 2D laser rangefinder.

John Leonard and Hans Feder tackle this problem in the underwater environment which is potentially non-planar. The algorithm directs the sonar sensor towards targets so as to provide optimal localization. Experimental results from tank tests are presented.

PIC

# Navigation of a Martian Rover in Very Rough Terrain

by Ray Jarvis
Intelligent Robotics Research Centre
Monash University
Wellington Road, Clayton   vic 3168
Australia
Ph: +61 3 9905 3470   Fax: +61 3 9905 3454
e-mail: Ray.Jarvis@eng.monash.edu.au

## Abstract:

This paper presents initial experimental outcomes of research aimed at building an autonomous navigation system for a working scale model, Russian built, Marskohod Martian Rover capable of negotiating very rough terrain, including climbing over obstacles of up to the height of its own wheel diameter (25 cm). Whilst the final goal is to build a system capable of navigating the vehicle through initially unknown, very rough terrain, mainly by means of natural landmarks, the stage of research reported here concerns navigating only in known rough terrain.

## 1. Introduction:

Autonomous robot navigation has been an area of intense research in recent times [1,2]. The rich set of affordable sensors and computational resources now available to support such systems makes such an endeavour reasonably straightforward [3,4] in level floor indoor domains, whether or not the obstacle field is initially unknown and/or changing. Extending applicability to outdoor variable terrain situations is still a challenge [5] but one which is tractable due to the potential use of GPS localisation, optical gyroscopes, inertial navigation systems and laser range-finders etc.

In the context of potential (but hardly likely to be implemented in our case) planetary exploration, the use of GPS can not be contemplated and in any case, for some Earth based operations, GPS can be rendered inoperative due to obscurance by high-rise buildings, mountains, heavy foliage cover etc.

In very rough terrain there are still a number of fundamental but not completely answered questions concerning natural landmark based autonomous navigation in initially unknown environments which may be subject to change. Creating and maintaining 3D environmental maps and using them as the basis of localisation without unbounded, incrementally acquired errors whilst also accommodating the extent, geometry and climbing capability of the robot vehicle are still hard areas of research. There are several important advantages in using a natural landmark based approach to localisation. The first has already been referred to above as the avoidance of the need for prior site preparation. The second is that it is essentially a free-ranging method. Thirdly, if environmental mapping is needed for obstacle avoidance and path planning, the same sensory acquired data can be used for localisation. Fourthly, (and this is a point often overlooked), when obstacles themselves can be used as landmarks, their profusion (in cluttered situations) allows the position of the vehicle to be determined quite accurately (and particularly relative to the immediately proximate obstacles) just when such quality of information is critical for obstacle avoidance. In a complimentary manner, when obstacles used as natural landmarks are sparsely placed, the quality of absolute localisation may well degrade, but the need for accuracy is also reduced since the obstacle field is presumably less of a threat to the proper passage of the vehicle. Furthermore, even if the absolute localisation of the vehicle is subject to serious error, the relative placement of the vehicle within its immediate surroundings can still be sufficiently accurately known for obstacle avoidance to be carried out reliably.

There are two modes of natural landmark usage for localisation which can be identified; these are 'specific' landmark mode and the 'generalised' landmark mode. In many circumstances it may be advantageous to combine them. In 'specific' mode, landmarks are selected, identified and subsequently recognised as if they were beacons. A minimal set of previously selected and identified specific landmarks must be later recognised at measured bearing angles and/or range to determine the position and orientation of the robotic vehicle. It is not difficult to list the attributes of potential landmarks which make them suitable or ideal for this purpose. Specific landmarks should ideally be detectable in the sense of being easily extracted out of sensory data, unique in not being duplicated in the environment, temporally stable in that they should not move, viewpoint invariant in that they should appear similar from varying viewpoints, compact in that their position should be easily measured in a way which supports the geometrical accuracy of the fixing calculation, strategically well distributed in that no ill-posed groupings degrade the accuracy of localisation, prominent in that they remain observable over a large viewing area and unambiguously recognisable. To test such qualities, pattern recognition concepts such as feature extraction and classification can be adapted from the classical literature. Such schemes can provide 'filters' for selecting and maintaining a track of identifiable and otherwise suitable specific landmarks. Of course, continuity of observability can greatly assist in tracking reliability, since the search space for detection and matching candidates to previously collected data can be reduced through such constraints. Some terrains and particularly those that contain prominent man-made structures are excellent providers of ideal specific landmarks. In other cases the alternative (generalised) natural landmark mode may be more reliable and perhaps even essential.

In 'generalised' landmark mode, the concern is not to select, detect and recognise landmarks but to match raw data in the form of volumetric (shape) and visual (appearance) models of land masses, obstacles, undulating terrain etc. against previously sensed and represented data. This mode does not have 'recognition' as its key requirement and uniqueness of match is achieved through the complexity and scale of the matching process, appropriate exploitation of continuity constrains and use of dead reckoning estimation (e.g. odometry and/or inertial guidance) to provide local search starting points.   The advantage of this approach is that it does not require the rigour of selecting and tracking specific and recognisable landmarks but its chief disadvantage is that, whilst essentially structurally uncomplicated, it is computationally intense in its likely correlation based matching requirements over large correlation masks with multi-dimensional vector components.

For both modes, uncertainty can be accommodated through the exploitation of redundancy which is likely to be manifested within the complexities of the environment. The real 'trick' is to know precisely how to limit the computational requirements to meet both minimal reliability and accuracy needs in initially unknown and perhaps time-varying obstacle fields.

The above discussion on natural landmark based navigation provides a background for what is reported in this paper and represents the larger term aspirations of the project.

As a first step, a fully known environment is navigated and the landmarks used are not natural but contrived, but in such a way that the extension to natural landmarks does not require a drastic rethinking of strategy.

In the next section the model Marskohod M96 rover vehicle and its instrumentation are described.  Then follows a section on the artificially created laboratory environment used and the means employed to obtain an accurate 3D terrain map of it.  The next section deals with path planning in a very rough terrain environment.  Then a novel localisation method is described which, whilst using contrived landmarks, does not used identified beacons at pre-known locations. Descriptions of initial experiments carried out follow.   The paper closes with a section on future work, discussion and conclusions.

## 2.   Marskohod   M96   Martian   Rover   Model   and Instrumentation

Figure 1. Shows the unusual structure of the robot vehicle which makes it a superb climber.  This vehicle was built in St. Petersburg by the Rover Company (TRANSMARSH) using authentic designs relating back to earlier Soviet Space Agency activity in lunar and martian rover development.  Each of its titanium plate wheels contains an independent electric motor/gear set and shaft encoders.   The controller electronics were installed and tested at the Automation Laboratory at the Helsinki University of Technology.  The vehicle weighs approximately 30 kg and can carry a pay-load of about its weight.  Three passive rotational body joints (bend in the middle and swivels between the central wheel pair and those at each end) contribute to its ability to traverse very rough terrain and to climb over abrupt obstacles just higher than its own wheel diameter [See figure 5].  It can also climb loose surfaces of a slope gradient in excess of 30 degrees.  This vehicle is an ideal platform for the research proposed.   Its size makes it feasible to construct an artificial, indoor, very rough terrain 'sand pit' with sloping hills, plateaus, rocky outcrops and ridges, thus enabling the limits of the vehicle, sensors and artificial intelligence capabilities to be tested with convenience. Outdoor experiments can be carried out at a later stage in development.

The rover is richly instrumented. A four video camera frame on a pitch/roll (as distinct from a pan/tilt) mount is provided for possible multi-camera passive stereopsis ranging [6] and up to four other cameras (perhaps at the side and back) can be catered for. A 8 x 4 video cross bar switch is provided for video selection for radio transmission. Two output channels of the switch are fed into a field multiplexer so that a single video signal can be transmitted; at the receiving end the field sequential signal can be demultiplexed to provide a stereo pair of images for range analysis. The stereo cameras are mounted on a pitch/roll head so that the pose of the camera frame can be adjusted for the pitch and roll of the vehicle. A pitch/roll sensor mounted on the camera frame provides the feedback signals to permit this adjustment. In our initial experiments the camera sensor data is not used but longer term plans include stereo ranging and possibly optical flow ranging.

An Erwin-Sick sheet scanning laser time-of-flight range-finder which returns ± 3 cm accurate range data at 1/2 degree intervals of a 180 degree sweep is mounted on passive gimbal system which allows it to remain pointing forward with its beam horizontal. It is intended to motorise a weight which can move forward and backward to cause the scanner to look up or down but still under gravity so that the attitude of the vehicle is factored out. Also on-board is an Andrew optical gyroscope carefully calibrated over the working temperature range. It is used for steering control feedback.

The vehicle communicates with a Silicon Graphics workstation via a radio ethernet bridge; an on-board serial line server collects sensor data for transmission to the workstation and provides wheel drive and pitch/roll controls from the workstation to the vehicle. A video transmitter delivers camera data to the workstation vis a standard television receiver.

The vehicle [See Figure 2] has six hollow truncated cone wheels built of sheet titanium, each with a powerful motor/gear set inside. A set of six PID controllers is provided so that each wheel can be independently driven. In our initial experiments, however, the left set and the right set are driven in a ganged manner. The three axels can pivot at their mid points and the body of the vehicle has a passive link which allows the vehicle to pivot about a horizontal axis at its middle. This geometry contributes greatly to the vehicle's climbing skills.

## 3. Laboratory Marscape

The vehicle is small enough to navigate it indoors with the advantages of convenience and security but without invalidating the experiments in the context of potential outdoor repetition.

A 12m x 12m single room building was available for the project. After much thought it was discovered that the chute of a cement truck could deliver gravel, sand and scoria through the doors of this building and some 15 cubic metres of such materials were laid out within the building [See Figure 3]. The chief gardener at Monash University was kind enough to make available various volcanic looking rocks from his stockpile. Two upturned fountain pools served as larger structures to complete the 'sand pit', which itself is 10m x 10m, leaving some space around two sides for access and support equipment. The result is a very contemplative space.

The 3D terrain of the 'sand pit' was mapped using an Erwin Sick range-finder pointing downwards from a carriage mounted on a beam running the whole length of the ceiling at roof height. The carriage is accurately moved along the beam using a stepper motor driven drum with the sheet scan of the range-finder at right angles to the supporting beam. Over a two hour period, averaging multiple scans at each position, a 200 x 200 array of fairly accurate (± 1/2 cm) range measurements were obtained. From these a terrain map was built [See Figure 4]; the map has some flaws due to range shadows caused by rocks in the terrain, behind which the range-finder beam could not reach. These shadow flaws were interpolated over.

## 4. Path Planning

A Distance Transform [7], DT, path planning strategy devised by the author and proven robust in previous work [4] was used in this project. In previous applications the DT strategy would begin by creating a binary map indicating occupied and unoccupied cells on a rectangularly tessellated grid. This map could pre-exist or be developed incrementally using on-board sensor data with replanning at every modification of the map. The vehicle is absolutely not able to penetrate any obstacle space.

The situation for this current work requires a different approach, since there is a continuum of tractability represented by the combination of the 3D terrain map and versatile climbing capability of the vehicle. As a start, two simple ideas were tried and seem to be very promising. Firstly, the size of the rover was accommodated by propagating (at high points in the terrain) the highest local point height in the terrain over a specified distance corresponding to the turning circle radius of the vehicle. This has the effect of broadening the protrusions in the landscape. More appropriate constructions will be applied at a later stage. Then a terrain cost map was built by regarding the tractability of each cell of the 200 x 200 terrain map as proportional to the square of the localised average absolute slope of the terrain. Now the DT was applied in the normal way after nominating a goal point. The method propagates accumulating costs away from the goal to all parts of the map. A steepest descent path from any starting point leads to the goal in an optimal path that has no local entrapment flaws. The results were very pleasing [See Figure 5] since they produced paths that humans might sensibly choose. Note that the one cost propagation process serves all possible starting points.

For simplicity's sake, it was felt that an optimal path consisting of cell to cell transitions from a starting point to goal was not as useful as a sequence of straight line segments for the purpose of controlling the vehicle. A sequence of rotation and straight line movements of the vehicle are not only easy to execute but the localisation procedure yet to be described is simplified in this way.

Rather than developing some formal straight line approximation strategy a direct and effective, intuitively pleasing method was successfully devised.

The method is as follows:

1. Step forward from a current starting point cell until any point along the way is greater than some specified tolerance distance from the straight line from the current starting point to the current stepping point.
2. Step back one point and establish a new starting point.
3. Repeat 1 and 2 until the goal is reached.

The straight line segments between the sequence of starting points is the piecewise linear path

which can be navigated by pairs of turns and straight line movements. This strategy was slightly modified by also nominating a minimal length of straight line, but this was not a significant change, it merely rides over a number of small kinks.

Figure 5 shows the terrain map, the optimal path and the straight line breakpoints (spikes) for the path for a typical tolerance figure. Setting a larger tolerance figure increase the average length of the straight line segments but with increasing risk that the path is hazardous.

## 5. Localisation

The longer term plans for this project include fully autonomous navigation in very rough, initially unknown (and possibly changing) terrain using only natural landmarks, knowledge about which must be extracted from on-board sensor data. The current work deals with a known environment and contrived landmarks. Wanting to avoid using beacons such as laser or ultrasound strobes or bar codes as being just too far away from the natural landmark approach and requiring too much prior site preparation, a novel alternative localisation method involving 'null beacon' was devised.

It was discovered, after much experimentation that a particular shiny gift wrapping paper with a golden sheen returned very little energy to the Sick range-finder when the material was curved and sloped away from the range-finder. This curious finding led to the building of one metre high cones covered with the selected wrapping paper. When these cones are put into the terrain they return the maximum range readings (set at 50 metres) from the range-finder, that is, they look like holes in the environment, hence the term 'null beacons'. Examples of these 'null beacons' can be clearly seen in Figure 3. These holes are so unique that nothing natural in the limited environment can be mistaken for them. The range measured to non reflective sheets or panels directly behind the cones but protruding beyond them give estimates of how far away the cones are. The holes signify cones and the back board gives their distance; thus a set of bearing/distance pair of values can be used as a basis of localisation, with a minimum of two detectable 'holes' being sufficient for this purpose. 'Holes' in the range data can be clearly seen in Figure 6.

The implementation of this scheme is nearly complete.

The procedure for localisation is as follows:

1. From a known starting point and orientation a set of bearing/range values to several cone 'null beacons' is collected. The offset from the range sensors to

the centre of rotation of the rover is compensated for and a set of bearing/range pairs relative to the centre of rotation and orientation of the robot is derived.

2 . The robot is directed to rotate a specified angle and then to move a specified small distance in a straight line along a previously determined path to the specified goal.

3 . The dead reconning new position and orientation of the robot is calculated using the gyroscope for the turn angle and estimated linear movement (currently using timed constant speed) for distance.

4 . A new set of bearing/range pairs for observable beacons are determined using the range-finder and range-finder/centre of rotation offset adjustments made.

5 . Using the dead reconning estimates from 3, above, the location of the beacons are geometrically rotated and translated to where they should be relative to the starting reference if the dead reconning estimation were accurate.

6 . Beacons are matched using nearest neighbour distance measures, excluding any pairs which don't meet a specified distance tolerance.

7 . Adjustments are made to the estimation position and orientation of the vehicle and steps 5, 6 and 7 repeated until convergence. A first try adjustment strategy is to calculate the bearing and lateral shift which would allow each separate beacon pair to be exactly matched and add a fraction of the average rotations and lateral shifts to the previous estimates.

Provided at least two of the beacons discovered in 4, above, can be correctly matched with those found in 1, above, this method should be acceptable and not accumulate errors. Testing of this procedure is not yet complete, unfortunately.

## 6. Experiments and Current Project Status

All individual sensors and their communication links to an off-board Silicon Graphics workstation have been tested successfully. The optical gyroscope, tilt sensor and Erwin Sick range-finder are serviced by software daemons which collect data and store it at shared memory locations. Each sensor is also served by a graphics window which can be opened to display tits data. A steering wheel allows manual control of the robot in teleoperation mode. A vector arrow can be dragged around a dial and its length changed with the mouse. Various differential speeds of the right three wheels and left three wheels can be altered at will to drive the vehicle forward, in reverse, turn on the spot or to follow a curved path. The video multiplexer can be switched using a selector panel which appears on screen when required. Field sequential video collected from the on-board video transmitter by a standard TV receiver can be connected to the video input port of the workstation and displayed. If the multiplexing carry the same signal a normal interfaced image can be viewed on the workstation. In either case the odd and even interlaced scans can be separated and displayed separately.

Terrain data from an overhead scan of the laboratory can be collected when needed, perhaps when changes have been made to the environment.

Path planning and piecewise linear paths can be generated using this terrain data and specifying goal and start points.

The code for the 'null beacon' localisation scheme has been written and is currently being debugged. Once this component is operational, all that remains is to forge the links between the movement control, the planned path and the localising scheme, all of which are separately in place ready for linking, which should be a trivial exercise.

## 7. Discussion and Conclusion

This paper has presented the first stage of a project to develop an autonomous navigation system for a working model Martian rover in very rough terrain, initially unknown and subject to change, using natural landmarks. The current stage of the project is that navigation in known rough terrain using contrived but not identified landmarks is almost complete.

In the near future, natural landmarks which can be found using on-board sensor data will replace the contrived landmarks and it will be of interest to see if a variation of the contrived landmark localisation scheme will operate under this new regime.

One of the real difficulties of 3D terrain mapping using on-board sensors alone is the mutual uncertainty of the localisation/mapping inter-relationship when the mapping provides the basis of localisation, as it does in the natural landmark process. Just how to maximally exploit the intrinsic redundancy of these processes as a weapon against uncertainty is really the main game. If this conundrum can be practically resolved one has gone a long way towards a total solution to the natural landmark based navigation problem.

## Bibliography:

1. Haist, A., Siméon, T. and Taix, M. A Landmark-based Planner for Rough Terrain Navigation, Experimental Robotics V, The Fifth International Symposium, Barcelona, Catalonia, June 15-18, 1997, pp.216-226.
2. Suzuki, S., Kato, T., Asada, M. and Hosoda, K. Behaviour Learning for a Mobile Robot with Omnidirectional Vision Enhanced by an Active Zoom Mechanism, Intelligent Autonomous Systems, edited by Kakazu, Y., Wada, M. And Sato.T., IOS Press, 1998, pp.242-249.
3. Montano, L. And Asensio, J.R. Real-Time Robot Navigation in Unstructured Environments Using a 3D Laser Rangefinder, Proc. IROS 97, Genoble, France, Sept. 7-11, 1997, pp.526-532.
4. Jarvis, R.A. Etherbot - An Autonomous Mobile on a Local Area Network Radio Tether, Proc. Fifth International Symposium on Experimental Robotics, Barcelona, Catalonia, June 15-18, 1997, pp.151-163.
5. Jarvis, R.A. An Autonomous Heavy Duty Outdoor Robotic Tracked Vehicle, International Conference on Robots and Systems, Grenoble, France, Sept. 8-12, 1997, pp.352-359.

6 . Jarvis, R.A. Quad - Vision Ranging for Applications, Proc. 4th Australian Joint Conference on Artificial Intelligence, Nov. 21-23, 1990, Perth, Western Australia, pp.682-698.

7 . Jarvis, R.A. On Distance Transform Based Collision - Free Path Planning for Robot Navigation in Known, Unknown and Time-Varying Environments, invited chapter for a book entitled 'Advanced Mobile Robots' edited by Professor Yuan F. Zang World Scientific Publishing Co. Pty. Ltd. 1994, pp.3-31.

Figure 1. Marskohod M96 Martian Rover Vehicle.

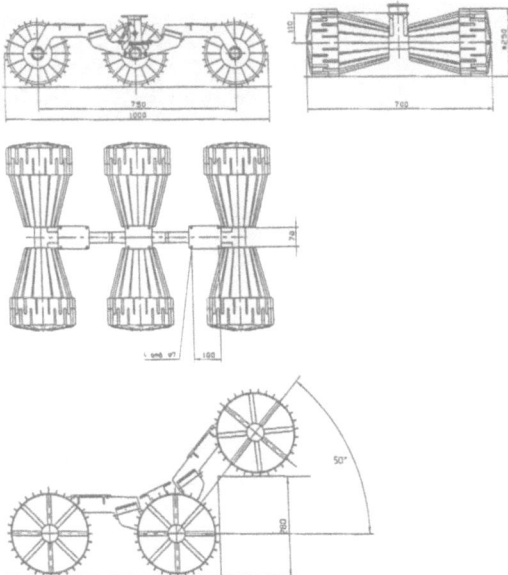

Figure 2. Rover Geometry.

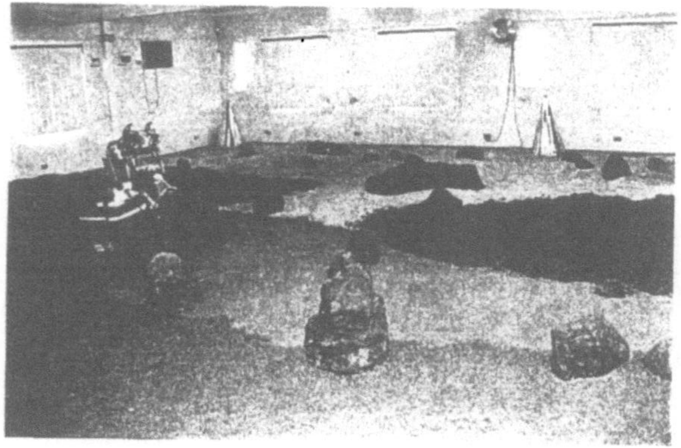

Figure 5. Terrain Map with Optimal Path (and breakpoints for piecewise straight line approximation).

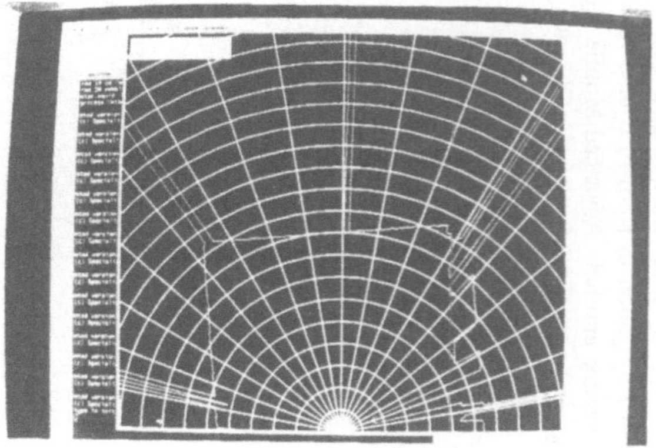

Figure 6. 'Holes' in the Range Data.

Figure 3. Indoor 'Martian' Terrain 'Sand Pit'.

Figure 4. Terrain Map of 'Sand Pit'.

# An Experimental and Theoretical Investigation into Simultaneous Localisation and Map Building

M.W.M.G. Dissanayake, P.Newman, H.F. Durrant-Whyte,
S. Clark and M. Csorba
Australian Centre for Field Robotics
Department of Mechanical and Mechatronic Engineering
The University of Sydney
NSW 2006, Australia

### Abstract

The simultaneous localisation and map building (SLAM) problem asks if it is possible for an autonomous vehicle to start in an unknown location in an unknown environment and then to incrementally build a map of this environment while simultaneously using this map to compute absolute vehicle location. Starting from the estimation-theoretic foundations of this problem developed in [5, 4, 2], this paper proves that a solution to the SLAM problem is indeed possible. The underlying structure of the SLAM problem is first elucidated. A proof that the estimated map converges monotonically to a relative map with zero uncertainty is then developed. It is then shown that the absolute accuracy of the map and the vehicle location reach a lower bound defined only by the initial vehicle uncertainty. Together, these results show that it is possible for an autonomous vehicle to start in an unknown location in an unknown environment and, using relative observations only, incrementally build a perfect map of the world and simultaneously to compute a bounded estimate of vehicle location.

## 1   Introduction

The solution to the simultaneous localisation and map building (SLAM) problem is, in many respects, a "Holy Grail" of the autonomous vehicle research community. The ability to place an autonomous vehicle at an unknown location in an unknown environment and then have it build a map, using only relative observations of the environment, and then simultaneously to use this map to navigate would indeed make such a robot "autonomous". A solution to the SLAM problem would be of inestimable value in a range of applications where absolute position or precise map information is unobtainable, including, amongst others, autonomous planetary exploration, subsea autonomous vehicles, autonomous air-borne vehicles, and autonomous all-terrain vehicles in tasks such as mining and construction.

The general SLAM problem has been the subject of substantial research since the inception of a robotics research community and indeed before this in areas such as manned vehicle navigation systems and geophysical surveying. A number of approaches have been proposed to address both the SLAM problem and also more simplified navigation problems where additional map or vehicle location information is made available. The most popular of these is the estimation-theoretic or Kalman-filter based approach. A major advantage of this approach is that it is possible to develop a complete proof of the various properties of the SLAM problem and to study systematically the evolution of the map and the uncertainty in the map and vehicle location. A proof of existence and convergence for a solution of the SLAM problem within a formal estimation-theoretic framework also encompasses the widest possible range of navigation problems and implies that solutions to the problem using other approaches are possible.

This paper starts from the original estimation-theoretic work of Smith, Self and Cheeseman [5]. It assumes an autonomous vehicle (mobile robot) equipped with a sensor capable of making measurements of the location of landmarks relative to the vehicle. The vehicle starts at an unknown location with no knowledge of the location of landmarks in the environment. As the vehicle moves through the environment (in a stochastic manner) it makes relative observations of the location of individual landmarks. This paper then proves the following three results:

1. The determinant of any submatrix of the map covariance matrix decreases monotonically as observations are successively made.

2. In the limit as the number of observations increases, the landmark estimates become fully correlated.

3. In the limit, the covariance associated with any single landmark location estimate is determined only by the initial covariance in the vehicle location estimate.

These three results describe, in full, the convergence properties of the map and its steady state behaviour.

This paper makes three principal contributions to the solution of the SLAM problem. Firstly, it proves three key convergence properties of the full SLAM filter. Secondly, it elucidates the true structure of the SLAM problem and shows how this can be used in developing consistent SLAM algorithms.

# 2   The Estimation-Theoretic SLAM Problem

This section establishes the mathematical framework employed in the study of the SLAM problem. This framework is identical in all respects to that used in Smith et. al. [5] and uses the same notation as that adopted in Leonard and Durrant-Whyte [2].

## 2.1 Vehicle and Land-Mark Models

The setting for the SLAM problem is that of a vehicle with a known kinematic model, starting at an unknown location, moving through an environment containing a population of features or landmarks. The terms feature and landmark will be used synonymously. The vehicle is equipped with a sensor that can take measurements of the relative location between any individual landmark and the vehicle itself as shown in Figure 1. The absolute locations of the landmarks are not available. Without prejudice, a linear (synchronous) discrete-time model of the evolution of the vehicle and the observations of landmarks is adopted. Although vehicle motion and the observation of landmarks is almost always non-linear and asynchronous in any real navigation problem, the use of linear synchronous models does not affect the validity of the proofs in Section 3 other than to require the same linearisation assumptions as those normally employed in the development of an extended Kalman filter. The state of the system of

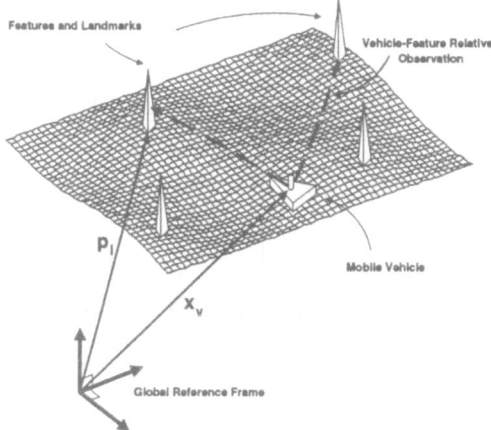

Figure 1: The Structure of the SLAM problem. A vehicle takes relative observations of environment landmarks. The absolute location of landmarks and vehicle are unknown.

interest consists of the position and orientation of the vehicle together with the position of all landmarks. The state of the vehicle at a time step $k$ is denoted $\mathbf{x}_v(k)$. The motion of the vehicle through the environment is modeled by a conventional linear discrete-time state transition equation or process model of the form

$$\mathbf{x}_v(k+1) = \mathbf{F}_v(k)\mathbf{x}_v(k) + \mathbf{u}_v(k+1) + \mathbf{v}_v(k+1), \qquad (1)$$

where $\mathbf{F}_v(k)$ is the state transition matrix, $\mathbf{u}_v(k)$ a vector of control inputs, and $\mathbf{v}_v(k)$ a vector of temporally uncorrelated process noise errors with zero mean and covariance $\mathbf{Q}_v(k)$ ( [3] for further details). The location of the $i^{th}$ landmark is denoted $\mathbf{p}_i$. The "state transition equation" for the $i^{th}$ landmark is

$$\mathbf{p}_i(k+1) = \mathbf{p}_i(k) = \mathbf{p}_i, \tag{2}$$

since landmarks are assumed stationary. Without loss of generality the number of landmarks in the environment is arbitrarily set at $N$. The vector of all $N$ landmarks is denoted

$$\mathbf{p} = \begin{bmatrix} \mathbf{p}_1^T & \cdots & \mathbf{p}_N^T \end{bmatrix}^T \tag{3}$$

The augmented state vector containing both the state of the vehicle and the state of all landmark locations is denoted

$$\mathbf{x}(k) = \begin{bmatrix} \mathbf{x}_v^T(k) & \mathbf{p}_1^T & \cdots & \mathbf{p}_N^T \end{bmatrix}^T. \tag{4}$$

The augmented state transition model for the complete system may now be written as

$$\begin{bmatrix} \mathbf{x}_v(k+1) \\ \mathbf{p}_1 \\ \vdots \\ \mathbf{p}_N \end{bmatrix} = \begin{bmatrix} \mathbf{F}_v(k) & 0 & \cdots & 0 \\ 0 & \mathbf{I}_{p_1} & \cdots & 0 \\ \vdots & \vdots & \ddots & 0 \\ 0 & 0 & 0 & \mathbf{I}_{p_N} \end{bmatrix} \begin{bmatrix} \mathbf{x}_v(k) \\ \mathbf{p}_1 \\ \vdots \\ \mathbf{p}_N \end{bmatrix} \tag{5}$$

$$+ \begin{bmatrix} \mathbf{u}_v(k+1) \\ \mathbf{0}_{p_1} \\ \vdots \\ \mathbf{0}_{p_N} \end{bmatrix} + \begin{bmatrix} \mathbf{v}_v(k+1) \\ \mathbf{0}_{p_1} \\ \vdots \\ \mathbf{0}_{p_N} \end{bmatrix} \tag{6}$$

$$\mathbf{x}(k+1) = \mathbf{F}(k)\mathbf{x}(k) + \mathbf{u}(k+1) + \mathbf{v}(k+1) \tag{7}$$

where $\mathbf{I}_{p_i}$ is the $\dim(p_i) \times \dim(p_i)$ identity matrix and $\mathbf{0}_{p_i}$ is the $\dim(p_i) \times \dim(p_i)$ null matrix.

## 2.2 The Observation Model

The vehicle is equipped with a sensor that can obtain observations of the relative location of landmarks with respect to the vehicle. Again, without prejudice, observations are assumed to be linear and synchronous. The observation model for the $i^{th}$ landmark is written in the form

$$\mathbf{z}_i(k) = \mathbf{H}_i\mathbf{x}(k) + \mathbf{w}_i(k) \tag{8}$$

$$= \mathbf{H}_{p_i}\mathbf{p} - \mathbf{H}_v\mathbf{x}_v(k) + \mathbf{w}_i(k) \tag{9}$$

$$\tag{10}$$

where $\mathbf{w}_i(k)$ is a vector of temporally uncorrelated observation errors with zero mean and variance $\mathbf{R}_i(k)$. It is important to note that the observation model for the $i^{th}$ landmark is written in the form

$$\mathbf{H}_i = [-\mathbf{H}_v, 0 \cdots 0, \mathbf{H}_{p_i}, 0 \cdots 0]$$

$$= [-\mathbf{H}_v, \mathbf{H}_{mi}]$$

This structure reflects the fact that the observations are "relative" between the vehicle and the landmark, often in the form of relative location, or relative range and bearing.

## 2.3 The Estimation Process

In the estimation-theoretic formulation of the SLAM problem, the Kalman filter is used to provide estimates of vehicle and landmark location [3] and [2]. The Kalman filter recursively computes estimates for a state $\mathbf{x}(k)$ which is evolving according to the process model in Equation 5 and which is being observed according to the observation model in Equation 8. The Kalman filter algorithm proceeds recursively in three stages:

- Prediction: The algorithm first generates a prediction for the state estimate, the observation (relative to the $i^{th}$ landmark) and the state estimate covariance at time $k + 1$ according to

$$\hat{\mathbf{x}}(k + 1|k) = \mathbf{F}(k)\hat{\mathbf{x}}(k|k) + \mathbf{u}(k) \tag{11}$$

$$\hat{\mathbf{z}}_i(k + 1|k) = \mathbf{H}_i(k)\hat{\mathbf{x}}(k + 1|k) \tag{12}$$

$$\mathbf{P}(k + 1|k) = \mathbf{F}(k)\mathbf{P}(k|k)\mathbf{F}^T(k) + \mathbf{Q}(k), \tag{13}$$

respectively.

- Observation: Following the prediction, an observation $\mathbf{z}_i(k + 1)$ of the $i^{th}$ landmark of the true state $\mathbf{x}(k + 1)$ is made according to Equation 8. Assuming correct landmark association, an innovation is calculated as follows

$$\nu_i(k + 1) = \mathbf{z}_i(k + 1) - \hat{\mathbf{z}}_i(k + 1|k) \tag{14}$$

together with an associated innovation covariance matrix given by

$$\mathbf{S}_i(k + 1) = \mathbf{H}_i(k)\mathbf{P}(k + 1|k)\mathbf{H}_i^T(k) + \mathbf{R}_i(k + 1). \tag{15}$$

- Update: The state estimate and corresponding state estimate covariance are then updated according to:

$$\hat{\mathbf{x}}(k + 1|k + 1) = \hat{\mathbf{x}}(k + 1|k) + \mathbf{W}_i(k + 1)\nu_i(k + 1) \tag{16}$$

$$\mathbf{P}(k + 1|k + 1) = \mathbf{P}(k + 1|k) - \mathbf{W}_i(k + 1)\mathbf{S}_i(k + 1)\mathbf{W}_i^T(k + 1) \tag{17}$$

Where the gain matrix $\mathbf{W}_i(k + 1)$ is given by

$$\mathbf{W}_i(k + 1) = \mathbf{P}(k + 1|k)\mathbf{H}_i^T(k)\mathbf{S}_i^{-1}(k + 1) \tag{18}$$

The update of the state estimate covariance matrix is of paramount importance to the SLAM problem. Understanding the structure and evolution of the state covariance matrix is the key component to this solution of the SLAM problem.

# 3  Structure of the SLAM problem

In this section proofs for the three key results underlying the structure of the SLAM problem are provided. The implications of these results will also be examined in detail. The properties of positive semi-definite matrices that are invoked implicitly in the following proofs.

## 3.1  Convergence of the map covariance matrix

The state covariance matrix may be written in block form as

$$\mathbf{P}(i|j) = \begin{bmatrix} \mathbf{P}_{vv}(i|j) & \mathbf{P}_{vm}(i|j) \\ \mathbf{P}_{vm}^T(i|j) & \mathbf{P}_m(i|j) \end{bmatrix}$$

where $\mathbf{P}_{vv}(i|j)$ is the error covariance matrix associated with the vehicle state estimate, $\mathbf{P}_m(i|j)$ is the map covariance matrix associated with the landmark state estimates, and $\mathbf{P}_{vm}(i|j)$ is the cross-covariance matrix between vehicle and landmark states.

**Theorem 1** *The determinant of any submatrix of the map covariance matrix decreases monotonically as successive observations are made.*

The algorithm is initialised using a positive semi-definite (*psd*) state covariance matrix $\mathbf{P}(0|0)$. The matrices $\mathbf{Q}$ and $\mathbf{R}_i$ are both *psd*, and consequently the matrices $\mathbf{P}(k+1|k)$, $\mathbf{S}_i(k+1)$, $\mathbf{W}_i(k+1)\mathbf{S}_i(k+1)\mathbf{W}_i^T(k+1)$ and $\mathbf{P}(k+1|k+1)$ are all *psd*. From Equation 17, and for any landmark $i$,

$$\det \mathbf{P}(k+1|k+1) = \det(\mathbf{P}(k+1|k) - \mathbf{W}_i(k+1)\mathbf{S}_i(k+1)\mathbf{W}^T(k+1))$$
$$\leq \det \mathbf{P}(k+1|k). \tag{19}$$

The determinant of the state covariance matrix is a measure of the volume of the uncertainty ellipsoid associated with the state estimate. Equation 19 states that the total uncertainty of the state estimate does not increase during an update.

Any principle submatrix of a *psd* matrix is also *psd*. Thus, from Equation 19 the map covariance matrix also has the property

$$\det \mathbf{P}_m(k+1|k+1) \leq \det \mathbf{P}_m(k+1|k) \tag{20}$$

From Equation 13, the full state covariance prediction equation may be written in the form

$$\begin{bmatrix} \mathbf{P}_{vv}(k+1|k) & \mathbf{P}_{vm}(k+1|k) \\ \mathbf{P}_{vm}^T(k+1|k) & \mathbf{P}_m(k+1|k) \end{bmatrix} = \begin{bmatrix} \mathbf{F}_v\mathbf{P}_{vv}(k|k)\mathbf{F}_v^T + \mathbf{Q}_{vv} & \mathbf{F}_v\mathbf{P}_{vm}(k|k) \\ \mathbf{P}_{vm}^T(k|k)\mathbf{F}_v^T & \mathbf{P}_m(k|k). \end{bmatrix}$$

Thus, as landmarks are assumed stationary, no process noise is injected in to the predicted map states. Consequently, the map covariance matrix and any principle submatrix of the map covariance matrix has the property that

$$\mathbf{P}_m(k+1|k) = \mathbf{P}_m(k|k). \tag{21}$$

Note, that this is clearly not true for the full covariance matrix as process noise is injected in to the vehicle location predictions and so the prediction covariance grows during the prediction step.

It follows from Equations 20 and 21 that the map covariance matrix has the property that

$$\det \mathbf{P}_m(k+1|k+1) \leq \det \mathbf{P}_m(k|k). \tag{22}$$

Furthermore, the general properties of *psd* matrices ensure that this inequality holds for *any* submatrix of the map covariance matrix. In particular, for any diagonal element $\sigma_{ii}^2$ of the map covariance matrix (state variance),

$$\sigma_{ii}^2(k+1|k+1) \leq \sigma_{ii}^2(k|k).$$

Thus the error in the estimate of the absolute location of every landmark also does not increase.

**Theorem 2** *In the limit as successive observations are made, the errors in estimated landmark location become fully correlated.*

As the map covariance matrix does not increase, in the limit it will reach a steady-state lower bound such that

$$\lim_{k \to \infty} [\mathbf{P}_m(k+1|k+1) - \mathbf{P}_m(k|k)] = 0 \tag{23}$$

The update stage for the SLAM algorithm can be written as

$$
\begin{aligned}
\mathbf{P}(k+1|k+1) &= \mathbf{P}(k+1|k) - \mathbf{W}_i(k+1)\mathbf{S}_i\mathbf{W}_i^T(k+1) \\
&= \mathbf{P}(k+1|k) - \mathbf{P}(k+1|k)\mathbf{H}_i^T\mathbf{S}_i^{-1}\mathbf{H}_i\mathbf{P}(k+1|k) \\
&= \mathbf{P}(k+1|k) + \begin{bmatrix} \mathbf{M}_1 \\ \mathbf{M}_2 \end{bmatrix} \mathbf{S}_i^{-1} \begin{bmatrix} \mathbf{M}_1^T & \mathbf{M}_2^T \end{bmatrix} \\
&= \mathbf{P}(k+1|k) - \begin{bmatrix} \mathbf{M}_1\mathbf{S}_i^{-1}\mathbf{M}_1^T & \mathbf{M}_1\mathbf{S}_i^{-1}\mathbf{M}_2^T \\ \mathbf{M}_2\mathbf{S}_i^{-1}\mathbf{M}_1^T & \mathbf{M}_2\mathbf{S}_i^{-1}\mathbf{M}_2^T \end{bmatrix} \tag{24}
\end{aligned}
$$

where

$$\mathbf{M}_1 = -\mathbf{P}_{vv}(k+1|k)\mathbf{H}_v + \mathbf{P}_{vm}(k+1|k)\mathbf{H}_{mi}^T$$
$$\mathbf{M}_2 = -\mathbf{P}_{vm}(k+1|k)^T\mathbf{H}_v + \mathbf{P}_m(k+1|k)\mathbf{H}_{mi}^T$$

The update of the map covariance matrix $\mathbf{P}_m$ can now be written as

$$
\begin{aligned}
\mathbf{P}_m(k+1|k+1) &= \mathbf{P}_m(k+1|k) - \mathbf{M}_2\mathbf{S}_i^{-1}\mathbf{M}_2^T \\
&= \mathbf{P}_m(k|k) - \mathbf{M}_2\mathbf{S}_i^{-1}\mathbf{M}_2^T \tag{25}
\end{aligned}
$$

Together, equations 23 and 25 require that the matrix $\mathbf{M}_2\mathbf{S}_i^{-1}\mathbf{M}_2^T = 0$. As the inverse of the innovation covariance matrix $\mathbf{S}_i^{-1}$ is always *psd*, this requires that $\mathbf{M}_2 = 0$ or

$$\mathbf{P}_m(k|k)\mathbf{H}_{mi}^T = \mathbf{P}_{vm}(k|k)^T\mathbf{H}_v \tag{26}$$

Equation 26 holds for all landmark observation models $i$ and thus, in the limit, the block columns of $\mathbf{P}_m(k|k)$ are linearly dependent.

A consequence of this fact is that in the limit the determinant of any submap of the map covariance matrix, containing at least two landmarks, tends to zero.

$$\lim_{k \to \infty} \left[ \det \mathbf{P}_m(k|k) \right] = 0 \tag{27}$$

This implies that the landmarks become progressively more correlated as successive observations are made. In the limit, given the exact location of one landmark the location of all other landmarks can be deduced with absolute certainty and the map is fully correlated.

In the specific case where landmarks are similar (all points for example), then the observation models will be the same $\mathbf{H}_{pi} = \mathbf{H}_{pj}$ and so Equation 26 requires that the block columns of $\mathbf{P}_m(k|k)$ are also identical. Furthermore, because $\mathbf{P}_m(k|k)$ is symmetric it follows that for any two landmarks $i$ and $j$, the elements of the joint covariance matrix must satisfy

$$\mathbf{P}_{ii}(k|k) = \mathbf{P}_{jj}(k|k) = \mathbf{P}_{ij}(k|k) = \mathbf{P}_{ij}^T(k|k) \tag{28}$$

A consequence of this is that the covariance in the estimated relative location of landmarks tends to zero: Define $\hat{\mathbf{d}}_{ij}(k|k)$ to be the estimated relative location of two landmarks:

$$\hat{\mathbf{d}}_{ij}(k|k) = \hat{\mathbf{p}}_i(k|k) - \hat{\mathbf{p}}_j(k|k)$$
$$= \mathbf{G}_{ij}\hat{\mathbf{x}}(k|k)$$

The estimated covariance $\mathbf{P}_d(k|k)$ of $\hat{\mathbf{d}}(k|k)$ is computed as

$$\mathbf{P}_d(k|k) = \mathbf{G}_{ij}\mathbf{P}(k|k)\mathbf{G}_{ij}^T$$
$$= \mathbf{P}_{ii}(k|k) + \mathbf{P}_{jj}(k|k) - \mathbf{P}_{ij}(k|k) - \mathbf{P}_{ij}^T(k|k) = 0.$$

Thus, in the limit, $\mathbf{P}_d(k|k) = 0$ and the relationship between the landmarks is known with complete certainty.

## 3.2 Lower bounds on the map covariance matrix

The two map convergence theorems are concerned only with the relations between landmarks. It has been demonstrated that the uncertainty in the relative locations of landmarks decreases monotonically to zero as successive observations are made. A consequence of Theorem 1 is that the absolute landmark covariance also do not increase. However, Theorem 2 *does not* imply that the absolute landmark covariances also go to zero. Theorem 3 shows that the absolute landmark location covariances do reach a (non zero) lower bound.

**Theorem 3** *In the limit, the lower bound on the covariance matrix associated with any single landmark estimate is determined only by the initial covariance in the vehicle estimate $\mathbf{P}_{0v}$ at the time of the first sighting of the first landmark.*

It is convenient to use the information form of the Kalman filter to examine the limiting behaviour of the state covariance matrix [3]. For a single landmark, the state covariance update equation may be written as

$$P(k|k)^{-1} = P(k|k-1)^{-1} + \begin{bmatrix} -H_v^T \\ H_{pl}^T \end{bmatrix}^T R_1^{-1} \begin{bmatrix} -H_v & H_{pl} \end{bmatrix} \qquad (29)$$

Consider first the case when $Q_v(k) = 0$ so that, for map elements,

$$P^{-1}(k|k-1) = P^{-1}(k-1|k-1). \qquad (30)$$

Applying Equations 29 and 30 successively for $k$ observations of a single landmark results in

$$P(k|k)^{-1} = \begin{bmatrix} P_{0v}^{-1} & 0 \\ 0 & 0 \end{bmatrix} + \begin{bmatrix} kH_v^T R_1^{-1} H_v & -kH_v^T R_1^{-1} H_{pl} \\ -kH_{pl}^T R_1^{-1} H_v & kH_{pl}^T R_1^{-1} H_{pi}. \end{bmatrix}$$

Invoking the matrix inversion lemma for partitioned matrices,

$$P(k|k) = \begin{bmatrix} P_{0v}^{-1} & P_{0v}^{-1}H_v^T H_{pl}^{-T} \\ H_{pl}^{-1}H_v P_{0v} & H_{pl}^{-1}H_v P_{0v} \left[ H_{pl}^{-1}H_v \right]^T + \frac{H_{pl}^{-1}R_1 H_{pl}^{-T}}{k} \end{bmatrix},$$

where $H_{pl}^{-1}$ is taken to be the appropriate generalised inverse. In the limit,

$$\lim_{k \to \infty} P(k|k) = \begin{bmatrix} P_{0v}^{-1} & P_{0v}^{-1}H_v^T \left[ H_{pl}^T \right]^{-1} \\ H_{pl}^{-1}H_v P_{0v} & H_{pl}^{-1}H_v P_{0v} \left[ H_{pl}^{-1}H_v \right]^T \end{bmatrix} \qquad (31)$$

Equation 31 gives a lower bound on the solitary landmark state estimate variance as

$$P_{ii}(\infty) = H_{pl}^{-1}H_v P_{0v} \left[ H_{pl}^{-1}H_v \right]^T. \qquad (32)$$

Theorem 2 requires that the absolute map variances become equal in the limit. Thus Equation 32 also provides a limit for all absolute landmark variances.

Equation 32 shows that, in the case where $Q_v(k) = 0$, the limiting map covariance depends only on the initial vehicle location uncertainty $P_{0v}$. The term $H_{pl}^{-1}H_v$ simply transforms covariance information from the vehicle state space to landmark state space.

In the case where $Q_v(k) \neq 0$, the two competing effects of loss of information due to process noise injection and the increase in information content through observations, determine the limiting covariance. Determining the limit in this case is analytically intractable. The limiting covariance of the map will generally depend on $P_{0v}, Q$ and $R$, but can never be below the limit given in Equation 31.

# 4 Discussion and Conclusions

The three theorems derived describe, in full, the convergence properties of the map and its steady state behaviour. As the vehicle progresses through an environment the uncertainty in the estimates of relative landmark locations reduces monotonically. In the limit, errors in the estimates of any pair of landmarks become fully correlated. This means that given the exact location of any one landmark, the location of any other landmark in the map can also be determined with absolute certainty. As the map converges the error in the absolute location estimate of every landmark (and thus the whole map) reaches a lower bound determined only by the error that existed when the first observation was made.

The authors have implemented the simultaneous localisation and map building (SLAM) algorithm on a standard road vehicle. The vehicle is equipped with a millimeter-wave radar (MMWR) which provides observations of the location of landmarks with respect to the vehicle. The implementation demonstrates key properties of the SLAM algorithm; convergence, consistency and boundedness of the map error.

The implementation also highlights a number of key landmarks of the SLAM algorithm and its practical development. In particular, the implementation shows how generally non-linear vehicle and observation models may be incorporated in the algorithm, how the issue of data association can be dealt with, and how landmarks are initialised and tracked as the algorithm proceeds.

It is not possible to provide further details of the implementation here, due to space limitations. However, a technical report [1] on this work will be made available during the conference.

# References

[1] M.W.M.G Dissanayake, P. Newman, H.F. Durrant-Whyte, S. Clark, and M. Csorba. A solution to the simulataneous localisation and map building (slam) problem. Internal Tech Report ACFR-TR-01-99, Australian Center for Field Robotics, University of Sydney, Australia, January 1999.

[2] J. Leonard and H.F. Durrant-Whyte. *Directed Sonar Sensing for Mobile Robot Navigation*. Kluwer Academic Publishers, 1992.

[3] P. Maybeck. *Stochastic Models, Estimation and Control*, volume 1. Academic Press, 1982.

[4] P. Moutarlier and R. Chatila. Stochastic multisensor data fusion for mobile robot localization and environment modelling. In *Fifth Int. Symp. Robotics Research*, pages 85–94, 1989.

[5] P. Smith, M. Self, and R. Cheeseman. Estimating uncertain spatial relationships in robotics. In I.J.Cox and G.T. Wilfon, editors, *Autonomous Robot Vehicles*. Springer-Verlag, 1990.

# Continuous Probabilistic Mapping by Autonomous Robots

Jesús Salido Tercero[1], Christiaan J. J. Paredis[2], and Pradeep K. Khosla[2]

[1] Universidad de Castilla-La Mancha, ES de Informática,
Ronda de Calatrava 5, 13071 Ciudad Real, Spain
[2] Carnegie Mellon University,
Institute for Complex Engineered Systems,
5000 Forbes Avenue, Pittsburgh, PA 15213, USA

**Abstract.** In this paper, we present a new approach for continuous probabilistic mapping. The objective is to build metric maps of unknown environments through cooperation between multiple autonomous mobile robots. The approach is based on a Bayesian update rule that can be used to integrate the range sensing data coming from multiple sensors on multiple robots. In addition, the algorithm is fast and computationally inexpensive so that it can be implemented on small robots with limited computation resources. The paper describes the algorithm and illustrates it with experiments in simulation and on real robots.

## 1 Introduction

For efficient navigation in cluttered environments, maps are an indispensable tool. Maps allow mobile robots to autonomously plan trajectories through the environment while avoiding obstacles. Maps are often pre-recorded or generated by human operators. However, in unknown or rapidly changing environments, these maps are usually unavailable or outdated. In order to maintain autonomous operation in these situations, robots should record their own experiences and build internal maps themselves.

In recent research results, two different approaches can be identified for representing the environment:

1. *Metric or grid-based representations* [5][9]. In this approach, the world is divided into evenly spaced cells in a grid. Each cell contains information related to the corresponding region in the environment.
2. *Topological representations* [4] Here the world is represented as a graph in which nodes represent states, places, or landmarks. If there is a direct path between two nodes, they are connected with an arc.

Both approaches have advantages and disadvantages. Metric maps are easy to build and maintain, but they require accurate measurements of the robot position. Furthermore, building metric maps is in general time-consuming and may require a large amount of storage space. Topological maps, on the other hand, are much more difficult to build but do not depend as much on accurate position measurements. In addition, their compactness makes them highly suitable for fast global planning. Overall, metric maps have been used most often for sensor-based map-building.

In this paper, we present a novel approach to building grid based metric maps. Its novelty resides in its generality and simplicity. It requires only limited computational resources and can be easily implemented in real-time. It allows multiple sensing modalities to be combined (sonar, laser-range-finder, camera, etc.) and it can combine maps from multiple robots into a single global map.

## 2  Previous related work

As we mentioned in the introduction, metric maps have been most successfully applied until now. In particular, *grid based occupancy maps* [5][9][10] have shown to be convenient and useful. In an occupancy map, the world is represented by a grid in which each cell contains the probability of occupancy; a value of zero corresponds to free space, a value of one corresponds to an obstacle.[1]

In grid based maps, Bayes' rule [11] is used to update occupancy values over time. The updated occupancy value *(a posteriori probability)* is based on the occupancy value in the current map *(a priori probability)* and the current sensor reading *(conditional probability)*. The update process requires a sensor model defining the occupancy values of the grid cells around the sensor, given the current sensor reading. Several solutions have been proposed for the sensor modeling problem:

1. **Probabilistic models.** A probability density function is used to model the sensor. Commonly, Gaussian probability density functions are used. [5][10].
2. **Histograms.** This approach is usually applied for *local planning* (i.e. obstacle avoidance), for which a fast response is critical. To avoid the time consuming evaluation of a probability density functions, a simple histogram is used. [2].
3. **Learned models.** Instead of using a Gaussian approximation, the probability density function is learned [15]. This provides greater accuracy but at a the cost of greater computational requirements at run-time as well as for training. A neural network trained in simulation has been used to determine the probability density function [15].

The type of sensor model has a significant impact on the applicability of the method. For applications in which very accurate global plans need to be constructed and computational resources are readily available, the most accurate sensor model is appropriate, namely a learned model, even if this requires a high computation cost. On the other hand, if information for local obstacle avoidance is the goal of the algorithm, a very simple sensor model can be used with adequate results. Because such a simple model can be evaluated very quickly, a high sample rate can be achieved, which allows the robot to move more quickly while still avoiding collisions.

The method that we propose in this article, makes a compromise between these two extremes. It uses a sensor model that is slightly more expensive to evaluate than in the Histogramic In-Motion Mapping approach [2]. Yet, the method provides maps that are sufficiently accurate to perform global planning, at a fraction of the cost of learned models.

The whole map-building process consists of the following three steps:

---

[1] For visual purposes that likelihood is usually represented by gray levels where 'black' means a completely occupied cell and 'white', free space.

1. Sensing.
2. Computation of conditional occupancy probabilities based on the sensor model.
3. Update the occupancy values in the grid using Bayes' rule.

An important requirement for building metric maps is the accurate determination for the robot position and orientation. That does not pose a big problem for outdoor applications where global positioning systems (GPS and compass) are available. However, for indoor applications or space applications these global positioning systems are usually not available.

Early results used dead-reckoning and filtering to locate the robot in its environment [1][5][10] Recently, a new set of probabilistic approaches has emerged based on *Markov Decision Processes* [13]. Instead of using a single position estimate, a probability distribution is used to represent the robot's belief of being in a particular position/orientation. This method requires an 'a priori' map to be supplied to the robot, which may cause some undesired cyclic behavior. Moreover, the basic assumptions for a Markov process are not satisfied in highly dynamic environments. To overcome these problems, researchers have introduced modifications to the basic MDP, such as the *(Partially Observable Markov Decision Process)* [3] and the work in [7]. Instead of combining user supplied maps with sensor readings, recent work at Carnegie Mellon University has focussed on combining map-based localization and autonomous map-building [6][14][15].

As part of the CyberScout project, we are also developing a sonar-based indoor localization system. In this system, each robot is equipped with a sonar beacon and an RF transmitter, as illustrated in Figure 1. A conical reflector is mounted above the sonar transceiver to reflect the sound into a horizontal plane and provide 360° coverage. Periodically, each robot will transmit a sonar and an RF pulse simultaneously. All the other robots in the vicinity will receive the RF pulse almost instantaneously and the sonar pulse after a delay proportional to the distance from the transmitting robot. When the distances between all the robots are known, one can use trilateralization to determine their relative positions. This localization system will allow us in the future to implement the mapping algorithm described in this article on the small robot platforms shown in Figure 1.

# 3 Continuous probabilistic mapping — CPM

This section describes the details of our approach for building metric maps called Continuous Probabilistic Mapping (CPM). By *continuous* we emphasize the fact that CPM allows one to update the map continuously—every execution cycle. The approach requires only limited computation each time-step and can easily be implemented on the small robot platforms with limited computational resources that we use in the lab. The applications that we consider are cooperative exploration and reconnaissance by a team of autonomous robots in unknown and possibly dynamic environments.

In CPM, the environment is divided into uniform cells $c_{i,j}$ (with center at $[x_i, y_j]$). We define a metric $\text{Occup}(c)$ that defines the occupancy likelihood of the cell $c$:

In the further derivation of the CPM algorithm, we assume that the robot's position $(x, y, \theta)$ can be measured sufficiently accurately.

**Fig. 1.** A Millibot (size 6 × 6 × 6cm) with a sonar-beacon.

**Table 1.** Occupancy metric

| Value | Description | | |
|-------|-------------|---|---|
| 0.0 | minimum occupancy | free space | no uncertainty |
| 0.5 | medium occupancy | unknown | maximum uncertainty |
| 1.0 | maximum occupancy | obstacle | no uncertainty |

We further assume that the sensor readings consist of proximity measurements (e.g. sonar or laser-range-finder) and that the characteristics of the sensor are well understood (e.g. field of view and maximum range).

### 3.1 The model for 'a priory' occupancy likelihood

CPM uses a simplified 'a priori' occupancy likelihood based on the concept of *field of view* (FOV) of a sensor. By FOV we mean the *spatial region where a sensor is able to perceive*. As is illustrated in Figure 2, the FOV can be characterized by two parameters: the maximum range $(r)$, and the perception angle $(\alpha)$. For example, for the sonar used in our lab (Polaroid 6500 series) the maximum range is 2 meters and the perception angle is 20 degrees.

Based on the FOV, we define a function $Occup_0(s, c_{i,j})$ for each cell $c_{i,j}$ around the sensor $s$. If a cell $c$ covers only a fraction $f$ of the FOV of the sensor $s$, the function $Occup_0$ is proportional to $(1 - f)$ or

$$Occup_0(s, c_{i,j}) = \frac{1}{2}\left(1 - \frac{\text{Area}(FOV \cap c)}{\text{Area}(c)}\right) \tag{1}$$

One can interpret this function as the probability that a cell is occupied when no obstacles are detected. As a result, any cell that does not overlap with the FOV of the sensor is assigned a value of 0.5 meaning that the occupancy is completely unknown.

For the further development of the algorithm, we only need to consider the cells that overlap with the FOV. As indicated in Figure 2, we call this set of cells the *Sensor Occupancy Pattern* or SOP. The $Occup_0(s, c_{i,j})$ values for the SOP are independent of the sensor readings and can be computed beforehand. An SOP requires little storage and can be applied to any type of range sensor.

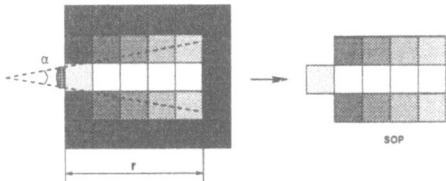

**Fig. 2.** The a priori sensor occupancy values (r=range, $\alpha$=perception angle), and the Sensor Occupancy Pattern (SOP)

## 3.2 Computation of the occupancy function

In this section, we will explain how the SOP is combined with a sensor reading to obtain the occupancy value for each cell in the FOV. These values will later be combined with the occupancy probabilities in the global map to update the robot's world view.

The computation of the occupancy function consists of two steps. In a first step, we classify the cells in the field of view into three categories, depending on whether they are within the range of the current sensor reading or not. Next, we combine this information with the values in the SOP to obtain the final occupancy value of sensor.

The classification process divides the cells in the FOV into three categories: *occupied, free or unknown*. Consider the example in Figure 3. Each cell is a square of dimension $C_s$. The distance $d = D_c - S$ is the difference between the sensor reading $S$ and the distance to the center of the cell $D_c$. The classification function $h(c, S)$ is than computed as:

$$h(c, S) = \begin{cases} \textbf{Occupied,} & \text{if} \quad |d| < Cs/2 \\ \textbf{Free,} & \text{if} \quad S = r \quad \vee \quad (|d| \geq Cs/2 \wedge d < 0) \\ \textbf{Uknown,} & \text{if} \quad |d| \geq Cs/2 \quad \wedge \quad d \geq 0 \end{cases} \qquad (2)$$

The classification is conservative in that it overestimates the probability of occupation. From the range measurement, we know that somewhere at a distance $r$ there is an obstacle in the FOV. However we do not know in which cell exactly. Therefore, we label every cell at distance $r$ as occupied, even though it is not very likely that all these cells are actually occupied. Our conservative approach still provides accurate results in the long term because unoccupied cells that were

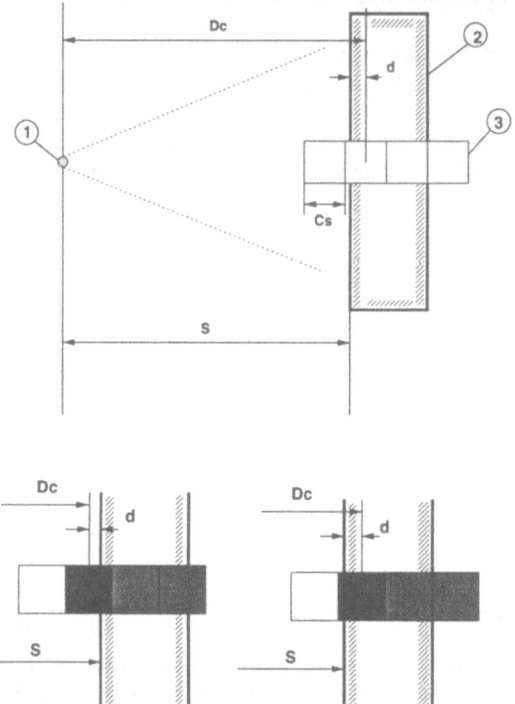

**Fig. 3.** Classification of cells: (1) sensor, (2) obstacle, and (3) cell

wrongly labeled as occupied based on one sensor reading will most likely be labeled correctly when the robot changes positions.

In the classification, we do not take into account the extent to which a cell is inside the FOV of the sensor. Therefore, the classification value is now combined with the a priori occupancy to obtain the actual occupancy value $\text{Occup}(c, S)$ of each cell:

$$\text{Occup}(c, S) = \begin{cases} Occup_0 & \text{if } h(c, S) = \textbf{Free} \\ 0.5 & \text{if } h(c, S) = \textbf{Unknown} \\ 1 - Occup_0 & \text{if } h(c, S) = \textbf{Occupied} \end{cases} \tag{3}$$

It is clear that the computation of the occupancy values requires minimal computation. The total number of operations needed is proportional to the total number of cells in the SOPs of all sensors. Compared to other occupancy grid methods, the computations per cell are much simpler and the number of cells is much reduced by only considering the cells in the FOV. In the next section, these occupancy values will be used to update the global map of the environment using a Bayes' update rule.

## 3.3 Spatial-temporal integration

So far, we have illustrated how to obtain an approximate probability distribution for a single range sensor based on a single sensor reading. However, mapping is a dynamic process in which sensor readings from different locations and at different time instants are fused into a single global representation. Bayesian probability theory has been applied successfully to the mapping problem [5][10][9][15]. In this section, we describe how Bayes' rule is applied in our approach.

After having computed the function $Occup(c)$ for each of the sensors, there are two steps remaining. In a first step, the cells in the grid around each sensor are mapped to cells in the global map using the position and orientation information from the robot. Then, in a second step, the occupancy value of each sensor cell is used to update the estimated occupancy of the corresponding cell in the global map. Because the robot moves, over time cells in the global map are covered multiple times by different sensors from different sensing locations and most likely.

Using Bayes' rule, all these sensor readings are integrated into the global map, improving the estimates for the occupancy of each cell. This integration is accomplished using the following equation [15]:

$$Occup(c \mid S^{(1)}, S^{(2)} \cdots S^{(T)}) = 1 - \left(1 + \prod_{\tau=1}^{T} \frac{Occup(c \mid S^{(\tau)})}{1 - Occup(c \mid S^{(\tau)})}\right)^{-1} \quad (4)$$

This equation expresses the a posteriori probability of occupation of cell $c$ given a sequence of readings $(S^{(1)}, S^{(2)} \cdots S^{(T)})$. In the equation, we assume an initial occupancy value of $Occup(c) = 0.5$ for all the cells. The sequence of readings may include readings at different time instants as well as readings from different sensing locations, different sensors on the same robot, or sensors on multiple robots.

Notice that Equation 4 can also be evaluated iteratively, which reduces the store requirements significantly; only one value per cell needs to be stored.

$$Occup(c \mid S^{(1)}, S^{(2)} \cdots S^{(T)}) =$$
$$1 - \left(1 + \frac{Occup(c \mid S^{(T)})}{1 - Occup(c \mid S^{(T)})} \frac{Occup(c \mid S^{(1)} \cdots S^{(T-1)})}{1 - Occup(c \mid S^{(1)} \cdots S^{(T-1)})}\right)^{-1} \quad (5)$$

Finally, the same equation can also be used to integrate maps from two different robots. One can write Equation 4 as

$$Occup(c \mid S^{(1)}, S^{(2)} \cdots S^{(N)}) =$$
$$1 - \left(1 + \frac{Occup(c \mid S^{(1)} \cdots S^{(k)})}{1 - Occup(c \mid S^{(1)} \cdots S^{(k)})} \frac{Occup(c \mid S^{(k+1)} \cdots S^{(N)})}{1 - Occup(c \mid S^{(k+1)} \cdots S^{(N)})}\right)^{-1} \quad (6)$$

This allows one to combine maps from different robots (sensor readings 1 through $k$ from robot 1; sensor readings $k + 1$ through $N$ from robot 2).

**Implementational considerations.** The main purpose for the integration process discussed in this section is to produce a map in a global frame of reference. However, the sensor occupancy values are obtained in the robot's frame of reference. Although the local cells can be mapped to the global frame of reference by applying a translation and rotation transformation,the cells do not line up with the cells of the global map in general. This problem is solved by using a simple interpolation scheme that determines the sensor occupancy value at each of the centers of the cells in the global map.

A second important implementational issues has to do with dynamically changing environments. When a cell in the global map is occupied by an obstacle, the occupancy value of that cell will move closer and closer to one as more and more sensor reading are incorporated. Due to the finite precision of floating point computation, the occupancy value may actually become equal to one at some point in time. However, this causes problems in the evaluation of Equation 5. Even with a careful implementation of Equation 5, there still remains a problem with dynamic environments. Once, a cell reaches a value of either zero or one, the value cannot change anymore, regardless of the current sensor readings. Therefore, in dynamic environments, where the position of obstacles may change over time, the occupancy value should never be allowed to reach zero or one completely. In our implementation we limit the occupancy value to remain inside the interval $[10^{-6}, 1 - 10^{-6}]$.

## 4 Experiments and results

We performed experiments in simulation as well as with real robots. All the experiments were executed using the CyberRAVE environment developed at Carnegie Mellon University [8] [12]. CyberRAVE is a general purpose framework for simulation and command and control of multiple robots. It allows algorithms developed in a simulated environment to be transfered transparently to the real robot, and even allows real and simulated robots to interact with each other in a single experiment. CyberRAVE allows us to quickly develop algorithms for multi-robot systems, evaluate different sensor configurations in simulation, and transition the software to real hardware one robot at a time.

### 4.1 Experiments in simulated environments

The main purpose for the simulation experiments is to determine how well CPM performs in terms of real-time computation and its ability to integrate maps in a multi-robot system.

The experiment consists of a collaborative mapping task with at team of five robots. The simulated robots are modeled after the RC-tank based robots developed in our lab (Figure 5). They are equipped with 8 sonars: 5 in the front, 1 left, 1 right, and 1 in the back. The sonars are modeled to have a sensing cone with $r = 2$m. and $\alpha = 20°$). The environment the robots need to map consists of a flat terrain with polygonal obstacles.

Figure 4 illustrates some of the results. Initially, the robots start without any knowledge of the environment: the complete map is initialized with occupancy values of 0.5. While the user maneuvers the robots around using a simulated joystick,

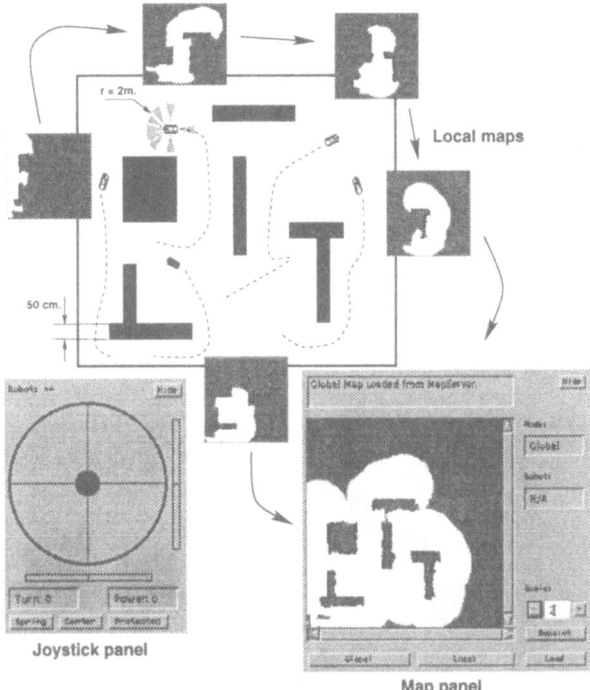

**Fig. 4.** Cooperative mapping results in simulation using CyberRAVE

the robots build up information about the environment through their sonar sensors. This results in occupancy values moving towards one (black) in cells containing obstacles, and toward zero (white) in free space. Each robot builds up a local map containing occupancy information for the area it has traversed. Periodically, the local maps for each of the robots is transmitted to a "Team Leader" who uses CPM to integrate the local maps into a single global map.

## 4.2 Experiments with real robots

To perform mapping experiments with real robots, we have equipped our lab with a vision-based localization system, as is illustrated in Figure 6. The CPM algorithm requires that the global position of the robot be known. Since GPS sensors cannot be used inside, we use two overhead cameras and a cognachrome tracking system to measure the position and orientation of the robots with an accuracy of about 5 cm and 5 degrees. Relative to the size of the robots (30x50cm), the tracking system provides sufficient accuracy to build maps collaboratively.

The robots shown in Figure 5 are small RC-tanks that have been retrofitted with a PC-104 stack containing a 486 computer running Linux OS, a digital IO board for sensor and actuator interaction, a wireless Ethernet connection, and a hard disk. They are further equipped with 8 sonar sensors, 5 IR obstacle detectors, a B/W camera mounted on a pan/tilt mechanism, and stereo microphones. For the

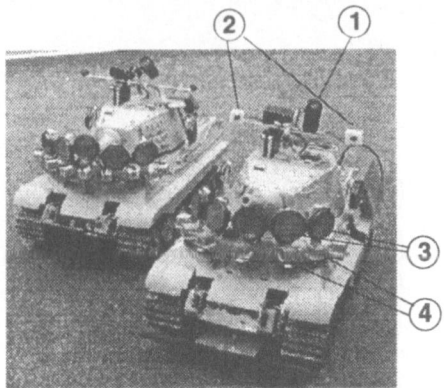

**Fig. 5.** Scouts: (1) B/W camera, (2) microphones, (3) sonars, and (4) IRs.

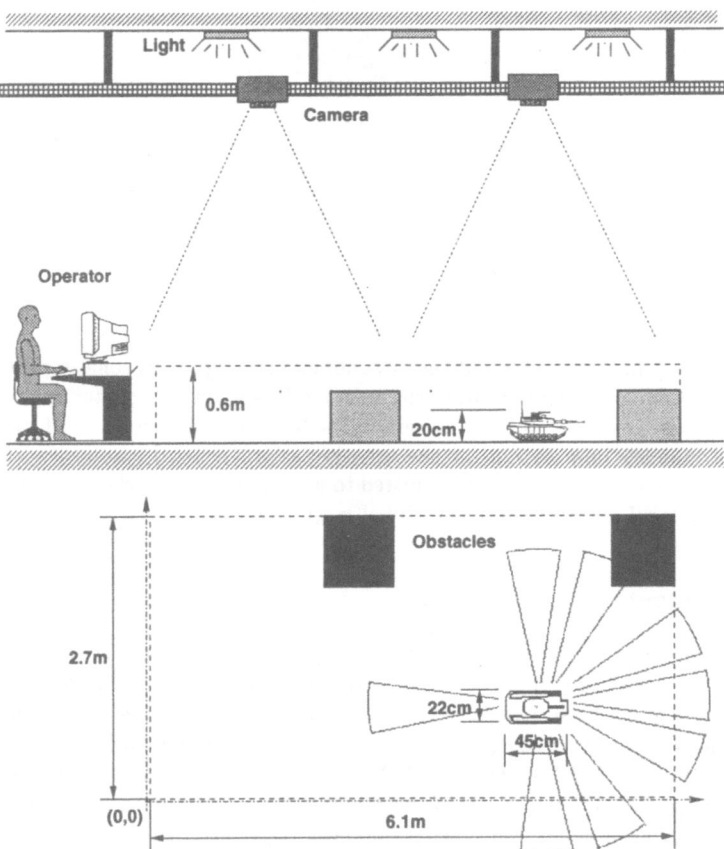

**Fig. 6.** The physical layout of the testing area.

mapping experiment, we only use the 8 Polaroid sonars (series 6500) with a sampling rate of about 5Hz. In their current configuration the sonars have a reliable sensing range of 0.50m to 2.0m with a precision of ±0.02m.

The area in which the experiments are conducted is delimited by fiberboard panels. When the robot is at an oblique angle with respect to the paneling, the sonar pulse may not reflect back to the robot directly but bounce back via one of the other four panels. This multi-path effect may erroneously cause the robot to sense obstacles further than they really are or even stop the robot from sensing the obstacle at all. However, from a mapping perspective, infrequent erroneous measurements pose no problem. Over time, only the correct measurements contain enough correlated information to show up in the map.

Figure 7 shows the results obtained in a mapping experiment compared to the actual obstacle layout. The quality of the map is not only determined by the size of the grid cells, but also by the trajectory followed by the robot and the the number of multi-path readings. In this experiment, the robot moved through the environment relatively quickly and experienced a several number of multi-path readings. The result is that the map marks the inside of the obstacles and even areas outside the boundary of the experimental area as unoccupied. The obstacles on the right side of the map are not marked, because the area has not yet been explored by the robot.

The second and third experiment were produced with 10 × 10cm cells, and b

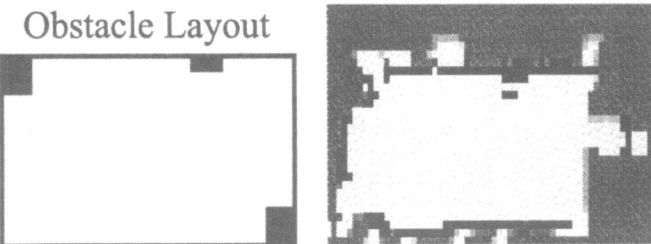

**Fig. 7.** Mapping result with a real robot

## 5  Summary

In this paper, we have described a novel method for continuous mapping in unknown environments: Continuous Probabilistic Mapping. The novelty of CPM resides in two properties: simplicity for real-time computation, and scalability for multiple robots each with multiple sensors.

The algorithm requires only limited computational resources so that it can be executed at a high frequency, which allows the robot to move quickly while still avoiding obstacles. This simplicity is also important for our future work in small distributed robotic systems in which computation is at a premium.

With respect to distributed robotics, the scalability of the algorithm is also very important. The algorithm allows robots to integrate their sensing data locally for collision avoidance, but at the same time to combine this local information with that of other robots to obtain a global view of the environment.

# Acknowledgement

This research was supported in part by DARPA/ETO under contract DABT63-97-1-0003, and by the Institute for Complex Engineered Systems at Carnegie Mellon University.

# References

1. J. Borenstein, H. R. Everett, and L. Feng. *Navigating Mobile Robots: Sensors and Techniques.* A. K. Peters, Ltd., first edition, 1996.
2. J. Borenstein and Y. Koren. Histogramic in-motion mapping for mobile robot abstacle avoidance. *IEEE Trans. on Robotics and Automation*, 7(4):535–539, 1991.
3. Anthony R. Cassandra, L. Kaelbling, and J. Kurien. Acting under uncertainty: Discrete bayesian models for mobile-robot navigation. In *IEEE/IROS Int. Conf. on Intelligent Robotics and Systems*, 1996.
4. H. Choset. *Sensor based motion planning: The hierarchical generalized Voronoi graph.* PhD thesis, California Institute of Technology, USA, 1996.
5. Alberto Elfes. *Occupancy grids: A probabilistic framework for mobile robot perception and navigation.* PhD thesis, Electrical and Computer Engineering Dept., Carnegie Mellon Univ., 1989.
6. W. Burgard et al. The interactive museum tour-guide robot. In *AAAI, Nat. Conf. on Artificial Intelligence*, 1998.
7. Dieter Fox, W. Burgard, S. Thrun, and A. Cremers. Position estimation for mobile robots in dynamic environments. In *AAAI, Nat. Conf. on Artificial Intelligence*, 1998.
8. Wes Huang et al. CyberRAVE: A real and virtual environment for CyberScout, August 1998.
9. Martin C. Martin and Hans P. Moravec. Robot evidence grids. Technical Report CMU-RI-TR-96-06, Carnegie Mellon University, March 1996.
10. H.P. Moravec. Sensor fusion in certainty grids for mobile robots. *AI Magazine*, Summer:116–121, 1988.
11. Athanosios Papoulis. *Probability, Random Variables, and Stochastic Processes.* McGraw-Hill, third edition, 1991.
12. J. Salido, J.M. Dolan, J. Hampshire, and P. Khosla. A modified reactive control framework for cooperative mobile robots. In *SPIE International Symposium on Sensor Fusion and Decentralized Control in Autonomous Robotic Systems*, volume 3209, pages 90–100, October 1997.
13. Reid Simmons and S. Koenig. Probabilistic robot navigation in partially observable environments. In *International Joint Conference on Artificial Intelligence*, pages 1080–1087, August 1995.
14. S. Thrun, D. Fox, and W. Burgard. Probabilistic mapping of a environment by a mobile robot. In *AAAI, Nat. Conf. on Artificial Intelligence*, 1998.
15. Sebastian Thrun. Learning maps for indoor mobile robot navigation. *AI Magazine*, 1997.

# Sensor Influence in the Performance of Simultaneous Mobile Robot Localization and Map Building[*]

J.A. Castellanos    J.M.M. Montiel    J. Neira    J.D. Tardós

Departamento de Informática e Ingeniería de Sistemas
Universidad de Zaragoza, Spain
E-mail: {jacaste, josemari, jneira, tardos}@posta.unizar.es

**Abstract:** Mobile robot navigation in unknown environments requires the concurrent estimation of the mobile robot localization with respect to a base reference and the construction of a global map of the navigation area. In this paper we present a comparative study of the performance of the localization and map building processes using two distinct sensorial systems: a rotating 2D laser rangefinder, and a trinocular stereo vision system.

## 1. Introduction

Simultaneous localization and map building is one of the key problems in autonomous mobile robot navigation. Different approaches have been reported in the literature after the initial theoretical contributions of Smith et al. [1] and the early experiments of Chatila et al. [2] and Leonard et al. [3]. An important application in which we are interested is the development of the sensorial system for autonomous wheel-chairs for handicapped people. Information gathered by the sensorial system would be used for navigation purposes. Both the localization of the vehicle and the construction of a map of its surroundings are required. The main goal of this paper is to discuss the issues concerning the type of sensor used by the system. We present a comparative study of the performance of the localization and map building processes using two distinct sensorial systems: a 2D laser rangefinder, and a trinocular vision system.

In our approach we use a probabilistic model, the Symmetries and Perturbation Model (SPmodel) [4, 5], to represent uncertain geometric information given by any sensor in a general and systematic way. As reported in the literature, the solution to the simultaneous robot localization and map building problem requires maintaining a representation of the relationships between the location estimations of the robot and the features included in the map [6, 7], which in our work are represented by the cross-correlations between their estimations. In this paper we briefly present the formulation of the Symmetries and Perturbations Map (SPmap), a probabilistic framework for the simultaneous

---
[*]This work has been supported by spanish CICYT project TAP97-0992-C02-01

localization and map building problem whose main advantage is its generality, i.e. sensor and feature independence.

The rest of the paper is structured as follows. In section 2 we describe our approach to the simultaneous localization and map building problem. Section 3 presents the experiment performed and the sensor data processing algorithms used. Experimental results obtained with a 2D laser rangefinder and a trinocular vision system are discussed in section 4, whilst the main conclusions of our work are drawn in the last section.

## 2. Simultaneous Localization and Map Building

### 2.1. The Symmetries and Perturbation Model

In our feature-based approach, uncertain geometric information is represented using a probabilistic model: the Symmetries and Perturbation Model (SPmodel) [4, 5] which combines the use of probability theory to represent the imprecision in the location of a geometric element, and the theory of symmetries to represent the partiallity due to characteristics of each type of geometric element.

In the SPmodel, the location of a geometric element $E$ with respect to a base reference $W$ is given by a *location vector* $\mathbf{x}_{WE} = (x, y, \phi)^T$. The estimation of the location of an element is denoted by $\hat{\mathbf{x}}_{WE}$, and the estimation error is represented locally by a *differential location vector* $\mathbf{d}_E$ relative to the reference attached to the element. Thus, the true location of the element is:

$$\mathbf{x}_{WE} = \hat{\mathbf{x}}_{WE} \oplus \mathbf{d}_E \tag{1}$$

where $\oplus$ represents the composition of location vectors. To account for the symmetries of the geometric element, we assign in $\mathbf{d}_E$ a null value to the degrees of freedom corresponding to them, because they do not represent an effective location error. We call *perturbation vector* the vector $\mathbf{p}_E$ formed by the non null elements of $\mathbf{d}_E$. Both vectors can be related by a row selection matrix $\mathbf{B}_E$ that we call *self-binding matrix* of the geometric element:

$$\mathbf{d}_E = \mathbf{B}_E^T \mathbf{p}_E \quad ; \quad \mathbf{p}_E = \mathbf{B}_E \mathbf{d}_E \tag{2}$$

Then, the *uncertain location* of every geometric entity is represented in the SPmodel by a quadruple $\mathbf{L}_{WE} = (\hat{\mathbf{x}}_{WE}, \hat{\mathbf{p}}_E, \mathbf{C}_E, \mathbf{B}_E)$, where the transformation $\hat{\mathbf{x}}_{WE}$ is an estimation taken as base for perturbations, $\hat{\mathbf{p}}_E$ is the estimated value of the perturbation vector, and $\mathbf{C}_E$ its covariance.

### 2.2. The Symmetries and Perturbation Map

The Symmetries and Perturbation Map (SPmap) is a complete representation of the environment of the robot which includes the uncertain location of the mobile robot $\mathbf{L}_{WR}$, the uncertain locations of the features obtained from sensor observations $\mathbf{L}_{WF_i}, i \in \{1 \dots N_F\}$ and their interdependencies. The SPmap can be defined as a quadruple:

$$\mathbf{SPmap} = (\hat{\mathbf{x}}^W, \hat{\mathbf{p}}^W, \mathbf{C}^W, \mathbf{B}^W) \tag{3}$$

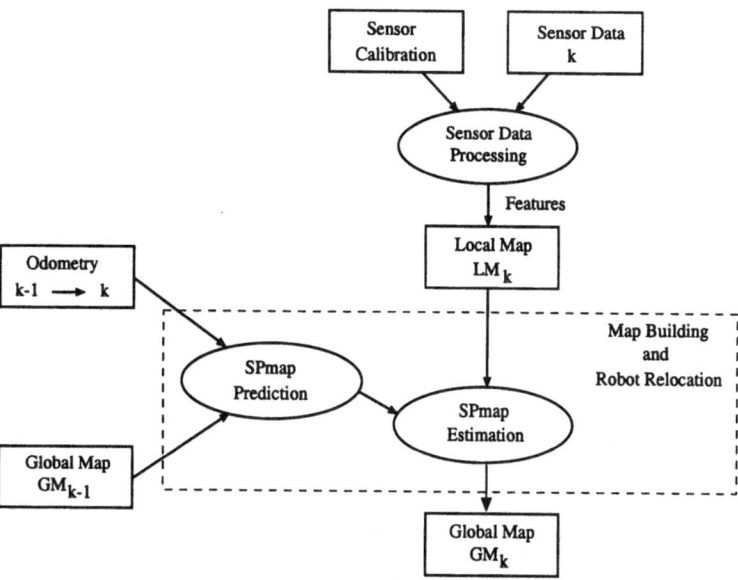

*Figure 1. Simultaneous mobile robot localization and map building.*

where $\hat{\mathbf{x}}^W$ is the *estimated location vector* of the SPmap and $\mathbf{p}^W$ is the *perturbation vector* of the SPmap:

$$\hat{\mathbf{x}}^W = \begin{bmatrix} \hat{\mathbf{x}}_{WR} \\ \hat{\mathbf{x}}_{WF_1} \\ \vdots \\ \hat{\mathbf{x}}_{WF_{N_F}} \end{bmatrix} \quad ; \quad \mathbf{p}^W = \begin{bmatrix} \mathbf{d}_R \\ \mathbf{p}_{F_1} \\ \vdots \\ \mathbf{p}_{F_{N_F}} \end{bmatrix} \tag{4}$$

The true location of the robot and the map features is:

$$\mathbf{x}^W = \hat{\mathbf{x}}^W \oplus (\mathbf{B}^W)^T \mathbf{p}^W \tag{5}$$

where the composition operator $\oplus$ applies in this case to each of the components of the vectors, and $\mathbf{B}^W$ is the *binding matrix* of the SPmap, a diagonal matrix formed by the self-binding matrix of the robot and the self-binding matrices of the map features:

$$\mathbf{B}^W = \mathrm{diag}\left( \mathbf{B}_R, \ \mathbf{B}_{F_1}, \ \ldots, \ \mathbf{B}_{F_{N_F}} \right) \tag{6}$$

The covariance matrix of the SPmap represents the covariance of the estimation of the robot and the map feature locations, the cross-covariances between the robot and the map features, and finally, the cross-covariances between the map features themselves:

$$\mathbf{C}^W = \begin{pmatrix} \mathbf{C}_R & \mathbf{C}_{RF_1} & \cdots & \mathbf{C}_{RF_{N_F}} \\ \mathbf{C}_{RF_1}^T & \mathbf{C}_{F_1} & \cdots & \mathbf{C}_{F_1 F_{N_F}} \\ \vdots & \vdots & \ddots & \vdots \\ \mathbf{C}_{RF_{N_F}}^T & \mathbf{C}_{F_1 F_{N_F}}^T & \cdots & \mathbf{C}_{F_{N_F}} \end{pmatrix} \tag{7}$$

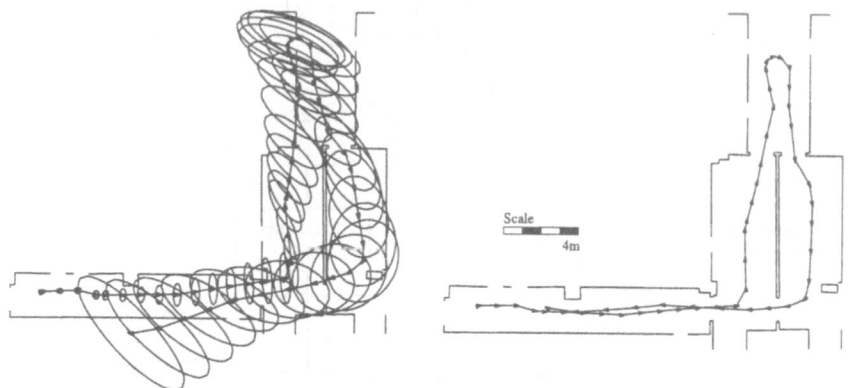

*Figure 2. Robot trajectory according to odometry (left) and true trajectory measured with the theodolites (right).*

Note that we may represent any type of geometric entity within this general framework. The proposed approach can include different types of features obtained by different types of sensors, and thus it is suitable to deal with a multisensor system. In our experiment, either 2D laser segments or trinocular vertical edges are fused in the estimation of both the robot and the feature locations using an EKF-based algorithm. Figure 1 describes the incremental construction of the SPmap. More details can be found in [7].

## 3. Experiment Design

### 3.1. Measurement of the Vehicle Trajectory

The experiment was carried out using a Labmate$^{TM}$ mobile robot, a 2D laser rangefinder and a trinocular vision system, both mounted on the mobile robot, and a pair of theodolites, used as a precise and independent location measurement equipment. At each step of the programmed robot trajectory, the robot location according both to odometry, and measured with the theodolites, were obtained (figure 2). The *environment model* was a set of vertical edges, corresponding to wall corners and door frames, whose location was measured with the theodolites. The location of vertical walls was calculated using this information. This model is used as ground-truth to evaluate the precision of the map building processes. Also at each step of the robot trajectory, the environment was sensed using both sensors as explained below.

### 3.2. Processing of 2D Laser Readings

A set of 2D points, expressed in polar coordinates with respect to the sensor, were gathered from the surroundings of the vehicle. A maximum range of 6.5 m. and an angular resolution of 0.5 deg. were used during experimentation. Measurement error for each 2D laser reading was modelled by white-gaussian noise mainly characterized by the variances in range and azimuth angle [8]. A two-step segmentation algorithm was applied [9]. First, by application of a tracking-like algorithm, the set of 2D laser readings was divided into groups

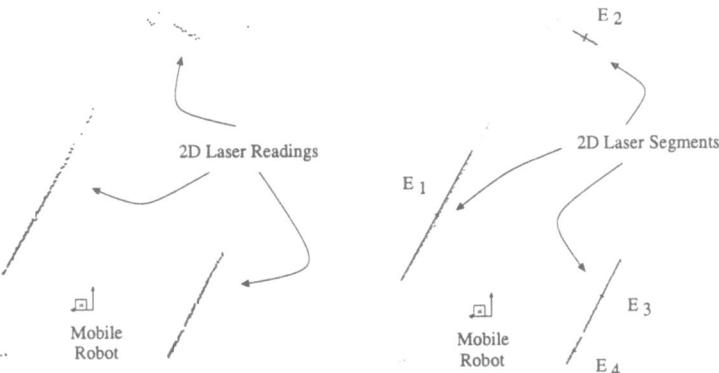

*Figure 3. Laser scan and 2D segments obtained at step 38 of the robot trajectory.*
*Lateral uncertainty has been represented for each detected segment.*

of points to which a polygonal line without gaps could be fitted. Then, an
iterative line fitting technique was considered to detect the endpoints of each
2D segment included in one of those groups. Finally, suboptimal estimation
was used to obtain the final representation associated to each laser segment.
Also, an estimation of the segment length was computed from the endpoints.
Segments shorter than 30 cm. were discarded from further processing. Figure
3 describes an example of segmentation of 2D laser readings gathered during
our experimentation.

### 3.3. Processing of Trinocular Vision

In the case of trinocular vision, large vertical edges were extracted from the
three images and matched to obtain the location of trinocular vertical edges
corresponding to corners and door frames. The trinocular vision system was
composed of three B&W CCD cameras, with a lens focal of $f = 6$ mm., image
resolution of $512 \times 512$, and camera baselines of 113.7 mm., 110.8 mm. and
186.4 mm. In this work only vertical image segments were considered, allowing
deviations from verticallity up to 22.5 deg.

Trinocular stereo processing followed these steps: 1) *image segment detection:* The three images were processed using Burns' algorithm [10]. Three conditions were considered to filter out segments: segments shorter than 100 pixels,
non-vertical segments, and segments whose grey-level gradient was smaller than
8 grey levels per pixel; 2) *image segment matching:* using the computed image segments and the calibration information of the cameras (calibration was
performed using Tsai's algorithm [11]), images were sequentially processed in
a EKF-based predict-match-update loop [12] (matching was performed in 3D
to profit from segment overlapping in the image); 3) *image segment projection
to 2D:* from the vertical projection of the midpoint projecting ray of each of
the matched 3D image segments, the projection ray in 2D was obtained (the
standard deviation of the orientation for each projection ray was modelled as
0.1 deg.); 4) *2D vertical edge location:* the 2D location of the vertical segment
was computed by fusing the three 2D projection rays. Figure 4 shows the ver-

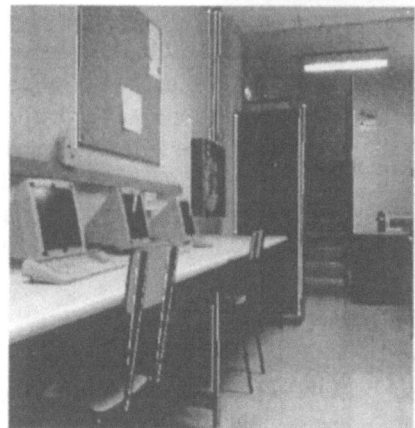

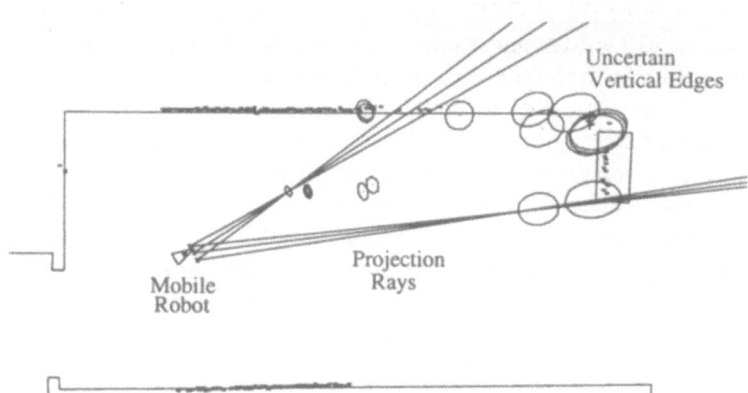

*Figure 4. Trinocular image set and vertical edge locations computed at step 38 of the trajectory (only the projection rays for two edges are shown). Elipses show the uncertainty (95% bounds) for the edge absolute location in 2D.*

tical edges obtained at step 38 of the robot trajectory, and the 2D local map obtained from the edges matched in the three images.

## 4. Experimental Results

In the first place, our experiment has shown that the SPmap approach to the simultaneous localization and map building problem is *appropriate* and *general* (figure 5). A statistical test based on the squared Mahalanobis distance was performed to validate the uncertainty model of the robot location for both the 2D laser range finder and the trinocular system solutions. Results showed that the laser-based solution passed the test in more than 95% of the cases, whilst the trinocular-based solution attained 87%. The trinocular-based solution was slightly optimistic due to the fact that the data association problem is more complex and some spurious matchings were accepted. Thus, this approach requires a more ellaborate solution than the closest neighbour used in our experiment. Nevertheless, with both sensors the robot adequately solved the *revisiting problem*, i.e. it recognized previously learned features of the environment.

With respect to *precision*, in our experiments we profited from ground-truth to obtain a precise estimation of the error in the vehicle location computed by each approach. Figure 6 presents the errors obtained at each point of the vehicle trajectory. The laser-based solution was bounded by a 20 cm. error in position and 1.5 deg. in orientation for the location of the vehicle. On the other hand, the trinocular-based approach obtained a maximum error of 35 cm. for the vehicle position and 2.5 deg. in orientation. Thus, the sensorial system based on laser rangefinder increased the accuracy of the solution. The environment map (figure 5) constructed in both cases proved to be adequate for navigation, although the map obtained by the laser-based approach may also be used for path planning purposes, which is not the case for the map obtained with trinocular vision.

The *complexity* of the approach based on trinocular vision is high, because this system requires an ellaborate calibration process, and there is much more computational effort involved. However, laser has been used to its maximum potential. On the contrary, only vertical edges, a minimum potential of the trinocular vision system have been considered. There are many more possibilities with the use of trinocular vision, such as extracting horizontal edges and planar surfaces, to be explored in the future.

## 5. Conclusions

In this paper we have compared the performance of two distinct sensorial systems on the simultaneous localization and map building problem for a mobile robot navigating indoors: on one hand, a 2D laser range finder, and on the other hand, a trinocular stereo system. The main conclusions derived from our experimentation can be summarized as follows: 1) The simultaneous localization and map building problem, using a non separable state-vector approach is a tractable problem for a reduced number of features. More structured representations are required for large-scale environments; 2) Using a 2D laser range

294

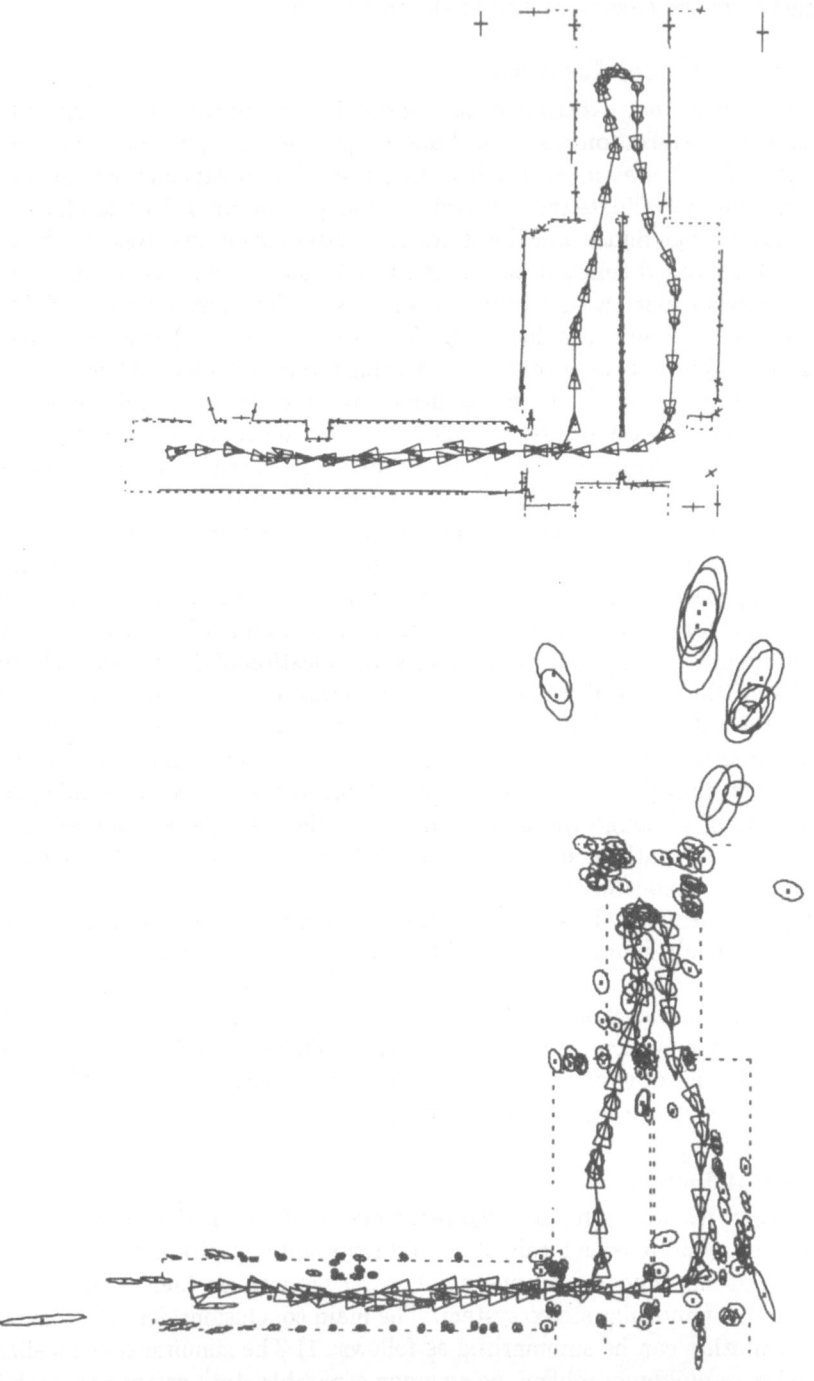

*Figure 5. Solutions for the simultaneous localization and map building problem obtained by laser (top) and trinocular (bottom). A hand-measured model map has been drawn for reference purposes.*

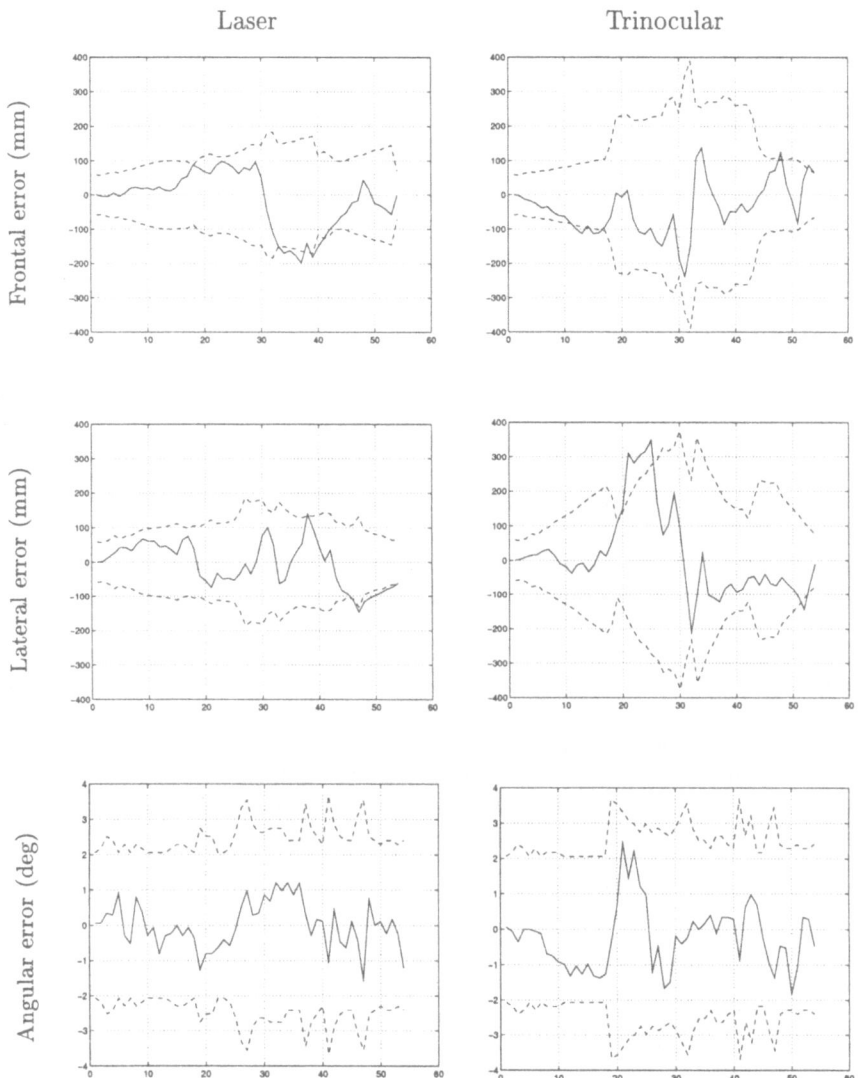

*Figure 6. Errors in the robot location estimation and 95% uncertainty bounds obtained by laser (left) and trinocular (right) along the trajectory of the vehicle.*

finder we obtained a more precise estimated localization for the vehicle than using trinocular vertical edges. One of the reasons was the higher semantical content of laser data as compared with the trinocular data. Also, data association by using the nearest-neighbour technique produced a degradation of the solution obtained by trinocular stereo as compared to the laser-based solution; 3) The laser-based approach allows the structuration of the navigation area towards the topological representation of the environment. However, such an structuration is difficult from the global map obtained by the trinocular-based approach.

Further work within our group is planned towards increasing the structuration of the navigation area and to profit from multisensor fusion using our general framework for the simultaneous localization and map building problem.

# References

[1] R. Smith, M. Self, and P. Cheeseman. Estimating uncertain spatial relationships in robotics. In J.F. Lemmer and L.N. Kanal, editors, *Uncertainty in Artificial Intelligence 2*, pages 435–461. Elsevier Science Pub., 1988.

[2] P. Moutarlier and R. Chatila. Stochastic multisensory data fusion for mobile robot location and environment modeling. In *5th Int. Symposium on Robotics Research*, Tokyo, Japan, 1989.

[3] J.J. Leonard and H.F. Durrant-Whyte. Simultaneous map building and localization for an autonomous mobile robot. In *Proc. 1991 IEEE/RSJ Int. Conf. on Intelligent Robots and Systems*, pages 1442–1447, Osaka, Japan, 1991.

[4] J.D. Tardós. Representing partial and uncertain sensorial information using the theory of symmetries. In *Proc. 1992 IEEE Int. Conf. on Robotics and Automation*, pages 1799–1804, Nice, France, 1992.

[5] J. Neira, J.D. Tardós, J. Horn, and G. Schmidt. Fusing range and intensity images for mobile robot localization. *IEEE Transactions on Robotics and Automation*, 15(1):76–84, 1999.

[6] P. Hébert, S. Betgé-Brezetz, and R. Chatila. Decoupling odometry and exteroceptive perception in building a global world map of a mobile robot: The use of local maps. In *Proc. 1996 IEEE Int. Conf. Robotics and Automation*, pages 757–764, Minneapolis, Minnesota, 1996.

[7] J.A. Castellanos, J.D. Tardós, and G. Schmidt. Building a global map of the environment of a mobile robot: The importance of correlations. In *Proc. 1997 IEEE Int. Conf. on Robotics and Automation*, pages 1053–1059, Albuquerque, NM, U.S.A, April 1997.

[8] J.A. Castellanos. *Mobile Robot Localization and Map Building: A Multisensor Fusion Approach*. PhD thesis, Dpto. de Ingeniería Eléctrica e Informática, University of Zaragoza, Spain, May 1998.

[9] J.A. Castellanos and J.D. Tardós. Laser-based segmentation and localization for a mobile robot. In F. Pin, M. Jamshidi, and P. Dauchez, editors, *Robotics and Manufacturing vol. 6 - Procs. of the 6th Int. Symposium (ISRAM)*, pages 101–108. ASME Press, New York, NY, 1996.

[10] J.B. Burns, A.R. Hanson, and E.M. Riseman. Extracting straight lines. *IEEE Trans. on Pattern Analysis and Machine Intelligence*, 8(4):425–455, 1986.

[11] R.Y. Tsai. A versatile camera calibration technique for high-accuracy 3d machine vision metrology using off-the-self tv cameras and lenses. *IEEE Journal of Robotics and Automation*, 4(3):323–344, 1987.

[12] J.M Martínez, Z. Zhang, and L. Montano. Segment-based structure from an imprecisely located moving camera. In *IEEE Int. Symposium on Computer Vision*, pages 182–187, Florida, November 1995.

# Experimental analysis of adaptive concurrent mapping and localization using sonar

John J. Leonard and Hans Jacob S. Feder
Department of Ocean Engineering
Massachusetts Institute of Technology
Cambridge, MA 02139, USA
jleonard@mit.edu,feder@deslab.mit.edu

## Abstract

This paper describes the experimental verification a method for performing concurrent mapping and localization (CML) adaptively using sonar. Predicted sensor readings and expected dead-reckoning errors are used to evaluate different potential actions of the robot. The action that yields the maximum information is selected. The performance of this approach to CML is investigated via experiments with a dynamic underwater sonar sensing system in a 9 meter by 3 meter by 1 meter testing tank. Increased navigation and mapping accuracy are demonstrated in comparison to results obtained with non-adaptive sensor motion.

# 1   Introduction

This paper describes the experimental verification a method for performing concurrent mapping and localization (CML) adaptively using sonar. The goal of CML is to enable a mobile robot to build a map of an unknown environment, while concurrently using that map for navigation. CML has been a central problem in robotics research due to its theoretical challenges and critical importance to many mobile robot applications [1, 2, 3]. The motivation for our experiments is to test new methods for CML that can enable autonomous underwater vehicles (AUVs) to navigate in unknown environments without relying on a priori maps or acoustic beacons [4].

Navigation is essential for successful operation of underwater vehicles in a range of scientific, commercial, and military applications [4]. Accurate positioning information is vital for the safety of the vehicle and for the utility of the data it collects. The error growth rates of inertial and dead-reckoning systems available for AUVs are usually unacceptable. Underwater vehicles cannot use the Global Positioning System (GPS) to reset dead-reckoning error growth unless they surface, which is usually undesirable. Positioning systems that use acoustic transponders [5] are often expensive and impractical to deploy and calibrate. Navigation algorithms based on existing maps have been proposed [4], but sufficiently accurate a priori maps are often unobtainable.

Adaptive sensing strategies have the potential to save time and maximize the efficiency of the concurrent mapping and localization process for an AUV. Energy efficiency is one of the most challenging issues in the design of underwater vehicle systems [6]. Techniques for building a map of sufficient resolution as quickly as possible would be highly beneficial. Many survey class AUVs must maintain forward motion for controllability, hence the ability to adaptively choose a sensing and motion strategy which obtains the most information about the environment is especially important.

Sonar is the principle sensor for AUV navigation. Possible sonar systems include mechanically and electronically scanned sonars, side-scan sonar and multibeam bathymetric mapping sonars [7]. The rate of information obtained from a mechanically scanned sonar is low, making adaptive strategies especially beneficial. Electronically scanned sonars can provide information at very high data rates, but enormous processing loads make real-time implementations difficult. Adaptive techniques can be used to limit sensing to selected regions of interest, dramatically reducing computational requirements.

In our experiments, adaptive sensing is formulated as the evaluation of different actions that the robot can take and the selection of the action that maximizes the amount of information acquired. This general problem has been considered in a variety of contexts [8, 9] but not specifically for concurrent mapping and localization. CML provides an interesting context within which to address adaptive sensing because of the trade-off between dead-reckoning error and sensor data acquisition. The information gained by observing an environmental feature from multiple vantage points must counteract the rise in uncertainty that results from the motion of the vehicle.

## 2 Theory

The algorithm used in our adaptive CML experiments is an enhanced version of the stochastic mapping method introduced by Smith, Self and Cheeseman [1], which we have coined augmented stochastic mapping (ASM). ASM differs from stochastic mapping in that it includes a logic based track initiator, a logic based track deleter and a nearest neighbor data association strategy[10]. In stochastic mapping, a single state vector $\mathbf{x}[k] = [\mathbf{x}_r^T \ \mathbf{x}_1^T ... \mathbf{x}_N^T]^T$ contains the vehicle's state, $\mathbf{x}_r$, as well as all the features states (the map), $\mathbf{x}_i$, that are being modeled at time step $k$. Further, the covariance for the system is estimated and denoted by $\mathbf{P}$. The estimated state vector and associated covariance is obtained through an extended Kalman filter (EKF) with a state transition model for the vehicle's dynamics given by

$$\mathbf{x}[k+1] = \mathbf{f}(\mathbf{x}[k], \mathbf{u}[k]) + \mathbf{d_x}(\mathbf{u}[k]), \qquad (1)$$

where $\mathbf{u}[k]$ is the control input at time step $k$ and $\mathbf{d_x}$ is a white stochastic process. Measurements of features in the environment are taken through range and bearing measurements relative to the vehicle's current position and stacked into a vector $\mathbf{z}$. The measurements are modeled by an observation model, $\mathbf{h}$, that is

$$z[k] = \mathbf{h}(\mathbf{x}[k]) + \mathbf{d_z} \tag{2}$$

where $\mathbf{d_z}$ is a white stochastic process.

In order to perform adaptive CML, we are interested in the information that can be gained from taking the measurement $z[k]$. This can be quantified using the Fisher information [10]

$$\mathbf{I_z}(\mathbf{x})[k|k] = -\mathop{E}_{\mathbf{z}}\{\nabla_{\mathbf{x}}\nabla_{\mathbf{x}}^T \ln p(\mathbf{Z}^k|\mathbf{x}[k])\} - \mathop{E}_{\mathbf{z}}\{\nabla_{\mathbf{x}}\nabla_{\mathbf{x}}^T \ln p(\mathbf{x})\}. \tag{3}$$

The information $\mathbf{I_z}(\mathbf{x})[k|k]$ quantifies the information of all the measurements $z[0]...z[k]$ as well as the information in the prior $p(\mathbf{x})$, such as, knowledge about the uncertainty of our vehicle and sonar model.

Under the assumption that the EKF is a good approximation to an efficient estimator, the information $\mathbf{I_z}(\mathbf{x})[k|k]$ is well approximated by the inverse of the error covariance $\mathbf{P}$ of the system. Thus, the information of the system at time step $k$ given the prior and all past measurements is readily available to us directly from the EKF. Following the method described in [11], we obtain the transformation that relates the current information $\mathbf{I_z}(\mathbf{x})[k|k]$ to the predicted resulting information $\mathbf{I_z}(\mathbf{x})[k+1|k+1]$ due to an action $\mathbf{u}[k]$. The action that maximizes the information gained can be expressed as:

$$\mathbf{u}[k] = \max_{\mathbf{u}} \arg \mathbf{I}_{k+1|k+1} = \min_{\mathbf{u}} \arg \mathbf{P}_{k+1|k+1}. \tag{4}$$

The information is a matrix and we require a metric to quantify the information. Further, it is desired that this metric have a simple physical interpretation.

The question of defining a metric for adaptive sensing has been considered by previous researchers in different contexts. Manyika utilized a metric based on entropy to define a scalar cost function that was used in a multi-sensor robot localization system [8]. Singh has developed an entropic measure for grid-based mapping and implemented it with the Autonomous Benthic Explorer [12].

For feature-based CML, it is desirable to use a metric that makes explicit the tradeoff between uncertainty in feature locations and uncertainty in the vehicle position estimate. To accomplish this, we define the metric by a cost function $C(\mathbf{P})$, which gives the total area of all the error ellipses, (i.e., highest probability density regions) and is thus a measure of our confidence in our map and robot position [11]. That is,

$$C(\mathbf{P}) = \pi \prod_j \sqrt{\lambda_j(\mathbf{P}_{rr})} + \pi \sum_{i=1}^{N} \prod_j \sqrt{\lambda_j(\mathbf{P}_{ii})}, \tag{5}$$

where $\mathbf{P}_{rr}$ signifies the estimated covariance of the vehicle and $\mathbf{P}_{ii}$ signifies the estimated covariance of feature $i$; $\lambda_j(\cdot)$ is the $j$-th eigenvalue of its argument. The action to take is obtained by evaluating Equation (5) over the action space of the robot. Using this mettric, an algorithm for performing adaptive augmented stochastic mapping is shown in Figure 1.

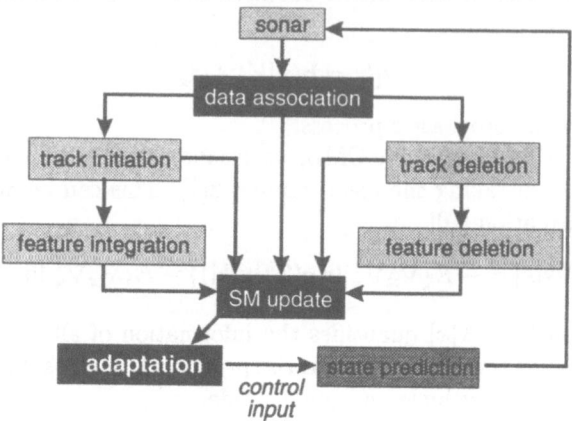

Figure 1: Structure of the adaptive augmented stochastic mapping (adaptive ASM) algorithm.

# 3 Results

Experiments were conducted to test the adaptive stochastic mapping algorithm using a narrow-beam 675 kHz sector scan sonar mounted on a planar robotic positioning system, as shown in Figure 2. The positioning system was controlled by a Compumotor AT6450 controller card. The system was mounted on a 3.0 by 9.0 by 1.0 meter testing tank. The stochastic mapping algorithm was implemented and integrated with code to control the AT6450 controllers, resulting in a closed loop system for performing CML. At each time step, Equation (5) was minimized over the action space of the robot to choose the motion and scanning angles of the sensor.

The experiments were designed to simulate an underwater vehicle equipped with a sonar that can scan at any direction relative to the vehicle at each time step. Conducting complete 360° scans of the environment at every time step is slow with a mechanically scanned sonar and computationally expensive with an electronically scanned sonar. For these experiments, we envisioned a vehicle mounted with two sonars, one looking forward for obstacle avoidance, and one that can be scanned in any direction for localization purposes. The forward looking sonar was assumed to scan an angle of ±30°. The scanning sonar was limited to scan over an angle of ±15° at each time step. The scanning was performed in intervals of 0.15°. The sonars were modeled to have a standard deviation in bearing of 10.0° and 2.0 centimeters in range. Between each scan by the sonar, the vehicle could move between 15 cm and 30 cm. The lower limit signifies a minimum speed that the vehicle could move at before loosing controllability, while the upper limit signifies the maximum speed of the vehicle. The vehicle was constrained to only turn in increments of 22.5°. Further, we assumed that the vehicle was equipped with a dead-reckoning system with an accuracy of 10% of distance traveled and an accuracy of 1.0° in heading.

Figure 3 shows a typical scan taken by the sonar from the origin. The

Figure 2: The planar robotic positioning system and sector-scan sonar used in the underwater sonar experiments. The water in the tank is approximately 1 meter deep. The transducer was translated and rotated in a horizontal plane mid-way through the water column.

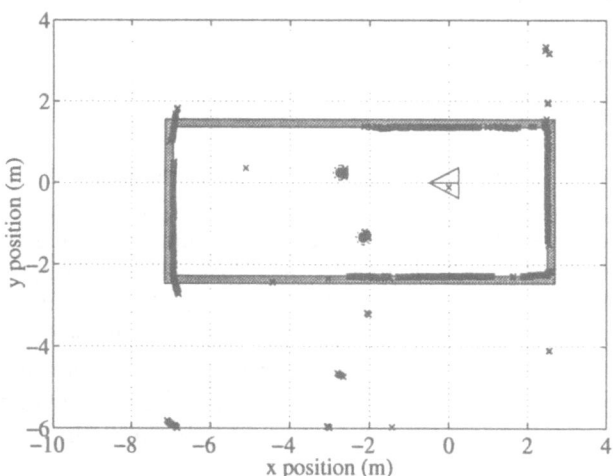

Figure 3: The returns from the underwater sonar for a 360° scan of the tank from the origin. The crosses shows individual returns. The small circles identify the position of the features (PVC tubes), with a dotted 5 cm outside circle drawn around them to signify the minimum allowable distance between the sonar and the features. The sonar was mounted on the carriage of the positioning system, which served as a "simulated AUV". The location of the sonar is shown by a triangle. The outline of the tank is shown in gray.

crosses show individual sonar returns. The circles show the features (PVC tubes). The dotted circles around the features signify the minimum allowable distance between the vehicle and the features. The triangle shows the position of the sensor. Circular arc features were extracted from the sonar scans using the technique described in [13].

Figure 4 shows the sensor trajectory for a representative underwater sonar experiment performing adaptive ASM. The sensor started at the origin and moved around the tank using adaptive ASM to decide where to move and where to scan. Based on minimizing of the cost function in Equation (5), the vehicle selected one target to scan to provide localization information. In addition, at each time step, the sonar was also scanned in front of the vehicle for obstacle avoidance. Figure 5 shows the $x$ and $y$ errors for the experiment and associated $3\sigma$ error bounds (99% highest confidence area). No sonar measurements were obtained from approximately time step 75 to time step 85 due to a communication error between the sonar head and the host PC. The left figure of Figure 6 shows the cost as a function of time. Solid vertical lines in Figures 5 and Figure 6 indicate the time steps when features of the environment were removed, the dashed-dotted vertical lines indicate the period in which the communication error occurred. The right figure in Figure 6 plots the vehicle position error versus time for the stochastic mapping algorithm in comparison to dead-reckoning.

Figures 7 and 8 show the results of a non-adaptive experiment in which the vehicle moved in a straight line in the negative $x$ direction. Without adaptive motion, the observability of the features was degraded and the $y$ estimate diverged.

# 4 Conclusion

The performance of adaptive augmented stochastic mapping (adaptive ASM) for performing concurrent mapping and localization (CML) was investigated via experiments with a dynamic underwater sonar sensing system in a 9 meter by 3 meter by 1 meter testing tank. The experiments show the superior performance of adaptive ASM versus non-adaptive ASM or systems based on dead-reckoning alone. Further, an intuitively appealing "exploratory" behavior emerges, with the result that different features of the environment are selectively explored. The algorithm also incorporates addition and removal of features from the environment and is robust to communication failure between the sonar and the host PC.

In related work, we have supplemented our experiments with a simulation study of an AUV performing long duration missions to map large areas using forward look sonar data [14]. Simulations with external disturbances such as ocean currents demonstrate convincingly the superior performance of ASM over dead-reckoning alone [15].

To address the issue of map scaling in CML [3], we have developed the technique of Decoupled Stochastic Mapping [14]. This method reduces the $\mathcal{O}(n^3)$ computational complexity of stochastic mapping (where $n$ is the number of fea-

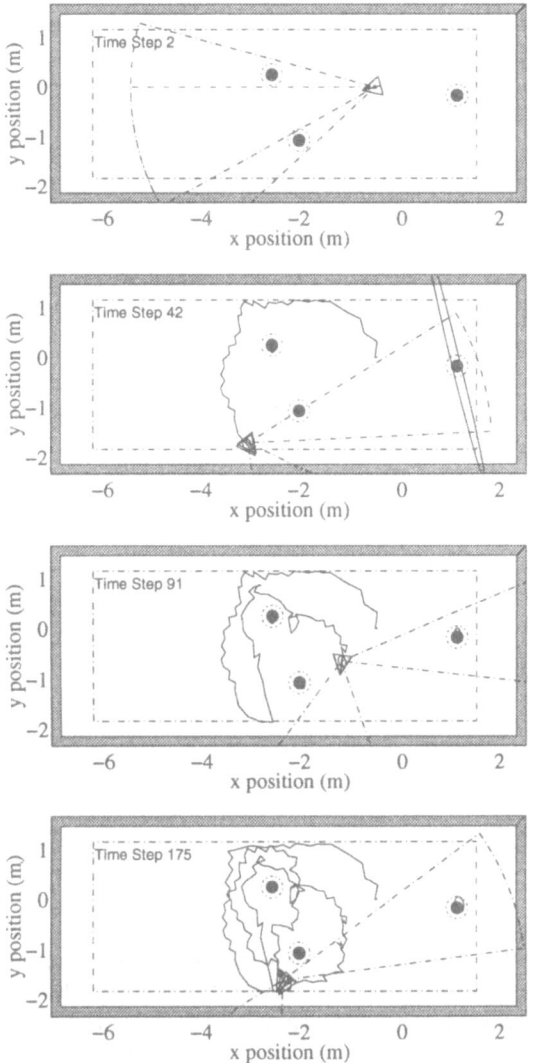

Figure 4: Time evolution of the sensor trajectory for an adaptive stochastic mapping experiment with the underwater sonar. The feature on the right was added at the 40th time step and the two features on the left were removed towards the end of the experiment. The vehicle started at the origin and moved adaptively through the environment to investigate different features in turn, maximizing Equation (5) at each time step. The filled circles designate the feature locations, and are surrounded by dotted circles which designate the "stand off" distance used by an obstacle avoidance routine. The dashed-dotted line represents the scanning region selected by the vehicle at the last time step. The triangle designates the vehicle's position. The vehicle was constrained from moving outside the dashed-dotted rectangle to avoid collisions with the walls of the tank. Sonar returns originating from outside the dashed-dotted rectangle were rejected.

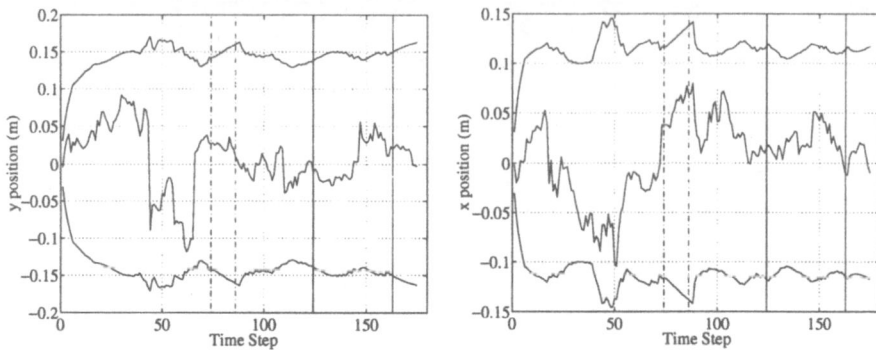

Figure 5: Position errors in the $x$ and $y$ directions and 3-$\sigma$ confidence bounds for adaptive experiment.

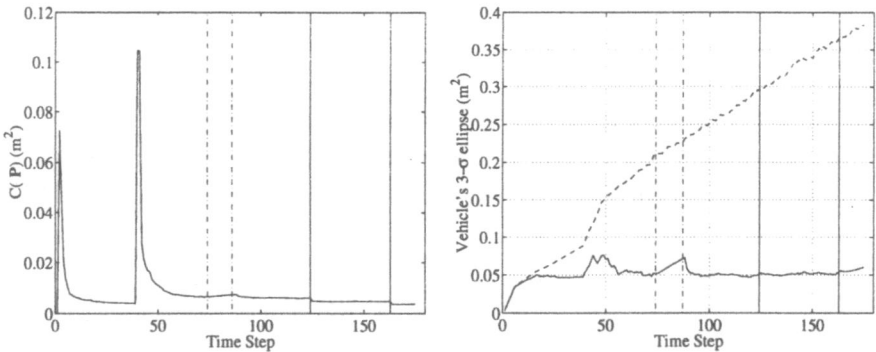

Figure 6: **Left figure:** Cost, $C(\mathbf{P})$ of Equation (5), as a function of time for adaptive underwater experiment. The cost increased at approximately the 40th time step when the third object was inserted into the tank and was observed for the first time. During the time interval between the two dashed vertical lines, no sonar data was obtained due to a serial communications problem between the PC and the sonar head. The two solid vertical lines designate the time steps during which features were removed from the tank, to simulate a dynamic environment. **Right figure:** Vehicle position error versus time for dead-reckoning (dashed line) and the adaptive ASM algorithm (solid line). The two solid vertical lines designate the time steps when features were removed from the tank, to simulate a dynamic environment.

tures), to an algorithm which is constant in computation time while maintaining consistent, globally-referenced error bounds.

In general, we have seen excellent agreement between simulations and real data. We are collaborating with the Naval Undersea Warfare Center in New-

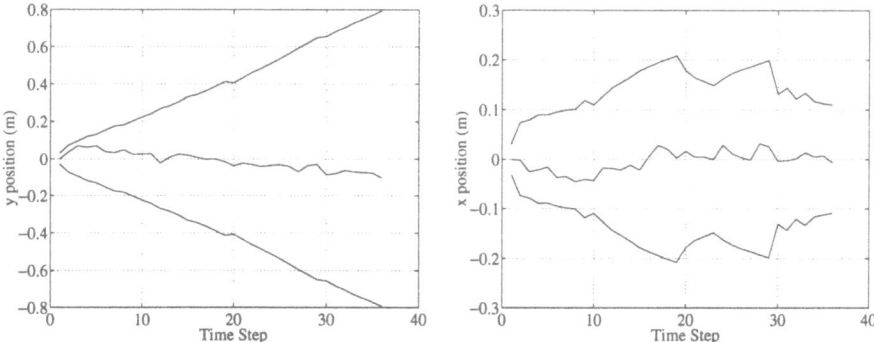

Figure 7: Position errors in the $x$ and $y$ directions and 3-$\sigma$ confidence bounds for non-adaptive underwater experiment. While accurate location information was obtained in the $x$ direction, the CML process diverged in the $y$ direction.

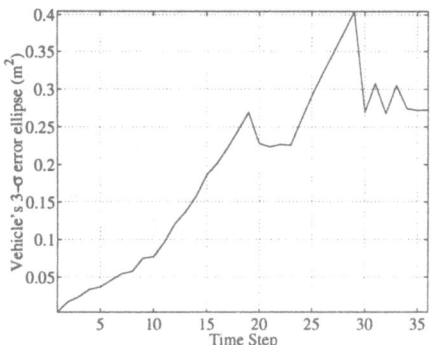

Figure 8: Vehicle error as a function of time for non-adaptive experiment.

port, Rhode Island to implement CML on-board a US Navy unmanned underwater vehicle using data from a high-resolution forward-look sonar array [16]. Preliminary results have been obtained that demonstrate successful CML postprocessing of a small set of real data acquired in Narragansett Bay, Rhode Island. [4].

Experimental work in progress is testing new methods for adaptive echolocation using a broad beam, wide-band binaural sonar system that mimics the sonar characteristics of a bottle-nose dolphin [17].

### Acknowledgments
This research has been funded in part by the National Science Foundation under Career Award BES-9733040, the Naval Undersea Warfare Center, Newport, RI, U.S.A, and the C. S. Draper Laboratories under University IRD Project number 98-1002. H.J.S.F. acknowledges the support of NFR (Norwegian Research Council) through grant 109338/410. J.J.L. acknowledges the support of the Henry L. and Grace Doherty Assistant Professorship in Ocean Utilization. The authors would like to thank Seth

Lloyd and Chris Smith for their helpful suggestions and Paul Newman for his assistance in interfacing the Imagenex 881 sonar.

# References

[1] R. Smith, M. Self, and P. Cheeseman. Estimating uncertain spatial relationships in robotics. In I. Cox and G. Wilfong, editors, *Autonomous Robot Vehicles*. Springer-Verlag, 1990.

[2] P. Moutarlier and R. Chatila. Stochastic multisensory data fusion for mobile robot location and environment modeling. In *5th Int. Symposium on Robotics Research*, Tokyo, 1989.

[3] J. J. Leonard and H. F. Durrant-Whyte. Simultaneous map building and localization for an autonomous mobile robot. In *Proc. IEEE Int. Workshop on Intelligent Robots and Systems*, pages 1442–1447, Osaka, Japan, 1991.

[4] J. J. Leonard, A. A. Bennett, C. M. Smith, and H. J. S. Feder. Autonomous underwater vehicle navigation. In *IEEE ICRA Workshop on Navigation of Outdoor Autonomous Vehicles*, Leuven, Belguim, May 1998.

[5] P. H. Milne. *Underwater Acoustic Positioning Systems*. London: E. F. N. Spon, 1983.

[6] J. G. Bellingham and J. S. Willcox. Optimizing AUV oceanographic surveys. In *IEEE Conference on Autonomous Underwater Vehicles*, Monterey, CA, 1996.

[7] R. Urick. *Principles of Underwater Sound*. New York: McGraw-Hill, 1983.

[8] J. S. Manyika and H. F. Durrant-Whyte. *Data Fusion and Sensor Management: A decentralized information-theoretic approach*. New York: Ellis Horwoord, 1994.

[9] S. Russell and P. Norvig. *Artificial Intelligence A Modern Approach*. Prentice Hall, 1995.

[10] Y. Bar-Shalom and T. E. Fortmann. *Tracking and Data Association*. Academic Press, 1988.

[11] H. J. S. Feder, J. J. Leonard, and C. M. Smith. Adaptive concurrent mapping and localization using sonar. In *Proc. IEEE Int. Workshop on Intelligent Robots and Systems*, Victoria, B.C., Canada, October 1998.

[12] H. Singh. *An entropic framework for AUV sensor modelling*. PhD thesis, Massachusetts Institute of Technology, 1995.

[13] B. A. Moran, J. J. Leonard, and C. Chryssostomidis. Curved shape reconstruction using multiple hypothesis tracking. *IEEE J. Ocean Engineering*, 22(4):625–638, 1997.

[14] J. J. Leonard and H. J. S. Feder. Decoupled stochastic mapping. Technical Report Marine Robotics Laboratory Technical Memorandum 99-1, Massachusetts Institute of Technology, 1999.

[15] H. J. S. Feder, C. M. Smith, and J. J. Leonard. Incorporating environmental measurements in navigation. In *IEEE AUV*, Cambridge, MA, August 1998.

[16] R. N. Carpenter. Concurrent mapping and localization using forward look sonar. In *IEEE AUV*, 1998.

[17] W. Au. *The Sonar of Dolphins*. New York: Springer-Verlag, 1993.

# Chapter 8

# Planning and Navigation

The previous chapter discussed how mobile robots can determine their location in an unknown environment. This chapter is concerned with how robots can plan and execute paths so as to reach a goal location.

Ray Jarvis describes the problem of navigating a Martian rover through rough, but known terrain. A terrain 'tractability' map is generated from laser scanner data and an achievable path to the goal is found using a modified distance transform method.

Karl Iagnemma and colleagues describe manipulator and traction control systems for a Martian rover. The manipulator control uses feedback from a base force/torque sensor to compensate for friction due to the small motors and high-reduction ratio gearboxes used in the lightweight design. The vehicle has 6 independently powered wheels in a reconfigurable rocker-bogie arrangement. A fuzzy logic controller is shown to achieve optimal online traction control.

Continuing on the Martian theme is the paper by Laubach and Burdick on a path planning system for future Martian rovers which must be able to perform more autonomously than Sojourner. In particular they require the vehicle to be able to progress toward a goal even if it is obscured for part of the mission. Their algorithm makes local terrain maps from mast based stereo cameras, generates subpaths and executes them until the goal is achieved.

Brock and Khatib present the elastic strips conceptual framework which integrates path planning and execution in the presence of changing obstacles.

# Quasi Real-time Walking Control of a Bipedal Humanoid Robot
# Based on Walking Pattern Synthesis

*Samuel Agus SETIAWAN  **Sang Ho HYON
***Jin'ichi YAMAGUCHI  * ***Atsuo TAKANISHI
*Department of Mechanical Engineering, Waseda University
**Department of Control and Systems Engineering, Tokyo Institute of Technology
***Humanoid Research Laboratory, Advanced Research Institute for Science and
Engineering, Waseda University

## Abstract

This research is aimed at the development of bipedal humanoid robots working in human living space, with a focus on its physical construction and motion control method. At the initial stage, we developed a 35 DOF bipedal humanoid robot WABIAN (WAseda BIpedal humANoid)[1] in 1996, and proposed a control method for dynamic cooperative biped walking. By this control method, we have realized a various number of patterns, such as dynamic forward or backward walking, marching in place, dancing, and carrying a load. In this paper, we presented a quasi real-time walking control of a bipedal humanoid based on those walking pattern syntheses. Concretely, we developed a follow-walking control method with a patterns switching technique for a bipedal humanoid robot to follow human motion through hand contact, as an application to a human-robot physical interaction. At the same time, we adopted a 6-axis force-torque sensor system to WABIAN's hand. By using the control methods, WABIAN has been able to perform dynamic marching in place, walking forward and backward during a continuous time while someone is pushing or pulling its hand in such a way.

## 1. Introduction

Bipedal humanoid robots intended to share the same working space with humans have different functional abilities from robots in factories or hazardous environments. The ability to achieve various motion patterns is required by robots to behave freely in a human living environment. Further, they are strongly desired to have a flexible workability, such as moving along with human motion while in physical contact.

In this research, we intended to develop a control method to realize the generation of various motions by pattern selections and a switching algorithm in the real-time way. The patterns are precomputed and provided by the pattern generation software before walking. We applied the control method to realize human-robot interaction.

Physical interaction between humans may be realized by the action of shaking hands, walking together hand in hand, and even dancing. From these cases, it is reasonable to suppose that the hand has an important role in physical interactions with humans. Thus, our object for this research is to realize locomotive tracking motion by a bipedal humanoid robot to human motion through hand contact using the newly developed control technique.

In this paper, we first describe a control method for dynamic cooperative biped walking. This walking control method is an improved version of a model based walking control with compensation for three-axis moments by trunk motion, which has been applied to our former bipedal robot WL-12RV[2]. It is used to stabilize the dynamic walking of the robot. Next, we present human-follow walking with the patterns switching control technique. Finally, we describe the results of the experiments performed using WABIAN.

## 2. Stabilization walking control method for a bipedal humanoid robot

The control method which only used the consideration of the upper-limb's model in its moment calculation algorithm, is a preliminary part of whole body cooperative walking, our newest and most recently proposed motion control method[3]. It is composed of two main components. The first component includes an algorithm for computing the compensatory trunk motion from the motion of the lower-limbs, the time trajectory of ZMP, and the time trajectory of the hands planned arbitrarily before walking. Further, this algorithm consists of the following four submain parts, i.e.: (1) modeling of the robot, (2) derivation of the ZMP equations, (3) computation of approximate trunk motion in frequency domain, and (4) computation of strict trunk motion by iteratively computing the approximate trunk motion. The other component of the control method is a program control for walking using preset walking patterns. The latter is one kind of a general open-loop method in robot control, so that will be not discussed in this paper. Below we describe the algorithm for computing the compensatory trunk motion.

For modeling the robot, first we assumed the walking system as follows:
(1) The robot is a system of particles.
(2) The floor for walking is rigid and not moved by any force or moment.
(3) The X-axis and the Y-axis in a Cartesian coordinate form a plane that is the same as that of the floor. The Z-axis is settled perpendicular to the floor.
(4) The contact region between the foot and the floor is a set of contact points.
(5) The coefficient of friction for rotation around the X, Y and Z-axes is zero at the contact point.

From the assumption that the upper-limb is one part of the trunk, we first define an

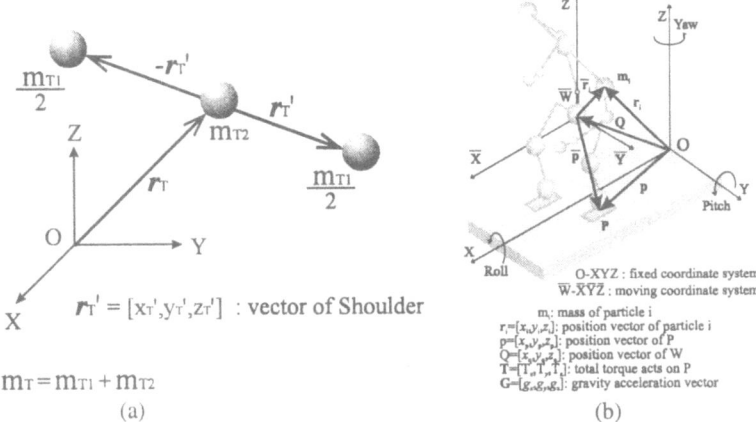

$r_T' = [x_T', y_T', z_T']$ : vector of Shoulder

$m_T = m_{T1} + m_{T2}$

(a)

O-XYZ : fixed coordinate system
$\overline{W}$-$\overline{XYZ}$ : moving coordinate system

$m_i$: mass of particle i
$r_i = [x_i, y_i, z_i]$: position vector of particle i
$p = [x_p, y_p, z_p]$: position vector of P
$Q = [x_q, y_q, z_q]$: position vector of W
$T = [T_x, T_y, T_z]$: total torque acts on P
$G = [g_x, g_y, g_z]$: gravity acceleration vector

(b)

**Fig. 1** Definition of position vectors for trunk

approximation model of the trunk and the position vectors as shown in **Fig.1**. Based on this model and the D'Lambert principle, the moment balance around point P on the floor

$$m_{T1}r_{T1}' \times \ddot{r}_{T1}' + \sum_i^{all\_particles} m_i(\mathbf{r}_i - p) \times (\ddot{\mathbf{r}}_i + \mathbf{G}) + \mathbf{T} = 0 \quad (1)$$

can be expressed as below:

Point P is defined as ZMP, so we denote the position vector of P as $P_{zmp}(x_{zmp}, y_{zmp}, 0)$. To consider the relative motion of each part, the translational moving coordinate $\overline{W}$ - $\overline{XYZ}$ is established on the waist of the robot on a parallel with the fixed coordinate O-XYZ (shown in Fig. 1). $Q(x_q, y_q, z_q)$ is defined as the origin of $\overline{W}$ - $\overline{XYZ}$ on the O-XYZ. Using the coordinate $\overline{W}$ - $\overline{XYZ}$, equation (1) can be modified and expanded into (2), (3), (4), (5) by putting the terms about the motion of the upper-limb particles on the left-hand side as unknown variables, and the terms about the moment generated by the lower-limb particles on the right-hand side as known parameters, named $M(Mx, My, Mz)$ respectively.

$$m_{T1}(z_T'\ddot{x}_T' - x_T'\ddot{z}_T') + m_T(\bar{z}_T + z_q)(\ddot{\bar{x}}_T + \ddot{x}_q)$$
$$- m_T(\ddot{\bar{z}}_T + \ddot{z}_q + g)(\bar{x}_T - \bar{x}_{zmp}) = -My(t) \quad \bullet(2)$$

$$m_{T1}(y_T'\ddot{z}_T' - z_T'\ddot{y}_T') - m_T(\bar{z}_T + z_q)(\ddot{\bar{y}}_T + \ddot{y}_q)$$
$$+ m_T(\ddot{\bar{z}}_T + \ddot{z}_q + g)(\bar{y}_T - \bar{y}_{zmp}) = -Mx(t) \quad \bullet(3)$$

$$m_{T1}(x_T'\ddot{y}_T' - y_T'\ddot{x}_T') + Mz_T(t) = -Mz(t) \quad \bullet\bullet(4)$$

$$Mz_T(t) = -m_T(\ddot{\overline{x}}_T + \ddot{x}_q)(\overline{y}_T - \overline{y}_{zmp})$$
$$+ m_T(\ddot{\overline{y}}_T + \ddot{y}_q)(\overline{x}_T - x_{zmp}) \quad \bullet\bullet (5)$$

By assuming that neither the waist nor the trunk particles move vertically, i.e., the trunk arm rotates on the horizontal plane only, the equations could be decoupled and linearized, and then indicated as the approximate model's equations (6), (7) and (8).

$$m_T(\overline{z}_T + z_q)(\ddot{\overline{x}}_T + \ddot{x}_q) - m_T g(\overline{x}_T - \overline{x}_{zmp}) = -My \quad (6)$$

$$-m_T(\overline{z}_T + z_q)(\ddot{\overline{y}}_T + \ddot{y}_q) - m_T g(\overline{y}_T - \overline{y}_{zmp}) = -Mx \quad (7)$$

$$m_{T1}R^2\theta_y = -Mz_T - Mz \quad (8)$$

$\theta_y$ is the rotational angle of the yaw-axis actuator, and R is the radius of the trunk's arm.

Each equation can be represented as a Fourier series, since in the case of steady walking, $My, Mx, Mz$ are periodic functions. Hence, by comparing the Fourier Transform coefficients of both sides of each equation, we can easily acquire the approximate periodic solution of the trunk motion. We determine the offset term in the equation of the yaw-axis moment as a value where the generated yaw-motion angle is within the range of the rotatable region.

Further, to obtain strict solutions the approximate solutions are computed again iteratively while the errors are accumulated to the side of known generated moments. The computation ends when the errors converge to a certain value within a tolerance margin.

**Fig.2** shows outline of the flowchart of the compensatory trunk motion computation algorithm described above. Further discussion of the algorithm and its simulation results are described in [1]. In the next chapter, we will discuss the effect of this control method in the follow-walking motion.

## 3. Human-follow Walking Control Technique

This method tries to realize a follow-walking motion by selecting and generating switchable unit patterns, based on a behaviour criterion for human-robot interaction. The upper-limb's trajectory is determined from the force information applied on the robot's hand and computed using the method of virtual compliance control[4]. The trajectory of the lower-limb is determined arbitrarily while its pattern's type is decided by judging the direction of the robot's tracking motion. The robot recognizes the guiding direction of human motion by detecting the position or displacement of its hand, and then decides the next walking pattern while synchronizing it with the present walking condition. In the case where no pattern is selected, we programmed to let the present condition be continued by the robot.

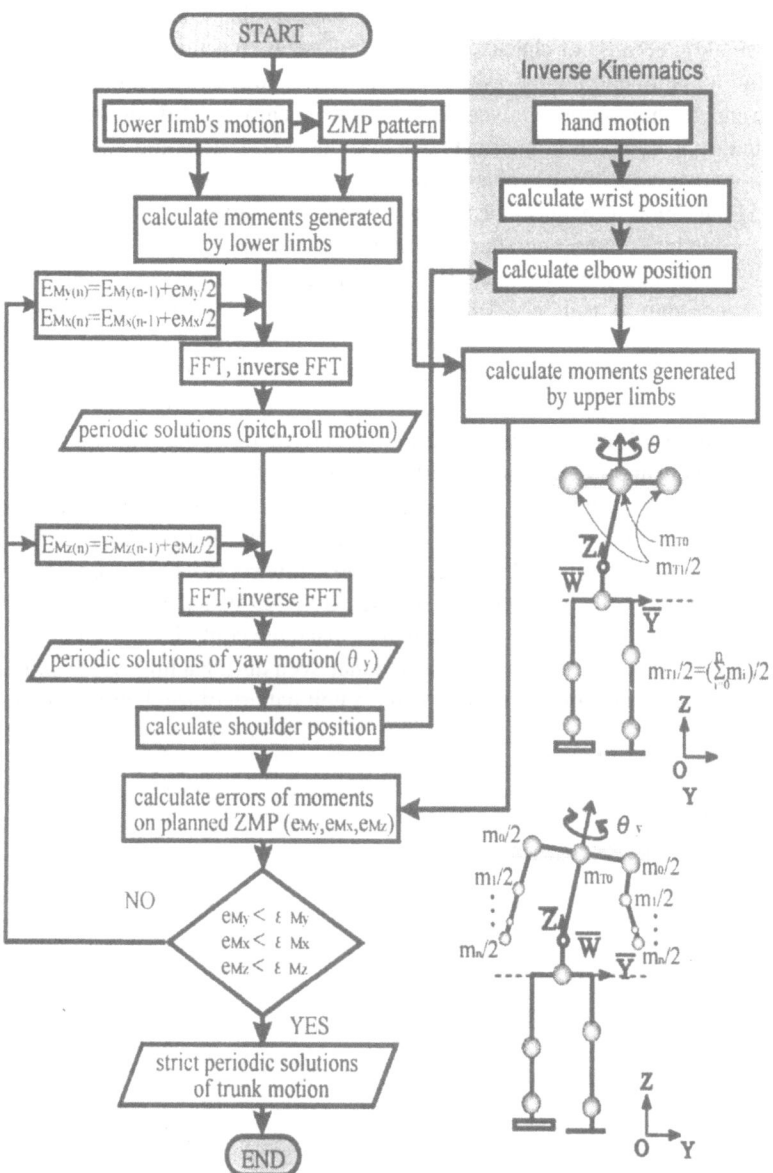

**Fig. 2** Flowchart to compute trunk compensatory motion

314

In addition, we know from the transfer functions of Eq. (6) and (7) that the above control method doesn't cope with initial value problem and doesn't satisfy the law of causality, because of characteristics of the Lorentz function. Also, from the simulation and experimental data as shown in **Fig.4**, we could conclude that the trunk compensation motion moved earlier than the shift of ZMP on the floor. This means that trunk compensation affects one or more steps before and after in a pattern time.

To overcome these problems, we proposed the solution methods described below.

(1) The motion patterns of the lower limbs are given as unit patterns, which have one before and one after gait attributes of a step pattern (in consideration of the trunk's motion dynamic). By combining some simple and selective patterns, it is possible to realize a following motion by a bipedal robot similar to a human. Unit patterns that are combined from three gait patterns (step forward, backward, marching in place) and are computed to have a continuous position, velocity and acceleration of its trajectory, are calculated offline.

(2) The trunk compensatory motion calculated from the moment generated by the lower-limbs is computed offline, and it is applied to the robot as sets of unit patterns (**Fig.5**) in the same way as the lower limbs' patterns. Through experiments using WABIAN, we could clarify that in circumstance of movable margin of the trunk's angle, the ZMP shift caused by the upper limbs' smooth motion could be ignored. This was done so that the change in attitude of the upper limbs could be disregarded to make the control technique simpler.

By a simulation we have confirmed that even during a high speed motion (±1.0[sec/step]), it is possible to make a unit pattern of trunk motion by taking into

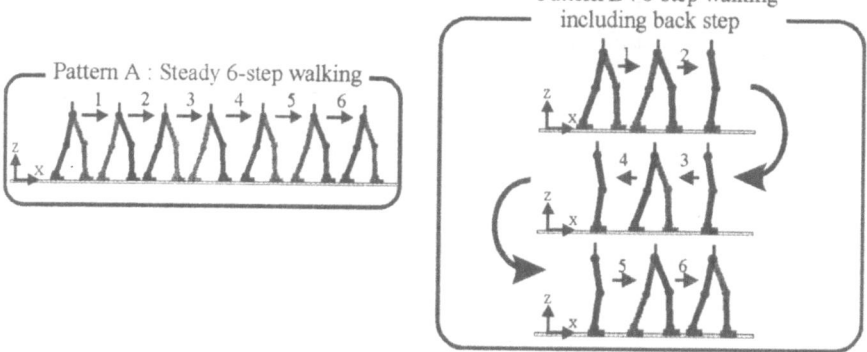

**Fig.3** Two patterns used to evaluate the effect of compensatory motion's dynamic

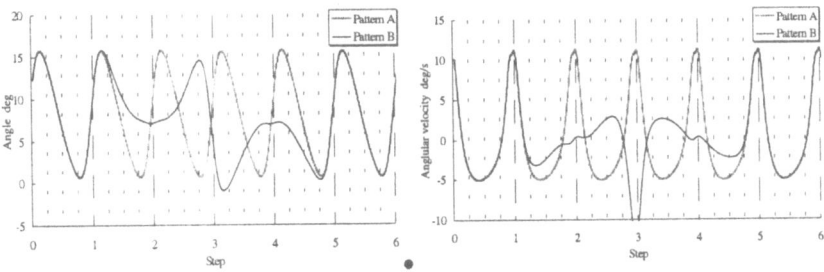

(a)   step time : 6.40[s], step length : 0.1[m]

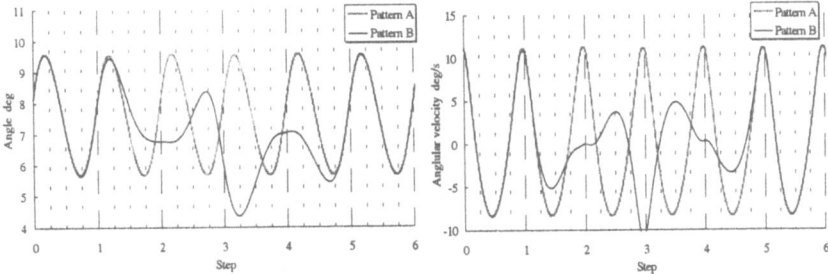

(b)   step time : 1.28[s], step length : 0.1[m]

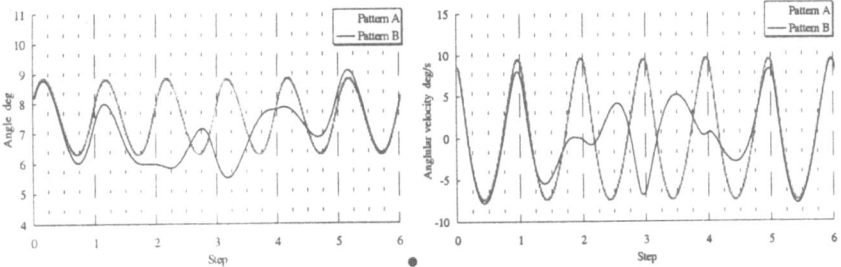

(d)   step time : 0.96[s], step length : 0.1[m]

**Fig. 4** Trunk trajectories of two walk patterns

account only one step before and after of a unit pattern as an attribute. The unit patterns were created using a software simulator, and include indexes for pattern searching and attributes of current, before and after steps. These unit patterns are preloaded as one step long angle data in the memory.

**A**

| # | prev | curr | next |
|---|------|------|------|
| 0 | FL | FR | FL |
| 1 | AFL | FR | FL |
| 2 | FL | FR | DFL |
| 3 | AFL | FR | DFL |

**B**

| # | prev | curr | next |
|---|------|------|------|
| 4 | SL | AFR | FL |
| 5 | DFL | AFR | FL |
| 6 | DBL | AFR | FL |
| 7 | IDLE | AFR | FL |
| 8 | SL | AFR | DFL |
| 9 | DFL | AFR | DFL |
| 10 | DBL | AFR | DFL |
| 11 | IDLE | AFR | DFL |

**C**

| # | prev | curr | next |
|---|------|------|------|
| 12 | FL | DFR | SL |
| 13 | AFL | DFR | SL |
| 14 | FL | DFR | AFL |
| 15 | AFL | DFR | AFL |
| 16 | FL | DFR | ABL |
| 17 | AFL | DFR | ABL |
| 18 | FL | DFR | IDLE |
| 19 | AFL | DFR | IDLE |

**D**

| # | prev | curr | next |
|---|------|------|------|
| 20 | BL | BR | BL |
| 21 | ABL | BR | BL |
| 22 | BL | BR | DBL |
| 23 | ABL | BR | DBL |

**E**

| # | prev | curr | next |
|---|------|------|------|
| 24 | SL | ABR | BL |
| 25 | DBL | ABR | BL |
| 26 | DFL | ABR | BL |
| 27 | IDLE | ABR | BL |
| 28 | SL | ABR | DBL |
| 29 | DBL | ABR | DBL |
| 30 | DFL | ABR | DBL |
| 31 | IDLE | ABR | DBL |

**F**

| # | prev | curr | next |
|---|------|------|------|
| 32 | BL | DBR | SL |
| 33 | ABL | DBR | SL |
| 34 | BL | DBR | ABL |
| 35 | ABL | DBR | ABL |
| 36 | BL | DBR | AFL |
| 37 | ABL | DBR | AFL |
| 38 | BL | DBR | IDLE |
| 39 | ABL | DBR | IDLE |

**G**

| # | prev | curr | next |
|---|------|------|------|
| 40 | FR | FL | FR |
| 41 | AFR | FL | FR |
| 42 | FR | FL | DFR |
| 43 | AFR | FL | DFR |

**H**

| # | prev | curr | next |
|---|------|------|------|
| 44 | SR | AFL | FR |
| 45 | DFR | AFL | FR |
| 46 | DBR | AFL | FR |
| 47 | IDLE | AFL | FR |
| 48 | SR | AFL | DFR |
| 49 | DFR | AFL | DFR |
| 50 | DBR | AFL | DFR |
| 51 | IDLE | AFL | DFR |

**I**

| # | prev | curr | next |
|---|------|------|------|
| 52 | FR | DFL | SR |
| 53 | AFR | DFL | SR |
| 54 | FR | DFL | AFR |
| 55 | AFR | DFL | AFR |
| 56 | FR | DFL | ABR |
| 57 | AFR | DFL | ABR |
| 58 | FR | DFL | IDLE |
| 59 | AFR | DFL | IDLE |

**J**

| # | prev | curr | next |
|---|------|------|------|
| 60 | BR | BL | BR |
| 61 | ABR | BL | BR |
| 62 | BR | BL | DBR |
| 63 | ABR | BL | DBR |

**K**

| # | prev | curr | next |
|---|------|------|------|
| 64 | SR | ABL | BR |
| 65 | DBR | ABL | BR |
| 66 | DFR | ABL | BR |
| 67 | IDLE | ABL | BR |
| 68 | SR | ABL | DBR |
| 69 | DBR | ABL | DBR |
| 70 | DFR | ABL | DBR |
| 71 | IDLE | ABL | DBR |

**L**

| # | prev | curr | next |
|---|------|------|------|
| 72 | BR | DBL | SR |
| 73 | ABR | DBL | SR |
| 74 | BR | DBL | ABR |
| 75 | ABR | DBL | ABR |
| 76 | BR | DBL | AFR |
| 77 | ABR | DBL | AFR |
| 78 | BR | DBL | IDLE |
| 79 | ABR | DBL | IDLE |

**M**

| # | prev | curr | next |
|---|------|------|------|
| 80 | SL | SR | SL |
| 81 | DFL | SR | SL |
| 82 | DBL | SR | SL |
| 83 | IDLE | SR | SL |
| 84 | SL | SR | AFL |
| 85 | DFL | SR | AFL |
| 86 | DBL | SR | AFL |
| 87 | IDLE | SR | AFL |
| 88 | SL | SR | ABL |
| 89 | DBL | SR | ABL |
| 90 | DFL | SR | ABL |
| 91 | IDLE | SR | ABL |
| 92 | SL | SR | IDLE |
| 93 | DFL | SR | IDLE |
| 94 | DBL | SR | IDLE |

**N**

| # | prev | curr | next |
|---|------|------|------|
| 95 | SR | SL | SR |
| 96 | DFR | SL | SR |
| 97 | DBR | SL | SR |
| 98 | IDLE | SL | SR |
| 99 | SR | SL | AFR |
| 100 | DFR | SL | AFR |
| 101 | DBR | SL | AFR |
| 102 | IDLE | SL | AFR |
| 103 | SR | SL | ABR |
| 104 | DBR | SL | ABR |
| 105 | DFR | SL | ABR |
| 106 | IDLE | SL | ABR |
| 107 | SR | SL | IDLE |
| 108 | DFR | SL | IDLE |
| 109 | DBR | SL | IDLE |

F : step forward at const speed
B : step back at const speed
S : marching in place
AF : step forward acceleration
AB : step back acceleration
DF : step forward deceleration
DB : step back deceleration
IDLE : idling (stand still)

R : right leg
L : left leg

**Fig.5** Unit Walking Patterns

## 4. Experiments

To let WABIAN be able to follow a human's guidance through hand, we installed a six-axis moment and force sensor on its wrist and applied a compliance control to it.

Tracking experiments were performed to determine the compliance parameters, and walking experiments were done to decide suitable step time and width under the condition of the follow-walking technique mentioned above. From these experiments, we could confirm that the effect of the upper limb's attitude is more dominant in a fast motion, as the frequency of the trunk compensatory moment becomes higher as the robot moves faster. Therefore the change of attitude of the upper limbs could not be disregarded anymore, and a more effective control technique is needed to realize a switching motion while walking.

**Fig.6** shows cut of scene of the human-follow walking experiment (16 continous number of steps). The step time is 1.28[s/step], the step width is 0.1[m], and the inverse viscosity coefficient for the arm compliance is 0.15[m/Ns].

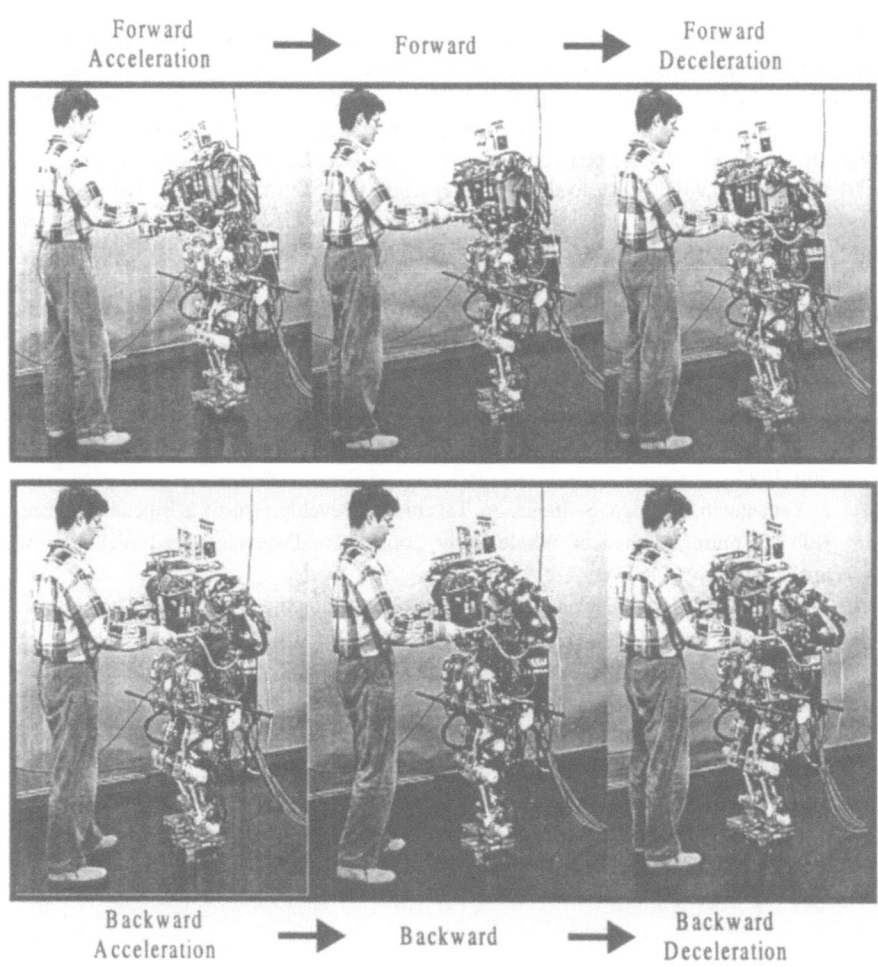

**Fig.6** Scene of Experiment with WABIAN

# 5. Conclusions

In this research, we presented a control technique to realize physical interaction between a human and a life-size humanoid robot based on a quasi real-time walking control of walking pattern synthesis. We revised our bipedal robot WABIAN by installing a new six-axis force sensor system on its wrist. We performed a human-follow walking experiment using this system and a newly developed control technique. We confirmed the effectiveness of both control methods by performing experiments using WABIAN.

**Acknowledgment**

This study has been conducted as a part of the project: Humanoid at HUREL (HUmanoid REsearch Laboratory), Advanced Research Institute for Science and Engineering, Waseda University. The authors would like to thank ATR, NAMCO Ltd., YASKAWA ELECTRIC Corp., NISSAN MOTOR CO., LTD., and NIPPON STEEL CORPORATION for their cooperation in this study. A part of this study was done by the Japanese Grant-in-Aid for Science Research (No. 07405012) and NEDO (New Energy and Industrial Technology Development Organization).

The authors would also like to thank NITTA Corp, OKINO Industries ltd., Harmonic Drive Systems, Inc., YKK Corp, Daisuke AOYAGI and Akihiro NAGAMATSU for supporting us in developing the hardware for the six-axis moment and force sensor system.

**References**

[1] J. Yamaguchi, S. Inoue, D. Nishino, A. Takanishi "Development of a Bipedal Robot Having Antagonistic Joints and Three DOF Trunk", Proc. of IROS'98, pp. 96-101, 1998

[2] J. Yamaguchi, A. Takanishi, I. Kato "Development of a Biped Walking Robot Compensating for Three-Axis Moment by Trunk Motion", Proc. of IROS'93, pp. 561-566, 1993

[3] J. Yamaguchi, E. Soga, S. Inoue, A. Takanishi "Development of a Bipedal Humanoid Robot: Control Method of Whole Body Cooperative Dynamic Biped Walking", will appear in Proc. of ICRA'99

[4] M. T. Mason "Compliance and Force Control for Computer Controlled Manipulators", IEEE Trans. on SMC, Vol.11, No.6, pp.418-432, 1981

[5] T. Yoshikawa "Analysis and Control of Robot Manipulators with Redundancy", Robotics Research, MIT Press, 1984

[6] Q. Huang, S. Sugano "Motion Planning of Stabilization and Cooperation of a Mobile Manipulator", Proc. of the 1996 ICRA, pp. 568-575, 1996

# Experimental Validation of Physics-Based Planning and Control Algorithms for Planetary Robotic Rovers

Karl Iagnemma   Robert Burn   Eric Wilhelm   Steven Dubowsky
Department of Mechanical Engineering
Massachusetts Institute of Technology
Cambridge, MA  02139 U.S.A.
dubowsky@mit.edu

**Abstract**: Robotic planetary exploration is a major component of the United States' NASA space science program. The focus of our research is to develop rover planning and control algorithms for high-performance robotic planetary explorers based on the physics of these systems. Experimental evaluation is essential to ensure that unmodeled effects do not degrade algorithm performance. To perform this evaluation a low-cost rover test-bed has been developed. It consists of a rocker-bogie type rover with an on-board manipulator operating in a rough-terrain environment. In this paper the design and fabrication of an experimental rover system is described, and the experimental validation of several rover control algorithms is presented. The experimental results obtained are key to the evaluation and validation of our research.

## 1. Introduction

On July 4, 1997, the Mars Pathfinder mission landed the Sojourner telerobotic rover on the surface of Mars [1]. The United States' NASA plans planetary robotic missions in the years 2001, 2003, and 2005 to Mars with far more ambitious performance objectives than those set for Sojourner. Future planetary explorers will need to navigate rugged terrain and travel substantial distances, while exercising a high degree of autonomy [2,3]. The focus of our research at the MIT Mechanical Engineering Field and Space Robotics Laboratory (FSRL) is on developing methods and algorithms for the design, planning and control of high-performance robotic planetary explorers based on the physics of these systems.

To validate the effectiveness of the developed algorithms, experimental evaluation is essential. To perform this evaluation a low-cost rover test-bed has been developed (see Figure 1). The FSRL rover is a 6-wheeled rocker-bogie vehicle that is similar kinematically to the JPL Lightweight Survivable Rover (LSR) (see Figure 2) [4]. The FSRL rover has a three degree-of-freedom manipulator mounted on the front of its chassis. It uses several prototype end-effector concepts for manipulation of rock samples. The experimental system chassis contains shape memory alloy

(SMA) actuated variable geometry mechanisms that re-configure the system to improve its ability to traverse difficult terrain. All power is provided by on-board batteries. Low-level control and planning are performed using an on-board PC/104 computing architecture. A wireless modem is used for external communication to obtain high-level commands from a task planner. The experimental system weighs 6.1 kg. and was constructed for less than $10,000.

*Figure 1: The Field and Space Robotics Laboratory (FSRL) Rover (without cover)*

In our research, algorithms have been developed that use rover models which are designed to be practical for on-board implementation. Based of these models, a fuzzy logic traction control algorithm has been developed [5,6]. Rough terrain path-planning algorithms have also been developed [5,7]. Finally, a high-performance manipulator control algorithm developed for fixed-base manipulators has been successfully extended to the rover-mounted manipulator case [8]. To demonstrate the effectiveness of these algorithms with the FSRL rover, an 8' by 10' laboratory terrain has been constructed with challenging Mars-like features, such as hills, rocks, and an extended ditch system.

*Figure 2: The JPL LSR (left) and Sojourner (right)*

## 2. Rover Mechanical Design

The rover mechanical system is based on a six-wheeled rocker-bogie design, which is widely utilized by NASA due to its excellent mobility characteristics (see Figure 3) [9]. The rover features six independently powered wheels driven by geared Portescap DC motors. A 501:1 gear reduction produces a peak output torque of 100 oz-in and maximum angular velocity of 12 rpm. The resulting maximum velocity of the rover is approximately 8 cm/sec. The rover is steered with skid-steering. The suspension of the rover is constructed from aluminum square tubing, and includes integral motor mounts. The body is made from a formed aluminum sheet and supports the system electronics and sensors. The body also serves as an attachment point for a manipulator arm and stereo cameras. A mechanical differential allows the body to "split the difference" of the two rocker angles. The total cost of construction of the mechanical structure was approximately $2500.

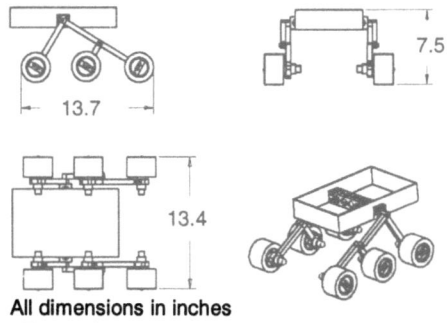

All dimensions in inches

*Figure 3: Experimental Rover Dimensions*

## 2.1. Manipulator

The three degree-of-freedom manipulator mounted on the front of the rover is shown in Figure 4. The manipulator's light weight (approximately 16 ounces) is achieved by using low-weight aluminum members and small, highly geared motors. The joints are driven by MicroMo DC motors with gear ratios of 2961:1, 3092:1, and 944:1 at the trunk, shoulder, and elbow joints, respectively. With the high gear reduction, the manipulator is capable of exerting large forces. In some configurations, it can exert a force equal to one-half the rover weight. This high-force capability could be useful for manipulator-aided mobility or trap recovery. The base of the arm is mounted to a six degree-of-freedom force-torque sensor, which is used for control (see Section 4.1) [8].

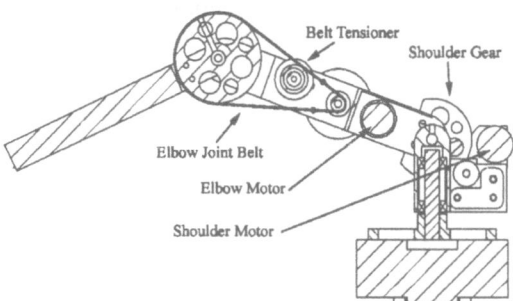

*Figure 4: Three Degree-of-Freedom Manipulator*

### 2.1.1. End-Effector

Several end-effector concepts have been developed for handling rock samples. The most effective has proven to be a lightweight three-fingered end-effector. It utilizes flexural joints and relies on shape-memory alloy (SMA) actuation (see Figure 5). Each of the three fingers are formed from 1/8" steel rod. A nylon mounting plate has integral flexures that allow motion without bearing surfaces, eliminating the need for lubrication and considerably simplifying design and fabrication. A 0.006" diameter Flexinol SMA wire provides retracting force to each finger from its normally closed position. The wires are connected to the ends of the fingers, and run along the bottom of the mounting plate to increase their working length, and thus allow greater finger travel.

The three-fingered gripper is designed to be able to grasp rocks up to 2 ½" in diameter, and be able to support the weight of a typical volcanic-type rock of that size. The prototype end-effector weighs 2 ounces.

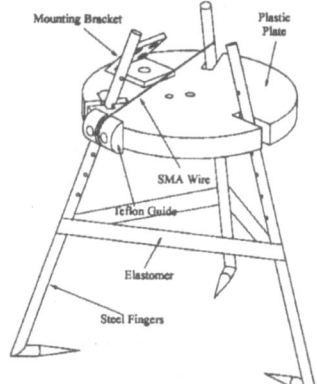

*Figure 5: Three-Fingered End-Effector*

## 2.2. Reconfigurability

An SMA-actuated reconfigurable rocker-bogie suspension concept has been developed which allows the rover to modify its geometry to improve mobility and avoid failure situations. The mechanism allows the rover to squat one or both sides of its suspension, and thus increase its stability margin when required (for example, when the rover is on a transverse slope). Additionally, reconfigurability allows the rover to reposition its center of mass when performing traction control [5,6].

An illustration of a reconfigurable rocker mechanism is shown in Figure 6. A Flexinol SMA wire provides retracting force for the rocker, and a multi-jaw coupling locks the rocker links in place, fixing the rocker angle. The SMA wire is routed over Delrin wire guides to increase the working length.

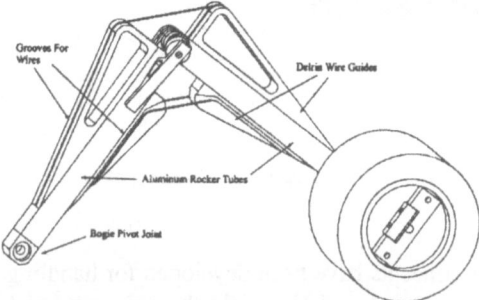

*Figure 6: Reconfigurable Rocker Concept*

# 3. Electronic and Power System

The rover electronics system was designed to be compact, low-cost and lightweight (see Figure 7). A block diagram of the system can be seen in Figure 8. The system is based on a PC/104 486 computing platform, with additional modules for digital and analog IO, encoder reading, and interface to a JR3 six-axis force/torque sensor which is mounted underneath the manipulator (see Section 2.1). NiCad rechargeable batteries power the rover. The rover is outfitted with a full suite of sensors including motor tachometers and encoders, rocker/bogie angle potentiometers, and a three-axis accelerometer. Communication with an operator is accomplished via a National Semiconductor AirShare wireless modem operating at 9600 baud. The total cost of the electronic system was less than $6000.

*Figure 7: Rover Electronics Packaging*

## 3.1. Computation

Computation is performed on a 486DX2 66 Mhz PC/104 single-board computer. A PC/104 computing platform was chosen due to its small size, light weight, and low power consumption [10]. The motherboard is 203 mm by 146 mm, with a functional depth of 40 mm, and weighs 15.5 ounces. It operates on a single power supply of 5 volts at 2 amps. It is interfaced via the PC/104 bus to modules which perform A/D conversion, D/A conversion, and digital input and output.

The system can support 8 differential A/D channels, 16 single-ended A/D channels, 16 D/A channels, 80 digital IO lines, and 8 quadrature encoders. It can also power 12 motors via pulse-width modulation.

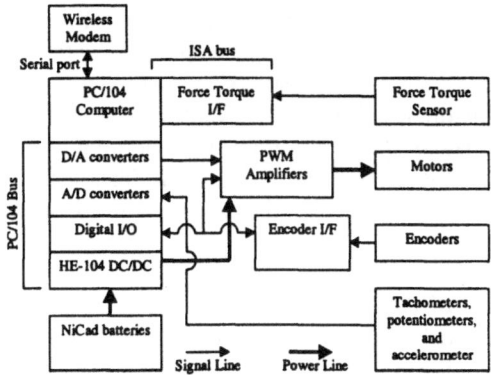

*Figure 8: Electronic System Block Diagram*

## 3.2. Power

The rover is powered by NiCad batteries. Each battery pack is rated at 7.2 volts and has a capacity of 1700 mAh, and the rover uses two battery packs connected in series. A Tri-M Engineering Systems PC/104 HE-104 DC/DC converter provides power to the system. The DC/DC converter can provide 50 watts of continuous filtered power, and is very tolerant of shocks and vibration. Motor control of the drive motors and manipulator motors is accomplished through PWM amplifiers.

## 3.3. Sensing

The rover is equipped with numerous sensors to monitor its performance and determine its state. The six drive wheels are equipped with tachometers, which are read by differential A/D lines. The three manipulator motors are equipped with magnetic encoders, and are read by a custom interface based on a US Digital LS7266R1 dual quadrature encoder chip. A three-axis Crossbow CXL04M3 accelerometer is mounted on the body of the rover, to determine its roll and pitch relative to an inertial frame. A JR3 67M25A six-axis force/torque sensor is mounted under the base of the manipulator. A dedicated ISA bus board provides interface support for the force/torque sensor. Finally, potentiometers are mounted on the rocker and bogie joints to determine their angular position.

# 4. Rover Analysis and Control

The Field and Space Robotics Laboratory has been actively researching control algorithms for planetary rovers in recent years. Recently, a high-performance manipulator control algorithm developed for fixed-base manipulators has been successfully extended to the rover-mounted manipulator case [8]. A fuzzy logic traction control algorithm has been developed [5,6]. Rough terrain path-planning algorithms have also been developed [5,7].

Two of these algorithms are described briefly in the following subsections. Experimental results using the rover testbed are presented in Section 5.

## 4.1. Rover-Mounted BSC Control

Planetary rover-mounted manipulators are expected to perform high-precision tasks such as instrument placement. These lightweight manipulators have large gear ratios, which leads to high drivetrain friction and makes high-precision control difficult. A control algorithm called Base-Sensor Control (BSC) has been developed for high-precision control of fixed-base manipulators with high joint friction [8]. This method utilizes feedback from a six-axis force/torque sensor mounted under the base of the manipulator, which is used to estimate the torque at each manipulator joint. With an estimate of the joint torque, accurate joint torque control is possible and disturbances such as friction can be rejected. This leads to improved friction compensation, which in turn improves the execution of fine-motion tasks.

### 4.1.1. BSC Control Overview

A simplified version of the BSC control algorithm has been developed for slow manipulator motions such as might be required in planetary scientific tasks [8]. This simplified algorithm relies only on feedback from the force/torque sensor and knowledge of the manipulator configuration and kinematic parameters. In this simplified algorithm, the torque at a manipulator's n joints is estimated as:

$$\tau = \mathbf{A}(q)\big(\mathbf{W}_{base} - \mathbf{Y}(q)\phi\big) \tag{1}$$

where $\mathbf{A}(q)$ is an nx6 matrix that depends on the manipulator joint configurations and kinematic parameters, $\mathbf{W}_{base}$ is a 6x1 vector containing the wrench measured by the force/torque sensor, $\mathbf{Y}(q)$ is a 6xm matrix that relies on the manipulator joint configurations, and $\phi$ is an mx1 vector of manipulator mass parameters. Note that the quantity $(\mathbf{Y}(q)\cdot\phi)$ represents gravity compensation.

The estimated joint torque is used in a control scheme comprised of an inner torque-control loop and an outer position-control loop (see Figure 9). An inner loop integral compensator provides low-pass filtering and zero steady-state error. A simple proportional-derivative controller is employed in the outer loop. This simple control architecture has been shown to provide very precise position control during small, slow motions [8].

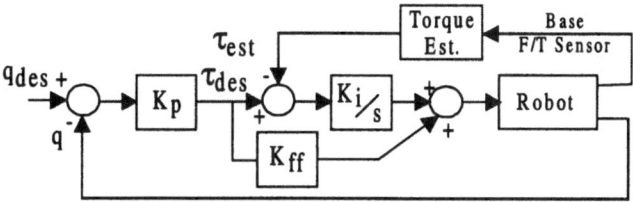

*Figure 9: BSC Control Block Diagram*

### 4.1.2. Rover-Mounted BSC Control

The above control algorithm can be modified for application to rover-mounted manipulators by assuming that the motion of the rover is slow enough that dynamic forces applied to the force/torque sensor by rover motion are negligible. Then Equation 1 becomes:

$$\tau = \mathbf{A}(q)\big(\mathbf{W}_{base} - \mathbf{Y}(q,\Psi,\Theta)\phi\big) \tag{2}$$

The gravity compensation matrix $\mathbf{Y}$ is now a function of both the manipulator configuration and the rover body pitch ($\Psi$) and roll ($\Theta$). Note also that manipulator mass parameters which previously did not appear in $\phi$ (i.e. were "unidentifiable") may now appear uniquely [11].

Equation (2) is used to estimate the joint torque of a rover-mounted manipulator, and a control scheme identical to that shown in Figure 9 can be implemented. Current work in the Field and Space Robotics Laboratory involves extending this control method to allow for rover dynamic effects.

### 4.2. Rough-Terrain Traction Control

Future planetary exploration  missions will require rovers to traverse challenging terrain in order to achieve scientific objectives. Traction control reduces unplanned wheel slip, and thus improves traversability in rough terrain. Traction control also improves localization accuracy, and reduces power consumption.

An analysis of a planar rocker-bogie type rover has been performed with the goal of developing a traction control system [6]. It shows that the force balance equations of the system shown in Figure 10 can be written in the form:

$$\mathbf{M} \cdot \begin{bmatrix} F_x^i \\ N_1 \\ N_2 \\ N_3 \end{bmatrix} = \mathbf{X} \tag{3}$$

where $\mathbf{M}$ is a 4x4 matrix of ground contact angles and kinematic parameters, and $\mathbf{X}$ is a 4x1 vector of the rover kinematics and the input (traction) forces $T_1$, $T_2$, and $T_3$. The system is solvable by inverting $\mathbf{M}$, either analytically or numerically.

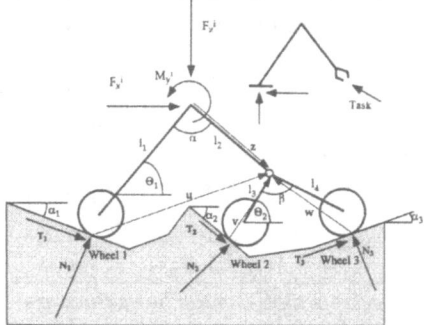

*Figure 10: Traction Control Force Analysis*

From this analysis, a control scheme could be developed for on-line optimal traction control. However, this requires knowledge of wheel-ground contact angles, which are difficult to measure in practice. To circumvent this problem, a set of heuristic rules have been developed based on the preceding analysis [6]. Based on these heuristics, a fuzzy-logic controller has been developed that systematically exploits the heuristics. The controller utilizes the desired wheel angular velocity, the measured angular velocities of the wheels, and an estimate of the slip at each wheel.

## 5. Experimental Results

The control methods described above have been tested on the FSRL rover operating in our Mars-like laboratory terrain.

### 5.1. Rover-Mounted BSC Control Results

Figures 11a and 11b show typical experimental results of the rover-mounted BSC algorithm compared with standard PID control. Here, the first and second (torso and shoulder) joints of the manipulator were commanded to track 0.2° amplitude 0.05 hz sinusoids. At these very slow speeds, nonlinear friction effects can dominate manipulator tracking performance. The tracking performance of joints one and two have been plotted against one another for compactness. The BSC-controlled manipulator had an RMS error of 0.0138°, compared to 0.0297° for PID control. The maximum error was reduced from 0.0749° for PID control to 0.0322° for BSC.

Essentially, BSC control reduced the position errors of the manipulator by approximately 50%, a substantial performance improvement. This improvement was difficult to predict with simulation due to system compliance and complex backlash in the rover differential. Experimentation showed that BSC control is robust to these unmodeled phenomena.

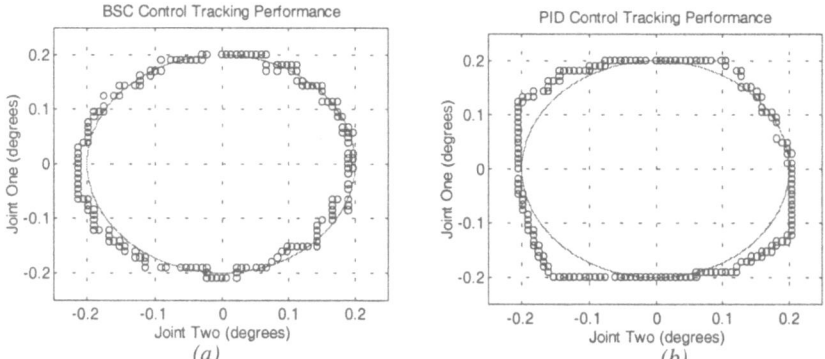

Figure 11: BSC Tracking Performance (Figure 11a) vs. PID (Figure 11b)

## 5.2. Rough-Terrain Traction Control Results

For experimental evaluation of the fuzzy-logic control algorithm the rover was commanded to perform slope and step climbing tasks under various loading conditions. In the slope climbing task the rover was commanded to traverse a slope with a grade of approximately 18 degrees, which created a near-slip situation. In the step climbing task the rover was commanded to traverse a sharp step followed by a slope. Results of the success rate of the fuzzy logic controller is summarized in Table I and compared to conventional PI velocity control.

Table I: Summary of Results of Slope and Step Climbing Tasks

|  | Fuzzy Logic | PI Velocity Control |
|---|---|---|
| Slope Task | 85% | 60% |
| Step Task | 63% | 27% |

The fuzzy logic controller performed better than the PI controller in the slope climbing task, where it improved the success rate by 42%. The fuzzy logic controller significantly improved performance in the step climbing task, where it improved the success rate by 133%. Current work in the Field and Space Robotics Laboratory involves refining the heuristics of the fuzzy logic controller, and improving the slip-detection algorithm. These factors should further improve the performance of the fuzzy logic controller.

# 6. Conclusions

A low-cost rover test-bed has been developed, which consists of a rocker-bogie type rover with an on-board manipulator operating in a rough-terrain environment. It has a three degree-of-freedom manipulator mounted on the front of the rover chassis, which uses a prototype end-effector for manipulation of rock samples. The experimental system chassis contains shape memory alloy (SMA) actuated variable geometry mechanisms that re-configure the system to improve its ability to traverse difficult terrain. Power is provided by on-board batteries, and control is performed using an on-board PC 104 computing architecture.

The rover has proven very useful in the study of advanced control algorithms for planetary robotic explorers. Examples of experimental validation of a high-performance rover-mounted manipulator control algorithm and a fuzzy logic traction control algorithm have been presented.

## 7. Acknowledgements

This work is supported by the NASA Jet Propulsion Laboratory under contract number 960456. The authors would like to acknowledge the support of Dr. Paul Schenker and Dr. Eric Baumgartner at JPL. The support and encouragement of Guillermo Rodriguez is also acknowledged.

## 8. References

[1] Golombek, M. P. 1998 Mars Pathfinder Mission and Science Results. In: *Proceedings of the 29th Lunar and Planetary Science Conference*

[2] Matijevic, J., 1997 01 Surveyor Rover: Requirements, Constraints and Challenges (www.cs.cmu.edu/~meteorite/2001rover/)

[3] Hayati, S., Volpe, R., Backes, P., Balaram, J., and Welch, W. 1996 Microrover Research for Exploration of Mars. In: *AIAA Forum on Advanced. Developments in Space Robotics*

[4] Schenker, P., Sword, L., Ganino, et al. 1997 Lightweight Rovers for Mars Science Exploration and Sample Return. in: *Proceedings of SPIE XVI Intelligent Robots and Computer Vision Conference*

[5] Farritor, S., Hacot, H., and Dubowsky, S. 1998 Physics-Based Planning for Planetary Exploration. In: *1998 IEEE International Conference on Robotics and Automation*

[6] Hacot, H. 1998 *Analysis and Traction Control of a Rocker-Bogie Planetary Rover*. M.S. Thesis, Massachusetts Institute of Technology

[7] Iagnemma, K., Genot, F. and Dubowsky, S. 1999 Rapid Physics-Based Rough-Terrain Rover Planning with Sensor and Control Uncertainty. In: *1999 IEEE International Conference on Robotics and Automation*

[8] Morel, G., Iagnemma, K., and Dubowsky, S. The Precise Control of Manipulators with High Joint Friction Using Base Force/Torque Sensing. Submitted to *Automatica: The Journal of the International Federation of Automatic Control*

[9] Bickler, D. 1992 The New Family of JPL Planetary Surface Vehicles. In: Proceedings of the International Symposium On Missions, Technologies, and Design of Planetary Mobile Vehicles

[10] Wilhelm, E. 1998 *Design of a Wireless Control System for a Laboratory Planetary Rover*. B.S. Thesis, Massachusetts Institute of Technology

[11] West, H., Papadopoulos, E., Dubowsky, S., and Cheah, H. 1989 A Method For Estimating the Mass Properties of a Manipulator by Measuring the Reaction Moments at its Base. In: *IEEE International Conference on Robotics and Automation*

# Elastic Strips: A Framework for Integrated Planning and Execution

Oliver Brock            Oussama Khatib

Robotics Laboratory, Department of Computer Science
Stanford University, Stanford, California 94305, U.S.A.
email: {oli,khatib}@cs.stanford.edu

**Abstract:** The execution of robotic tasks in dynamic, unstructured environments requires the generation of motion plans that respect global constraints imposed by the task while avoiding collisions with stationary, moving, and unforeseen obstacles. This paper presents the elastic strip framework, which addresses this problem by integrating global motion planning methods with a reactive motion execution approach. To maintain a collision-free trajectory, a given motion plan is incrementally modified to reflect changes in the environment. This modification can be performed without suspending task behavior. The elastic strip framework is computationally efficient and can be applied to robots with many degrees of freedom. The paper also presents experimental results obtained by the implementation of this framework on the the Stanford Mobile Manipulator.

## 1. Introduction

The automated execution of tasks by robots in dynamic, unstructured, and potentially populated environments requires sophisticated motion capabilities. In such environments unforeseen or moving obstacles may repeatedly invalidate a previously planned motion. The successful completion of a task will then require the frequent generation of revised motion plans that accommodate changes in the environment. In highly structured environments this problem can be avoided by imposing constraints on the motion of obstacles: it is common for motion planning algorithms to assume that obstacles are stationary or moving on predetermined trajectories [13]. These assumptions are unrealistic for environments outside the factory floor or the laboratory environment, like those encountered in robotic applications such as service or field robotics.

The necessity of frequent regeneration of motion plans for task execution in unstructured environments entails a computational difficulty for the generation of robot motion. Planning algorithms generally take on the order of minutes to determine a plan, under ideal circumstances still on the order of seconds [8]. Hence, using planning algorithms the avoidance of moving or unforeseen obstacles cannot be guaranteed. Reactive motion schemes, like the potential field approach [9], on the other hand, are able to avoid moving obstacles in real-time but might fail to achieve the task by getting trapped in local minima.

This dichotomy of complete or resolution complete but inefficient global planning algorithms and incomplete but efficient local, reactive execution approaches has been the starting point for various efforts to improve the performance of robot motion generation algorithms [11]. In one approach the concepts of potential field-based obstacle avoidance and approximate cell decomposition motion planning were used in conjunction to yield a reactive framework for planning and execution of robot motion [5]. Only partial knowledge of the environment is required and small, unforeseen obstacles or minor changes in the environment can be tolerated.

Whereas the previously mentioned approach can only be applied to mobile robots, the elastic band framework [14] is also suited for robot manipulators. In this approach planning and reactive execution are used in sequence: first a plan is generated using a conventional motion planner; subsequently this plan is modified incrementally as a reaction to changes in the environment. Real-time obstacle avoidance for robots with few degrees of freedom has been demonstrated.

For mobile robots the global dynamic window approach integrates planning and execution by incorporating an efficient global path planner into the control loop of a dynamics-based execution scheme [3]. This approach allows for high-speed navigation of a mobile base in an unknown and dynamic environment.

In another approach to integrated planning and control a plan is converted into a trajectory using a path-based parameterization, rather than a time-based one [15]. This allows the corresponding path-based controller to interrupt the execution of the trajectory, if an unforeseen obstacle is detected. Once the obstacle has been removed or an evasion maneuver has been specified by a human operator, the execution of the original trajectory is resumed.

The elastic strip framework [2] presented in this paper integrates planning and execution for robots with many degrees of freedom. This approach allows for reactive obstacle avoidance behavior that does not suspend task execution, enabling robots with many degrees of freedom to perform tasks in dynamic and unstructured environments. The paper also presents the experimental validation of this framework on the Stanford Mobile Manipulator, a six degree-of-freedom (DOF) PUMA 560 mounted on a 3 DOF holonomic mobile base.

## 2. Elastic Strip Framework

The elastic *strip* framework is very similar in spirit to the elastic *band* framework [14]. In the elastic *band* framework a previously planned robot motion is modeled as elastic material. A path between an initial and a final configuration can be imagined as a rubber band spanning the gap between two points in space. Obstacles exert a repulsive force on the trajectory, resulting in an incremental modification. This can be imagined as a moving obstacle pushing and deforming the rubber band. When the obstacle is removed, the trajectory will return to its initial configuration, just as the rubber band would. An elastic *band* is represented as a one-dimensional curve in configuration space. This leads to high computational complexity for high-dimensional configuration spaces. Furthermore, the specification of tasks for robots is most naturally done in workspace. Elastic bands, however, represent a path in the configuration space.

The elastic *strip* framework operates entirely in the workspace in order to avoid aforementioned problems. The characterization of free space becomes more accurate in the workspace than that in configuration space, resulting in a more efficient description of trajectories. In addition, by avoiding configuration space computation, the framework becomes applicable to robots with many degrees of freedom. The trajectory and the task are both described in workspace. In the elastic strip framework a trajectory can be imagined as elastic material filling the volume swept by the robot along the trajectory. This strip of elastic material deforms when obstacles approach and regains its shape as they retract.

## 2.1. Free Space Representation

To guarantee that the current trajectory is entirely in free space or to modify it due to the motion of an obstacle, the free space around the volume swept by the robot along the trajectory must be known. The simplest representation of free space around a point $p$ in the workspace is a sphere: it is described by four parameters and can be computed with just one distance computation. Such a sphere is called *bubble* [14] and is defined as

$$\mathcal{B}(p) = \{\, q : \|p - q\| < \rho(p)\},$$

where $\rho(p)$ computes the minimum distance from $p$ to an obstacle.

A set of bubbles is used to describe the local free space around a configuration $q$ of a robot $\mathcal{R}$. This set is called *protective hull* $\mathcal{P}_q^{\mathcal{R}}$ and is defined as

$$\mathcal{P}_q^{\mathcal{R}} = \bigcup_{p \in \mathcal{R}} \mathcal{B}(p).$$

Not every point $p$ needs to be covered by a bubble. A heuristic is used for selecting a small set of points yielding an accurate description of the free space around configuration $q$. An example of a protective hull around the Stanford Mobile Manipulator is shown in Figure 1 a).

Along the trajectory $\mathcal{U}$ a sequence of configurations $q_0, \cdots, q_n$ is chosen. This sequence is called an elastic strip $\mathcal{S}_{\mathcal{U}}^{\mathcal{R}}$ if the union of the protective hulls $\mathcal{P}_i^{\mathcal{R}}$ of the configurations $q_i, 1 \leq i \leq n$ fulfills the condition

$$V_{\mathcal{U}}^{\mathcal{R}} \subseteq \mathcal{T}_{\mathcal{S}}^{\mathcal{R}} = \bigcup_{0 \leq i \leq n} \mathcal{P}_i^{\mathcal{R}}, \tag{1}$$

where $V_{\mathcal{U}}^{\mathcal{R}}$ is the workspace volume of robot $\mathcal{R}$ swept along the trajectory $\mathcal{U}$. The union $\mathcal{T}_{\mathcal{S}}^{\mathcal{R}}$ of protective hulls is called *elastic tunnel*. It can be imagined as a tunnel of free space within which the trajectory can be modified without colliding with obstacles. An example of an elastic tunnel is shown in Figure 1 b). Three configurations represent snapshots of the motion along a trajectory. The union of the protective hulls around those configurations form an elastic tunnel. It contains the volume swept by the robot along the trajectory.

This representation of free space is the key to the performance of the elastic strip framework. It can be computed very efficiently, while providing a good approximation of the the actual free space.

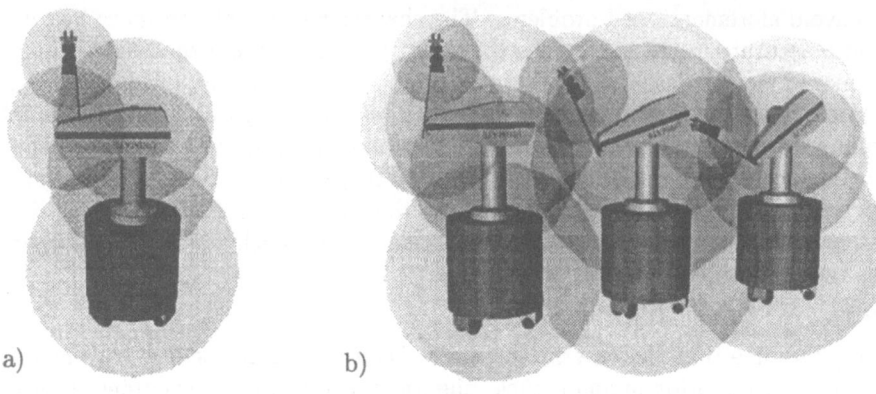

Figure 1. a) A protective hull around the Stanford Mobile Manipulator    b) An elastic tunnel formed by several overlapping protective hulls

## 2.2. Elastic Strip Modification

The elastic strip $S$ is subjected to external forces that keep the trajectory free of collision and to internal forces that result in a short and smooth trajectory. External forces are caused by a repulsive potential associated with obstacles. For a point $p$, this potential function is defined as

$$V_{ext}(p) = \begin{cases} \frac{1}{2}k_r(\rho_0 - \rho(p))^2 & \text{if } \rho(p) < \rho_0 \\ 0 & \text{otherwise} \end{cases},$$

where $\rho_0$ defines the region of influence around obstacles and $k_r$ is the repulsion gain. Internal forces are caused by virtual springs attached to control points on consecutive configurations along the elastic strip. How external and internal forces are used to modify the trajectory in accordance with the task specification is described in section 2.3.

As obstacles approach the trajectory, the size of protective hulls decreases. If this leads to the violation of equation 1, intermediate configurations are inserted into the elastic strip, until the swept volume of the robot is again entirely covered by protective hulls. The retraction of obstacles, on the other hand, leads to the enlargement of protective hulls. In this case redundant configurations are removed.

The integration of planning and execution paradigms in the elastic strip framework consists of the incremental and reactive modification of a global plan. As the forces acting on the elastic strip do not change the topological properties of the represented trajectory, the global properties of the plan are maintained. The robot can be guaranteed to reach the goal as long as the plan remains topologically feasible. Hence, an efficient, reactive scheme has been integrated with a global plan.

## 2.3. Motion Behavior

A task can consist of different subtasks, each potentially requiring different motion behavior. Using the elastic strip framework these subtasks can be described in a very intuitive way. A force $F_{task}$ is specified in operational space [10] at the end-effector. This force can be derived from the potential $V_{task}$ associated with the task. Joint torques $\Gamma_{task}$ required to accomplish the task can be computed by a simple mapping of $F_{task} = -\nabla V_{task}$ acting at the end-effector point $e$ using the Jacobian $J(q)$ at that point for configuration $q$:

$$\Gamma_{task} = J^T(q)F_{task}. \tag{2}$$

Operational space control of redundant manipulators can accommodate different kinds of motion behavior. In the simplest case, a part has to be moved on any trajectory between two locations. To implement obstacle avoidance, the existing trajectory can be modified to accommodate unforeseen or moving obstacles. No particular motion behavior is required to accomplish the task and the joint torques $\Gamma_{task}$ can be computed by adding the torques resulting from internal and external forces to equation 2:

$$\Gamma_{task} = J^T(q)F_{task} + \sum_{p \in \mathcal{R}} J_p^T(q)F_p,$$

where $F_p$ is the sum of internal and external forces action at point $p$ and $J_p(q)$ is the Jacobian at that point in configuration $q$. Internal forces are caused by the potential function $V_{int}$ associated with the virtual springs of the simulated elastic material and external forces can be derived from the potential function $V_{ext}$ resulting from the proximity of obstacles: $F_p = -\nabla V_{int} - \nabla V_{ext}$.

Redundancy of a robot with respect to its task can be exploited to integrate task behavior and obstacle avoidance behavior. Some tasks may require the end-effector to remain stationary or to move along a given trajectory without deviation. This entails that obstacle avoidance cannot alter the motion of the end-effector.

For redundant systems the relationship between joint torques and operational forces is given by

$$\Gamma_{task} = J^T(q)F_{task} + \left[I - J^T(q)\overline{J}^T(q)\right]\Gamma_0, \tag{3}$$

with $$\Gamma_0 = \sum_{p \in \mathcal{R}} J_p^T(q)F_p \quad \text{and} \quad \overline{J}(q) = A^{-1}(q)J^T(q)\Lambda(q),$$

where $\overline{J}(q)$ is the dynamically consistent generalized inverse [10]. The term $[I - J^T(q)\overline{J}^T(q)]$ corresponds to the null space of the Jacobian $J(q)$. This relationship provides a decomposition of joint torques into two dynamically decoupled control vectors: joint torques resulting in forces acting at the end-effector $(J^T(q)F_{task})$ and joint torques that only affect internal motions $\left(\left[I - J^T(q)\overline{J}^T(q)\right]\Gamma_0\right)$ [12]. This allows to control the end-effector by a force $(F_{task})$ in operational space, whereas internal motions can be independently controlled by joint torques $(\Gamma_0)$ that do not alter the end-effector's dynamic behavior.

This framework is integrated with the elastic strip approach by simulating the effect of internal and external forces on the robot at the configurations $\{q_0, \cdots, q_n\}$ representing the elastic strip. The elastic strip can now be modified in a way that avoids obstacles by exploiting the redundant degrees of freedom of the robot but leaves the end-effector motion unaltered. If kinematic constraints of the robot make obstacle avoidance impossible the task needs to be aborted or modified. This can occur when the robot reaches the border of its workspace, for example.

## 3. Trajectory Generation and Execution

An elastic strip $S$ represents a sequence of discrete configurations $\{q_0, \ldots, q_n\}$ along a path from $q_0$ to $q_n$. The conversion of such a sequence into a time-parameterized trajectory is a well-studied problem [1, 6]. In the elastic strip framework, however, this sequence is changing in discrete steps as the elastic strip is modified in reaction to changes in the environment. Let $q_t$ and $\dot{q}_t$ be the vectors of joint positions and joint velocities of the robot at time $t$. They resulted from the execution of the trajectory described by the elastic strip $S_{t-1}$ at time $t-1$. The modified elastic strip $S_t$ will represent a different trajectory with a desired velocity of $\dot{q}'_t$ at time $t$ and a potentially unattainable position $q'_{t+1}$ and velocity $\dot{q}'_{t+1}$ at time $t+1$. Due to the comparatively slow rate at which the update of the elastic strip occurs, the difference between the current and desired position and velocity, $\|q_t - q'_t\|$ and $\|\dot{q}_t - \dot{q}'_t\|$, can be large. Hence, the application of conventional approaches to trajectory execution would not result in desirable behavior.

This problem could be solved by either requiring the initial portion of the elastic strip to remain constant or by limiting modification to those trajectories achievable, given the dynamic constraints of the robot. These solutions have two disadvantages: For one, invalidating the path represented by the elastic strip requires a computationally expensive replanning operation, which should be avoided if at all possible. Hence, it is desirable for the elastic strip to represent a valid path, even if inconsistent with the current state of the robot. Secondly, when executing a motion on a mobile base, such as the Stanford Mobile Manipulator, execution errors accumulate for the mobile base due to slippage of the wheels. To reduce this error various relocalization schemes can be employed. The error accumulated between different relocalizations can be expected to be large enough to invalidate a trajectory originally incorporating dynamic constraints imposed by the robot. Insufficient actuator capabilities could also be regarded as a source of error.

To avoid these disadvantages a novel approach to the execution of changing trajectories is presented in this section. The general idea is to maintain a valid trajectory that corresponds to the elastic strip but ignores the dynamic state of the robot. The initial configuration $q_0$ of the strip will be constrained to coincide with the configuration of the robot. Hence, the trajectory represented by the strip will be almost correct, only ignoring dynamics. The final trajectory then results from merging the robot's current motion with the previously computed trajectory. These two steps are detailed in the two following subsections for a single degree of freedom. The extension to many degrees of freedom is trivial.

## 3.1. Generating the Trajectory

The elastic strip represents a sequence of configurations $\{q_0, \ldots, q_n\}$ that are connected by straight-line segments in joint space. The discontinuous velocity change that can occur at a configuration $q_i$ along the piecewise linear trajectory cannot be executed by the robot without coming to rest at $q_i$. As this is not desirable, an interpolation technique is applied to convert the piecewise linear trajectory into one with a continuous first derivative. As we can expect frequent velocity changes during the execution of a changing trajectory, this interpolation is performed with cubic polynomials, which also guarantee a continuous second derivative within a given segment of the trajectory.

When using a standard scheme for trajectory generation with cubic polynomials [6], the trajectory passes through a set of via points, corresponding to the configurations $q_i$ along the elastic strip. This may results in a large deviation from the straight-line trajectory between two adjacent configurations. Due to the free space description with protective hulls, however, this is the portion of the trajectory where the free space description is most narrow. It is hence desirable for the robot to follow this portion of the trajectory as closely as possible. To achieve this, the straight-line segments will be connected by *cubic turns*. The maximum allowed duration of a turn at configuration $q_i$ can be inferred from the local free space and determines the velocity along the adjacent line segments from $q_{i-1}$ to $q_i$ to $q_{i+1}$.

To allow the execution of trajectories in tight spaces, a *turning cubic* will always begin by accelerating with the maximum acceleration $\ddot{q}_{max}$. Taking into account that the acceleration during a cubic can be described by a line, the duration $d$ of the turning cubic is computed as follows:

$$\Delta\dot{q} = |\dot{q}_{i-1} - \dot{q}_i| = \int_0^d \ddot{q}(t)\,dt = \frac{d\,\ddot{q}_{max}}{2} \quad \rightarrow \quad d = \frac{2 \cdot |\Delta\dot{q}|}{\ddot{q}_{max}},$$

where $\dot{q}_i$ denotes the velocity along the line segment between $q_i$ and $q_{i+1}$. The points at which the turning cubic connects with the adjacent line segments can be computed by equating their motion equations:

$$q_i + (d_i - d_{\Delta i})\dot{q}_i + \Delta q = q_{i+1} + (d - d_{\Delta i})\dot{q}_{i+1},$$

where $d_i$ denotes the duration of the line segment between $q_i$ and $q_{i+1}$ assuming constant velocity $\dot{q}_i$, and $\Delta q$ stands for the change in joint position between beginning and ending of the turning cubic. Solving for $d_{\Delta i}$, the parameters of the turning cubic are computed. Its execution will begin at time $t_{i+1} - d_{\Delta i}$ and end at time $t_{i+1} + (d - d_{\Delta i})$, where $t_i$ denotes the time at which execution of the $i$th segment begins. Using those values the starting and ending point of the turning cubic with respect to the line segments can be computed.

## 3.2. Merging the Robot's Motion with the Trajectory

The trajectory resulting from the method described in the previous subsection does not take into account the current velocity of the robot. Also, the robot's position might have changed due to the correction of accumulated execution error. To connect the robot to this trajectory, an iterative optimization algorithm could be used [7]. This algorithm retains the original time parameterization of the trajectory, which causes the robot to "catch up" with the trajectory. In dynamic environments it is unreasonable to impose a time constraint for the robot to reach the goal configuration. Therefore we will reconnect the robot's current state to the trajectory according to its dynamic capabilities and then adapt the time-parameterization of the remainder of the trajectory.

The cubic segment that merges the current state of the robot with a segment on the trajectory can be computed by equating the position and the velocity equations of the merging segment and the segment on the trajectory. When merging with a line segment the following equations result:

$$q_0 + d\,\dot{q}_0 = q_r + \dot{q}_r\,d + a_1\,d^2 + a_2\,d^3$$
$$\dot{q}_0 = \dot{q}_r + 2\,a_1\,d + 3\,a_2\,d^2,$$

where $q_r$ and $\dot{q}_r$ are the position and the velocity of the robot, $d$ is the duration of the merging cubic, and $a_1, a_2$ are its coefficients. When attempting to merge with another cubic segment of the original trajectory, a solution can be found in a similar fashion. The duration $d$ of the merger has to be determined such that the acceleration constraints of the robot are not violated.

## 4. Experimental Results

The elastic strip framework was implemented and tested on the Stanford Mobile Manipulator, a 9 degree-of-freedom (DOF) robotic system, consisting of a PUMA 560 arm mounted on a holonomic mobile base. The robot was controlled using the Robotics Library [4] running on a dedicated on-board 90 MHz Pentium PC. The algorithms presented in this paper were executed on a 400 MHz Pentium PC. In the example shown below, update rates of the elastic strip varied between 10 and 100 Hz.

To demonstrate the elastic strip framework, the Stanford Mobile Manipulator was commanded to move five meters along the $x$-axis of the global coordinate frame, while keeping the arm's posture. During the execution of this plan an unforeseen obstacle, another Stanford Mobile Manipulator, forces the first robot to deviate from the original plan. It is modified using the elastic strip framework to avoid collision while achieving the desired goal configuration. Two different perspectives of the simulated modification of the trajectory are shown in Figure 2. A sequence of snapshots from the execution on the real robot can be seen in Figure 3. The plot of the base trajectory, as well as the trajectory for the first three joints of the PUMA 560 manipulator are shown in Figure 4.

## 5. Conclusion

The elastic strip framework is an efficient method for motion execution for robots with many degrees of freedom in dynamic environments. It allows obstacle avoidance behavior without the suspension of task execution. Hence,

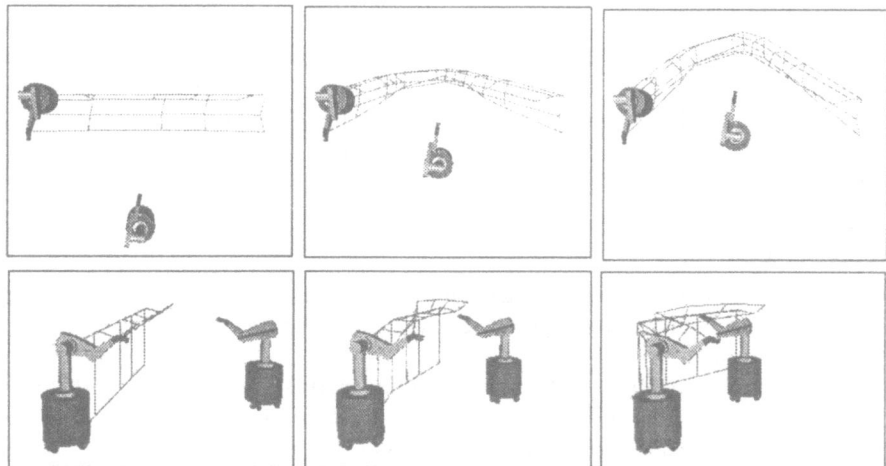

Figure 2. The elastic strip, shown as gray lines, is modified incrementally in order to maintain a valid path while avoiding a moving obstacle

Figure 3. Execution of a plan using the elastic strip framework; the path is modified in real-time to avoid the obstacle

this approach is well suited for mobile manipulation. The effectiveness of the proposed algorithm is derived from an efficient integration of planning and execution methods. The validity of the algorithm has been experimentally verified using the Stanford Mobile Manipulator, a 9 DOF robotic system, consisting of a mobile base and a manipulator arm.

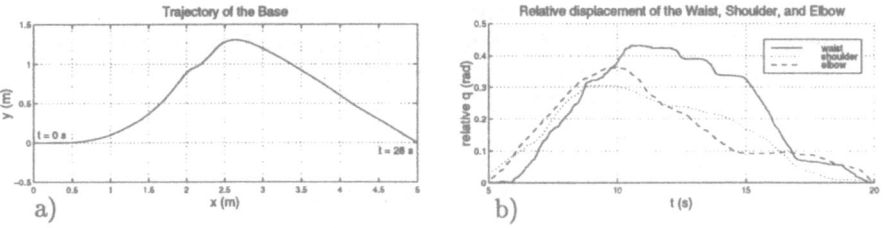

Figure 4. a) Trajectory of the base    b) Relative displacement of the arm

338

# Acknowledgments

The authors would like to thank Kyong-Sok Chang, Bob Holmberg, and Diego Ruspini for their helpful insights and discussion in preparing this paper. In particular we would like to thank Kyong-Sok Chang for helping to gather the experimental data. The financial support of Boeing and NSF (grants IRI-9320017 and CAD-9320419) is gratefully acknowledged.

# References

[1] James. E. Bobrow, S. Dubowsky, and J. S. Gibson. Time-optimal control of robotic manipulators along specified paths. *Int. J. of Robotics Research*, 4(3):3–17, 1985.

[2] Oliver Brock and Oussama Khatib. Executing motion plans for robots with many degrees of freedom in dynamic environments. In *Proc. Int. Conf. on Robotics and Automation*, volume 1, pages 1–6, 1998.

[3] Oliver Brock and Oussama Khatib. High-speed navigation using the global dynamic window approach. In *Proc. Int. Conf. on Robotics and Automation*, 1999. To appear.

[4] Kyong-Sok Chang. *Robotics Library*. Stanford University, 1998. Dynamic control and simulation library.

[5] Wonyun Choi and Jean-Claude Latombe. A reactive architecture for planning and executing robot motions with incomplete knowledge. In *Proc. Int. Conf. on Intelligent Robots and Systems*, volume 1, pages 24–29, 1991.

[6] John J. Craig. *Introduction to Robotics: Mechanics and Control*. Addison-Wesley, 2nd edition, 1989.

[7] Elisabeth A. Croft, Robert G. Fenton, and Beno Benhabib. Time-optimal interception of objects moving along topologically varying paths. In *Proc. Int. Conf. on Systems, Man, and Cybernetics*, volume 5, pages 4089–94, 1995.

[8] Lydia E. Kavraki, Peter Švestka, Jean-Claude Latombe, and Mark H. Overmars. Probabilistic roadmaps for path planning in high-dimensional configuration spaces. *IEEE Trans. on Robotics and Automation*, 12(4):566–580, 1996.

[9] Oussama Khatib. Real-time obstacle avoidance for manipulators and mobile robots. *Int. J. of Robotics Research*, 5(1):90–8, 1986.

[10] Oussama Khatib. A unified approach to motion and force control of robot manipulators: The operational space formulation. *Int. J. of Robotics Research*, 3(1):43–53, 1987.

[11] Oussama Khatib. Towards integrated planning and control. In *Proceedings of IFAC Symposium on Robot Control*, volume 1, pages 305–13, 1994.

[12] Oussama Khatib, Kazu Yokoi, Kyong-Sok Chang, Diego Ruspini, Robert Holmberg, and Arancha Casal. Vehicle/arm coordination and multiple mobile manipulator decentralized cooperation. In *Proc. Int. Conf. on Intelligent Robots and Systems*, volume 2, pages 546–53, 1996.

[13] Jean-Claude Latombe. *Robot Motion Planning*. Kluwer Academic Publishers, Boston, 1991.

[14] Sean Quinlan and Oussama Khatib. Elastic bands: Connecting path planning and control. In *Proc. Int. Conf. on Robotics and Automation*, volume 2, pages 802–7, 1993.

[15] Tzyh-Jong Tarn. Path-based approach to integrated planning and control for robotic systems. *Automatica*, 32(12):1675–87, 1996.

# RoverBug: Long Range Navigation for Mars Rovers

Sharon Laubach and Joel Burdick

California Institute of Technology

Pasadena, CA, USA

Sharon.Laubach@jpl.nasa.gov / jwb@robby.caltech.edu

**Abstract:** After Mars Pathfinder's success, a demand for new mobile robots for planetary exploration arose. These robots must be able to autonomously traverse long distances over rough, unknown terrain, under severe resource constraints. We present the "RoverBug" algorithm, which is complete, correct, requires minimal memory, and uses only on-board sensors, which are guided by the algorithm to sense only the data needed for navigation. The implementation of RoverBug on the Rocky7 Mars Rover prototype at the Jet Propulsion Laboratory (JPL) is described, and experimental results from operation in simulated martian terrain are presented.

## 1. Introduction

The Mars Pathfinder mission illustrated the benefits of including a mobile robotic explorer on a planetary mission. While previous forays allowed only remote exploration or were limited to a small site by an immobile lander, the Sojourner rover was able to roam and to place its instruments on objects of interest. Mars missions currently being planned call for new rovers capable of operation for up to a year, compared to the 83 sols (martian days) of operation for the Sojourner rover. The rovers are also required to traverse vastly greater distances: up to 100m/sol, versus Sojourner's 104m/83 sols. Lessons learned from Mars Pathfinder indicate a need for significantly increased rover autonomy in order to meet mission criteria within severe constraints including limited communication opportunities with Earth, power, and computational capacity.

### 1.1. Motion Planning on Mars

A key advance in functionality required for planetary rovers is greater navigational autonomy. Given a goal which cannot be seen from the rover's location, the rover must use its sensors to navigate safely and accurately through unknown, rough terrain to that goal, autonomously. This will require, in particular, improved motion planning and localisation algorithms.

Navigation techniques for planetary rovers must assume no prior knowledge of the environment and must be sensor-based and robust. They must also operate under severe constraints of power, computational resources, and memory, due to the high cost of flight components. Due to dead reckoning errors, slippage, nonholonomic fine-positioning constraints, and constraints on mission time available, using rover motion to augment sensing is costly. Simultaneously, limited memory, computational capacity, power and time available all argue for minimising the amount of data sensed and processed. Thus, practical navigation techniques must utilise the available sensing array in a scheme which efficiently senses only the data needed for navigation, requires minimal memory to store salient features of the environs, and conserves rover motion.

*Figure 1. Typical terrain encountered on Mars by the Sojourner rover.*

## 2. Relevant Work

Much of the work in motion planning can be divided into three major categories: "classical" path planners, heuristic planners, and "complete and correct" sensor-based motion planners. "Classical" planners assume full knowledge of the environment, and are complete. Heuristic planners, generally based upon a set of "behaviours," can be used in unknown environments but may generate long paths and do not guarantee the goal will ever be reached, nor that the algorithm will halt. (A more detailed discussion is given in [1].) The third category, which relies solely upon the rover's sensors and yet guarantees completeness, is most relevant to the problem of autonomous planetary navigation.

Two approaches to such planners adapt classical methods to local sensed areas. One technique builds "roadmaps" using data from the visible region, such as Choset's HGVG [2], Rimon's adaptation of Canny's OPP [3], and the TangentBug algorithm of Kamon, Rivlin, and Rimon [4]. The second approach springs from approximate cell decomposition, filling in a grid-based world model as information is gathered, exemplified by Stentz' D* algorithm [5]. These methods differ in their level of development for real systems. The sensor-based version of OPP is currently strictly theoretical, owing to the difficult-to-implement sensors required. The HGVG has been implemented on a mobile robot using sonar sensors. This planner produces paths which are maximally distant from obstacles, a plus for rover safety, but it works best in a contained area—a description not applicable to the typical martian environment (Fig. 1).

The D* algorithm and TangentBug are both useful in unbounded environments, and both produce "locally optimal" solutions: the resultant paths are the shortest length possible using only local information. D* has been implemented on an autonomous HMMWV; however, its grid-based world model requires a significant amount of memory for storage, and the algorithm's completeness depends upon the cell granularity of its world model.

TangentBug motivates the work presented here. Its world model is streamlined, comprising primarily the sensed obstacle endpoints. The planner consists of two "modes", motion-to-goal and boundary following, which interact to ensure global convergence if the goal is reachable, and which "fail gracefully" if the goal is unreachable. The algorithm is memory-efficient, fairly robust, and conserves robot motion. However, some of its assumptions do not apply to the "rover problem" of navigating in planetary terrain: TangentBug assumes that the robot is a point, that obstacles block both motion and sensing, and that the robot's sensors provide an omnidirectional view.

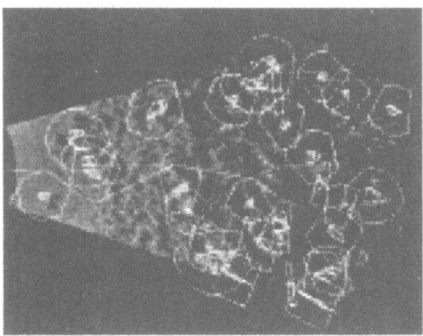

*Figure 2. Rangemap data from a stereo pair. This image also shows detected obstacles, and a path generated by the RoverBug algorithm (see Section 4).*

The current plan for a rover sensing system consists of a mast-mounted stereo pair of cameras that can pan and tilt. These cameras have a 30° to 45° field of view (FOV), and the "visible region" associated with these sensors sweeps out roughly a wedge, with limited downrange radius $R$ due to both the tilt angle and camera resolution. (Fig. 2 shows data from such a sensing array.) The rovers also feature chassis-mounted stereo pairs on the front and back. Given the constraints described above, we cannot simply pan the mast-mounted sensor array and combine many views to obtain a 360° view at each step. Rather, the planner should be able to identify the minimal number of sensor readings needed (and which specific areas to sense) to proceed at each step, while avoiding unnecessary rover motion. Thus, we have developed the "Wedgebug" algorithm and its extension, "RoverBug", to address these issues for flight microrovers. Wedgebug deals with the limited FOV of rovers in an efficient manner, minimising the need to sense and store data, using autonomous gaze control. The RoverBug implementation discussed in Section 4 relaxes the assumptions that the rover is a point robot, and that obstacles block sensing.

## 3. The Wedgebug Algorithm

Wedgebug assumes the following: The rover is modelled as a point robot in a 2D world where every point is either contained within an impassable obstacle or lies in freespace ($\mathfrak{F}$). Obstacles block both motion and sensing. The rover's sensors, from position $x$, detect ranges within a wedge $W_i = W(x, \vec{v}_i)$ which sweeps out an angle $2\alpha$ and is centered on the direction $\vec{v}_i$, where $\angle(\overrightarrow{xT}, \vec{v}_i) = 2i\alpha$ ($T$ is the goal). Define $C$ as the arc boundary of $W_i$ at radius $R$, and $\partial W_i$ as the union of the two bounding rays of $W_i$ (Fig. 3). The "interior" of $W_i$ is defined as $\text{int}(W_i) = \overline{W_i} - \partial W_i$ (an "interior" point may lie on $C$). Let $d(a, b)$ be the Euclidean distance between points $a$ and $b$.

Wedgebug, like TangentBug, has two modes which interact to ensure global convergence: *motion-to-goal* (*MtG*) and *boundary following* (*BF*). Each mode is further divided into components to improve efficiency and handle the limited FOV. At the robot's initial position $A$, an initialisation step records the parameter $d_{\text{LEAVE}} = d(A, T)$. This parameter marks the farthest the robot can stray from $T$ during an *MtG* segment. Thereafter, *MtG* is typically the dominant

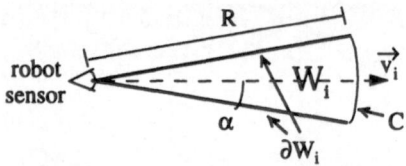

*Figure 3. Anatomy of a wedge.*

behaviour. It directs the robot to move toward the goal using a local version of the tangent graph, restricted to the visible area (Fig. 4). The robot (at position $x$) first senses a wedge, $W_0$, directed toward the goal. The tangent graph (or "reduced visibility graph") is constructed, consisting of all line segments in $\mathfrak{F}$ connecting $x$, $T$, and all obstacle vertices such that the segments are tangent to any obstacles they encounter [6]. The *local tangent graph* within the wedge $W$, LTG($W$), is defined as the tangent graph restricted to $W$. (Obstacle boundaries appear as continuous contours in the range data; the endpoints of these contours are called the "obstacle vertices". Each endpoint $e$ corresponds to a discontinuity in the range data or to an intersection of a contour with $\partial W$ or $C$.) The planner constructs LTG($W_0$). The planner optionally adds a node, the projection of $T$ onto $C$, so LTG($W_0$) contains a path directly towards $T$. The planner then searches a subgraph, $G1(W_0)$, consisting of those nodes closer to $T$ than both the robot's current position and $d_{\text{LEAVE}}$, for the optimal local subpath. Using the criteria in Section 3.1, the robot may scan additional wedges, construct the LTG in the conglomerate wedge $\overline{W}(x)$ (see Section 3.1), and search for a new subpath. After executing the resultant subpath, *MtG* begins anew. This behaviour is continued until either the goal is reached, $T$ is deemed unreachable, or the robot encounters a local minimum in d($\cdot, T$). In the latter case, the planner switches to *BF*. The objective of this mode is to skirt the boundary of the obstacle which contains the local minimum, still calculating LTG($W_0$), until one of two events occur: either the robot completes a loop, in which case $T$ is unreachable and the algorithm halts; or LTG($W_0$) contains a new subpath toward $T$ and the planner switches back to *MtG*. Based upon the two operational modes, *MtG* and *BF*, it can be proved that Wedgebug is complete and correct [7]. Practically, by implementing a form of gaze control, the algorithm also deals with the limited FOV of flight rovers in a manner which is efficient and minimises the need to sense and store data. Furthermore, Wedgebug produces locally optimal paths. Hence, it is well suited to conserving rover energy. The next two sections describe the *MtG* and *BF* modes in more detail. (See [7] for a thorough description.)

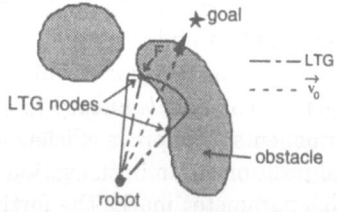

*Figure 4. LTG calculated within $W(x, \vec{v}_0)$.*

## 3.1. Motion-to-Goal

The goal of *MtG* is to move so the robot's distance to $T$ is nonincreasing. During *MtG*, the robot can either move through $\mathfrak{F}$ toward the goal ("direct mode"), or it must skirt the boundary of an obstacle while moving toward $T$ ("sliding mode"). Further, "sliding mode" contains a submode, "virtual *MtG*," to improve efficiency. That is, during normal *MtG*, the planner scans a single wedge toward $T$ to determine whether a path exists. If it is apparent that more information could lead to a shortcut, "virtual *MtG*" scans additional wedges in order to determine the appropriate path.

The first actions taken in a new *MtG* step are to scan $W_0$, construct $\text{LTG}(W_0)$, and search $G1(W_0)$. The shortest local path $P$ will either aim directly toward $T$, or will pass through an endpoint $e$ of a *blocking obstacle* (which lies directly between $x$ and $T$). Call the point through which $P$ passes—either $e$, or the projection of $T$ on $C$ ($T_g$)—the *focus point*, $F$ (Fig. 4). The focus point (fixed for each step) is the goal for each *MtG* step: the subpath for the current *MtG* step is precisely $\overline{xF}$. Its position within the robot's FOV determines whether additional wedge views are needed. Basically, if the subpath moves the robot through the interior of the visible region, the robot executes this subpath and begins the next *MtG* step. If, on the other hand, $F$ lies on the obstacle boundary, an additional view in this direction could produce a shortcut around the blocking obstacle—that is, the robot could "virtually circumnavigate" a portion of the boundary without moving (see Fig. 5). The "virtual *MtG*" mode ends when 1) the robot has found a suitable shortcut, so the robot moves along this subpath and begins the next *MtG* step; 2) the rover detects that it is sensing part of a region not useful for *MtG*, i.e. farther than the rover from $T$; or 3) the robot detects that the obstacle boundary is curving away from $T$, so the robot can no longer "virtually slide" in this direction without losing ground. In the latter cases, if the rover has not yet established a traversal direction, the robot attempts to round the obstacle in the opposing direction. If this attempt fails, the robot has encountered a local minimum in $d(\cdot, T)$, and the planner switches to *BF*.

In order to prevent backtracking while circumnavigating an obstacle $O$ (both "virtually" and while moving), the algorithm establishes a traversal direction—call it *positive* ($\rho^+$)—upon first sensing $O$. Thereafter, the robot must round the obstacle in only the positive direction, until (1) the blocking obstacle is changed (including the case when $F = T_g$), (2) the robot detects

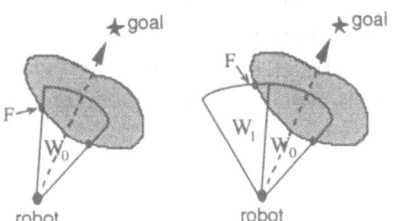

*Figure 5.* "*Virtual MtG.*" *The figure on the left shows the first part of an MtG step. The nodes of LTG(W₀) are marked. F satisfies the conditions for "virtual MtG," so the robot scans $W_1$ (right). Now, $F \in int(W_0 \cup W_1)$, so "virtual MtG" ends.*

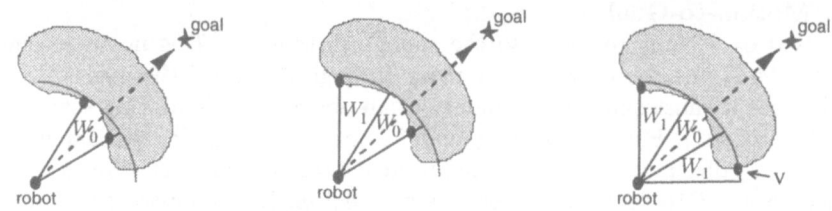

*Figure 6.* "*Virtual BF.*" *The figure on the left depicts the first part of a "virtual BF" step. The nodes of $LTG(W_0)$ are marked. Since $\nexists V \in int(LTG(W_0))$, the robot scans $W_1$ (center). Again, $\nexists V \in int(LTG(W_0 \cup W_1))$, so the robot scans $W_{-1}$ (right). Now, $V \in int(W_0 \cup W_1 \cup W_{-1})$, so "virtual BF" ends.*

that it has completely circumnavigated $O$ and the algorithm halts, or (3) the planner switches to $BF$. The latter case occurs when the robot can no longer decrease its distance to $T$—i.e, it has entered the basin of attraction of a local minimum in $d(\cdot, T)$.

## 3.2. Boundary Following

The goal of $BF$ is to skirt the blocking obstacle $O$ until progress can be made once more toward $T$, thereby escaping a local minimum. As with $MtG$, $BF$ is split into two submodes. Normal $BF$ uses two wedge views, one toward $T$ and one in the direction of travel around the obstacle boundary ($\rho^+$), to determine when a path towards $T$ appears while the robot circumnavigates $O$. Immediately after a switch from $MtG$ to $BF$, however, $\rho^+$ may not be defined. In this case, "virtual $BF$" is used to take advantage of the information from the current distance from $O$ to choose this direction wisely. (The motivation for "virtual $BF$," as for "virtual $MtG$," is the idea that it is less costly for the robot to swivel its sensors than for the robot to move.) In essence, the robot swings its sensors back and forth in a prescribed manner to search for the "best" place to move and begin normal $BF$.

More precisely, the robot initially scans $W_1 = W(x, \vec{v}_1)$, where the positive direction is chosen by comparing the tangents to $\partial O$ at the wedge boundary. The "steepest" side is chosen with the idea that the rover may be able to progress further "virtually" in this direction. As before, let $\overline{W} = \bigcup^{sensed} W_k(x)$. The planner computes $LTG(\overline{W})$. Similarly to $MtG$, if a shortcut is found through the interior of $\overline{W}$, the robot moves along this path and begins normal $BF$, first recording $d_{reach}$, the closest distance to $T$ found along $\partial O$, $\rho^+$, and $V_{loop}$, a marker used to detect whether the robot has looped around $O$. Otherwise, the planner directs the sensor to scan $W_{-1} = W(x, \vec{v}_{-1})$, constructs $\overline{W} = W_0 \cup W_1 \cup W_{-1}$, and searches the freshly expanded $LTG(\overline{W})$. In this manner, the robot scans back and forth until a suitable shortcut is found, then travels it to begin normal $BF$. "Virtual $BF$" ends when one of three events are detected: 1) a suitable shortcut is found, so the robot moves along this subpath and begins normal $BF$; 2) the latest wedge overlaps a previously scanned region—the robot is trapped by an encircling obstacle, and the algorithm halts; and finally 3) "virtual $BF$" is no longer useful, since a second obstacle obscures the blocking obstacle; the point where the two boundaries "meet" is called the "framing point" $V_f$. If one such point is visible, the robot scans once more in the opposing direction and "virtual *BF*" ends. If a shortcut as in (1) is found, the robot moves there to begin normal *BF*. Otherwise, the rover moves to the point on $\partial O$ adjacent to $V_f$. If there are two such points, $V_l$ and $V_r$, the rover moves to the point which is "farthest around" the obstacle, then begins normal *BF*.

To start a normal *BF* step, the robot senses $W_0$ and searches $G1(W_0)$. *BF* exits if: (1) $T \in W_0$, in which case the robot moves to $T$ and the algorithm halts, or (2) $\exists V \in G1(W_0)$ such that $d(V,T) < d_{reach}$, the *leaving condition*, in which case the planner sets $d_{\text{LEAVE}}$ to $d(V,T)$ and begins a new *MtG* segment. If neither of these conditions hold, the planner directs the sensor to scan a wedge $W_x$ along the tangent to the obstacle boundary. If $V_{loop} \in W_x$ and it is in the sensed portion of $\partial O$ containing $x$, the robot has executed a loop—the goal is unreachable, and the algorithm halts. Otherwise, the planner computes the farthest the robot can traverse along $\partial O$ in $W_x$. The robot records $d_{reach}$, moves, and begins a new *BF* step.

### 3.3. Sketch of Proof of Convergence

The proof of convergence for Wedgebug is similar to TangentBug's proof [4]. Each robot motion can be characterised as a particular type of motion segment; in turn, each type of segment can be shown to have finite length. Following [4], it can be shown that there are a finite number of each type of segment, so the path terminates after finite length. Due to space limitations, we detail only the proof that *BF* segments have finite length. The proofs for the other types of motion segments are analogous.

Define $S_i$ to be the point where the planner switches from *MtG* to *BF* at obstacle $O_i$; $L_i$ the point where the *leaving condition* is met on $O_i$; and finally $Q_i$, the point where a loop is detected on $O_i$. There are two types of *BF* segments: $[S_i, L_i]$, and $[S_i, Q_i]$. Let $P_i$ be the perimeter of obstacle $O_i$.

**Lemma.** *BF* segments are finite length.

*Proof.* a) $[S_i, L_i]$. Let $N$ be the point where the robot first touches $\partial O_i$. The path then consists of two pieces: $[S_i, N]$ and $[N, L_i]$. Since $N$ and $L_i$ both lie on $\partial O_i$, the robot is traversing $O_i$ in a fixed direction, and the robot has not detected a loop, we have length($[N, L_i]$) $\leq$ length($[N, Q_i]$) $\leq$ length($[N, V_{loop}]$). Further, since $V_{loop}$ is in the opposite direction from traversal, we know that the segments $\overline{NV_{loop}}$ and $\overline{V_{loop}N}$ do not overlap. Therefore, since $P_i$ is finite, we have length($[N, L_i]$) $\leq$ length($[N, V_{loop}]$) $\leq P_i < \infty$. Further, $d(S_i, N) \leq R$. Thus, length($[S_i, L_i]$) $\leq d(S_i, N) +$ length($[N, L_i]$) $\leq R + P_i < \infty$.

b) $[S_i, Q_i]$. Similarly, length($[S_i, Q_i]$) $\leq d(S_i, N) + P_i \leq R + P_i < \infty$. $\square$

## 4. Implementation and Results

An extended version of the Wedgebug algorithm, called "RoverBug," has been implemented on the JPL Rocky7 prototype microrover (Fig. 7), a research vehicle designed to test technologies for future missions [8]. Rocky7, which at 60cm x 50cm x 35cm is roughly the same size as the Sojourner rover, has a few important differences relevant to future rovers, including carrying three stereo pairs of cameras for navigation: two body mounted, and one on a deployable 1.2m mast. In addition, the rover software features a localisation algorithm utilising mast imagery to aid in dead-reckoning [9].

Although the Wedgebug algorithm is an important step, it still does not capture the complexities of the real world. For instance, the rover is not a point robot; a problem addressed in the RoverBug implementation by calculating the obstacles' "silhouettes": the smallest polygon bounding the projection of each $SE(2)$ obstacle onto $\Re^2$. Another issue arises since the mast imagery can "see over" many obstacles: the resulting visible region is not a star-shaped set, and the LTG is much richer than in the development in Section 3. Also, the mast is limited in its ability to sense obstacles within 1m of the rover, since the obstacle detection algorithm searches for steps in elevation, not easy to detect while looking straight down on the tops of rocks. Thus, care must be taken while executing the subpaths.

The experimental scenario is as follows: Rocky7 is situated in unknown, rough terrain. The remote human operator views panoramic imagery returned by the rover, or orbiter and/or descent imagery to designate a goal coordinate. The operator then transmits the command to navigate to the goal, which sets in motion the autonomous planner. RoverBug begins by directing the mast to image towards the goal. Software on-board produces a rangemap, detects obstacles, and computes the obstacles' convex hulls. RoverBug then computes the obstacles' silhouettes, and searches the resulting LTG—which now truly looks the part of a local tangent graph—to produce the first subpath (see Fig. 2). The planner directs the mast to look in the appropriate direction(s), and incrementally builds and executes each subpath until the goal is reached.

As before, the motion-to-goal ($MtG$) mode is the dominant behaviour, moving the rover toward $T$ monotonically, and boundary following ($BF$) is used to escape local minima in $d(\cdot, T)$. However, rather than endowing each of these modes with a "virtual" submode, we combine the two submodes into a single "virtual sliding" ($VS$) mode in the interest of reducing the implementation's footprint, helping further minimise memory requirements. Upon detecting an obstacle chain which prevents progress through the single wedge view used by $MtG$ to move toward the goal, the robot invokes $VS$ mode. The purpose of

*Figure 7. The Rocky7 Prototype Microrover. It is pictured in the JPL MarsYard, an outdoor testing arena featuring simulated martian terrain.*

*VS* mode is to scan one wedge at a time, back and forth, until enough of the (visible) extent of the blocking obstacle has been seen[1] in order to determine which direction the robot should circumnavigate the obstacle chain. (Such a chain can be seen spanning the end of the wedge in Fig. 2.) At this point, RoverBug plans a path through the cumulative LTG (at $x$) to the appropriate edge of the blocking obstacle (chain) boundary, and enters *BF* mode.

*BF* mode, in turn, relies upon the body-mounted stereo cameras to gather information about obstacles, for two reasons: 1) the rover is likely too close for mast imagery to be useful, and 2) since the mast should not be raised while the rover moves, it would adversely affect the time required to navigate if the mast were raised and lowered between each (small) *BF* step. Since these cameras are fixed to the rover chassis, it is necessary to rotate the rover itself to ensure views in the necessary directions. As with Wedgebug, each *BF* step begins with a view toward the goal. If no obstacles are detected in this direction, then the rover raises its mast to determine whether the *leaving condition* (same as for Wedgebug) is satisfied. If the leaving condition holds, the rover switches back to *MtG*. Otherwise, the rover turns in the direction of circumnavigation, images the new region (with the body-mounted cameras), and determines whether another view is required to advance as far as possible around the obstacle boundary. In this wise, the rover uses its body cameras to skirt the obstacle chain, taking shortcuts where possible, until either the rover can again progress toward the goal, or until the robot has detected a loop. As before, a loop indicates that the goal is unreachable, and the algorithm halts.

Finally, in order to cope with the limitations of dead reckoning, the localisation algorithm described in [9] brackets each *MtG* step, to update the rover's knowledge of its position after executing each subpath, and each *BF* segment, to relocalise the rover after it has skirted an obstacle boundary. (If the boundary is particularly long, the rover may execute several localisation events along the way, due to the short accuracy range of the localisation algorithm.)

The *MtG* mode of RoverBug has been tested extensively in the JPL MarsYard, as well as in natural arroyo terrain, including traverses for tens of meters requiring multiple iterations of the motion planning and localisation algorithms. Figure 8 shows the processed experimental data from a 21m traverse in the MarsYard. Each "wedge" in the figure depicts a projected view of the rangemap computed from a stereo pair of (mast) images taken by the rover during its traverse. That is, each wedge represents the rover's FOV, containing terrain height information. The polygons represent the inaccessible regions around rocks deemed too large for rover traversal. The path superimposed on the figure is the path taken by the rover to avoid the rock obstacles and reach the goal. The "jags" in the path represent localisation events where dead reckoning is updated using the algorithm described in [9]. As in Fig. 2, the silhouettes are computed within each wedge view, and a subpath generated, which is executed before the next wedge view is taken. The resultant multi-step path runs from left to right.

---

[1] Note that in the RoverBug scenario, the frequency of "framing nodes" is significantly reduced, since the rover can "see over" most obstacles.

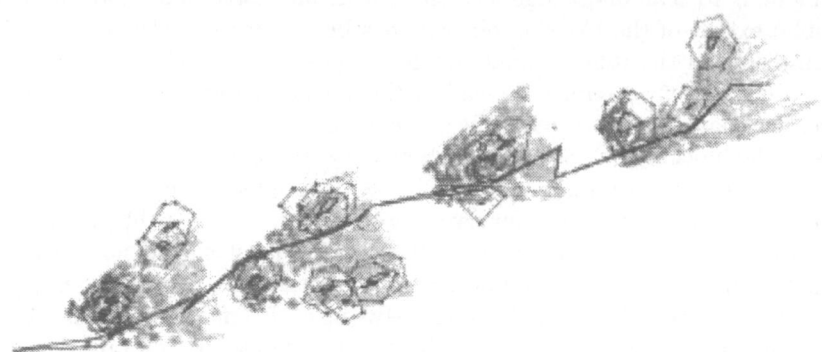

*Figure 8. Results from the JPL Mars Yard. The path runs left to right. Each wedge is a rangemap from mast imagery, extending roughly 5m from the imaging position.*

## 5. Conclusion

The requirements for autonomous flight rovers for planetary exploration provide compelling motivation for work in sensor-based navigation. This paper continues the work begun in [1] to develop, implement, and test a robust, practical path planner for the Rocky7 prototype microrover. We believe that the RoverBug planner will significantly augment microrovers' autonomous navigation ability, which in turn will aid in producing successful mobile robot missions.

### Acknowledgments
The work described here was carried out at the Jet Propulsion Laboratory, California Institute of Technology, under a contract with the National Aeronautics and Space Administration. The authors acknowledge the Long Range Science Rover team, particularly Samad Hayati, and the Mars Pathfinder Rover team, for help, inspiration, and flight experience with a rover.

## References
[1] Laubach S L, Burdick J W, Matthies L H 1998 An autonomous path-planner implemented on the Rocky7 prototype microrover. *Proc IEEE Conf Rob. Automat.*
[2] Choset H 1996 Sensor based motion planning: the Hierarchical Generalized Voronoi Graph. Ph.D. thesis, California Inst of Tech
[3] Rimon E, Canny J 1994 Construction of c-space roadmaps from local sensory data: what should the sensors look for? *Proc IEEE Conf Rob. Automat.*
[4] Kamon I, Rimon E, Rivlin E 1995 A new range-sensor based globally convergent navigation algorithm for mobile robots. CIS–Center of Intelligent Systems 9517, Computer Science Dept, Technion, Israel
[5] Stentz A 1994 Optimal and efficient path planning for partially-known environments. *Proc IEEE Conf Rob. Automat.*
[6] Latombe J-C 1991 *Robot Motion Planning*. Kluwer Academic Publishers, Boston
[7] Laubach S L 1999 A practical autonomous sensor-based path planner for flight planetary microrovers. Ph.D. thesis, California Inst of Tech
[8] Volpe R, Balaram J, Ohm T, Ivlev R 1996 The Rocky7 Mars Rover prototype. *Proc IEEE/RSJ Conf Intelligent Robots and Sys.*
[9] Olson C, Matthies L 1998 Maximum likelihood rover localisation by matching range maps. *Proc IEEE Conf Rob. Automat.*

# Chapter 9

# Programming and Learning

Perhaps the biggest problem with robots today is having to teach them, that is, encode the task to be performed. For industrial robots the teach box is still the most common way of entering spatial data and program steps — a method that is functionally unchanged in nearly 20 years. Research into CAD-based programming is ongoing but has not yet made significant inroads on the shop floor. Learning (from others and from mistakes) is a key aspect of intelligence, and robots will never be considered intelligent until they have some learning capability.

The first paper by Tsuneo Yoshikawa and Kazuyuki Henmi captures the force and position data required for a human skill, in this case calligraphy. That information is subsequently displayed to a student using haptics, see Chapter 10, and vision.

James Trevelyan and his colleagues at The University of Western Australia are pioneers in web telerobotics, having operated a system online continuously since 1994. The large number of willing users are actually participating in a long-term experiment into ways in which humans and robots interact. In this paper James Trevelyan and Barney Dalton describe the web telerobot system, the user interface and conclusions about user behaviour.

RoboCup is an international robotic soccer tournament. The game was originally conceived as a 'grand challenge' for robotics but has now taken off as a form of entertainment (which is perhaps the future for robotics). Minoru Asada et al. describe the history of the game and provide a summary of activity in the three leagues. There is substantial international activity and the long-term goal is to take on the human world-cup team.

The software packages ORCADD and MAESTRO have been developed to provide an integrated framework for simulating and implementing complex continuous-time and discrete-event systems. Eve Coste-Maniere, Nicolas Turro and Oussama Khatib describe an application of this technology to the Stanford ROMEO and JULIET robots for a safety critical human assistance task.

Finally, Dinesh Pai and colleagues describe ACME, an internet accessible shared facility for acquiring physical data about objects. This 15 degree-of-freedom system provides accurately registered range, color and surface properties that can be used to build accurate physical models of the object.

PIC

# Human Skill Transfer Using Haptic Virtual Reality Technology

Tsuneo Yoshikawa and Kazuyuki Henmi*
Department of Mechanical Engineering
Kyoto University
Kyoto, 606-8501 Japan
(* Currently with IBM Japan, Ltd.)
yoshi@mech.kyoto-u.ac.jp

**Abstract:** One potential field of application of haptic virtual reality technology will be skill training. An approach for transferring a teacher's skill to a student using haptic virtual reality technology is proposed and, as an example of its application, a skill transfer system for calligraphy is developed. In many of skilled tasks, both position and force trajectories of the teacher's body and/or hand play crucial role. Recognizing this fact, in our system, the teacher's skill is first recorded in terms of position and force and then this recorded skill consisting of position and force trajectories is displayed to the student through a haptic display device and a secondary display device. We have implemented two methods of skill display: one is to use the haptic display device for displaying the position information, and the other is to use it for displaying the force information. The remaining information is displayed using the secondary display device, i.e., visual display in the case of our calligraphy system. A preliminary experimental result of the calligraphy system is also presented.

## 1. Introduction

Virtual reality is a technology that makes it possible for a person to feel as if he/she is really surrounded by a virtual world built by a computer software. The importance of haptic sensation in virtual reality has also been recognized and various haptic display devices have been developed to provide the feel of touch and force to the operator in the virtual world. [1]-[9]. Control algorithms for displaying dynamics of virtual objects have also been developed [10] [11] [12].

One potential application field of this haptic virtual reality technology is skill education. Although several systems for experiencing sports, medical operations, and so on have been developed [13] [14], displaying and transferring a teacher's skill to a student using haptic display device has not been paid much attention so far.

In the present paper, we are mainly interested in human skill in manipulating some tool or object, such as writing brush, pencil, knife, tweezers, grinder, peg, probe, ball, bat, tennis racket, etc. This kind of human skill involves not only the motion (or position) of the human's body but also the force human's

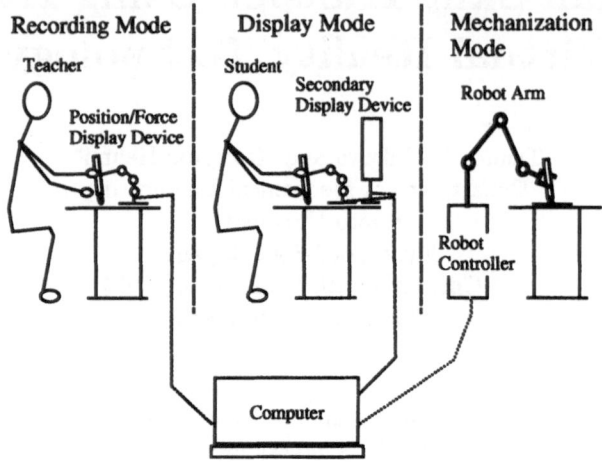

Figure 1. Conceptual diagram of skill transfer system

body applies to the tool or object. We further assume that the skill under consideration can be represented only by the position and force trajectories of the tool or object. For this class of skills, we study an approach to skill transfer using the haptic virtual reality technology. Usefulness of this approach and two methods of displaying the teacher's skill to the student are discussed. As an example of this approach, a virtual calligraphy system [15] is developed and a preliminary experimental result is presented.

## 2. Skill Transfer System Using Haptic Interface

### 2.1. Basic Concept

Fig.1 shows schematically the basic structure of the skill transfer system that uses haptic interface. The purpose of the system is to develop a virtual environment where the position and force trajectories of the teacher's skillful behavior are recorded first (recording mode or teacher mode), and then they are displayed to the student using haptic virtual reality technology (display mode or student mode).

Conventional way a teacher often adopts for teaching a skill to a student is to show his motion many times, to explain the motion orally, and to hold and move the student's hand or some other part to show a desired motion. However, this way may not be ideal, for example, from the viewpoint of showing precisely the force accompanying this motion. Also this method requires that the teacher and the student must be together at the time of teaching. The proposed skill transfer system is expected to be able to overcome these difficulties. Moreover, since the teacher's skill is recorded in the computer, it is possible to modify the recorded skill in terms of its scale, speed, and so on, to adapt the skill to each student considering his/her size, strength, and so on. Various display methods may also be used depending on which aspect of the skill should be emphasized.

Note that, since many skills are related to the dynamics of human body, tool, and environment, we usually have to take the dynamics of virtual objects and the display device into consideration [10] [12] for applying the haptic virtual reality technology to skill training.

Another kind of skill transfer we can think of is from human to machine or robot. This is shown by Mechanization Mode in Fig.1. This mode will play an important role in realizing Artificial Skill [16] using the virtual reality technology.

## 2.2. Algorithms for Displaying Teacher's Skill

When we display the teacher's skill to the student, it is important to display both the position and the force trajectories. However, it is impossible, in principle, to display these two information at the same time through one haptic display device. Based on this recognition, we propose to display one information directly to the student through the haptic display device, and another information indirectly to the student through a secondary display device. The method of displaying the position information directly through the haptic display device and displaying the force information through the secondary display device is called the Direct Position Display Method, and the method of displaying the force information through the haptic display device and displaying the position information through the secondary display device is called the Direct Force Display Method.

In many skills, the directions in which the force is important are rather limited. For example, in writing letters and /or drawing pictures on a flat smooth surface by a pencil or writing brush, the force applied in normal to the surface is more important than that in other directions. Hence we may want to display force in this normal direction whereas the position information is dominant in other directions. In robotics it is sometimes required that the end-effector position of a robot manipulator should be controlled in some directions and the force applied by the end-effector on an object should be controlled in other directions. The control algorithm realizing this requirement is called the hybrid position/force control. Especially, dynamic hybrid control algorithm [17] [18] realizes the desired position and force trajectories precisely by taking into account the dynamics of the robot arm itself. This appraoch will be useful for implementing the above display methods.

# 3. Virtual Calligraphy System

## 3.1. System Structure

As an example of skill transfer system, we have developed a virtual calligraphy system. The basic structure of the system and its overview are shown in Fig.2.

The haptic display device is composed of 2 three-joint miniature robot arms (each arm consists of a four-bar parallel link mechanism connected to the base through a rotational joint) and an aluminum pipe (198mm long and 5mm diameter) connected to the tips of the robot arms through ball joints. This pipe represents the brush.

The joints of the robot arms are driven by DC servo-motors (nominal torque 3600gmm) through reduction gears with reduction ratio of 1:8. Joint

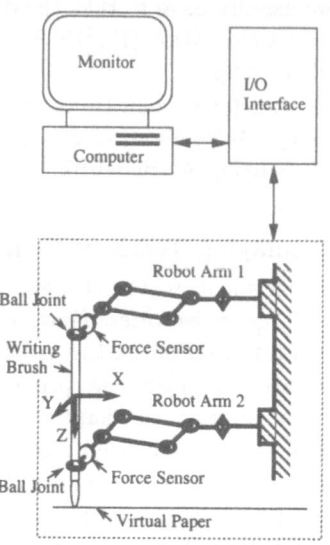

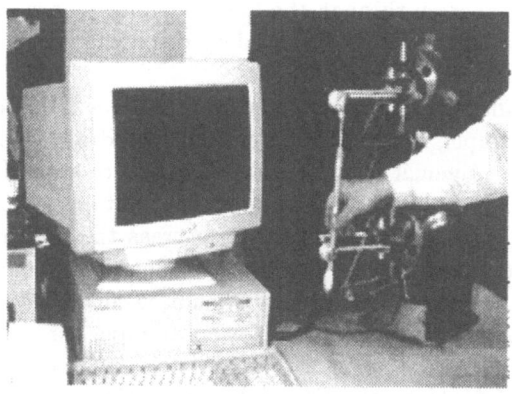

Figure 2. Virtual calligraphy system

angles are measured by optical encoders, and the joint velocities are obtained by calculating difference of the measured angles. Force sensors are attached to the tips of the robot arms. The host computer is a personal computer (NEC, i486DX2,66MHz). The C-language is used and the sampling period is 5ms. A CRT of the personal computer is used as the visual display device of position and force information of the teacher.

## 3.2. Control Algorithm

Both for recording teacher's skill and for displaying the recorded motion to students, the position and force of the robot arms are measured and the position and force control is performed.

The control algorithm is given as follows. Assume that the dynamics of the robot arm $j$ $(j = 1, 2)$ is given by

$$\tau_j - J^T(q_j)f_j = M(q_j)\ddot{q}_j + \hat{h}(q_j, \dot{q}_j) \tag{1}$$

where $q_j$ is the joint angle vector, $\tau_j$ is the joint driving torque, $J(q_j)$ is the Jacobian matrix, $M(q_j)$ is the inertia matrix, $\hat{h}(q_j, \dot{q}_j)$ is the term representing the centrifugal, Coriolis, viscous friction, and gravity forces, and $f_j$ is the external force applied to the tip of the arm. Superscript $T$ denotes the transpose of matrix or vector. Note that, since the arms are connected to the pipe through ball joints, no external moment is applied to the arms. Also note that $J, M$, and $\hat{h}$ are the same for the two arms.

Denoting the endpoint position of the arm $j$ as $r_j$, the relation between the endpoint velocity $\dot{r}_j$ and the joint velocity $\dot{q}_j$ is given by

$$\dot{r}_j = J(q_j)\dot{q}_j \tag{2}$$

Differentiating this, we have

$$\ddot{r}_j = \dot{J}(q_j)\dot{q}_j + J(q_j)\ddot{q}_j \tag{3}$$

The position control algorithm of the arms is given as follows. First from (1) and (3), we obtain

$$\tau_j = M(q_j)J^{-1}(q_j)(\ddot{r}_j - \dot{J}(q_j)\dot{q}_j) + \hat{h}(q_j, \dot{q}_j) - J^T(q_j)f_j \tag{4}$$

Then we replace $\ddot{r}_j$ in the above equation by

$$\ddot{r}_j = \ddot{r}_{jd} + K_v(\dot{r}_{jd} - \dot{r}_j) + K_p(r_{jd} - r_j) \tag{5}$$

where $r_{jd}$ is the desired position, $K_v$ and $K_p$ are the feedback gains for the velocity error and the position error.

As for the force control, we replace $f_j$ in (4) by

$$f_j = f_{jd} + K_f(f_{jd} - f_j) \tag{6}$$

where $f_{jd}$ is the desired force, $K_f$ is the feedback gain for the force error.

### 3.2.1. Teacher Mode

We have two modes of operation of the system: teacher mode and student mode. The teacher mode is described in this section.

We define the reference coordinate frame as shown in Fig.2. That is, choosing an upright position of the brush, we define the center of the brush as the origin of the frame. We then select $Z$ axis in the vertical direction and $X, Y$ axes in the horizontal plane.

In the teacher mode, the control is done so that the teacher feels the vertical reaction force from the virtual paper and the friction force in the horizontal plane when the tip of the brush is in touch with the paper. The character written by the teacher is shown on the visual display by overlapping circles whose radius is proportional to the teacher's force applied in $Z$ direction.

Considering the character of the brush, the vertical reaction force from the paper $f_p$ is assumed to be given by

$$f_p = K(z - z_0) \tag{7}$$

$$K = \begin{cases} g(z - z_0)^2 + K_0, & \text{if} z \leq z_0 \\ 0, & \text{if} z > z_0 \end{cases}$$

where $K$ is the position feedback gain, $g$ is a constant, $z$ is the $Z$ coordinate value of the brush-tip position, $z_o$ is the $Z$ coordinate value of the paper position, and $K_0$ is the gain for the case of $z_0 = z$. This reaction force $f_p$ is displayed to the teacher as the desired value of the $Z$ component of $f_2$.

The friction force in the horizontal plane is calculated in the same way as in [11].

### 3.2.2. Student Mode

The student mode is now described in this section. In the horizontal plane the brush is position-controlled to follow the recorded position trajectory of the teacher. In the vertical direction, we want to display to the student (i) the relative position between the brush-tip and the paper and (ii) the pushing force against the paper. Note that we decided not to consider the display of force trajectory in the horizontal direction and to focus only on the vertical direction .

In displaying the teacher's horizontal trajectory, the two robot arms are position-controlled to follow the teacher's position trajectory. Hence the brush is forced to follow the teacher's brush motion. This way the student can learn the teacher's motion in the horizontal direction. This also allows the student to experience the posture and speed of the brush of the teacher's brush.

For displaying the teacher's vertical position and force trajectory, there are two methods to display the position and force information, as stated in section 3, These methods for the case of calligraphy system will be explained in more detail.

**(A) Direct position display method** This is a method in which the student learns the teacher's vertical brush position from the haptic display device and the pushing force from the visual display device. As shown in Fig.3, the pushing force at each point is represented by a circle with radius proportional to the magnitude of pushing force and its continuous overlapping representation gives a rough idea of force trajectory. The student's pushing force is also displayed in the same way using a different color on the same screen. In Fig.3 the teacher's force information is shown in white and the student's force information in black. On the left hand side, a continuous representation of the circle with radius proportional to the difference between the two pushing forces is also shown. The thinner this character is, the closer the student's pushing force is to the teacher's, meaning that the student's skill is closer to the teacher's skill in terms of the pushing force.

The desired trajectories of arms 1 and 2 in the direct position display method is given as follows. Let the endpoint trajectories of arm $j$ $(j = 1, 2)$

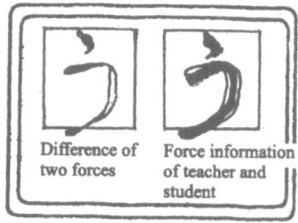

Difference of
two forces

Force information
of teacher and
student

Figure 3. Visual display of force trajectory in direct position display method

recorded in the teacher mode be denoted by $[x_{jt}, y_{jt}, z_{jt}]^T$ and the current position and force be $r_j = [x_j, y_j, z_j]^T$ and $f_j = [f_{jx}, f_{jy}, f_{jz}]^T$, respectively.

The desired position and force for arm 1 are given by

$$r_{1d} = \begin{pmatrix} x_{1d} \\ y_{1d} \\ z_{1d} \end{pmatrix} = \begin{pmatrix} x_{1t} \\ y_{1t} \\ z_1 \end{pmatrix}, \quad f_{1d} = \begin{pmatrix} f_{1xd} \\ f_{1yd} \\ f_{1zd} \end{pmatrix} = \begin{pmatrix} 0 \\ 0 \\ 0 \end{pmatrix} \tag{8}$$

These desired values mean that arm 1 is controlled in such a way that the position control is performed to display the teacher's position trajectory in the horizontal direction, and the force control with desired force 0 is performed in the vertical direction.

The desired position and force for arm 2 are given by

$$r_{2d} = \begin{pmatrix} x_{2d} \\ y_{2d} \\ z_{2d} \end{pmatrix} = \begin{pmatrix} x_{2t} \\ y_{2t} \\ z_{2t} \end{pmatrix}, \quad f_{2d} = \begin{pmatrix} f_{2xd} \\ f_{2yd} \\ f_{2zd} \end{pmatrix} = \begin{pmatrix} f_{2x} \\ f_{2y} \\ f_{2z} \end{pmatrix} \tag{9}$$

These values mean that arm 2 is controlled in such a way that the position control is performed to display the teacher's position trajectory in the horizontal direction, and the position control to the teacher's brush-tip trajectory is performed in the vertical direction.

**(B) Direct force display method** This is a method in which the student learns the teacher's vertical pushing force from the haptic display device and the brush-tip position from the visual display device.

In a fashion similar to that in Fig.3, the distances between the brush-tip and the paper for the teacher and for the student are displayed by using contures of circles with radius proportional to the distance.

The desired trajectories of arms 1 and 2 in the direct force display method is given in the following. Arm 1 is controlled using (8) and (9) as in the direct position display method. The desired position and force for arm 2 are given by

$$r_{2d} = \begin{pmatrix} x_{2t} \\ y_{2t} \\ z_2 \end{pmatrix}, \quad f_{2d} = \begin{pmatrix} f_{2x} \\ f_{2y} \\ f_{2zt} \end{pmatrix} \tag{10}$$

where $f_{2zt}$ is the $Z$ axis component of the teacher's force trajectory. These desired values mean that the teacher's position trajectory is displayed in the horizontal direction, and the pushing force trajectory is displayed in the vertical direction.

## 4. Experiment

We need to know how well the student can learn the skill of the teacher. A rough and intuitive way to evaluate this is to have the student write the character in free state (that is, in the teacher's mode) after having some exercise in the direct position (or direct force) display method and observe the change of the character.

The following procedure was taken for both the direct position and direct force display methods.

1. The teacher and the student write Japanese character "u" and the position and force trajectories are recorded.
2. Using the teacher's record, the student learns once by using the direct position (force) display method.
3. The student writes the character in free state.

The student repeats procedures 2 and 3 two more times.

Experimental results are shown in Figs.4(A) and (B). Fig.4(A) is for the direct position display method, and Fig.4(B) is for the direct force display method. In the figures, (a) is the teacher's trajectory, (b) through (e) are the student's trajectories before exercise, after one exercise, after two exercises, and after three exercises, respectively.

From these results, we can see that, in both display methods, the student's position and force trajectories resemble more and more to those of the teacher's as the exercise advances from (b) to (e). Hence we judge that there was some positive effect of skill learning using the haptic calligraphy system.

The above evaluation of the experimental result is, of course, not objective. To examine the effectiveness of various teaching and learning methods of human skill, we will need to evaluate the student's level of skill in more objective and quantitative way. This will be a future research topic.

## 5. Conclusion

An approach for transferring a teacher's skill to a student using haptic virtual reality technology has been proposed and, as an example of its application, a skill transfer system for calligraphy has been developed. Recognizing that both position and force trajectories of the teacher's body and/or hand play crucial role in many of skilled tasks, the teacher's skill is first recorded in terms of position and force trajectories in our system, and then this recorded skill is displayed to the student through a haptic display device and a secondary display device. We have implemented two methods of skill display: one is to use the haptic display device for displaying the position information, and the other is to use it for displaying the force information. The remaining information is displayed using the secondary display device, i.e., visual display in the case of our calligraphy system. A preliminary experiment of the calligraphy system has been done and some tendency of improvement of student's skill has been

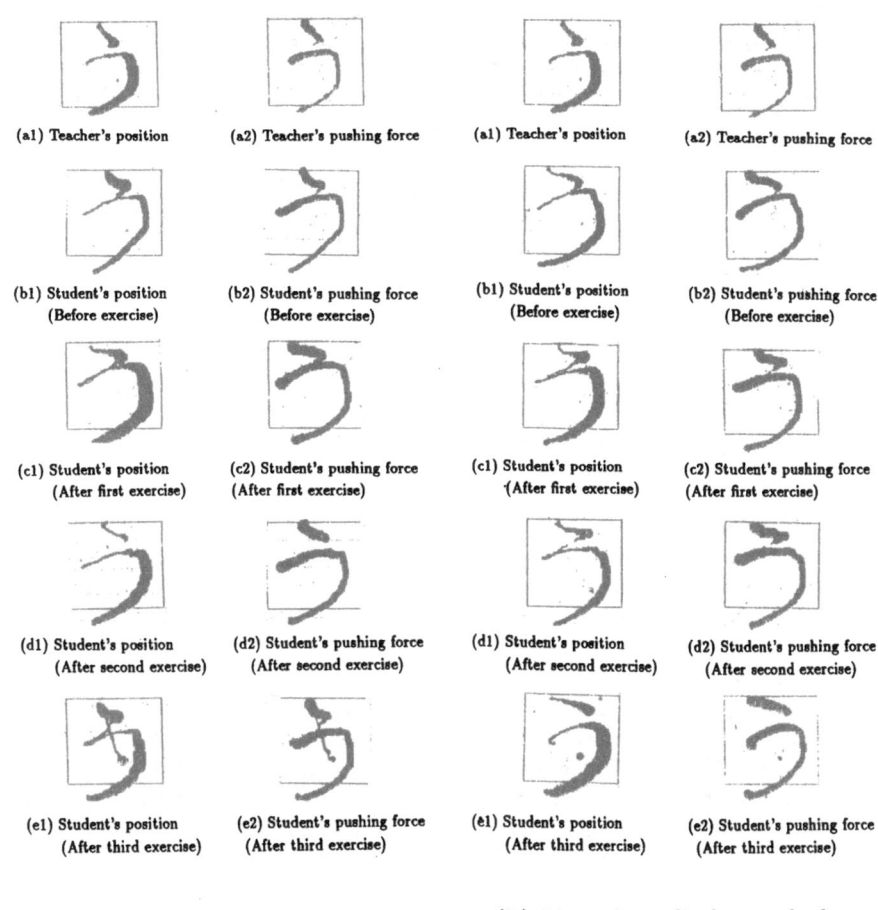

(a1) Teacher's position

(a2) Teacher's pushing force

(a1) Teacher's position

(a2) Teacher's pushing force

(b1) Student's position
(Before exercise)

(b2) Student's pushing force
(Before exercise)

(b1) Student's position
(Before exercise)

(b2) Student's pushing force
(Before exercise)

(c1) Student's position
(After first exercise)

(c2) Student's pushing force
(After first exercise)

(c1) Student's position
(After first exercise)

(c2) Student's pushing force
(After first exercise)

(d1) Student's position
(After second exercise)

(d2) Student's pushing force
(After second exercise)

(d1) Student's position
(After second exercise)

(d2) Student's pushing force
(After second exercise)

(e1) Student's position
(After third exercise)

(e2) Student's pushing force
(After third exercise)

(e1) Student's position
(After third exercise)

(e2) Student's pushing force
(After third exercise)

(A) Direct position display method

(B) Direct force display method

Figure 4. Experimental results

observed in the experiment. However, more detailed quantitative evaluation should be done to examine the effectiveness of the system.

Combination or mixture of the two display methods and/or utilization of some other method such as voice, sound, or vibration for presenting the information that is not directly displayed may also be effective. These will be future research topics.

## References

[1] M. Ouh-young, M. Pique, J. Hughes, N. Srinivasan, and F.P. Brooks, Jr., "Using a Manipulator For Force Display in Molecular Docking," Proc. of the 1988 IEEE International Conference on Robotics and Automation, pp.1824–1829, 1988.

[2] M.L. Agronin, "The Design of a Nine-String Six-Degree-of-Freedom Force-Feedback Joystick for Telemanipulation," Proc. NASA Workshop on Space Telerobotics, pp.341–348, 1987.

[3] H. Iwata, "Artificial Reality with Force-Feedback: Development of Desktop Virtual Space with Compact Master Manipulator," Comput. Graphics, vol.24, no.4, pp.165–170, 1990.

[4] M. Minsky, M. Ouh-young, O. Steele, F.P. Brooks, Jr., and M. Behensky, "Feeling and Seeing: Issues in Force Display," Comput. Graphics, vol.24, no.2, pp.235–243, 1990.

[5] R.D. Howe, "A Force-Reflecting Teleoperated Hand System for the Study of Tactile Sensing in Precision Manipulation," Proc. of the 1992 IEEE International Conference on Robotics and Automation, pp.1321-1326, 1992.

[6] T.H. Massie, "Initial Haptic Explorations with the Phantom: Virtual Touch Through Point Interaction," Master Thesis, Department of Mechanical Engineering, MIT, Feb. 1996.

[7] P. Buttolo and B. Hannaford, "Pen-Based Force Display for Prehension in Virtual Environment," Proc. VRAIS '95, pp.217-224, 1995.

[8] T. Yoshikawa and A. Nagura, "A Touch and Force Display System for Haptic Interface", Proc. of the 1997 IEEE International Conference on Robotics and Automation, pp.3018-3026, 1997.

[9] T. Yoshikawa and H. Ueda, "Haptic Virtual Reality: Display of Operating Feel of Dynamic Virtual Objects," Proc. of the 7th International Symposium of Robotics Research, pp.191-198, 1995.

[10] T. Yoshikawa, Y. Yokokohji, T. Matsumoto, and X-Z. Zheng, "Display of Feel for the Manipulation of Dynamic Virtual Objects," Trans. ASME J. DSMC, vol.117, no.4, pp.554–558, 1995.

[11] T. Yoshikawa, X-Z. Zheng, and T. Moriguchi, "Display of Operating Feel of Dynamic Virtual Objects with Frictional Surface," Transactions of the Japan Society of Mechanical Engineers, vol.60, no.578, pp.103-110, 1994.

[12] T. Yoshikawa and H. Ueda, "Construction of Virtual World Using Dynamic Modules and Interaction Modules," Proc. of the 1996 IEEE International Conference on Robotics and Automation, pp.2358-2364, 1996.

[13] S. Kawamura, M. Ida, T. Wada, and J-L. Wu, "Development of a Virtual Sports Machine Using a Wire Driven System–A Trial of Virtual Tennis," Proc. of the 1995 IEEE/RSJ International Conference on Intelligent Robots and Systems, vol.1, pp.111-116, 1995.

[14] N. Ayache, S. Cotin, and H. Delingette, "Surgery Simulation with Visual and Haptic Feedback," Preprints of the Eighth International Symposium of Robotics Research, pp.92–97, 1997.

[15] K. Henmi and T. Yoshikawa, "Virtual Lesson and Its Application to Virtual Calligraphy System," Proc. of the 1998 IEEE International Conference on Robotics and Automation, pp.1275-1280, 1998.

[16] T. Yoshikawa, "Artificial Skill —Understanding and Mechanization of Skill—," Journal of the Society of Instrument and Control Engineers, vol.37, no.7, pp.465-470, 1998 (in Japanese).

[17] T. Yoshikawa, Foundations of Robotics: Analysis and Control, MIT Press, Cambridge, Mass.1990.

[18] T. Yoshikawa, "Dynamic Hybrid Position/Force Control of Robot Manipulators —Description of Hand Constrtaints and Calculation of Joint Driving Force," IEEE Journal of Robotics and Automation, vol.3, no.5, pp.386-392, 1987.

# Experiments with Web-Based Telerobots

James Trevelyan, Barney Dalton
University of Western Australia
Perth, Australia
jamest@mech.uwa.edu.au, barney@mech.uwa.edu.au

**Abstract:**

This paper reviews some specific developments on which our research into web-based telerobotics has been focussed.

We have operated a web-based telerobot continuously since 1994 (except for a short period for installing a new robot). Remote users operate the robot to stack blocks on a flat table: we estimate that about 250,000 users have controlled the robot since then. Although this is a relatively simple and well structured task, we have had to fill many gaps in our understanding of robotics to make this work reliably.

With access to such a large number of casual users, we can draw some interesting conclusions about user behaviour and the reliability of our equipment and software.

Key aspects which have required considerable research effort have been the means by which the robot software interacts with the user's web browser and improving the usability of the interface. A simplified approach to kinematic calibration using a simple laser arrangement has also helped. Perhaps the most interesting issue is why the robot still attracts so many web users where the concept of flow helps to explain some of the user fascination.

## 1. Web Software

One key technology component is the way in which a telerobot is connected to the internet, most commonly using a world wide web browser for the user interface.

The original web-based robots pioneered simultaneously by Ken Goldberg[1] at the University of Southern California and Ken Taylor[2] at the University of Western Australia relied on CGI (common gateway interface) scripts which are widely used for web pages with dynamic content (eg. search engines). In the early stages the reliability of web servers and some operating systems was a significant weakness, but by early 1996 the web software components had become stable and reliable.

HTTP extended using CGI, and HTML are extremely popular for providing interaction on the Internet, the main reasons being they're simplicity, and complete portability. However they have two significant limitations. HTML provides some basic interface widgets, namely images, input boxes and buttons, but it is not a full GUI environment, and thus is a limitation for telerobotic interfaces. HTTP[3] is a request/response protocol, and thus provides no provision for server initiated communication. Server initiated communication is important in a remote control scenario, where users must receive asynchronous updates of changes at the remote site. A web telerobot requires an environment that supports server initiated communication and full GUI design capabilities, while remaining portable across platforms. In the early days of web telerobotics there was no simple solution, however now the development of JAVA means that applets incorporated within a web page can provide this functionality.

A web-based telerobot requires interactions between one or more users and at least two hardware devices: a robot and a camera to provide images or video to remote user(s) controlling the robot. In practice the number of hardware devices is usually greater: the robot may consist of an arm and related but physically distinct devices such as fixtures, special tools etc. Also, several cameras may be needed, and other sensors such as microphones or force sensors may be used in certain tasks.

With large numbers of devices and users, all potentially connected together by the Internet, there are opportunities for collaborative systems involving users and devices all over the world. There is also the potential to create a large, complex, inflexible, and unmaintainable system. To keep the software simple and maintain flexibility, it is beneficial to treat each device and user as separate software tasks or processes, and not necessarily expect these tasks to run on the same physical computer or operating system.

Given the distributed nature of the system, a communication architecture and protocol must be implemented. For two "peers"[1] in the system to communicate with each other, they must either have a direct connection, or be connected via other routers or proxies that are capable of forwarding there request to the right destination. The direct connection solution requires knowledge of the location of each peer and results in a $n^2$ number of connections. If a single "router"[4] is used for all peers to connect through, then only $n$ connections are required and peers only need know the location of the router.

Communication between peers is modelled as messages in an event delivery system, all requests, status updates, error reports, images etc are messages, this is often referred to as Message Oriented Middleware(MOM)[5]. MOM is designed to support deferred asynchronous communication. This contrasts with the remote procedure call architecture of CORBA[6] and RMI[7] which support synchronous communication. Telerobotics is an application where non-blocking asynchronous behaviour is actually desirable. Remote execution of commands can take a number of seconds and there are likely to be a number of intermediate results. This behaviour is only readily available in a MOM based system. MOM's semantics fit with those of supervisory control, where control signals and results are seen as messages between controller and manipulator. It is important that messages are executed in order, and that each is dependent on the last - this again fits well with the MOM framework. MOM is also especially well suited to situations in which clients collaborate as peers.

---

[1]The term peer is used for both clients and servers

The central MOM router is a simple Java process that maintains a connection to each peer. The router has a number of "channels" which peers publish/subscribe to. The router has a priority queues of pending messages for each peer that runs in their own threads. Access to a particular resource can be managed by the router, with the use of tokens. The router provides a number of services such as, authentication, peer lookup, time synchronisation, and token control.

Peers communicate with each other via the router using a simple message wrapper protocol. This message header contains the message size, the address, channel, priority, message id, reference id, and time the message was created. The rest of the message can contain application specific data and other headers. The router reads the size, address, channel, and priority of the message, replaces the address with that of the senders and places the message in each addressed peers queue. The message protocol is shown in figure 1.

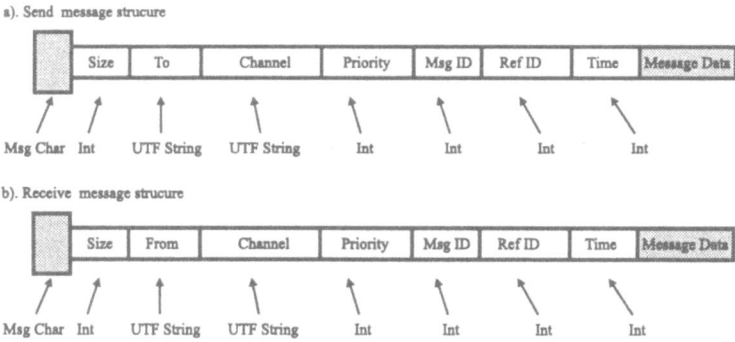

**Figure 1:** *The message wrapper used by all peers to communicate via the router. Message header fields are fixed length to minimise processing time within the router. Before forwarding a message, the router replaces the recipient's address (the 'to' field) with that of the sender.*

If any of the peer processes fail, the router still keeps running and maintains a list of pending messages. The router handles the result of a software reset or other reconnection process and re-attaches the message queue. If the user's computer hangs, or an applet fails, again the router handles the problem so the robot and its associated processes wait for the user to re-connect (for a limited time). Therefore the problems of unreliable software and connections are isolated from each peer, providing consistency and easier implementation.

The architecture of a peer, has the same low level structure for both control of physical devices and client user interfaces. A client for each channel and a transceiver implementation are all that is required. The client can be derived from a generic base class in either Java or C++, this provides the concrete message handling for a given channel. The transceiver remains unchanged across applications (a version has been written in both Java and C++) and is required to marshal/unmarshal data and send/receive data through the physical connection to the router. An example of this architecture applied to the UWA telerobot is shown in figure 2.

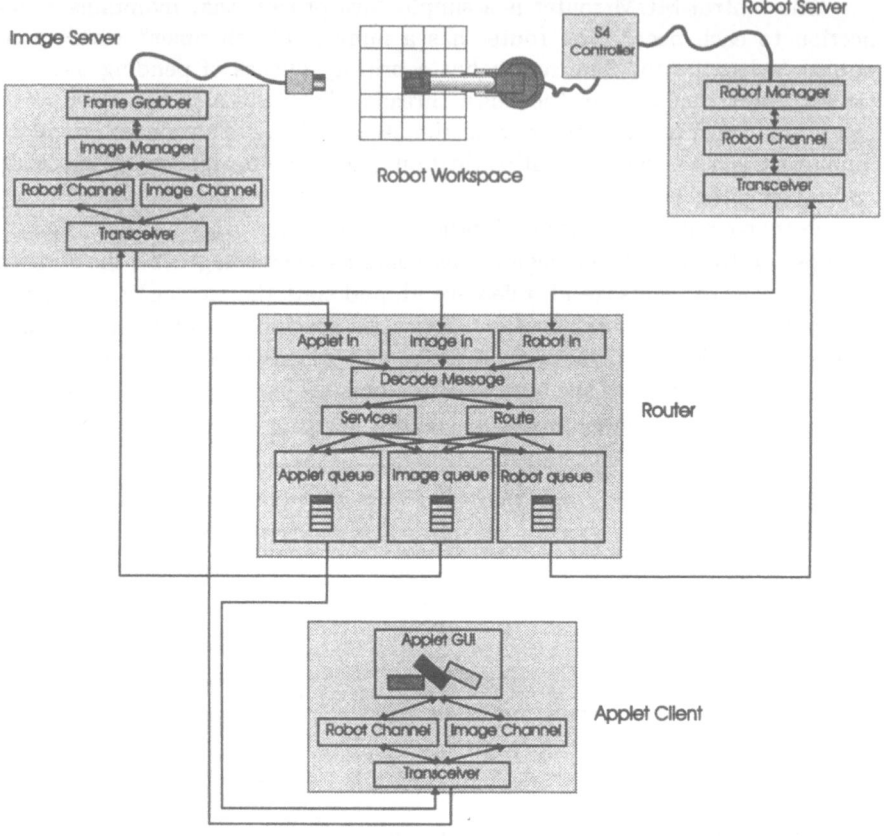

**Figure 2:** *The peer and router architectures shown for a single JAVA client, robot, and image capture device. The robot server subscribes to the robot channel and executes move requests as they are received, access control is managed by the router. New robot states are broadcast by the robot server as the robot is moved. The image server subscribes to the robot channel, and takes images for each new robot state, these images are then broadcast on the image channel. The applet subscribes to the robot and image channels to receive updates and, to submit control requests.*

## 2. Usability

Most user interaction with the robot has required the users to specify a full robot pose with numerical values. The coordinate values have to be estimated with reference to grid lines at 100 mm spacing on the table top.

We first simplified this by reducing three orientation coordinates to two: picking up a block with a simple gripper requires only two degrees of freedom.

We provided users with the option of clicking points in two or more camera images to specify positions, and a primitive virtual reality interface as a further option to make the task easier. Surprisingly few users were interested.

Recently a visiting student devised and built an "augmented reality"[8] interface which superimposes a symbolic representation of the gripper on the camera images. The user can move the gripper symbols in position or orientation by dragging with the mouse[9]. For the time being only a few web browsers can support the level of Java required for this interface so it will take some time to see how popular it becomes. The augmented reality device has been further developed to allow to allow interactive modelling of the environment[10]. Combined with collaborative architecture outlined above this will allow multiple users/agents to work together and model the remote scene. The interactive modelling interface is shown in figure 3, the user is adjusting the orientation of an overlaid block to align with a block in the workspace image.

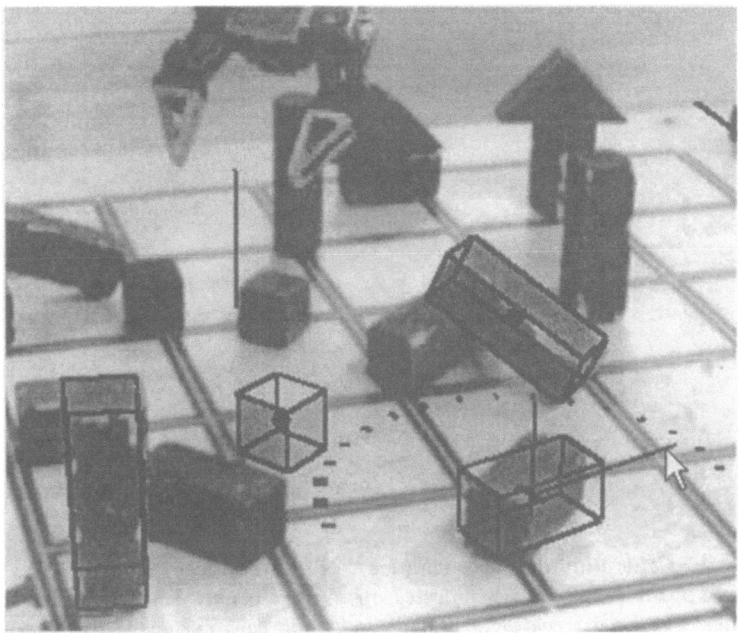

**Figure 3:** *Interactive modelling using augmented reality. This is a Java applet running in a web browser, the user is able to drag and rotate wireframe models of blocks, building up a model of the environment.*

## 3. Calibration

Another key technology is some form of calibration to exploit the repeatability of the robot. Cleary[11] developed an elegant laser calibration device which does not require precision components, yet provides great accuracy in measuring robot pose errors. Positioning errors were reduced to 1.3 times the specified repeatability of the arm over the work table used for the telerobot experiments and 3 times the specified repeatability outside this space.

The calibration device uses a plano-convex lens with reflective coatings arranged such that an incident laser beam produces two reflections: one off the flat front surface, and another off the curved rear surface. In perfect alignment, the two reflections align with the incident beam. Small deviations in position affect only the rear surface reflection, whereas small orientation deviations affect both beams. Four pose errors can be inferred by measuring the deflections of the two reflections.

The laser is arranged such that the beam passes through a small hole in a screen so that the reflections appear on the screen as diffuse spots. An off-axis CCD camera records the spot positions - a computer analyses the images to calculate the measured pose errors. The lab setup is shown in figure 4.

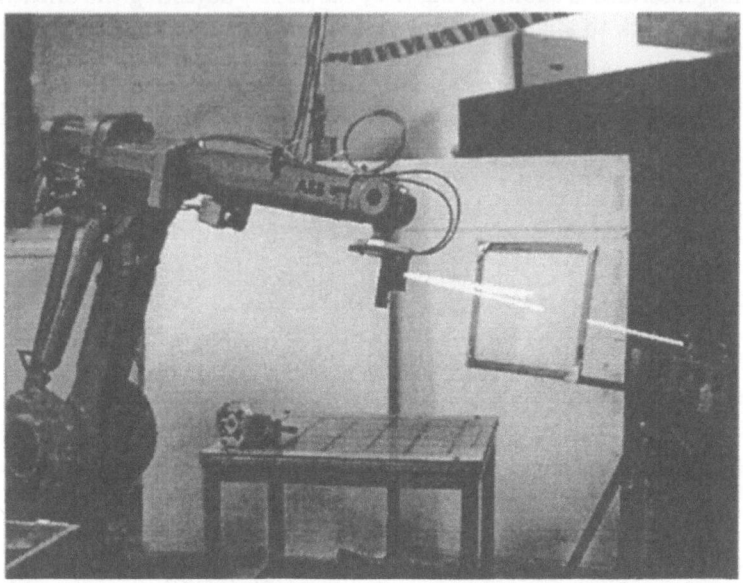

**Figure 4:** *Calibrating the robot with laser beam. The laser beam emerges through a pin hole in the screen and is reflected off a plano-convex lens on the robot gripper mounting. The lens is coated to provide front and back surface reflections. A CCD camera alongside the laser collects images to measure the positions of the reflections on the screen - the spot positions reveal pose errors in the wrist which lead to an improved kinematic model. Repeatable errors were reduced to less than 0.1mm using this technique.*

## 4. Entertainment or Fascination?

Analysis of user behaviour reveals some interesting results. The distribution of the number of robot moves made by users closely matches known statistical distributions[12]. The majority of users make just a few moves, perhaps to satisfy themselves that the robot actually responds to their commands. A tiny minority of users on the other hand use the robot for hours on end, constructing elaborate structures from the blocks. We have speculated that there may be similarities to visitor behaviour at interactive science and technology museums where many pass to another exhibit after a brief turn at the controls, but a few stay fascinated for hours.

The enduring attracting of our web site has been a pleasant surprise and we have speculated that telerobotic devices like this could provide an interesting form of recreation, perhaps entertainment in the future. The robot is in use about 70% of the time, night and day. A twin robot was installed (by Ken Taylor) at the Carnegie Science Centre in 1997 and shows similar usage statistics.

In attempting to explain this we have encountered the concept of "flow" which has been used to explain human behaviour in solitary fascination: intense concentration provides a very satisfying experience often accompanied by distorted time perception. Flow requires a degree of difficulty or challenge to a user to establish control. This might suggest that efforts to make the robot easier to use might be self-defeating unless the task is made correspondingly more difficult to provide sufficient challenge!

## References

[1] K. Goldberg, K. Mascha, M. Genter, N. Rothenberg, C. Sutter, and J. Wiegley. Desktop teleoperation via the world wide web. In *Proceedings of IEEE International Conference on Robotics and Automation*, May 1995.

[2] K. Taylor and J.Trevelyan. Australia's telerobot on the web. In *26th International Symposium on Industrial Robotics*, pages 39–44, Singapore, October 1995.

[3] Hypertext transfer protocol – http/1.1. `http://www.w3.org/Protocols/rfc2068/rfc2068`.

[4] B. Dalton and K. Taylor. A framework for internet robotics. In Roland Seigwart, editor, *IROS Workshop Robots on the Web*, pages 15–22, 1998.

[5] Message-orientated middleware. `http://sims.berkeley.edu/courses/is206/f97/GroupB/mom/`.

[6] The object management group. `http://www.omg.org`.

[7] Java remote method invocation. `http://java.sun.com/products/jdk/rmi/index.html`.

[8] Paul Milgram and Fumio Kishino. A taxonomy of mixed reality visual displays. *IEICE Transactions on Information Systems*, E77-D(12), 1994.

[9] H. Friz, B. Dalton, P. Elzer, and K.Taylor. Augmented reality in internet telerobotics using multiple monoscopic views. In *IEEE Transactions on Systems, Man, and Cybernetics*, pages 354–359, 1998.

[10] B. Dalton, H. Friz, and K. Taylor. Augmented reality based object modelling in internet robotics. In Matthew R. Stein, editor, *Telemanipulator and Telepresence Technologies V*, volume 3524, pages 210–217. SPIE, 1998.

[11] William Cleary. Robot calibration. Bachelors Thesis, Mechanical and Materials Engineering, University of Western Australia, 1997.

[12] K. Taylor and B. Dalton. Issues in Internet telerobotics. In *International Conference on Field and Service Robotics (FSR 97)*, pages 151–157, Canberra, Australia, December 1997.

# A Review of Robot World Cup Soccer Research Issues RoboCup: Today and Tomorrow

Minoru Asada
Adaptive Machine Systems, Osaka University
Suita, Osaka 565-0871, Japan
asada@ams.eng.osaka-u.ac.jp

Manuela Veloso
Computer Science Department, Carnegie Mellon University
Pittsburgh, PA 15213, USA
mmv@school.coral.cs.cmu.edu

Gerhard K. Kraetzschmar
Neural Information Processing, University of Ulm
Oberer Eselsberg, 89069 Ulm, Germany
gkk@neuro.informatik.uni-ulm.de

Hiroaki Kitano
Computer Science Lab, Sony Corp.
Tokyo 141-0022, Japan
and
ERATO Kitano Symbiotic Systems Project, JST
Tokyo 150-0001, Japan
kitano@csl.sony.co.jp

**Abstract:** RoboCup is an increasingly successful attempt to promote the full integration of robotics and AI research. The most prominent feature of RoboCup is that it provides the researchers with the opportunity to demonstrate their research results as a form of competition in a dynamically changing hostile environment, defined as the international standard game definition, in which the gamut of intelligent robotics research issues are naturally involved. This article describes what we have learned from the past RoboCup activities, mainly the first and the second RoboCups, and overview the future perspectives of RoboCup in the next century.

## 1. Introduction

RoboCup (The Robot World Cup Initiative) is an attempt to promote intelligent robotics research by providing a common task for evaluation of various theories, algorithms, and agent architectures [1]. RoboCup has currently chosen soccer as its standard task. In order for a robot (a physical robot or a software agent) to play a soccer game reasonably well, many technologies need to be integrated and a number of technical breakthroughs must be accomplished. The range of technologies spans the gamut of intelligent robotics research, including design principles for autonomous agents, multi-agent collaboration, strategy acquisition, real-time reasoning and planning, robot learning, and sensor-fusion.

The First Robot World Cup Soccer Games and Conferences (RoboCup-97) was held during the International Joint Conference on Artificial Intelligence (IJCAI-97) at Nagoya, Japan with 37 teams around the world, and the Second Robot World Cup Soccer Games and Conferences (RoboCup-98) was held on July 2-9, 1998 at La Cite des Sciences et de I'Industrie (La Cite) in Paris with 61 teams. RoboCup-99 Stockholm will be held in conjunction with IJCAI-99 participated in by over 120 teams. A series of technical workshops and competitions have been planned for the future. While the competition part of RoboCup is highlighted in the media, other important RoboCup activities include technical workshops, the RoboCup Challenge program (which defines a series of benchmark problems), education, and infrastructure development. As of December 1998, RoboCup activity involves thousands of researchers from over 36 countries. Further information is available from the web site: **http://www.robocup.org/**

Figure 1. RoboCup-98 in Paris

RoboCup has currently three kinds of leagues: (1)the simulation league, (2) the real robot small-size league, and (3) the real robot middle-size league. (Please see Figure 1 (left) for the real robot middle-size league competition site)

In this article, we review the challenge issues of real robot leagues and analyze the results of RoboCup-98. We compare the architectural differences between the leagues, and overview which research issues have been solved and how, and which have been left unsolved and remain as future issues.

## 2. Research Issues and Approaches

In this section, we discuss several research issues involved in the development of robots for RoboCup. One of the major reasons why RoboCup attracts so many researchers is that it requires the integration of a broad range of technologies into a team of complete agents, as opposed to a task-specific functional module. The following is a partial list of research areas which RoboCup covers:

- Agent architecture in general

- Combining reactive approaches and modeling/planning approaches

- Real-time recognition, planning, and reasoning

- Reasoning and action in a dynamic environment

- Sensor fusion

- Multi-agent systems in general

- Behavior learning for complex tasks

- Strategy acquisition

- Cognitive modeling in general

Currently, each league has its own architectural constraints, and therefore research issues are slightly different from each other. We have published proposal papers [2, 3] about research issues in RoboCup initiative. For the robotic agents in the real robot leagues, for both the small and middle-size ones, the following issues are considered:

- Efficient real-time global or distributed perception possibly from different sensing sources.

- Individual mechanical skills of the physical robots, in particular target aim and ball control.

- Strategic navigation and action to allow for robotic teamwork, by passing, receiving and intercepting the ball, and shooting at the goal.

More strategic issues are dealt in the simulation league and in the small-size real robot league while acquiring more primitive behaviors of each player is the main concern of the middle-size real robot league.

We held the first RoboCup competitions in August 1997, in Nagoya, in conjunction with IJCAI-97 [4]. There were 28, 4, and 5 participating teams in the simulation, small-size robot, and middle-size robot leagues, respectively. The second RoboCup workshop and competitions took place in July 1998, in Paris [5] in conjunction with ICMAS-98 and AgentsWorld. The number of teams increased significantly from RoboCup-97 to 34, 11, and 16 participating teams in the simulation, small-size robot, and middle-size robot leagues respectively. More than twenty countries participated. Every team had its own features some of which have been exposed during their matches with different degrees of success.

## 2.1. Arcitectural Analysis

There are two kinds of aspects in designing a robot team for RoboCup:

1. Physical structure of robots: actuators for mobility, kicking devices, perceptual (cameras, sonar, bumper sensor, laser range finder) and computational (CPUs, microprocessors) facilities.

2. Architectural structure of control software.

In the simulation league, both of the above issues are fixed, and therefore more strategical structure as a team has been considered. On the other hand, in the real robot leagues, individual teams have devised, built, and arranged their robots. Although the small league and the middle one have their own architectural constraints, there are variations of resource assignment and control structure of their robots. Table 1 shows the variations in architectural structure in terms of number of CPUs and cameras, and their arrangement.

Table 1. Variations in architectural structure

| Type | CPU | Vision | league |
|------|-----|--------|--------|
| A | 1 | 1 global | small-size |
| B | $n$ | 1 global | small-size |
| C | 1 | 1 global + $n$ local | small-size |
| D | 1+n | $n$ local | middle-size |
| E | $n$ | $n$ local | middle-size & simulation |

Three types A, B, and C inidicate a variation adopoted in the real robot small-size league. Type A is a typical structure many teams used in this league: the centralized contorl of multiple bodies through a global vision. Type B is a kind of multiagent system in which decision making is distributed and independent from each other although they share the global vision. CMUnited-98 in the small-size league took this sort of architecture. Type C features sensor coordination of global and local views based on the centralized control with multiple bodies. I-space (a joint team of Utsunomiya Univ. and Univ. of Tokyo, Japan) in the small-size league adopted this type architecture.

On the other hand, type E is a typical architecute adopted in both the simulation league and the real robot middle-size league: a completely distributed multiagent system. Type D used C. S. Freiburg team in the middle-size league adopted a combination of types A and E utilizing laser range finders mounted on players which make it possible to reconstruct the global view and to localize observed obejcts (teammates, opponents, and ball) in the field. That is, they changed the problem in the middle-size league into one in the small-size league.

Communication between agents is possible in all of the leagues. The simulation league is the only that uses it except one team Uttori in the middle-size league. In the following, we attempt to analyze the achivemnets in RoboCup-97 and 98 in the real robot leagues.

## 3. Small-Size Real Robot League

The RoboCup-98 small-size real robot league provides a very interesting framework to investigate the full integration of action, perception, and high-level reasoning in a team of multiple agents. Therefore, three main aspects need to be addressed in the development of a small-size RoboCup team: (i) hardware of physical robots; (ii) efficient perception; and (iii) individual and team strategy.

Although all of the eleven RoboCup-98 teams included distinguishing features at some of these three levels, it showed crucial to have a *complete* team with robust hardware, perception, and strategy, in order to perform overall well. This was certainly the case for the four top teams in the competition, namely CMUnited-98 (USA), Roboroos (Australia), 5DPO (Portugal), and Cambridge (UK), who classified in first, second, third, and fourth place respectively.

Figure 1 (right) shows a scene from the final match between CMUnited-98 and Queensland Roboroos (Ausuralia). We overview now the characteristics of the RoboCup-98 teams and the research issues addressed.

**Hardware:** All of the eleven RoboCup-98 participating teams consisted of robots built by each participating group. The actual construction of robots within the strict size limitations offered a real challenge, but gave rise to a series of interesting physical and mechanical devices. Remarkably, the robots exhibited sensor-activated kicking devices (iXs and J-Star, Japan, Paris-6, France, and CMUnited-98, USA), sophisticated ball holding and shooting tools for the goalie robot (Cambridge, UK), and impressive compact and robust designs (Roboroos, Australia, and UVB, Belgium). All of the robots were autonomously controlled through radio communication by off-board computers.

**Perception:** Ten out of the eleven teams used a single camera overlooking the complete field. The ISpace (Japan) team included one robot with an onboard vision camera.

Global perception simplifies the sharing of information among multiple agents. However global perception presents at the same time a real challenging research opportunity for reliable and real-time detection of the multiple mobile objects – the ball, and five robots on each team. In fact, both detection of robot position and orientation and robot tracking need to be very effective. The frame rate of the vision processing algorithms clearly impacted the performance of the team. Frame rates reached 30 frames/sec as in the CMUnited-98 team.

In addition to the team color (blue or yellow), most of the teams used a second color to mark their own robots and provide orientation information, hence only about their own robots. Robot identification was achieved in general by greedy data association between frames. The 5DPO (Portugal) and the Paris-6 (France) teams had a robust vision processing algorithm that used patterns to discriminate among the robots and to find their orientation.

The environment in the small-size league is highly dynamic with robots and the ball moving at speeds between 1m/s and 2m/s. An interesting research issue consists of the prediction of the motion of the mobile objects to combine it with strategy. It was not clear which teams actually developed prediction algorithms. In the particular case of the CMUnited-98 team, prediction of the movement of the ball was successfully achieved and highly used for motion (e.g., ball interception) and strategic decisions (e.g., goaltender behavior and pass/shoot decisions).

**Motion:** In this RoboCup league, a foul should be called when robots push each other. This rule offers another interesting research problem, namely obstacle avoidance and path planning in a highly dynamic environment. The majority of the teams in RoboCup-98 successfully developed algorithms for such difficult obstacle avoidance and the semi final and final games showed smooth games that demonstrated impressive obstacle avoidance algorithms.

**Strategy:** Following up on several of the research solutions devised for RoboCup-97 both in simulation and in the small-size robots, at RoboCup-98, all of the small-size teams showed a role-based team structure. As expected, the goaltender played a very important role in each team. Similarly to the goaltender of CMUnited-97, the goaltender of most of the teams stayed parallel to the goal line and tried to stay aligned with or intercept the ball. The goaltender represented a very important and crucial role. To remark were the goaltenders of Roboroos, CMUnited-98, and Cambridge.

Apart for CMUnited-98 which had a single defender and three attackers, most of the other teams invested more heavily on defense, assigning two robots as defenders. In particular, defenders in the Belgium and in the Paris-8 teams occupied key positions in front of the goal making it difficult for other teams to path plan around them and to try to devise shots through the reduced open goal areas. Defending with polygonally-shaped robots proved to be hard, as the ball is not easily controlled at a fine grain. In fact a few goals for different teams were scored into their own goals due to small movements of the defenders or goaltender very close to the goal. It is clearly still an open research question how to control the ball more accurately.

Finally, it is interesting to note that one of the main features of the winning CMUnited-98 team is its ability to collaborate as a team. Attacking robots continuously evaluate (30 times per second) their actions, namely either to pass the ball to another attacking teammate or to shoot directly at the goal. A decision-theoretic algorithm is used to assign the heuristic and probabilistic based values to the different possible actions. The action with the maximum value is selected. Furthermore, in the CMUnited-98 team, a robot who was not the one actively pursuing the ball is not merely passive. Instead each attacking robot *anticipates* the needs of the team and it positions itself in the location that maximizes the probability of a successful pass. CMUnited-98 uses a multiple-objective optimization algorithm with constraints to determine this strategic positioning. The objective functions maximize repulsion from other robots and minimize attraction the ball and to the attacking goal.

## 4. Middle-Size Real Robot League

The performance of robot behaviors in RoboCup-98 was better than in RoboCup-97 although the number of teams in the middle-size league drastically increased from 5 to 16, more than three times. However, the level of skills is under development, mainly putting more focus on individual behavior acquisition than cooperative teamwork. Engineering issues such as precise robot control and robust object detection are still main issues in this league.

### 4.1. Technological State-of-the-Art:

1. **Platforms:** Many of the new teams used off-the-shelf platforms, such as Activmedia's Pioneer-1 and Pioneer-AT robots (one team in RoboCup-97 and six teams in RoboCup-98) or Nomadics' Scout robot (one team in RoboCup-98). There is a trade-off between the use of well-equipped off-the-shell platforms and the building of originally designed ones such as RMIT, Uttori and so on. The former seems is easy to use immediately but less flexible while the latter seems more flexible but to consume much more time to build. Several teams such as Osaka, USC, and NAIST used the same cheap platforms, radio-controlled toy cars, but fairly modified them in different ways.

2. **Sensors:** Vision remains a central research topic in RoboCup. Recent progress of PC-based image processors enabled many teams to install them on-board, owing to which the better performance of robot behaviors were shown in RoboCup-98 than in RoboCup-97. In addition to image processor, sonar and bumper sensors, a laser range finder was introduced to reconstruct a global view of the field. However, there are still many problems with the perceptual capabilities of the robots, especially detecting other agents.

3. **Kicking mechanisms:** Some teams such as Freiburg, Ulm, and Uttori introduced their kicking devices on their robots based on pneumatic or solenoid-based activation in RoboCup-98. The kicking devices produced much higher ball accelerations than pushing the ball, and as a result, such robots could move the ball significantly better, which is one of the research issues in this league.

### 4.2. Research Results:

Since it seems difficult to survey the research issues attacked by the all teams in the middle size league, we show some of them which seems essential for RoboCup.

#### 4.2.1. Behavior Acquisition

Since the engineering issues such as precise motion control and robust vision are still main concerns in this league, many teams explicitly specify robot behaviors as a form of IF-THEN rules from a viewpoint of designers view. Trackies (Osaka

Univ., Japan) has been focusing learning and evolutionary approaches, and the work in [6] has made reinforcement learning applicable to dynamically changing multiagent environment after state vector estimation based on a method of system identification. Further, they applied genetic programming [7] to emerge turning around the ball behavior which is difficult to acquire by reinforcement learning because of loss of ball sight.

Werger [8] (Ullanta Performance Robotics) adopted a simple behavior co-ordination based on the subsumption architecture and well-considered physical constraints on the relationship between robot motions and the environment. During the matches, both two teams showed some good performance based on their methods.

### 4.2.2. Vision

All teams used vision as a main source of external sensing for individual robots. Most on-board cameras are fixed to the body, that is no active vision. There-fore, many robots rotate their bodies to expand robots' views. Two alternatives are considered. One is to use active vision, that is, panning a camera. NAIST's players have on-board panning cameras, and their searching behaviors (obser-vation) apart from body motions seem to have been explicitly programmed. One of Osaka's players also has a panning camera and they attempted to emerge a behavior combined with camera and body motions based on rein-forcement learning, but still under development except for simple situations. Active camera control is one of the essential issues because it is closely related to attention mechanism. Although this problem has not been intensively ad-dressed yet, RoboCup can provide a good testbed on this issue, that is, the role of attention will be made clear in the global context.

The other one is to use multiple or omni-directional vision. USC's Dream team-98 [9] put two cameras orienting to the opposite directions each other onto one player to extend the single robot's view. Omni-directional vision was first used by Osaka in RoboCup-97 in order for a goal keeper both to track their position in front of the goal as well as ball position [10] In RoboCup-98, more teams such as Italian and Australian adopted this sort of vision system. Again, Osaka attacked to emerge goal-keeping behavior by reinforcement learning, and showed an aggressive behavior by not simply goal saving but also pushing the ball ahead expecting to pass the ball to one of its teammates in future.

### 4.2.3. Environment Model and Localization

Currently, many teams used the geometric field model to localize their robots. For example, USC's Dream team used encoder of the robots to roughly estimate its position in the field to switch its behavior. However, there is no feedback, which results in accumulation of position errors. Then, visual information may correct these errors by capturing landmarks such as goals.

C. S. Freiburg [11] used laser range finders which provides fast and accurate range data. Each robot sends its range data to the host computer, which reconstructs the global field view including teammates and opponents based on these data and makes a plan, then informs such global information with action plan to each robot. As a result, their approach seems to be much more based on global positioning by LRF and centralized control (Type D in Table 1) although each player has its own CPU to detect a ball and to control its body.

Aside from the use of the geometric model of the field, one can approach the issue how to represent the world as one of cognitive model. From sensorimotor mapping to higher level cognition process, the robot may not need to have an explicit world model like a geometric map. Instead, some sort of internal representation can be obtained, which implicitly represents the relationship between sensorimotor mapping and robot environment [12, 13].

### 4.2.4. Communication

Communication between players is definitely a very important part of multi-agent coordination. Explicit communication lines between players were used in only Uttori team [14] via WEB-LAN. Players exchange the information and action commands to cooperate each other such as passing or saving the goal together based on the pre-specified communication protocol.

More interesting approach can be considered following human players behavior. Observation and action can be regarded as message receiving and passing if the robot can predict other robot actions through visual information. In order to acquire such capability, Uchibe et al. [15] proposed a state vector estimation method so that reinforcement learning can be applied to multiagent behavior learning. That is, implicit communication.

## 5. Conclusion

As a grand challenge, RoboCup is definitely stimulating a wide variety of approaches, and has produced rapid advances in key technologies. With a growing number of participants RoboCup is set to continue this rapid expansion. Although we have addressed the issues in real robot leagu, RoboCup researchers face an unique opportunity to learn and share solutions in three different agent architectural platforms with its three leagues,

## References

[1] H. Kitano, M. Asada, Y. Kuniyoshi, I. Noda, E. Osawa, and H. Matsubara. "robocup: A challenge problem of ai". *AI Magazine*, 18:73–85, 1997.

[2] Hiroaki Kitano, Milind Tambe, Peter Stone, Manuela Veloso, Silvia Coradeschi, Eiichi Osawa, Hitoshi Matsubara, Itsuki Noda, and Minoru Asada. The robocup synthetic agent challenge 97. In Hiroaki Kitano, editor, *RoboCup-97: Robot Soccer World Cup I*, pages 62–73. Springer, Lecture Note in Artificail Intelligence 1395, 1998.

[3] Minoru Asada, Peter Stone, Hiroaki Kitano, Aalexis Drogoul, Dominique Duhaut, Manuela Veloso, Hajime Asama, and Shoji Suzuki. The robocup physical agent challenge: Goals and protocols for phase i. In Hiroaki Kitano, editor, *RoboCup-97: Robot Soccer World Cup I*, pages 41–61. Springer, Lecture Note in Artificail Intelligence 1395, 1998.

[4] Hiroaki Kitano, editor. *RoboCup-97: Robot Soccer World Cup I*. Springer, Lecture Note in Artificail Intelligence 1395, 1998.

[5] M. Asada, editor. *Proc. of the second RoboCup Workshop*. The RoboCup Federation, 1998.

[6] Minoru Asada, Eiji Uchibe, and Koh Hosoda. Cooperative behavior acquisition for mobile robots in dynamically changing real worlds via vision-based reinforcement learning and development. *Artificial Intelligence*, 77:(to appear), 1999.

[7] Eiji Uchibe, Masateru Nakamura, and Minoru Asada. Co-evolution for cooperative behavior acquisition in a multiple mobile robot environment. In *Proc. of IROS'98*, pages 425–430, 1998.

[8] Barry Werger. Cooperation without deliberation: Minimalism, stateless, and tolerance in multi-robot teams. *Artificial Intelligence*, 77:(to appear), 1999.

[9] Wei Min Shen, Jafar Adibi, Rogelio Adobbati, Srini Lanksham, Hadi Moradi Benham Salemi, and Sheila Tejada. Integrated reactive soccer agents. In *RoboCup-98:Proceedings of the second RoboCup Workshop*, pages 251–264, 1998.

[10] Shóji Suzuki, Tatsunori Katoh, and Minoru Asada. An application of vision-based learning for a real robot in robocup learning of goal keeping behaviour for a mobile robot with omnidirectional vision and embeded servoing. In *RoboCup-98:Proceedings of the second RoboCup Workshop*, pages 281–290, 1998.

[11] Steffen Gutmann and Bernhard Nebel. The cs freiburg team. In *RoboCup-98:Proceedings of the second RoboCup Workshop*, pages 451–458, 1998.

[12] Minoru Asada, Shoichi Noda, and Koh Hosoda. Action based sensor space segmentation for soccer robot learning. *Applied Artificial Intelligence*, 12(2-3):149–164, 1998.

[13] Y. Takahashi, M. Asada, and K. Hosoda. Reasonable performance in less learning time by real robot based on incremental state space segmentation. In *Proc. of IEEE/RSJ International Conference on Intelligent Robots and Systems 1996 (IROS96)*, pages 1518–1524, 1996.

[14] K.Yokota, K. Ozaki, N. Watanabe, A. Matsumoto, D. Koyama, T. Ishikawa, K. Kawabata, H. Kaetsu, and H. Asama. Cooperative team play based on communication. In *RoboCup-98:Proceedings of the second RoboCup Workshop*, pages 491–496, 1998.

[15] E. Uchibe, M. Asada, and K. Hosoda. "state space construction for behavior acquisition in multi agent environments with vision and action". In *Proc. of ICCV 98*, pages 870–875, 1998.

# A Portable Programming Framework

Eve Coste-Maniere, Nicolas Turro
INRIA Sophia Antipolis
BP93 - 06902 Sophia Antipolis Cedex - France
Eve.Coste-Maniere@inria.fr, Nicolas.Turro@inria.fr

Oussama Khatib
The Robotics Lab
Stanford University
khatib@cs.stanford.edu

**Abstract:**

Programming is a key aspect in the development and practical use of complex robotic systems. The work presented in this paper describes developments conducted within INRIA's ORCCAD and MAESTRO programming framework for a robotic application, in which the Stanford mobile platform provides assistance to humans for performing physical tasks. This type of applications underlines the need for a well-designed programming architecture, without which such development becomes easily inextricable. The objective of this work is to study the portability and generality of the ORCCAD and MAESTRO framework by an implementation on an unfamiliar field of application (robotic assistance) and on a non-native platform.

## 1. Introduction

Today, robots are making inroads into new applications that require a wide range of skills and can be envisioned among humans in their daily life. The development of competent, practical systems that are seamless in operation and easy to use in such context rely upon theoretical progress in mechanics, sensing, automatic control, programming, together with advances in computing and computer power. Managing such a complexity can no longer rely on classical programming approaches and demands the adoption of a programming framework to structure the application design and integrate all these skills.

The central issues of the paper are to demonstrate the adequacy, generality and portability of the ORCCAD [2] and MAESTRO [3] programming framework developed at INRIA by implementing an instance of this class of unfamiliar field of application on a non-native platform: SAMM, the Stanford Assistant Mobile Manipulator [1], involved in an assistance application in which it provides help a human to perform physical tasks.

When designing at large complex applications (with coordinated control of the arm and base of the robot, and a proper managing of the interaction with end-user), the programmer is faced with multiple issues belonging to various domains. To cope with the complexity of such an application, our effort has focused on expanding the features of the experimental platform with the design methodology and tools provided in ORCCAD and MAESTRO in the context of autonomous robotics.

The paper is organized as follows. The first part introduces the requirements that the programmer must consider when implementing such an application and describes the solutions (methods and tools) proposed to meet these requirements. They adopt *automatic control theory* to describe vehicle and arm motions and *reactive system theory* to model the logic of the application. The programming framework (ORCCAD and MAESTRO) that integrates both aspects in then thoroughly presented. The second part of the paper demonstrates the suitability of our programming framework to tackle an unfamiliar field of application on a non-native platform through the set up of a representative human friendly manipulation application. A discussion and trends for the future conclude the paper.

## 2. Methodology and tools

### 2.1. Requirements for application design

When dealing with the design of a robot complex application, the programmer must cope with various requirements that range from automatic control to software engineering, such as reuse, efficiency, extensibility, and validation [2]. We focus in this section on modeling robot/human interactions and safety issues which are of prime importance for human-friendly applications.

#### 2.1.1. Human interaction modeling

In term of robot-human interaction, we propose to define and clearly separate *continuous interactions* and *discrete interactions*.

**Continuous interactions**   A *continuous* exchange of information between a human and the robot for a long period of time with respect to the servoing frequency (i.e. generally lasting several seconds) qualifies a continuous interaction that can be perceived through the robot's sensors. For example, a human can guide the robot by applying a force on its end effector or the robot provides the human with a description of moving objects within the operating field.

This kind of interactions can result in a sensor information (force, distance, image features etc) that can be used within the control laws and leads to sensor-driven control laws (force and redundant control as mentioned in section 2.2.1). It may also require a data flow scheme dedicated to the processing and display of real time sensory information.

**Discrete interactions**   The human-robot interactions can also consist of a low-bandwidth (or *discrete*) exchange of data with a meaningful semantic: discrete orders can be given to the robot via a set of buttons, vocal orders, or gestures. This kind of information can be used either to switch from a control mode to another (e.g. from motionless to human following), or to slightly modify the control law currently performed by changing gains or parameters for example. On the other hand, the robot through its control system can express some discrete facts, with regards to its own monitoring, e.g. "insertion successfully performed," or in relation to the environment, e.g. moving target or obstacle detection. In both cases, capabilities for servoing tasks preemption and synchronizations (with end-user, between base and arm) are required.

The real time process in charge of the *detection/production* of this discrete "events" may either be the control law itself, or an external process not related to the robot servoing that makes use of a dedicated interface (computer control panel, or voice recognition).

### 2.1.2. Safety requirements

Although safety is a general concern of the robotic domain, it takes a new dimension when robots must be used among humans for which every effort should be made to ensure that the system is as safe as it possibly can be. In this context, the safety needs can be divided into *intrinsic* and *application dependent* properties:

- The *intrinsic properties* concern both *numerical values*, e.g. maximum velocities allowed for the robot, or control stability margin, and *logical conditions*, e.g. situations where the arm is controlled with two different control laws at the same time or not controlled at all must never occur.

- The *application dependent needs* may be expressed in terms of logical properties: we might want to check, for instance, that whenever a human is in the range of the robot, the robot's arm must be immobile. And this should be checked before launching the execution of the system.

### 2.1.3. Approach

From this analysis, the need of a hybrid framework merging continuous and discrete aspects arises: continuous aspects to allow the description of the automatic control laws and of the continuous interactions with the end user, and discrete aspects for modeling task phases and switching, task synchronization when running in parallel and discrete interaction with the end-user.

For best efficiency, the handling of those two aspects should be done thru domain-specific methods. Those methods should be integrated coherently in a dedicated tool that exploits domain specific knowledge to render the programming incremental and intuitive. In the following, we propose a design framework that fulfills those requirements.

### 2.2. The methodology

As explained here-after, the proposed methodology to tackle human friendly applications uses innovative control-laws, a hybrid formalization, and its associated tools. The formal grounds of our methodology are based upon:

- The *automatic control theory* to formalize the continuous servoings and he continuous interactions with the end user.

- The *reactive systems theory* to model the the phases of the tasks and switching, task synchronization when running in parallel and discrete interactions with the end-user. In other words to model the logical behavior of the application. Moreover, this theory opens the gate to program verification.

### 2.2.1. Control laws design

The Stanford's Manipulation Group has developed various methodologies for task-oriented control of mobile manipulator platforms. These include: arm-vehicle coordination with dynamically decoupled manipulation and posture control, human-guided compliant motion control, and multiple-platform cooperation. The dynamic behavior at the end-effector of a mobile manipulator platform is describe by

$$\Lambda(\mathbf{q})\ddot{\mathbf{x}} + \mu(\mathbf{q}, \dot{\mathbf{q}}) + \mathbf{p}(\mathbf{q}) = \mathbf{F}; \qquad (1)$$

where $\mathbf{q}$ is the $n$ joint coordinates, $\mathbf{x}$ is the $m$ operational coordinates, $\Lambda(\mathbf{x})$ is the $m \times m$ kinetic energy matrix associated with the operational space. $\mu(\mathbf{q}, \dot{\mathbf{q}})$, $\mathbf{p}(\mathbf{x})$, and $\mathbf{F}$ are respectively the centrifugal and Coriolis force vector, gravity force vector, and generalized force vector acting in operational space.

The overall control structure for the integration of mobility and manipulation is based on the following decomposition of joint torques

$$\Gamma = J^T(\mathbf{q})\mathbf{F} + N^T(\mathbf{q})\Gamma_{\text{posture}}, \qquad (2)$$

where

$$N(\mathbf{q}) = \left[I - \bar{J}(\mathbf{q})J(\mathbf{q})\right] \qquad (3)$$

with

$$\bar{J}(\mathbf{q}) = A^{-1}(\mathbf{q})J^T(\mathbf{q})\Lambda(\mathbf{q}); \qquad (4)$$

$\bar{J}(\mathbf{q})$ is the *dynamically consistent generalized inverse*, [1], which minimizes the robot kinetic energy. This relationship provides a decomposition of joint forces into two control vectors: joint forces corresponding to forces acting at the effector, $J^T\mathbf{F}$, and joint forces that only affect the robot posture, $N^T\Gamma_{\text{posture}}$. To control the robot for a desired posture, the vector $\Gamma_{\text{posture}}$ is selected as the gradient of a potential function $V_{\text{posture}}$ constructed to meet the desired posture specifications. The gradient of this function

$$\Gamma_{\text{posture}} = -\nabla V_{\text{posture}}; \qquad (5)$$

provides the required attraction to the desired posture of the manipulator. With this posture behavior, the explicit specification of the associated motions is avoided, since desired behaviors are simply encoded into specialized potential functions for various types of operations. The interference of this gradient with the end-effector dynamics is avoided by projecting it into the dynamically consistent null space of $J^T(\mathbf{q})$, i.e. $N^T(\mathbf{q})\Gamma_{\text{posture}}$.

The control of free motion, constrained motion, force control, and human guided motions are all achieved by the task control vector $\mathbf{F}$.

### 2.2.2. Application Programming with MAESTRO/ORCCAD

This approach sets up a two layer software architecture. At the lowest level, ORCCAD [2] is dedicated to the implementation of elementary robotic actions where *automatic control* issues are of prime importance and strongly connected with *discrete-event* aspects. At the highest level, MAESTRO [3], a robotic specific programming language, focuses on the logical and temporal arrangement of the actions in order to build complex applications in terms of events and actions with an underlying formalization that takes benefit of the discrete-event theory to tackle reactive systems. The common discrete event theory ensure the consistency of this architecture.

**Features of the low-level** ORCCAD defines a ROBOT-TASK as the elementary robot action to be specified and implemented at this level. Formally, a ROBOT-TASK is defined as the parameterized specification of an elementary control law, that is to say the activation of a control scheme structurally invariant along the task duration (continuous aspects) and of a logical behavior associated with a set of signals (events) which may occur just before, during, and just after the task execution (discrete aspects).

With a dedicated interface that insulates a non-expert user from the technical aspects related to the real-time using automatic code generation, the implementation of a control-law is organized around a *data-flow diagram*, where modules communicate through input/output data ports. Each box represents one processing step of the control law. Modules can belong to three classes: the *algorithm* class to implement the stages of the control law; the *physical resource* class which are used as an interface between the control-law implementation and the physical robot or a simulator. This kind of module usually encapsulate a driver and therefore is the only non-portable part of the control scheme; and finally the *robot task automaton* class parameterized with events the ROBOT-TASK's generic behavior.

To facilitate the specification of the associated logical part, the events and their subsequent handling are typed: *pre-conditions*, *exceptions* corresponding to three different types of recovery procedures and *post-conditions*. The reception of these events controls the evolution of the action according to a pre-defined scheme: roughly speaking, the satisfaction of the pre-conditions leads to the activation of the control law; during its execution, if a specified exception occurs, it is handled according to its severity (the exceptions may have a scope limited to the ROBOT-TASK itself, may require a task switching or may lead to the abortion of the application); the reception of post-conditions implies the ending of the action.

**Features of the high-level** At this level, MAESTRO has been designed as a domain specific language whose objectives are:

- to enable the specification of a robotic application in terms of elementary actions and events, using an adequate syntax
- to generate an effective reactive controller for the real time execution and control of the whole application
- to enable formal verification on the programs

Behind a user-friendly syntax, sophisticated protocols are implemented for task synchronization, parallelism and sequencing. They use an accurate modeling of the elementary actions and are consistent with ORCCAD's ROBOT-TASKS. This subtle action management aims to minimize the time during which the robot is uncontrolled when a control mode switch occurs (thus taking into account non negligible initialization time for tasks). To implement this logical behavior, we use the reactive systems theory, and more precisely the synchronous approach, which leads to complex automata. In order to ease the application programming, those automata are automatically generated from MAESTRO programs and hidden from the mission programmer. Furthermore, determinism is ensured. It permits the use of theorem proving and model checking methods to prove some logical properties on the application behavior. Some domain specific coherency verifications will be performed automatically (e.g. checking that each physical resource is not controlled more than once at a time). Some application specific verifications can also be led by the programmer, checking the event/action dependencies, in order to find behavior not consistent with the application requirements.

## 3. Experimental demonstrations

We now discuss the generality and portability of the programming framework by implementing this mission on an human-friendly robotics application.

### 3.1. Applicative context

In order to materialize our objectives, we define an experiment, ahead of any application ever carried out by our two teams.

#### 3.1.1. Platform capabilities

The so-called SAMM (Stanford Assistant Mobile Manipulator system) platform, is made of the two robots ROMEO and JULIET. These robots are identical and each consists of a holonomic mobile base equipped with two sonar belts. A PUMA 560 arm, equipped with a gripper and a wrist force sensor. The platforms are fully autonomous in terms of power, computing, and communication. Each base houses a set of batteries, radio Ethernet link, and two Pentium-PC. One processor is dedicated to real-time control and runs under the VxWORKS OS. The second one is used for communication and less critical control, under Linux OS (see [1] for further details).

#### 3.1.2. Typical scenario

We now describe the targeted human-friendly application suitable to these hardware platforms. The purpose of the robot is to help a human during heavy manipulations. It should be able, upon user's request, to take an object, carry it into a given direction and perform a precise placement, guided by the operator.

During our preliminary implementation, as strict safety requirement was imposed, we specified that whenever a human is in the vicinity of the robot, the base of the robot stays immobile, and the arm's velocity remains bounded.

### 3.2. Implementing the application with ORCCAD and MAESTRO

In this section, we illustrate the implementation of the above-defined concepts. The experiments were led bottom-up, first installing and testing the low-level tools (ORCCAD), then some simple ROBOT-TASKS where implemented, before envisioning more complex applications involving the MAESTRO language.

### 3.2.1. Portability of the environment

Since the plugging points of ORCCADare clearly identified, its setup in a new environment, here in Stanford's manipulation group, is eased:

First, concerning the hardware/OS capabilities, all the necessary operating system calls are gathered in a small library, and in our case, only minor changes had to be performed due to some hardware specific capabilities (e.g. real-time clock).

Secondly, concerning the connection to the robot itself, a minute analysis of the local architecture gave us natural entry-points to plug ORCCAD drivers functions. Once ORCCADis ready to operate, MAESTROinstallation is straightforward since its interface with the robot uses only ORCCAD's services.

### 3.2.2. Implementation of the low-level's elementary-actions

**Methodology to reuse Stanford's control laws** SAMM includes a programming framework that was used to individually validate control laws, but this software targeted only one control structure, and one robot, pushing the hardware to its limits. Thus, as a drawback, some software engineering concepts such as the ease of integration, or of reuse were let aside. Therefore, the next step is to reuse the existing automatic control code and more precisely to encapsulate it into ORCCAD modules. ORCCAD's modularity. The clarity of the original code renders the transcription of regular C functions in ORCCAD's module straightforward. The main added value during this phase consists in encapsulating with discrete events the data-flow computation of the control law.

We first added postconditions to signal that an elementary ROBOT-TASK has fulfilled its objective. For example, in the case of a robot movement from a position to another, we consider that the goal is reached when the position error (in the operational or joint space, depending on the control-law) is smaller than a given threshold (which also depends on the control-law). Hence, for each control-law whose termination can be established internally, from data appearing in the control-scheme, we added a module whose purpose was to check this data and produce a postcondition event accordingly. Similarly, during the implementation of some tasks we included *temporal* postconditions, to indicate a predefined duration for the task's execution. In some cases, the same control-law could be used to *maintain* the robot at a given position for an unspecified duration. In this case, the ROBOT-TASK would have no postcondition and the termination of the action will be synchronized with an external event, at the application level. We also included some abnormal execution events in ROBOT-TASKS to formally specify the appropriate handling. Finally, each ROBOT-TASK is encapsulated in an abstract view handled by the primitives of the MAESTRO language. This external view conveys all information necessary to individually control the execution of the actions within the whole mission controller.

**Detailed example** Figure 1 displays the implementation of the Goal-position control of the arm within ORCCAD. This control mode can be used to perform simple arm movements.

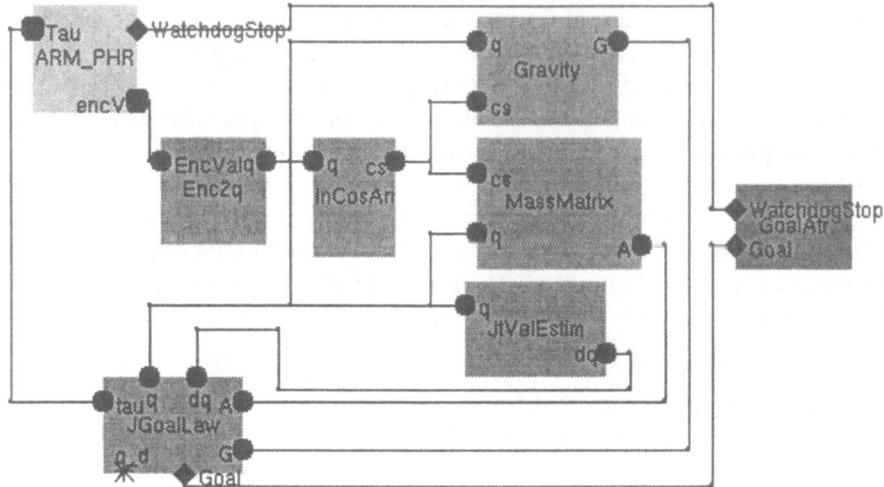

Figure 1. ORCCAD specification of a Goal-position control law

The different computational steps of the control law (Mass Matrix computation, Gravity computation, joint velocity estimation...) are clearly isolated in boxes. Those boxes belong to the algorithm class, and communicate with each others thru input/output ports expressing data dependency. The functions implementing those functionalities were reused form SAMM's original implementation.

The physical resource module (ARM_PHR) is on the upper-left. It represents the controlled part of the robot (SAMM's arm, in this case), so it has an output port providing the joint encoders values, and an input port receiving the torque to apply. The GoalAtr box on the right symbolizes the control law logical behavior. In this case, two events are taken into account:

- The use of the Emergency-button can be caught by software, and the detection of this event, WatchdogStop, was encapsulated in the physical resource module. This event is casted in a *T3 exception* in the logical behavior, and thus results in the software termination of the application. Although this is a severe error preventing the continuation off the application, some software clean-up, or data gathering must be performed anyway before the program termination.
- The Goal event is a *postcondition* which comes form the JGoalLaw module. This event will be produced when the current position of the arm, q, is close enough to a desired position, q_d. This desired position is a parameter of the control-law, set at the application level when the task is launched.

### 3.2.3. Implementation of the high level logical application

Once several ROBOT-TASKS have been designed using ORCCAD, they can be integrated in a procedure to implement a complete application. Figure 2 displays the most advanced application effectively experimented. It uses four different control modes: two for the arm and two for the base. Each physical

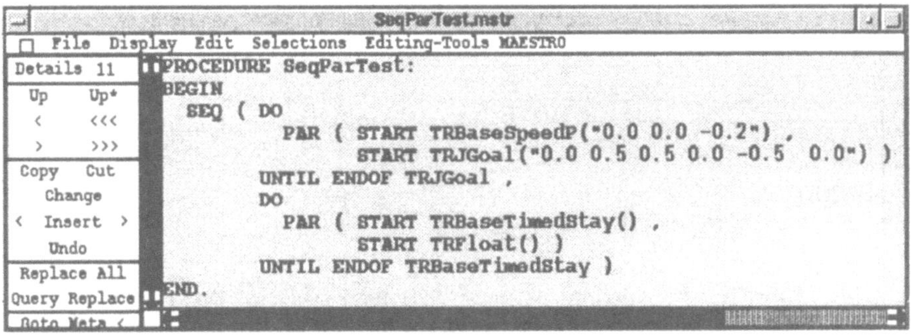

Figure 2. Specification in MAESTRO of the first part of the mission

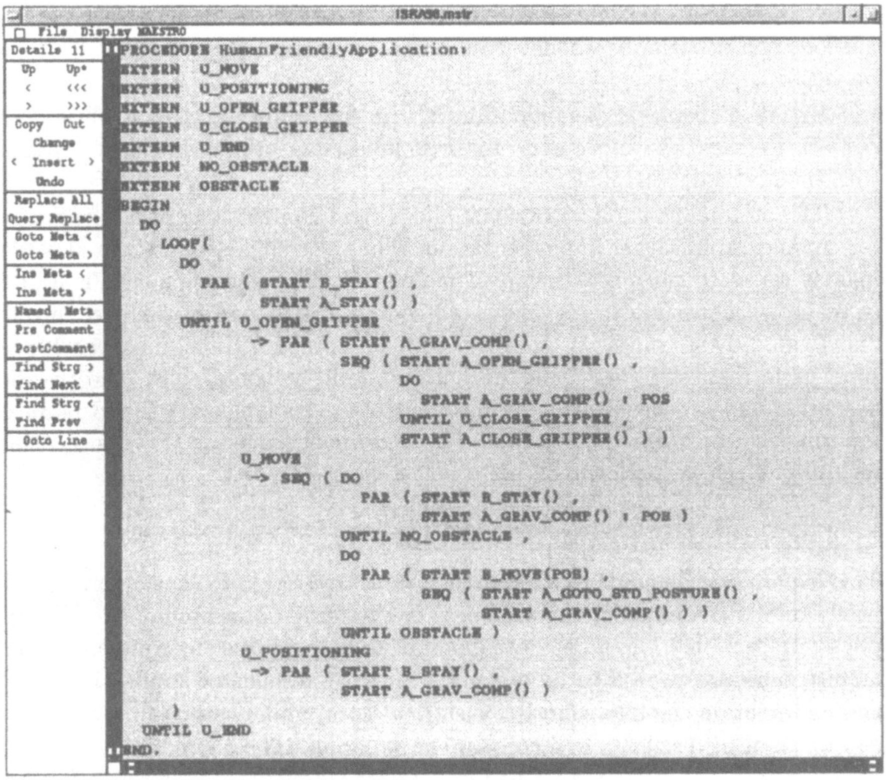

Figure 3. Specification in MAESTRO of the whole mission

resource (arm and base) is controlled independently and in parallel. Control-modes switching is synchronized upon the correct ending of one of the control laws (arm position reached then duration elapsed). In addition, external events such as U_OPEN_GRIPPER or Obstacle provide an easy interface for additional synchronization points.

This particular application was programmed using the MAESTRO (figure 3). Although experimentally not implemented, this example illustrates the verification of a logical property. From this specification, the requirement "base

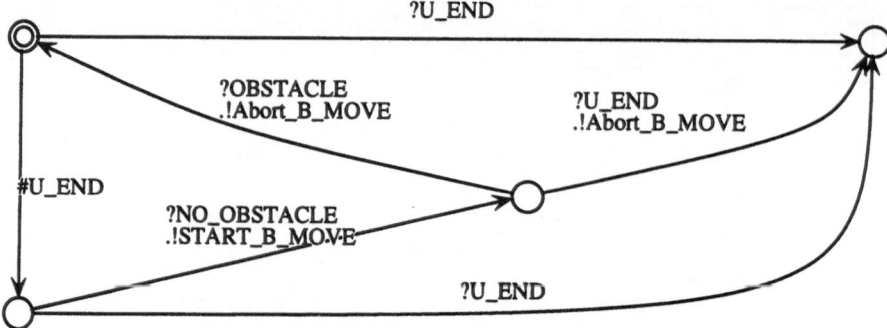

Figure 4. Verification of a logical property

motions can occur only whenever no human is in the range of the robot" can be verified:

Figure 4 displays the reduction of the full application controller, with respect to the base movement task (B_MOVE) and the presence of human (OBSTACLE or NO_OBSTACLE).

The original controller of SAMM had a rigid control structure, insufficient for complex application implementation. Thus, for single sophisticated servoings, it was very efficient, reaching 1 ms period. This efficiency level is harder to reach with the ORCCAD's generated code, but was met in the case of only one ROBOT-TASK running. However, during complex transitions, some real time performances issues could be encountered: at some point, the parallel execution and synchronization of four control modes is necessary and requires lowering the control frequency to 50 Hz (and in consequences, lowering gains), resulting in poorer performances.

## 4. Discussion and Future Trends

This experiment demonstrated that the ORCCAD/MAESTRO environment naturally fulfills the inherent requirements of the design of a human friendly application (section 2.1), and can easily adapt to new robotic environments. This architecture also proved to be adequate for several fields of applications such as manipulation, mobile robotics, electrical cars, underwater robotics, and a legged robot as discussed in a companion paper [4]. These experiments clearly demonstrated the effectiveness of ORCCAD/MAESTROand illustrated the impact of the properties built in this architecture.

The clear identification of the entities manipulated in the system (modules, tasks, events) facilitates the integration of new algorithms and control laws, the extensibility of the library of such control laws and the reuse of existing materials in new ROBOT-TASK or applications.

The automatic code generation from an abstract specification ensures some structural properties of the real-time controller; The use of formal methods allows program analyses to be performed with better tools such as theorem proving and model checking methods. Thus some logical properties on the application behavior can be proved and its logical correctness enforced.

Moreover this particular experiment brought the following contributions: The separation of continuous and discrete human-robot interaction is clearly matched by the MAESTRO/ORCCAD framework: discrete interactions are mapped into events formally handled within the framework of discrete event system while continuous interactions are integrated easily in control schemes using the sensors of the robot and their dedicated physical resource modules in ORCCAD. In addition, the availability of ORCCAD/MAESTRO for the new platforms is expected to widen the range of their applications.

However, these experiments have been exploratory in nature and several aspects remain to be investigated more thoroughly. First, multi-rate control-laws, with the inner loops running at higher frequencies than the outer, would be a very effective extension to our system.

Furthermore, safety is an essential aspect of human-friendly applications, and there is much to be considered regarding the level of safety that is acceptable with the various implications on the targeted applications. We mainly tackle safety from a the logical point of view. However, the safety issue in the context of human-friendly robots involves a number of other important considerations, e.g. mechanisms, actuators, sensors, and control structures, that must be fully addressed.

## References

[1] O. Khatib, K. Yokoi, K. Chang, D. Ruspini, R. Holmberg, and A. Casal. "Coordination and Decentralized Cooperation of Multiple Mobile Manipulators," *Journal of Robotic Systems*, vol. 13, 1996, pp. 755-64.

[2] Jean-Jacques Borrelly, Eve Coste-Maniere, Bernard Espiau, Konstantinos Kapellos, Roger Pissard-Gibollet, Daniel Simon, Nicolas Turro, *The* ORCCAD *Architecture*, The International Journal of Robotics Research, April 1998.

[3] Eve Coste-Maniere, Nicolas Turro, *The* MAESTRO *language and its environment: specification, validation and control of robotic missions*, IROS97, Grenoble 1997.

[4] Bernard Espiau, Isabelle Guigues, Roger Pissard-Gibollet, *an an Underactuated Leg with a Passive Spring at the Knee Achieve a Ballistic Step?*, this proceedings.

[5] Gerard Berry and Georges Gonthier, *The Synchronous Esterel Programming Language: Design, Semantics, Implementation*, Science of Computer Programming, November 1992, vol 19, p87-152.

# ACME,
# A Telerobotic Active Measurement Facility

Dinesh K. Pai, Jochen Lang, John Lloyd and Robert J. Woodham
Department of Computer Science
University of British Columbia
Vancouver, Canada
{pai|jlang|lloyd|woodham}@cs.ubc.ca

**Abstract:** We are developing a robotic measurement facility which makes it very easy to build "reality-based" models, i.e., computational models of existing, physical objects based on actual measurements. These include not only models of shape, but also reflectance, contact forces, and sound. Such realistic models are crucial in many applications, including telerobotics, virtual reality, computer-assisted medicine, computer animation, computer games, and training simulators.

The Active Measurement Facility (ACME) is an integrated robotic facility designed to acquire a rich set of measurements from objects of moderate size[1], for building accurate physical models. ACME can provide precise motions of a test object; acquire range measurements with a laser range finder; position a 3-CCD color video camera, a trinocular stereo vision system, and other sensors around the object; probe the test object with a robot arm equipped with a force/torque sensor; and acquire registered measurements from all these sensors.

ACME is a telerobotic system with fifteen degrees of freedom. Everything in ACME, from force controlled probing to camera settings and lighting, is under computer control. We have also developed an extensive teleprogramming system for ACME. ACME is designed to be a shared resource, and can be controlled from any remote location on the Internet.

## 1. Introduction

We would like to make it very easy to build "reality-based" models, i.e., effective computational models of real, physical objects based on actual measurements. Applications in telerobotics, virtual reality, training simulators and computer-assisted medicine require such realistic models. Computer animations and games could also profit from more realistic models of objects. Such models should be sufficiently accurate for meaningful simulation and analysis but also efficient for interaction using graphics, haptics, and auditory displays. Recently, there has been significant progress in some areas such as modeling

---

[1]Typically less than 0.5m in diameter. Larger objects can be accommodated depending on the measurements to be acquired.

geometric shape from measurements (e.g., using Cyberware[2], and Hymarc's Hyscan system[3]).

However, the current state of automation for building models is inadequate in important ways. There are no systems that allow integrated modeling of physical attributes such as surface texture and friction, elastic deformation and contact force response, surface reflectance and other radiometric properties, and the sound field due to impact. Such models are essential for realistic simulation, and for rich visual, haptic and auditory interaction in virtual environments. Acquiring the data to build such models remains extremely tedious. This has resulted in widespread reliance on simple, idealized, mathematical models of objects which are never validated by comparison with real world objects.

There is thus a need for automated systems which can measure a large number of properties with reasonable accuracy, and combine these measurements into computational models. The Active Measurement Facility (ACME) is an integrated robotic facility designed to acquire these measurements from small objects. While sensor-rich robotic systems have been previously developed elsewhere (e.g., [1]), we believe this is the first integrated, automated measurement facility for building comprehensive models of everyday objects.

The remainder of the paper is organized as follows. In Section 2 we give a brief overview of ACME hardware facilities. The control architecture, including its novel Internet-based user interface, is detailed in Section 3. In Section 4 we describe teleprogramming for ACME. We conclude with a summary and outlook in Section 5.

## 2. System Overview

The ACME hardware can be conceptually divided into the following subsystems (see also Figure 1):

- A 3-DOF *Test Station*, on which a test object is placed and can be moved for presentation to the sensing equipment. It consists of two translational motion stages mounted at right angles and a rotation stage mounted on top. (All stages are made by Daedal, with linear and rotary accuracies of $\pm0.00025$ in and $\pm10$ arc-min.)

- A *Field Measurement System* (FMS) for measuring the light and sound fields around the test object (see Figure 2). Light field measurements can be used, for instance, to tabulate the BRDF (bi-directional reflectance distribution function) for rendering [5], or simply to get images. Sound fields can be used to build sound synthesis models [2]. A key component of the field measurement system is a high quality 3-CCD color video camera with computer controlled zoom and focus (Sony DXC 950). The camera and other sensors can be positioned with a 5-DOF gantry robot. A trinocular stereo vision system is also included in the FMS and is described below.

---

[2]http://www.cyberware.com/
[3]http://www.hymarc.com/

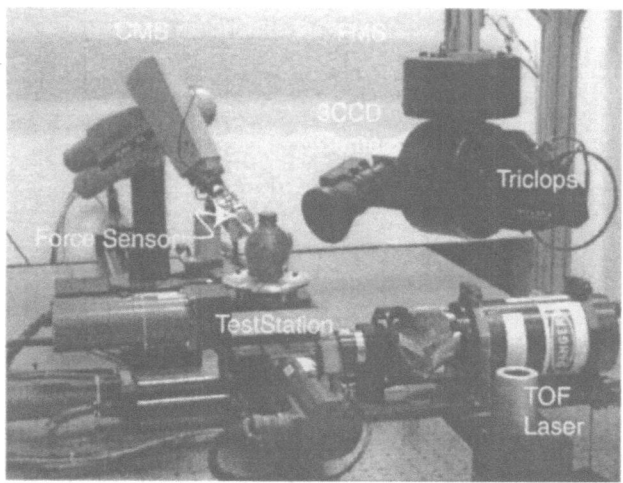

Figure 1. ACME Facility Overview

 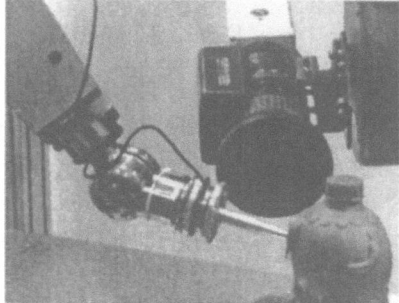

Field Measurement System (FMS)    Force/Position Measurement (CMS)

Figure 2. ACME Subsystems

- A *Contact Manipulation System* (CMS), which includes a Puma 260 robot arm with a wrist mounted 6-axis force/torque sensor (ATI Mini 40). The CMS is intended to measure properties such as friction and stiffness, and to make controlled impacts to generate sounds. The arm is mounted on a long linear motion stage to increase its work space. Contact forces with the test object are controlled using active compliance. Figure 2 shows this subsystem acquiring contact force data from a test object's surface.

- Range measurement systems. Currently, we have two range measurement devices: a time-of-flight laser range finder (Acuity AccuRange 3000 LIR) and a Color *Triclops* stereo vision system from Point Grey Research [10]. Other approaches to shape measurements may be incorporated in the future. We would like to experiment with photometric stereo [14] utilizing

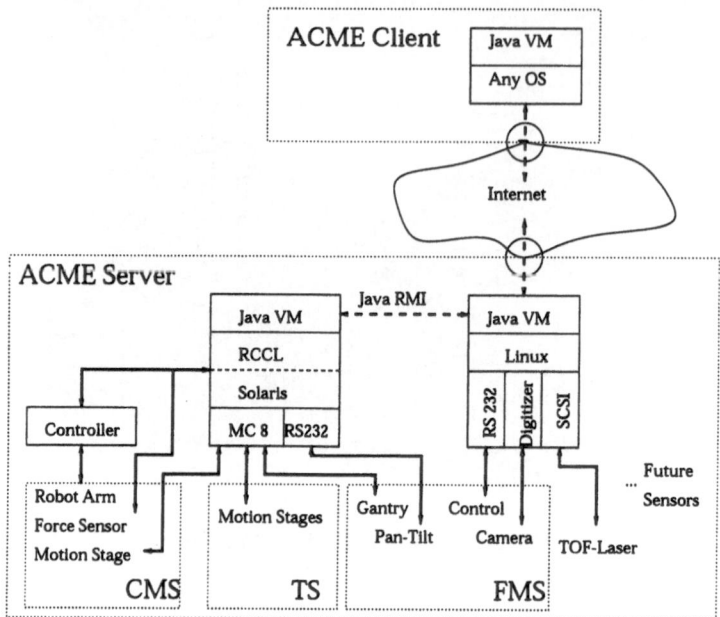

Figure 3. ACME Server: Overview of Control

the 3-CCD camera as the sensing device and appropriate light sources.

The design of ACME is directed at acquiring measurements for building models, and not the models themselves. However, ACME will provide nominal shape models, since these are necessary for obstacle avoidance, sensor planning, path planning, etc. The three-dimensional location and shape of objects on the test station are measured and registered relative to a facility-wide world coordinate frame.

- Miscellaneous subsystems, such as light sources, and control computers.

## 3. Control Architecture

In this section we describe the software and hardware architecture for controlling the ACME facility. The overall architecture is shown in Figure 3, and is divided into a server and a client. The robotic subsystems are controlled by the ACME Server (Section 3.1). Users interact with ACME using the client (Section 3.2). We discuss user level teleprogramming in Section 4.

### 3.1. ACME Server

The ACME server software consists of four layers. The highest layer is the user's Experiment which is described in the next section.

The next layer provides high level Java$^{TM}$[4] objects called Devices, which are in turn subclassed into Actuators and Sensors. Actuators are robotic

---

[4]Java is a trademark of Sun Microsystems Inc., MountainView, Ca., USA

subsystems which can be controlled as a unit: these include the Test Station, the CMS (implemented using a Puma 260 robot arm), and the 5-DOF gantry robot (x-y-z positioner plus 2-DOF pan-tilt head) comprising the FMS. Sensors are sources of measurement data, such as the time-of-flight laser range sensor, 24-bit color camera, and the CMS's 6 axis force/torque sensor. All Device objects keep track of their state and spatial position. Actuator objects also have methods to solve their inverse kinematics.

An important feature of our design is that motions are first-class objects which can be directly manipulated or stored like any other object in the language. For instance, a motion can be handed to a simulator for verification or to a server for execution on the actual hardware. This is in contrast with typical robot programming languages in which motions are implicitly defined through procedures. The smallest motion object is a Motion, which is a generalized way to specify a specific motion of an Actuator. Motions may also include a programmable impedance for those actuators capable of realizing this. Individual Motions may be assembled into a MotionPlan, which takes care of sequencing motions of a given actuator and starting motions of different actuators at the same time.

The next layer of software generates real-time trajectories for each actuator device from a MotionPlan. This layer is implemented using the Robot Control C Library (RCCL) [7] with a Java$^{TM}$ front-end. RCCL generates (at 100 Hz) the joint setpoints necessary to realize the specified position and force commands (basing the latter on input from a 6-DOF force sensor). These joint setpoints in turn form the input to various controllers which comprise the lowest layer: the 6 joint servos in the PUMA controller, an 8 axis motion control card (Precision Micro Dynamics MC-8) controlling the Test Station and x-y-z axes of the gantry, and a stepper motor interface for the gantry's pan-tilt head.

The grouping of physical devices into ACME subsystems has been described in 2, and it can also be seen from Figure 3. Physically, the system is distributed between several computers; separating low-level closed-loop control, trajectory generation, data acquisition, and networking.

## 3.2. ACME Client

The user interface of the ACME facility is through an ACME client. Figure 4 sketches the functionality of the client. The client serves as control terminal for the ACME facility; an important design feature is that the client can transparently connect to either a simulator of ACME, or to the real ACME facility.

The client consists of two major parts: the experiment window and the viewer. The experiment window is a simple development environment for designing ACME experiments. The viewer renders the state of the ACME facility (either real or simulated) and the state changes as an experiment proceeds. The complete ACME client is a 100% pure Java$^{TM}$ program enabling its use anywhere, on any system with an internet connection and Java$^{TM}$ virtual machine[5]

---

[5]However, the viewer utilizes the Java$^{TM}$3D API which is currently only available on a few systems.

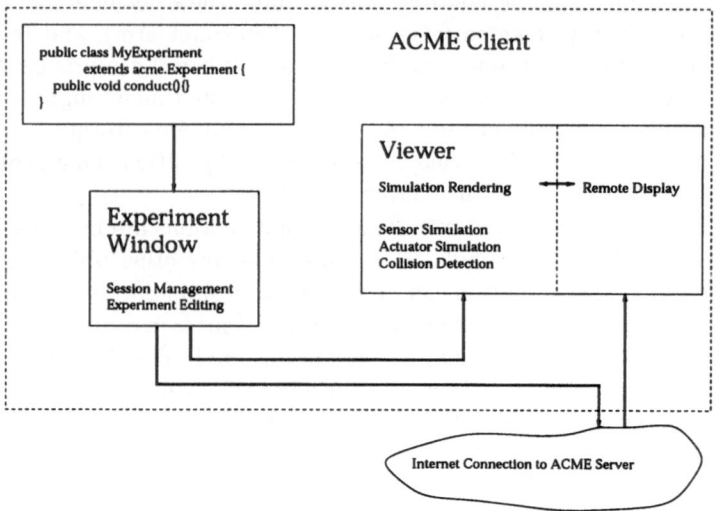

Figure 4. ACME Client: Functionality

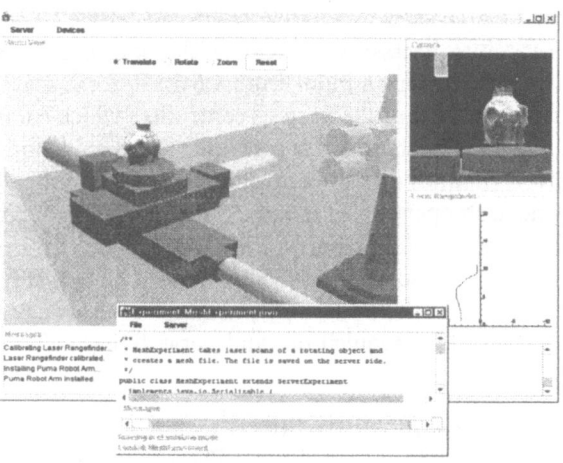

Figure 5. ACME Simulation on Client

The ACME client in stand-alone mode (connected to the simulator) is useful for design and verification of **Experiments**, with the aid of a three-dimensional graphical simulation of the motion commands. In addition to motion, some of the sensors in ACME are simulated as well, using an object model. The model may be available from earlier experiments with ACME, or in fact may be any simple shape description file. Currently, the only sensors simulated are the TOF-Laser and camera. Since ACME allows not just one-shot measurements, but adaptive measurement and on-line model construction,

ACME Experiments can be quite complex; the simulation provides a way to test these algorithms safely. As an example, the simulator can return laser scan data and images from a geometric model of a test object, so that user algorithms for shape reconstruction can be tested in great detail.

The ACME client is also an Internet terminal to the ACME robotic device. Once a user is satisfied with the results of an experiment in simulation, the client may be connected to the ACME server and run the experiment on the real ACME facility as described in Section 4. While the ACME server executes the experiment, it sends back state information and data to the ACME client. The data may have nearly any format from "raw" sensor output to the result of any computation specified with the user's experiment and its helper objects. The client's viewer renders the state updates as received from the server, providing visual feedback through the graphical display of ACME. Note that an ACME client on a system without the Java$^{TM}$3d API can still upload an experiment to the ACME server but can not use the viewer to provide visual feedback of an experiment in progress.

## 4. Teleprogramming ACME

ACME is a high degree of freedom robotic system, and therefore requires a good programming environment to use it effectively. Several teleprogramming systems have been demonstrated in the literature [6, 12, 3] from which we have adapted programming ideas for ACME, as well as from the Cornell Mobot Scheme environment [11]. Experiments can be run on ACME from almost any site on the Internet. See [9, 4, 13] for other work on telerobotics on the Internet. We describe the programming environment of ACME below.

ACME programs are called "Experiments" and are written in the Java$^{TM}$ programming language. Our design exploits dynamic class uploading, security, multi-threading, and networking provided by Java$^{TM}$.

A simple example using the ACME1.0 API is shown below. A user-defined class (called HelloRealWorld here) moves the test object to a desired configuration, sets camera parameters, and takes a series of images of the test object from different angles around the object by rotating the test station. The image data is then stored on the client's local disk. An experiment is required to implement a conduct method which is called by the ACME server to conduct the user's experiment.

```
import acme.*;

public class HelloRealWorld
  extends ExperimentBase
  implements java.io.Serializable {

  private DeviceCom devCom = null;
  private Camera cam = null;
  private TestStation ts = null;

  //... other methods
```

```
public void init() throws Exception {
  this.devCom = acme.getDeviceCom();
  this.ts = (TestStation) devCom.getDevice(TEST_STATION);
  this.cam = (Camera) devCom.getDevice(CAMERA);
  // set Camera state to desired settings
  CameraState camState = (CameraState) cam.getState();
  camState.zoom = 100; cam.setControl(camState);
  devCom.printMessage("Experiment initialized.");
  return;
}

public void conduct() {
  init();
  double[] jvals = new double[3];
  try {
    for (int deg = 0; deg < 360; deg += 45) {
      jvals[0] = 0; jvals[1] = 0;
      jvals[2] = deg / 180.0 * Math.PI;
      MotionPlan plan = new MotionPlan();
      Motion mot = plan.add(new Move (this.ts, jvals));
      plan.execute();
      while (mot.status() != Motion.COMPLETED) {
      Thread.sleep (1000);
      }
      ACMEImage image = this.cam.getImage();
      acme.saveData("image" + deg + ".ppm",
        ACMEImage.toBytePPM( image.getPixels(true), 640, 480));
    }
  }
  catch ( Exception e ) {
    devCom.printStatus( "Error in conduct()" );
    return;
  }
  return;
  }
}
```

---

The basic functionality for conducting experiments is defined by the interface **Experiment**, with some of the implementation provided by the abstract class **ExperimentBase**. In particular, it provides a reference to a special **acme** object. Various subsystems of ACME can be accessed through this reference; for instance the Test Station device is accessed invoking

```
acme.getDeviceCom().getDevice(TEST_STATION);
```

and an image can then be acquired by `cam.getImage()` and saved on the client's local disk using

```
acme.saveData("image" + deg + ".ppm",
 ACMEImage.toBytePPM(image.getPixels(true), 640, 480));
```

The client's Experiment class can be compiled on any computer. The experiment is conducted by uploading the class into a Java$^{TM}$ Virtual Machine (JVM) running on the ACME server. The ACME server provides native implementations of methods which can access and control the ACME hardware (see Figure 3).

Some features of our design are worth mentioning, since they make it easy to develop correct experiments and run them easily from a remote location.

First, while it would be possible, and conceptually simple, to use Remote Method Invocation (RMI) to directly teleoperate ACME from a remote site, the high latency and low bandwidth of the Internet makes this approach problematic, particularly for experiments involving contact with the test object. To address this problem, in previous work we have developed a system for model-based telerobotic control, and demonstrated its feasibility with a similar robot system [6, 8]. In ACME, we also use a teleprogramming approach, and send small robot programs (ACME Experiments) instead of low-level motion and sensing commands. This allows feedback loops to be closed at the remote ACME server and enables adaptive measurement techniques such as view-point planning.

Second, the ability to load classes at run time makes it possible to simulate the experiment by loading *exactly the same compiled bytecode* into a JVM on the user's computer. In this case, a different implementation of ACME classes and their methods are loaded into the JVM, and provide a graphical simulation of ACME (see Figure 5). Therefore, errors in the experiment can be caught early, and the simulated experiment can be immediately run on the real ACME facility, with no recompilation.

Third, dynamic class loading is well suited for this type of teleprogramming. The Experiment object is instantiated on the client, serialized, and sent to the ACME server, where it is re-instantiated. This has two benefits: (a) It allows interactive experimentation. The user can modify an Experiment class (for instance based on measurements returned from ACME), and reload it into a running ACME server. (b) It allows us to develop a library of idiomatic experiment objects which can be customized on the client (for instance based on user manipulation of the graphical simulation of ACME).

## 5. Conclusions

We are developing a telerobotic measurement facility, ACME, with goal of making it extremely easy to build more accurate and complete physical models of everyday objects. When completed, ACME will be able to acquire a large number of carefully registered measurements of a given object including shape, reflectance, sound, and contact forces. ACME is also a high degree of freedom robotic system with a complex array of sensors; we have developed an extensive teleprogramming architecture for programming this system from any location on the Internet.

### Acknowledgments

Several people have contributed significantly to the development of ACME at the University of British Columbia; without their efforts the system could not have been built. They are, in alphabetical order: R. Barman, C. Chiu, J. Fong, A. Fournier, L. Ke, S. Kingdon, and A. K. Mackworth. Financial support was provided in part by grants from NSERC and IRIS NCE. Lang and Lloyd were supported in part by NSERC fellowships. We would also like to thank Point Grey Research for their support and assistance with the Triclops system.

# References

[1] G. Hirzinger, B. Brunner, J. Dietrich, and J. Heindl, "Sensor-Based Space Robotics – ROTEX and Its Telerobotic Features". *IEEE Transactions on Robotics and Automation*, October 1993, pp. 649–663 (Vol. RA-9, No. 5).

[2] K. van den Doel and D. K. Pai, "The Sounds of Physical Shapes," *Presence*, The MIT Press, 1998, pp. 382-395, (Vol. 7, No. 4).

[3] J. Funda, T. S. Lindsay, and R. P. Paul, "Teleprogramming: Toward delay-invariant remote manipulation". *Presence*, Winter 1992, pp. 29–44 (Vol. 1, No. 1).

[4] K. Goldberg, M. Mascha, S. Gentner, N. Rothenberg, C. Sutter and J. Wiegley, "Desktop Teleoperation via the World Wide Web". *Proceedings 1995 IEEE International Conference on Robotics and Automation*, Nagoya, Japan, May, 1995, pp. 654–659.

[5] Paul Lalonde and Alain Fournier, "Generating Reflected Directions from BRDF Data". *Computer Graphics Forum, Special issue on Eurographics '97*, August 1997, pp. 293–300, (Vol. 16, No. 3).

[6] J. E. Lloyd, J. S. Beis, D. K. Pai, and D. G. Lowe, "Model-based Telerobotics with Vision". *Proceedings 1997 IEEE International Conference on Robotics and Automation* Albuquerque, NM, April 1997, pp. 1297–1304.

[7] John E. Lloyd and Vincent Hayward, "Multi-RCCL User's Guide". McGill CIM, April, 1992.

[8] J. E. Lloyd, and D. K. Pai, "Extracting Robotic Part-mating Programs from Operator Interaction with a Simulated Environment." In the proceedings of the *Fifth International Symposium on Experimental Robotics*, (Barcelona), June 1997.

[9] E. Paulos and J. Canny, "Delivering Real Reality to the World Wide Web via Telerobotics". *Proceedings 1996 IEEE International Conference on Robotics and Automation*, Minneapolis, Minnesota, April 1996, pp. 1694-1699.

[10] Triclops On-line Manual, *http://www.ptgrey.com/*, Point Grey Research, Vancouver, Canada.

[11] J. Rees and B. Donald. Program mobile robots in scheme. In *Proceedings 1992 IEEE International Conference on Robotics and Automation*, Nice, France, 1992, pp. 2681–2688.

[12] C. R. Sayers, "Operator Control of Telerobotic Systems for Real World Intervention". Ph. D. thesis, Department of Computer and Information Science, University of Pennsylvania, Philadelphia, PA 19104 USA, 1995.

[13] K. Taylor and J. Trevelyan, "Australia's Telerobot On The Web". *26th International Symposium On Industrial Robots*, Singapore, October 1995. http://telerobot.mech.uwa.edu.au/.

[14] R. J. Woodham, "Gradient and Curvature from the Photometric-Stereo Method, Including Local Confidence Estimation". *Journal of the Optical Society of America A*, November 1994, pp. 3050-3068, (Vol. 11, No. 11).

# Chapter 10

# Haptics

The word 'haptic' means relating to or based on the sense of touch. Haptic exploration is an important mechanism by which humans learn about their environment. Haptic 'displays' make it possible for a person to touch virtual computer objects as if they were real physical objects, or to perceive tactile information from a remote manipulator interacting with the environment. Haptics has matured considerably this decade and the following papers show the diversity of current research.

Friction is almost always present when two moving objects are in contact, so to add realism to haptic displays it is important to include a realistic friction model. Hayward and Armstrong introduce a new friction model which is able to replicate real world effects such as sticking, creeping and creaking.

Okamura et al. present a system that interprets human hand motion so as to guide stable robotic manipulation of the object to be explored. They then conduct experiments to determine the limits to detection of surface discontinuity features, surface roughness and friction.

Madhani, Niemeyer and Salisbury describe the design and control of a force-reflecting master-slave telerobot for minimally invasive surgery. The tendon driven slave has a macro-micro configuration and the overall force scaling of 50:1 allows the operator to perform delicate small-scale operations. Operating at an even smaller scale is the work by Nelson and colleagues on micro assembly. In such applications high relative positioning accuracy is required in an environment where atomic adhesive forces dominate inertial and viscous effects. They describe microforce sensing techniques that are used for gripping force control and contact detection.

# A New Computational Model of Friction Applied to Haptic Rendering

Vincent Hayward
Center for Intelligent Machines
McGill University, 3480 University Street
Montreal, Qc H3A 2A7, Canada
hayward@cim.mcgill.ca

Brian Armstrong
Department of Electrical Engineering and Computer Science
University of Wisconsin - Milwaukee, Post Office Box 784
Milwaukee, Wisconsin 53201-0784, U.S.A.
bsra@csd.uwm.edu

**Abstract:** A time-free, drift-free, multi-dimensional model of friction is introduced. A discrete implementation is developed which exhibits four solution regimes: sticking, creeping, oscillating, and sliding. Its computational solution is efficient to compute online and is robust to noise. It is applied to haptic rendering.

## 1. Introduction

Friction occurs almost everywhere. Many things, including human acts, depend on it. It is almost always present in machines. Usually friction is not wanted, so a great deal has been done to reduce it by design, or by control. In the present case, we want to synthesize it, so it can be presented under computer control to a subject using a haptic device [1]. The model introduced in this paper is also a possible contribution to the existing model-based compensation techniques. The model has the following properties:

1. It is time free (autonomous), only displacements enter in the formulation.
2. It neither drifts, nor relaxes.
3. It is robust to noise, input is not assumed to be noise-free.
4. A discrete formulation exists which is online and computationally efficient.
5. It has four regimes, one of them is a quasi-periodic oscillation.
6. It accounts for vector motions and forces (2D or 3D).
7. Its parameters have a simple physical interpretation.
8. It has a continuous counterpart.

## 2. Previous Models

Friction refers to the production of force as a result of relative movement between two objects in contact. The force must oppose motion when there is sliding (i.e. macroscopic motion). This forms the physical basis of dissipation. When there is adherence, microscopic motions can arise with contact compliance. To good approximation, the contact force must balance the net external forces. The force of adhesion in is not literally friction, but must be considered as part of a static friction model. Below a threshold of micro-displacement (or equivalently of force), the net macroscopic motion is truly identically zero. This forms the basis of energy storage. We can then speak of several states: relaxed, tense adhesion, and sliding. This basic model can be enhanced with the addition of dynamic friction effects which depend on time (rising static friction, frictional memory), or on state (friction irregularities, or dependencies as a function of space or of velocity as in hydrodynamic dependencies). For a complete survey, see [2]. In this paper, these model enhancements are not considered.

### 2.1. Presliding Displacement and Static Friction

Accounting for static friction and the transition to sliding motion is a considerable challenge for friction models, both in terms of the surface physics—the origin of adhesion was long debated and still cannot be predicted accurately—and in terms of modeling and simulation. Bodies in contact exhibit "presliding displacement" even when there is no true sliding. This motion arises with surface deformation in the contact. Over a small elastic regime energy is stored and a mass-spring behavior can be observed [3].

Dahl was the first to systematically study and model presliding displacement [4]. His model is further described below. A point addressed here is that Dahl's model, and more recent models based on it, exhibits drift when subjected to an arbitrarily small bias force and arbitrarily small vibrations. This drift is spurious: objects set on a slope and subjected to small vibrations do not continuously creep down the slope. Drift in the friction model is important for haptic interfaces because small vibrations originate from involuntary hand tremors, even in healthy individuals.

To define terms for modeling presliding displacement, consider first the one dimensional case of two objects with point contact, see the figure below. One object is termed the fixed object, the other is the moving object.

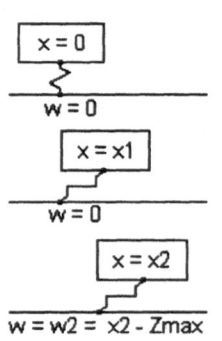 Friction is best understood by considering two points. One of them belongs to the moving object, call it $x$. The other defines an adhesion point, call it $w$. During adhesion, $w$ is attached to the fixed object, so $z = x - w$, a signed quantity, describes micromovements between the two objects. Most models, including Dahl's, define the friction force to be proportional to the strain $z$ as if the two objects were attached by a spring. The quantity $|z|$ is not allowed to exceed a small value called the breakaway distance (equivalently, the breakaway force), that we call here $z_{max} > 0$.

When $|z|$ reaches $z_{max}$ while $x$ monotonically increases (or decreases), the contact becomes fully tense and $w$ relocates so that at all times $|z| = z_{max}$. While the contact is fully tense, necessarily $\dot{x} = \dot{w}$ and $\dot{z} = 0$. This is called sliding. The magnitude of the friction force varies with the normal force, but this relationship is not considered here.

## 2.2. Computational Models

Karnopp considers computer simulations of friction using a model that switches from one dynamic equation to another [5]. It has been applied by Salcudean and Vlaar to haptic rendering [6]. It is similar in concept to the model just described but must use a velocity threshold to switch from sliding to adhesion. When sliding, the friction force is replaced by a viscous force.

Haessig and Friedland, discussing the original Dahl model, develop computational versions including a "bristle model" and a "reset integrator model" [7]. The bristle model has been applied to haptics by Chen et al. [8].

## 2.3. Dahl's Model

Dahl's formulation may be viewed as an attempt to account for the conceptual model of Section 2.1 using one single differential equation. The more general form proposed by Dahl is as follows [9, 10]:

$$\dot{f} = \sigma_0 v \left|1 - f/f_c \operatorname{sgn} v\right|^i \operatorname{sgn}(1 - f/f_c \operatorname{sgn} v), \tag{1}$$

where $v = \dot{x} = dx/dt$, $f$ is the friction force, $f_c$ the Coulomb force and $\sigma_0$ the assumed stiffness relating force to strain. The model may be formulated in terms of displacements, posing $f = \sigma_0 z$, $f_c = \sigma_0 z_{max}$, and $\alpha = 1/z_{max}$:

$$\dot{z} = v \left|1 - \alpha \operatorname{sgn} v \, z\right|^i \operatorname{sgn}(1 - \alpha \operatorname{sgn} v \, z).$$

This expression conveys that the rate of change of $z$ is zero when (i) $v$ is zero or when (ii) $v$ and $z = \pm z_{max}$ have the same signs. In both cases the contact remains tense, otherwise $|z|$ must decrease or increase so that $|z|$ can never exceed $z_{max}$. The exponent $i$ expresses how forcefully $z$ changes when $v$ is not zero and $|z| < z_{max}$. The $|.|$ and sign parts ensure the same behavior for any integer value of $i$. Many model enhancements introduce a dependency of $\alpha$ on time or state, as proposed by Canudas et al. [10].

From now on, we take $i = 1$ without affecting the discussion:

$$\dot{z} = v \left(1 - \alpha \operatorname{sgn} v \, z\right) = \dot{x} - \dot{w}.$$

Figure 1 shows $x(t)$ for a system governed by $\ddot{x} + \dot{x} = 0.5 + 0.1 \cos t - f(t)$, when $f(t)$ is given by Eq. (1), with $\sigma_0 = 100$ and $f_c = 1$.

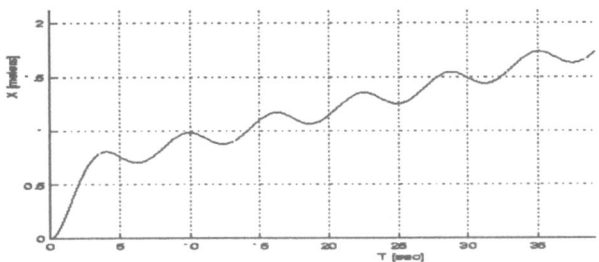

Figure 1. Solution obtained with Matlab$^{TM}$ ODE solver

Modeling the contact as a compliance should yield an oscillation for $x(t)$ as a filtered version of $f(t)$, since the applied force lies in the range $[0.4, 0.6]$, but a net average velocity of 0.03 meters per second is observed, although the breakaway force, or Coulomb friction level, is never applied.

### 2.3.1. Behavior at reversals

Of particular interest is the behavior of the solution at reversals, that is when $v$ changes sign. Because of the compliance in the contact, external rapidly varying applied forces will result in reversals of the microscopic motion.

To gain further insight, it is useful to rewrite the model in autonomous form, bearing in mind that in our conceptual account (Section 2.1), time was never needed, except to express that when there is sliding $\dot{x} = \dot{w}$ and $\dot{z} = 0$. The same idea can be expressed without time by writing that during sliding $x - w \pm z_{max}$ and $|z| = z_{max}$. The corresponding autonomous equation is obtained by eliminating time from Dahl's model:

$$dz/dx = 1 - \alpha \operatorname{sgn} dx \, z, \qquad (2)$$

thinking of $z$ as a function of $x$ instead of $t$. It is now easy to find the possible values of $dz/dx$. For a tense contact, this quantity has a value zero when $dx$ and $z$ have the same sign (as expected) and 2 when they have opposite signs.

### 2.3.2. The drift and relaxation ladders

When $z$ and $dx$ have opposite signs, $z$ changes up to twice as fast as $x$ does. In Section 2.1, we imagined $z$ to change at most at the rate of $x$, to keep $w$ invariant. Consider first that $x$ oscillates between two values. In Figure 2(a) the plot of $z$ against $x$ is shown, for all possible paths of $z(x)$ when $x$ varies between values further apart than $2z_{max}$. $z(x)$ traces closed major hysteresis loops which return to the point that they left since the ascending and the descending branches are symmetrical: we may substitute $z$ by $-z$ and $dx$ by $-dx$ leaving Eq. (2) invariant (a symmetry group). However, for small cycles around a non null value of $z$, the symmetry is broken: substitute $z$ by $z - a$ and Eq. (2) no longer is invariant. The minor paths no longer are symmetrical, they do not even trace loops but ladders, see Figure 2(b)(c), as shown now.

Suppose that $x$ cycles with a small amplitude $n$ around any location in the phase portrait. Call $\{z_m\}$ the sequence of extrema of $z$ at each reversal; $\Delta z^+$ and $\Delta z^-$ the increase of $z$ on the ascending and descending paths respectively.

$$\left. \begin{array}{l} dx > 0 : \Delta z^+ = n(+1 - \alpha z) \\ dx < 0 : \Delta z^- = n(-1 - \alpha z) \end{array} \right\} \frac{\Delta z}{z} = -2\alpha n.$$

$\{z_m\}$ converges (relaxes) to zero geometrically as in Figure 2(b). Now, an external force causing $z$ to oscillate around a nominal value yields a diverging drift ladder for $x$. Call $m$ the amplitude of the oscillation of $z$ around $z_{nom}$. Posing $\beta = \alpha z_{nom} < 1$, we find the progress of $x$ between two like signed extrema of $z$ to be: $2m\beta/(1 - \beta^2)$, see Figure 2(c). This is contrary to physical evidence. For small perturbations (say Brownian motion, permanent small tremors), a frictional contact neither relaxes nor drifts.

A drift-free friction model could either (i) have symmetrical minor hysteresis loops (this might not be easy to derive) or (ii) have no minor paths. In other terms, the ascending branches should overlap the descending ones to preserve symmetry. Memory would then be solely encoded in the sliding regime as in Figure 2(d). For the purpose of this paper, we opt for the second possibility.

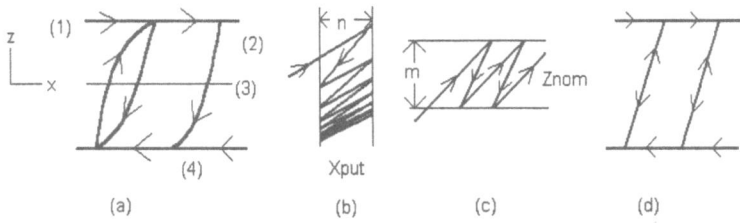

Figure 2. (a) Major hysteresis loops traced by Dahl's model. (1) sliding: $dz/dx = 0$: $dx > 0, z > 0$, (2) start relaxing: $dz/dx = 2$: $dx < 0, z > 0$, (3) relaxed $dz/dx = 1$: $dx < 0, z = 0$, (4) finish tensing $dz/dx = 0$: $dx < 0, z < 0$. (b) Converging drift ladder for small oscillations of $x$ around $x_{put}$. (c) Diverging ladder for small oscillations of $z$ around $z_{nom}$. (c) Drift free friction.

We now depart from the Dahl model, making $\alpha$ depend on $z$:

$$dz/dx = 1 - \alpha(z)\, \text{sgn}\, dx\; z. \tag{3}$$

We call $\alpha(z)$ the adhesion map. It is a nonlinear map which controls the rate of change of $z$, so that $w$ is allowed to move only when the contact is sufficiently tense. In general, $\alpha(z)$ must be identically zero for a range of values around the origin and have two asymptotes at $\alpha(z) = 1/z_{max}$ for large values of $z$. There are many possible choices for $\alpha(z)$, we explore two in Section 3.2.

## 3. An Online Discrete Scalar Model

The object is to compute a sequence $\{z_k\}$ given a sequence of position measurements $\{\bar{x}_k\}$. We first integrate (3) with respect to $x$ over an interval $h$ of any length between two samples:

$$z = x - \int_h \alpha(z)\, z\, |dx| = x - w \tag{4}$$

A sampled version does not require a clock, samples may arrive irregularly.

$$z_k = \bar{x}_k - w_k.$$

Tension at time $k$ is the measured position $\bar{x}_k$, the counterpart of the first term of the continuous equation (4), minus the discrete version of $w$. The $w_k$'s are the successive positions of the adhesion point: $w_k = w_{k-1} + \Delta w_k$. $\Delta w_k$ is the discrete counterpart of the integrand. Since samples are not necessarily equally spaced in time, simple filtering methods like discarding samples until a detectable displacement is found can be applied. Call $\bar{y}_k = |x_k - x_{k-1}|$ to emphasize the fact that the magnitude of small movements can be detected separately from the position measurements $\bar{x}_k$.

$$\Delta w_k = \alpha(z_k)\, \bar{y}_k\, z_k.$$

### 3.1. Stick-Slide

We may compute $z_k$ based on the desired properties of the solution.

$$w_k = \begin{cases} \bar{x}_k \pm z_{max}, & \text{if } |\bar{x}_k - w_{k-1}| > z_{max}; \\ w_{k-1}, & \text{otherwise.} \end{cases} \tag{5}$$

No integration is needed because we simply chose $\alpha(z) = 0$ for $|z| \leq z_{max}$ and $\alpha(z) = 1/z_{max}$ elsewhere. Note that we can choose $z_{max}$ to be very small, at the resolution limit of the position measurements. Calculations are indeed minimal. This formulation exhibits two regimes: stuck and sliding.

This model can be viewed as a filter which responds as shown on Figure 2(d). It can be used to detect extrema and signs of the derivative of $x$, robustly with respect to noise without ever estimating a derivative.

### 3.2. A More General Model

In an effort to find solutions for $z(x)$ which exhibit richer behaviors, we no longer take the computational shortcut of the previous section. The sequence of $\{w_k\}$ is now obtained from explicit Euler integration summing a sequence of estimated incremental displacements.

$$
w_k = \begin{cases} \bar{x}_k \pm z_{max}, & \text{if } \alpha(\bar{x}_k - w_{k-1})\,|\bar{x}_k - w_{k-1}| > 1; \\ w_{k-1} + \bar{y}_k\,\alpha(\bar{x}_k - w_{k-1})\,(\bar{x}_k - w_{k-1}), & \text{otherwise.} \end{cases}
$$

#### 3.2.1. Stick-slip-slide

We now take $\alpha(z)$ to be 0 for $|z| \leq z_{stick}$ and $\alpha(z) = 1/z_{max}$ elsewhere. The parameter $z_{stick}$ may be chosen to be equal to $z_{max}$ in which case this more elaborate model reverts to that of the previous section. Now, consider that $z_{stick}$ is larger than $z_{max}$. When $z$ is below $z_{max}$, the model behaves like in the previous section and does not drift. However, if the external force is sufficient for $z$ to reach $z_{stick}$, $w$ slips, the contact becomes stuck and the cycle starts again, yielding relaxation oscillations akin to a stick-slip behavior. Moreover, if the motion is sufficiently rapid, $z$ stays at $z_{max}$ and the oscillations vanish, yielding a sliding regime.

The sliding regime is entered and sustained when the condition $\alpha(z_k)z_k > 1$ is met. It is therefore dependent on sampling rate since $z_k = \bar{x}_k - w_k$ depends on the interval between two samples. This means that the higher the sampling rate, the "tighter" a contact can be simulated (further discussed in 3.2.3).

#### 3.2.2. Stick-creep-slip-slide

We may instead choose a smooth version of $\alpha(z)$, for example:

$$
\alpha(z) = \frac{1}{z_{max}}\,\frac{z^8}{z_{stick}^8 + z^8}
$$

which is nearly identical to zero for $|(z/z_{stick})| < 0.5$.

When the strain is sufficiently high, the model simulates an additional pre-sliding behavior because the model has a drift component (see discussion in [2], section 2.1.b.i). Let us call that creep to distinguish it from drift. If the motion of $x$ is sufficiently slow with respect to the drift and relaxation rates, the drift component consumes all of it so that $z$ remains below $z_{stick}$. On the other hand, if the motion of $x$ exceeds the drift rate, $z$ increases until it reaches $z_{stick}$. At this point, the behavior is the same as in the previous section. In summary, we have: stick, creep, oscillatory, and sliding regimes. See Figure 3.

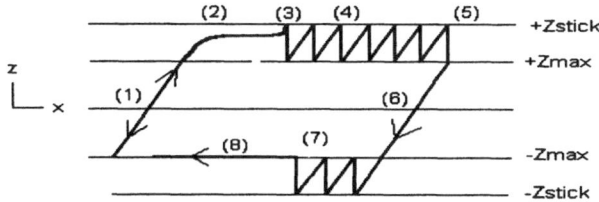

Figure 3. Trajectories when $z_{max} < z_{stick}$. Labels: (1) stuck, (2) slow motion, the model creeps along, (3) a faster motion, $z$ changes like $x$ does minus the creep, (4) oscillatory regime, (5) motion reversal, (7) motion cannot be fast immediately after a reversal, (8) fast motions slide.

### 3.2.3. Discussion

The reader will notice that implicitly, time was reintroduced in the discussion. The rates are function of time: the drift and relaxation rates depends on noise (which in turn depend on time), the motion rate, the rate of the unforced dynamics of the system during the return to $z_{max}$, and the sampling rate.

Additional care must be taken to determine the regime transition to sliding when $z$ is near $z_{max}$ and to determine the creep rate when $z$ is near $z_{stick}$, independently from the sampling rate. It is a balance between these rates that determines the regime, but each regime is time-free.

Convergence is guaranteed for any value of $\alpha$ and any value of $\bar{y}_k$ since $z_k = x_k - w_k$ is never allowed to exceed $z_{stick}$.

## 4. A Discrete Vectorial Model

We now turn our attention to a multi-dimension friction model. Points $x$ and $w$ are now associated to two or three coordinates, we denote them $X$ and $W$; $z$ is now a vector $Z = X - W$. In 3-D, we may think of $X$ as a particle and $W$ as the contact point with the ambient milieu.

It is helpful to look at Figure 4 to imagine the 2D counterpart of the scalar model just described.

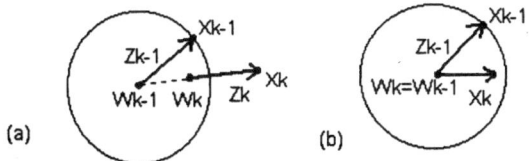

Figure 4. (a) Sliding: $X$ and $W$ move. (b) Stuck: $X$ moves, $W$ is invariant.

### 4.1. A Simple Model

The counterpart of (5) is straightforward since there is no integration:

$$W_k = \begin{cases} \bar{X}_k - \frac{\bar{X}_k - W_{k-1}}{|\bar{X}_k - W_{k-1}|} z_{max}, & \text{if } |\bar{X}_k - W_{k-1}| > z_{max}; \\ W_{k-1}, & \text{otherwise.} \end{cases}$$

This implements a robust motion direction detector, whereas the scalar version implements a sign detector. Since it is time-free, it is applicable to event-based input devices.

### 4.2. A More General Model

The most recent measurement $\bar{X}_k$ indicates the direction of movement, and $\bar{Y}_k = |\overline{X_k - X_{k-1}}|$. There are two choices for performing the integration.

$$\begin{aligned}
\Delta W_k &= \alpha \bar{Y}_k Z_{k-1} & Z_k &= \bar{X}_k - W_{k-1} \\
Z_k &= \bar{X}_k - W_k & \Delta W_k &= \alpha \bar{Y}_k Z_k
\end{aligned}$$

The second form is the only one that converges: the direction must be updated before the progress of $W$ (in the scalar case, the choice does not matter). This gives rise to the following discrete update law:

$$W_k = \begin{cases} \bar{X}_k - \dfrac{\bar{X}_k - W_{k-1}}{|\bar{X}_k - W_{k-1}|} z_{max}, & \text{if } \alpha(\bar{X}_k - W_{k-1})|\bar{X}_k - W_{k-1}| > 1; \\ W_{k-1} + \bar{Y}_k \alpha(\bar{X}_k - W_{k-1})(\bar{X}_k - W_{k-1}), & \text{otherwise.} \end{cases}$$

## 5. Experimental Results

Experiment were performed using the PenCat$^{\text{TM}}$ haptic device marketed by Haptic Technologies Inc.[1] It is hooked up to a host personal computer via a serial line which permits update rates up to 400 Hz, thanks to an embedded microprocessor.

In a first experiment, the friction model of Section 4.1 is used The thick line shows the trace of the "virtual pointer" on the screen as guided by the subject's hand, see Figure 5. The set of straight lines show $Z$ every 100 ms. In (1) and (2), $Z$ is in the direction of motion and approximates closely the tangent to the trace. In (3), the subject's hand applies a force low enough so that adhesion occurs, and $W$ becomes stationary. $Z$ is no longer tangent to the trace (which is a point) but is guided through a 360° counterclockwise sweep. In (4), motion stops also but $Z$ sweeps 90° clockwise before the motion resumes. The same maneuver is carried out in (5), counterclockwise this time. The experiment terminates in (6), where the pointer is driven around a fixed point. To indicate how slow the movement is, consider that the time of travel from (1) to (2) is about 2 seconds (the figure is to scale 1:1).

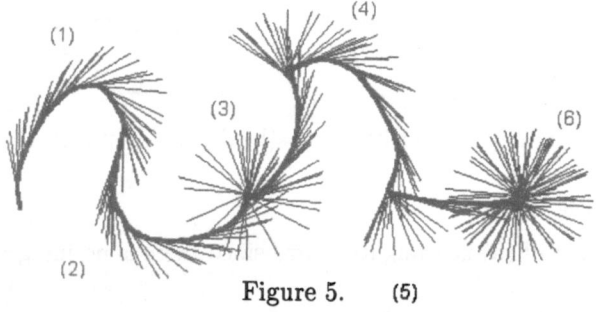

Figure 5.     (5)

---

[1] www.haptech.com

Here, the more general model of Section 4.2 is exemplified with $\alpha(z)$ as in Section 3.2.2. Figure 6 show the plot of $Z$ ($z_{max} \pm .25$ mm, $z_{stick} \pm .5$ mm) against $X$ ($\pm 50$ mm). The experiment starts in (0) where the virtual pointer is moved slowly to the right. It traverses the stuck zone (1), and continues in the creep zone before converging in (2) to an equilibrium between the motion rate and the creep rate. In (3), the motion is reversed over a small distance and so the ascending branch retraces exactly the descending one. In (4), the motion is reversed over a larger distance, so the model creeps for a while in the negative direction, eventually forming a hysteresis loop. Shortly after, a larger force is applied so $z$ climbs to reach $z_{stick}$, then returns to $z_{max}$. Because the "feed-rate" is high enough, $z$ quickly meets $z_{stick}$ again, yielding a limit cycle (5). In (6), the motion is reversed, attempting to move to the left now at higher speed. The subject's hand takes a while to reach a sufficient speed so an oscillatory regime (7) is entered before sliding in (8).

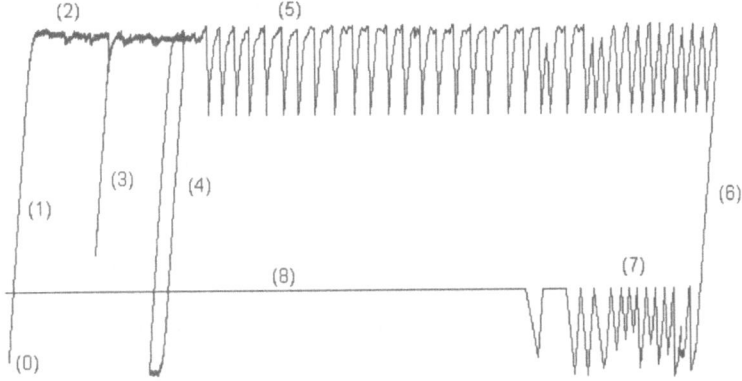

Figure 6.

Synthesized signals of this type of experiments have been recorded and are made available at www.cim.mcgill.ca/~haptic/squeaks.html.

## 6. Conclusion

It is shown that the Dahl friction model drifts when a small bias force is present with small oscillations of position. We hypothesize that this drift has not previously been identified because, in simulations, velocity during static friction can be exactly zero. In implementations where input is derived from measurements, such as haptic rendering, the drift becomes apparent.

The computational friction model proposed does not drift and provides several advantages for implementation. It is autonomous, and so suitable for event-based interfaces which may have nonuniform sampling. It is robust to noise, especially because it depends only on position. It is discrete in formulation, and so directly implementable by computer. It is extensible to 2D and 3D motions. Its parameters can be interpreted in terms of a surface deformation model, and it shows four regimes of behavior that are observed in physical systems.

The choice of a friction model depends on the application [2]. Here, the much simplified candidates for $\alpha(z)$ yield a variety of behaviors which are suitable for haptics. Various choices affect the haptic perception, so the question of perceptual relevance arises. For other applications, more direct association of the proposed friction model with tribophysics is needed, as is the introduction of dynamic friction effects. These are the subjects of continuing work.

## 7. Acknowledgments

The first author would like to thank McGill University for granting a sabbatical leave which made this research possible, and IRCAM, Paris, for generously hosting him. Thanks are due to X. Rodet and S. Tassart for many discussions on nonlinear oscillators. Special thanks to Joseph "Butch" Rovan, electronic music composer and performer for daring to use the reversal detector described in Section 3.1 during a public performance.

Funding for this research was also provided by the project "Core Issues in Haptic Interfaces for Virtual Environments and Communications" supported by IRIS (Phase 3), the Institute for Robotics and Intelligent Systems part of Canada's National Centers of Excellence program (NCE), and an operating grant "High Performance Robotic Devices" from NSERC, the National Science and Engineering Council of Canada.

## References

[1] Hayward V., Astley O. R. 1996. Performance measures for haptic interfaces. In: Giralt G., Hirzinger G., Eds., 1996 *Robotics Research: The 7th International Symposium*, Springer Verlag. pp. 195-207.

[2] Armstrong-Helouvry B., Dupont P., and Canudas De Wit C., 1994. Survey of models, analysis tools and compensation methods for the control of machines with friction. *Automatica.* 30(7):1083–1138.

[3] Polycarpou A., Soom A. 1992. Transitions between Sticking and Slipping. In: Ibrahim R. A., Soom, A., Eds., 1992. Proc. *Friction Induced Vibration, Chatter, Squeal, and Chaos, ASME Winter Annual Meeting, Anaheim,* DE-Vol. 49, New York: ASME, pp. 139-48.

[4] Dahl, P. R. 1968. A solid friction model. TOR-158(3107-18). The Aerospace Corporation, El-Secundo, California.

[5] Karnopp D. 1985. Computer Simulation of stick slip friction in mechanical dynamic systems. *ASME J. of Dyn. Systems, Meas. and Control.* 107:100-103.

[6] Salcudean S. E., Vlaar T. D. 1994. On the emulation of stiff walls and static friction with a magnetically levitated input/output device. Proc. *ASME Dynamic Systems and Control Division,* DSC-Vol. 55-1, New York: ASME, pp.303–310.

[7] Haessig D. A., Friedland B. 1991. On the modeling and simulation of friction. *ASME J. of Dyn. Systems, Meas. and Control.* 113:354–362.

[8] Chen, J., DiMattia, C., Taylor II, R. M., Falvo, M., Thiansathon, P., Superfine, R. 1997. Sticking to the point: A friction and adhesion model for simulated surfaces. Prod. *ASME Dynamics Systems and Control Division, DSC-Vol. 61,* pp. 167–171.

[9] Dahl, P. R. 1976. Solid friction damping of mechanical vibrations. *AIAA J.,* 14(2): 1675-82.

[10] Canudas de Witt C., Olsson H., Aström K. J. 1995. A new model for control of systems with friction. *IEEE T. on Automatic Control.* 40(3):419–425.

# Macro-Micro Control to Improve Bilateral Force-Reflecting Teleoperation

Akhil J. Madhani, Günter Niemeyer, and J. Kenneth Salisbury Jr.
Department of Mechanical Engineering and Artificial Intelligence Lab
Massachusetts Institute of Technology
Cambridge, MA 02139

## 1. Abstract

We study a simple teleoperator consisting of a single degree-of-freedom master and a two degree-of-freedom, macro-micro slave manipulator. Using an appropriate controller, the inertia and friction of the larger macro axis of the slave can be suppressed, thereby increasing the transparency of the teleoperator to the user. High-quality, stable, force-reflecting teleoperation is achieved with a 50:1 force scaling factor, without the use of explicit force sensing.

## 2. Introduction

When designing master and slave manipulators for bilateral force-reflecting teleoperation, it is desirable to make the dynamics of the teleoperator system transparent to the user. A good way to do this is to design manipulators with extremely low inertia and friction, [Hunter et al., 1989], [Buttolo and Hannaford, 1995], [Hwang and Hannaford, 1998]. Of course, this results in manipulators with small ranges of motion.

In this paper, we perform experiments with a simple master-slave system where the slave is a macro-micro manipulator, that is, it consists of a lightweight distal micro manipulator mounted on a heavier proximal macro manipulator. Macro-micro force control has been studied by a number of researchers, both theoretically and in the context of specific robotic hardware. For example, [Khatib, 1989] has studied the dynamics of macro/micro manipulators and proposed methods to improve their dynamic performance. [Sharon et al., 1988] applied the macro-micro concept to a hydraulic manipulator with external force sensing, and [Salcudean and Yan, 1994] applied the macro-micro concept to a six degree-of-freedom magnetically levitated wrist. Our goal was to develop a practical understanding using d.c. electric motors and cable-drive transmissions that could be directly applied to the design and control of our 7 and 8 degree-of-freedom slave manipulators for minimally invasive surgery. We designed these systems with small, cable-driven, dextrous wrists with low friction and inertia. The wrists act as micro manipulators and increase the surgeon's range of motion and mobility, while improving force feedback. We show that a slave can be designed with the range of motion of the macro manipulator, but with the sensitivity of the micro manipulator, thereby improving the quality of teleoperation. Descriptions and results for the more complex surgical teleoperators are given in [Madhani, 1998], [Madhani et al., 1998].

Figure 1. Photo of the macro-micro testbed. The master is on the left. The bar was added to increase the master inertia. The macro-micro slave is on the right.

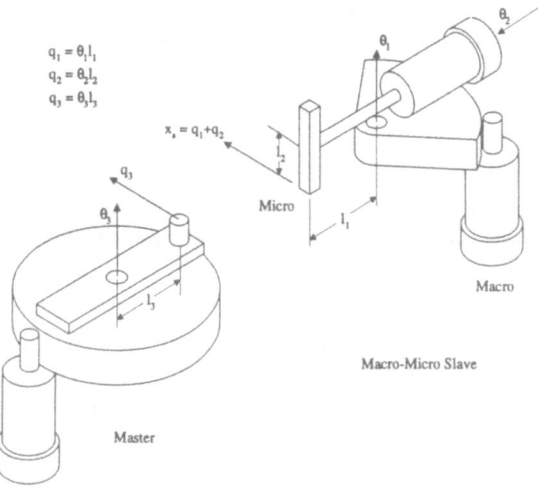

Figure 2. The slave consists of a single motor with a "T" shaped output shaft resting on a base turntable driven with a cable reduction. The master is a single turntable driven with a cable reduction.

## 3. Prototype Hardware

These ideas were studied using the testbed shown in Figure 1. A schematic drawing is shown in Figure 2 where system variables are defined. The slave has two degrees of freedom, a macro and a micro freedom. The master has one degree of freedom. All three axes are driven by brushed d.c. servomotors (Maxon Motors RE025-055-035EAA200A). In our discussion of the system, we map motor torques and rotations to endpoint forces and positions via the link lengths. In keeping with general notation for joint forces and positions, however, these are labeled $\tau_i$ and $q_i$ respectively, and have units of newtons and meters. Inertias are also measured at the endpoint of each joint and have units of kilograms.

The first joint position of the slave is labeled $q_1$ and the output force for this joint (force at the tip of the link) is $\tau_1$. This first axis is driven by a low friction, one stage cable reduction with a ratio of 13.87:1. The axis one encoder is a Hewlett-Packard optical encoder with 2000 counts per revolution after quadrature decoding. The output resolution is therefore 27,740 counts per revolution. At a link length of $l_1 = .0635$ m (2.5 in), this gives an endpoint resolution of $1.44 \times 10^{-5}$ m ($5.66 \times 10^{-4}$ in).

The second joint is mounted to the first joint such that its axis is orthogonal to and intersecting with the first. The second joint position is labeled $q_2$, and the output force of this second joint is labeled $\tau_2$. Mounted to the shaft of the joint 2 motor is a "T" shaped output link. This second joint is direct drive and uses a Canon laser encoder (TR-36) with 14,400 counts per revolution after quadrature decoding. At a link length of $l_2 = 0.025$ m (1 in), this gives an endpoint resolution of $1.11 \times 10^{-5}$ m ($4.36 \times 10^{-4}$ in).

The master is a single rotary joint whose position is labeled $q_3$. The force for this joint is $\tau_3$. The master is driven by a single stage cable reduction with a ratio of 10.67:1. The encoder on this axis is a Hewlett-Packard optical encoder with 2000 counts per revolution after quadrature, giving a total resolution of 21,340 counts per revolution. At a link length (the point where the user holds the master) of $l_3 = .044$ m (1.75 in), this gives an endpoint resolution of $1.29 \times 10^{-5}$ m ($5.1 \times 10^{-4}$ in).

The inertia for joint 1, the macro axis, is 1.16 kg. The inertia of joint 2, the micro, is .006 kg, and the inertia for joint 3, the master, is 1.53 kg. These values were found by simulating a mass-spring-damper system where the spring constant was set to our position gain and the mass and damping values were adjusted until the simulated and experimental responses coincided. The static friction in each axis was measured using a digital force sensor (Mark 10 Model BG2). The friction in joint 1, the macro, is approximately 0.47 N, the friction in joint 2, the micro, is 0.02 N, and the friction in joint 3, the master, is 0.54 N.

Computed output torques (Dell P.C., 166 MHz Pentium Processor) at approximately 6000 Hz are sent to 12-bit D/A boards (SensAble Devices, Cambridge, MA) and PWM servo amplifiers, (Model 303, Copley Controls, Westwood, MA). Encoder counters are mounted on the same board as the D/A converters.

## 4. Controllers

Three different controllers are described below. In each case, we demonstrate teleoperation with 50:1 force scaling and unity position scaling. This scales stiffness by 50:1 and scales power transmitted from the slave to the master by 50 times as well.

### 4.1. A Basic Teleoperator

Here we apply a basic teleoperator control to the master and micro actuator alone. The macro actuator is locked in place with clamps. We label the endpoint forces for the master and slave $F_m$ and $F_s$, and the cartesian stiffness and damping gains as $K_m$, $B_m$, $K_s$, $B_s$. The controller is given by:

$$F_m = -K_m(q_3 - q_2) - B_m(\dot{q}_3 - \dot{q}_2) \tag{1a}$$

416

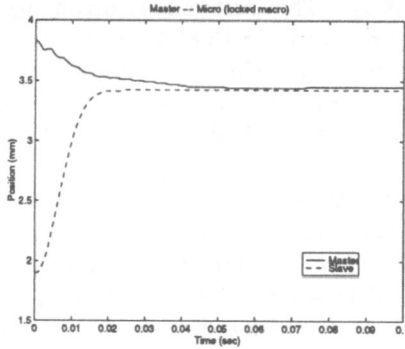
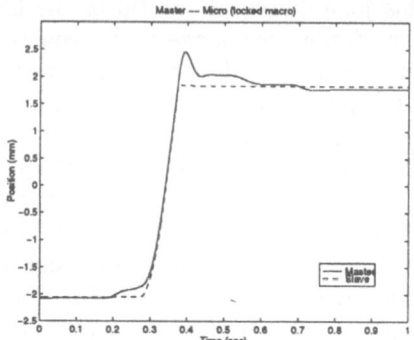

Figure 3. Step response for the master and micro axes slaved together using basic teleoperator control.

Figure 4. Contact response for the master and micro axes slaved together using basic teleoperator control.

$$F_s = K_s(q_3 - q_2) + B_s(\dot{q}_3 - \dot{q}_2) \tag{1b}$$

In order to achieve 50:1 force scaling, we set $K_m$ to be 8000 N/m, which is the highest we can reasonably achieve and set $K_s = K_m/50 = 160$ N/m. The damping gains were $B_m = 65$ Ns/m and $B_s = 1.3$ Ns/m. Because the master and slave have only one degree of freedom each, the joint forces are $\tau_2 = F_s$ and $\tau_3 = F_m$.

A 2 mm step response in the difference between endpoint positions is shown in Figure 3. Because the micro freedom, labelled "Slave" in the figure is so much lighter than the macro freedom, it moves considerably more than the master as they move together to achieve the same endpoint position. Figure 4 shows a contact plot for this system. In this test, the master is moved by hand until the slave makes contact with a fixed aluminum block. Here, the slave stops abruptly and the contact is well behaved and stable. The master overshoots slightly due to the finite stiffness connection between the master and slave. The force created by the resulting position error however quickly stops the user's hand and the master position settles to the slave position value as pressure is released by the user. The commanded output forces which the master applies to the user are 50 times that which the slave applies to the aluminum block. The actual output forces differ by the small amount of friction in the systems and by the forces required to decelerate the .006 kg micro output link inertia.

The actual feeling which the user experiences is fairly dramatic. Because the position has not been scaled between the master and slave, the effective stiffness of objects in the environment is multiplied by 50 when felt through the master. Contact with the aluminum block is felt sharply and clearly at the master. If the user holds the master in the right hand with a tight grasp, and uses the left hand to push the slave tip using soft foam or a piece of tissue, it is impossible for the right hand to avoid being pushed about. Trying to push a standard eraser across the table with the slave from the master side of the connection feels like pushing a laptop computer across the desk with one's finger.

The problem with this system is that the range of motion of the micro actuator is limited in this case to only a few centimeters. The price of having a very light, low friction slave is that it must also be small. As a result, motions of the master are limited to a small portion of it's workspace. The next two controllers represent different ways to use the macro actuator to increase the range of motion of the slave manipulator.

### 4.2. Jacobian Transpose Cartesian Control

A straightforward method of teleoperator control is to implement tracking between the endpoints of each manipulator. To do this, we calculate the endpoint position $\mathbf{x}$ using the manipulator forward kinematics and endpoint velocity $\dot{\mathbf{x}}$ using the manipulator Jacobian. We then apply a cartesian P.D. controller to find the resulting endpoint forces and torques, $\mathbf{F}$. These are then converted to joint torques $\tau$ using the transposed Jacobian and commanded to the master and slave motors.

Below, we apply this controller to our simple experimental system. Because our master has one-degree-of-freedom (it does not include orientation), we set the desired slave endpoint orientation to zero and apply an additional P.D. controller to maintain this orientation. The endpoint forces are given by:

$$F_m = -K_m(q_3 - x_s) - B_m(\dot{q}_3 - \dot{x}_s) \tag{2a}$$

$$F_s = K_s(q_3 - x_s) + B_s(\dot{q}_3 - \dot{x}_s) \tag{2b}$$

$$F_{q2} = -K_{orient}q_2 - B_{orient}\dot{q}_2 \tag{2c}$$

where: $x_s = q_1 + q_2$. Then our Jacobian is given by:

$$\begin{pmatrix} \dot{x}_s \\ \dot{q}_2 \end{pmatrix} = \begin{pmatrix} 1 & 1 \\ 0 & 1 \end{pmatrix} \begin{pmatrix} \dot{q}_1 \\ \dot{q}_2 \end{pmatrix} = J\dot{\mathbf{q}} \tag{3}$$

The joint torques are found using the transposed Jacobian:

$$\tau_3 = F_m \tag{4a}$$

$$\begin{pmatrix} \tau_1 \\ \tau_2 \end{pmatrix} = \begin{pmatrix} 1 & 0 \\ 1 & 1 \end{pmatrix} \begin{pmatrix} F_s \\ F_{q2} \end{pmatrix} = J^T \begin{pmatrix} F_s \\ F_{q2} \end{pmatrix} \tag{4b}$$

We find the controller viewed at the joint level is:

$$\tau_1 = K_s(q_3 - q_1 - q_2) + B_s(\dot{q}_3 - \dot{q}_1 - \dot{q}_2) \tag{5a}$$

$$\tau_2 = K_s(q_3 - q_1 - q_2) + B_s(\dot{q}_3 - \dot{q}_1 - \dot{q}_2) - K_{orient}q_2 - B_{orient}\dot{q}_2 \tag{5b}$$

$$\tau_3 = -K_m(q_3 - q_1 - q_2) - B_m(\dot{q}_3 - \dot{q}_1 - \dot{q}_2) \tag{5c}$$

The individual joint gains used were:

$$K_s = 160 \text{ N/m} \qquad B_s = 1.3 \text{ Ns/m}$$
$$K_m = 8000 \text{ N/m} \qquad B_m = 65 \text{ Ns/m}$$
$$K_{orient} = 50 \text{ N/m} \qquad B_{orient} = 0.5 \text{ Ns/m}$$

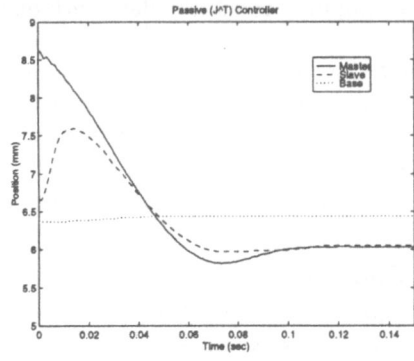

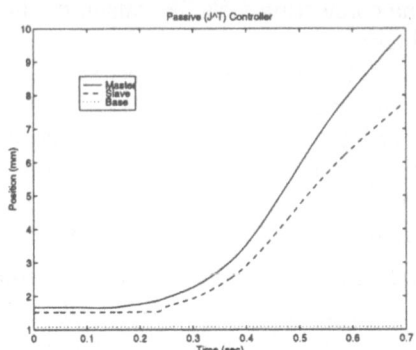

Figure 5. Step response for the master and slave using the Jacobian Transpose control.

Figure 6. Contact response for the master and slave using the Jacobian Transpose control.

The problem with this controller is that the effective joint gains are calculated based on kinematics alone, ignoring the large difference in inertia between the macro and micro axes. A gain of 160 N/m is simply too weak to move the macro axis given its inertia and friction. Figure 5 shows the 2 mm step response for the system with the Jacobian Transpose controller. Because of the small macro axis gain, it barely moves and all compensation is done by the master and micro axes. The endpoints come together and are also pulled towards the macro position. When trying to move the system by pushing on the master, the macro axis is again immovable, see Figure 6. The master motor reaches its torque limit after deflecting by about 9 mm and contact with the aluminum block is never made. At an 8000 N/m master stiffness, this means about 72 N of force was applied and the macro axis had yet to move.

### 4.3. Jacobian Inverse Controller

In this section, we describe a controller which suppresses the inertia of the macro actuator. This is in fact Jacobian Inverse manipulator control, as will be seen in section 4.3.1, which reduces to the controller presented below when applied to our system.

The controller is given as follows. First we simply command the master to follow the position of the tip of the slave manipulator:

$$\tau_3 = -K_3(q_3 - x_s) - B_3(\dot{q}_3 - \dot{x}_s) \tag{6}$$

$$x_s = q_1 + q_2 \tag{7}$$

Next we command the macro actuator to follow the position of the master without regards to the micro position:

$$\tau_1 = -K_1(q_1 - q_3) - B_1(\dot{q}_1 - \dot{q}_3) \tag{8}$$

Finally, we command the micro actuator to remain at zero:

$$\tau_2 = -K_2 q_2 - B_2 \dot{q}_2 \tag{9}$$

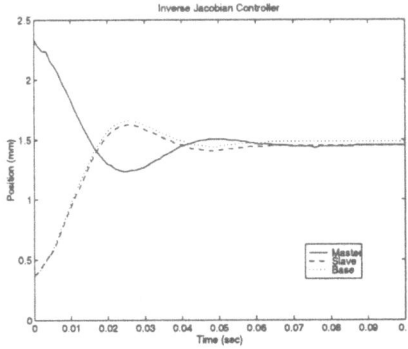

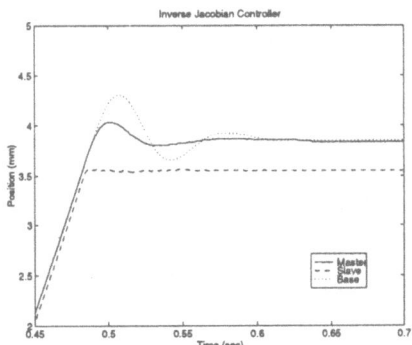

Figure 7. Step response for the master and slave using the Jacobian Inverse control.

Figure 8. Step response for the master and slave using the Jacobian Inverse control.

The gains are then tuned initially to provide as stiff a master and macro as possible i.e. optimized position control gains. The macro and micro, however, act as springs in series. To achieve 50:1 force scaling, the total endpoint stiffness of the slave must be 50 times less than that of the master. So the micro gains were retuned to achieve this. The gains were:

$$K_1 = 7000 \text{ N/m} \quad B_1 = 55 \text{ Ns/m}$$
$$K_2 = 163.74 \text{ N/m} \quad B_2 = 1.331 \text{ Ns/m}$$
$$K_3 = 8000 \text{ N/m} \quad B_3 = 65 \text{ Ns/m}$$

The macro and micro stiffnesses form springs in series, so that:

$$(163.74 \times 7000)/(7000 + 163.74) = 160 \text{ N/m}$$

It is important to note that the micro axis is decoupled from the macro axis. Regardless of the position of the master or macro axes, the micro axis simply tries to remain at its zero position.

Figure 7 shows the 2 mm step response for this controller. The master and macro respond to each other and come together in straightforward second order responses. Since the inertia and gains of these axes are similar they essentially mirror each others motions. The micro axis barely moves during the motion. The response appears well behaved and stable. Figure 8 shows a contact response plot for the Jacobian Inverse controller. Initially the master and slave are moving together towards an aluminum block. The slave hits the block and its tip position comes immediately to rest. The master actuator overshoots the slave tip position due to its finite servo stiffness, but quickly comes to rest. The macro (base) comes to rest at the position of the master. The roughly 0.25 mm offset between the master and slave represents a constant force being applied by the slave against the aluminum block and 50 times that force being applied to the user. The difference in the curves labeled Slave and Base (Macro) represent the relative motion between the micro and macro motions. While the micro stays in contact with the block, it deflects relative to the macro actuator.

This controller is well-behaved during contact and provides crisp feeling scaled force reflection. During freespace motions and during contact, the feel is very similar to that of the basic teleoperator above. It has the important advantage, however, that the range of motion is equal to that of the macro actuator.

Let us consider the inertia which the user feels when the system is moved through freespace. If we look at equations 7 and 8 and consider that in freespace motion, there is no disturbance applied to $q_2$, then $q_2 = 0$. These equations are then identical to those for the basic teleoperator, 1, with $q_1$ replacing $q_2$. However, the gain scale is $8000/7000$ so that the user feels the inertia of the slave macro freedom multiplied by $8/7$. If we could increase the gain of the macro freedom further, the macro freedom would feel lighter. So using this macro-micro approach, while moving through freespace, the user feels the inertia of the master, 1.53 kg, plus $8/7$ times the inertia of the macro base, or 1.33 kg. If our slave had consisted of only the macro base and we had implemented 50:1 force scaling, we would have felt a slave inertia of 50 times its actual inertia, or 58.7 kg. So the control effectively suppresses the inertia of the base by $58.7/1.33 = 44$ times.

### 4.3.1. Extension to Higher Degrees of Freedom

The Jacobian Inverse controller discussed earlier can be extended to higher degrees of freedom. The equations are given below.

In Jacobian Inverse control for a single manipulator, we calculate a desired velocity $\dot{x}_d$ and a desired incremental tip motion $\tilde{x}$ which will maintain our desired manipulator trajectory. We can combine these into a reference velocity $\dot{x}_r$ using a factor of $\lambda$ which defines the bandwidth of the trajectory command and should be selected to be within our positioning servo bandwidth:

$$\dot{x}_r = \dot{x}_d - \lambda\tilde{x}, \quad \tilde{x} = x - x_d \tag{10}$$

Note that if the cartesian position is lagging the desired cartesian position, $\dot{x}_r$ increases and if the cartesian position is leading the desired cartesian position, $\dot{x}_r$ decreases. Thus position feedback is implicitly included in $\dot{x}_r$. Then we convert our cartesian reference velocity into a desired joint velocity:

$$\dot{q}_r = J^{-1}\dot{x}_r = J^{-1}(\dot{x}_d - \lambda\tilde{x}) \tag{11}$$

Then we apply the following controller:

$$\tau = -B(\dot{q} - \dot{q}_r) \tag{12}$$

which is effectively a joint space P.D. controller with velocity gain $B$ and position gain $\lambda B$.

To show that these equations reduce to those given earlier for our two-degree-of-freedom macro-micro slave manipulator, let $x$ represent the cartesian output vector, let $x_s$ represent the output cartesian position and let $\theta$ represent the orientation of the output link. Then $\mathbf{x} = [x_s \ \theta]^T$. Let $\mathbf{q} = [q_1 \ q_2]^T$ represent a vector of joint positions, then:

$$\dot{\mathbf{x}} = J\dot{\mathbf{q}} = \begin{pmatrix} \dot{x}_s \\ \dot{\theta} \end{pmatrix} = \begin{pmatrix} 1 & 1 \\ 0 & 1/l_2 \end{pmatrix} \begin{pmatrix} \dot{q}_1 \\ \dot{q}_2 \end{pmatrix} \tag{13}$$

where $J$ is the manipulator Jacobian. Converting our cartesian reference velocity into a joint reference velocity we have:

$$\dot{\mathbf{q}}_r = J^{-1}\dot{\mathbf{x}}_r = J^{-1}(\dot{\mathbf{x}}_d - \lambda\tilde{\mathbf{x}}) = \begin{pmatrix} 1 & -l_2 \\ 0 & l_2 \end{pmatrix} \begin{pmatrix} \dot{x}_{sd} - \lambda\tilde{x}_s \\ \dot{\theta}_d - \lambda\tilde{\theta} \end{pmatrix} \tag{14}$$

We may substitute:

$$x_{sd} = q_3, \ \tilde{x}_s = q_1 + q_2 - q_3, \ \theta = q_2/l_2 \tag{15}$$

to get:

$$\dot{\mathbf{q}}_r = \begin{pmatrix} \dot{q}_3 - \lambda q_1 + \lambda q_3 \\ -\lambda q_2 \end{pmatrix} \tag{16}$$

Our slave controller is given by:

$$\tau = -B(\dot{\mathbf{q}} - \dot{\mathbf{q}}_r) \tag{17}$$
$$\tau_1 = -B(\dot{q}_1 - \dot{q}_3) - B\lambda(q_1 - q_3) \tag{18}$$
$$\tau_2 = -B\dot{q}_2 - \lambda B q_2 \tag{19}$$

The master has only one degree of freedom, so its controller is straightforward. Its desired position is simply the slave tip position so that its controller is given by:

$$\tau_3 = -K_3(q_3 - q_1 - q_2) - B_3(\dot{q}_3 - \dot{q}_1 - \dot{q}_2) \tag{20}$$

If we compare these equations to Equations 7, 8, and 9, we see that they are identical if we substitute $K_1 = \lambda B, K_2 = \lambda B, B_1 = B$, and $B_2 = B$.

A procedure for implementing force scaling can be laid out. We: 1) Design the master and micro inertias in a ratio equal (ideally) to the desired force scaling gains. The overall slave endpoint stiffness will typically be dominated by the micro stiffness. 2) Tune the gains of each axis individually for optimal position response. If bandwidth limitations were uniform across all axes, then inertia would be the only factor affecting stiffness gains, and our gains would now be in the ratio of our inertias and hence the desired force scale factor. In fact achieving the exact inertia ratio is difficult and other factors such as sensor resolution and friction will result in optimal gains not giving the desired force scale factor. So we must 3) retune (soften) certain gains to achieve our desired force scale factor.

## 4.4. Conclusions

Using a slave manipulator with macro and micro degrees of freedom can reduce endpoint inertia and friction when implementing bilateral force-relfecting teleoperation. Of the methods that we tried, (other types of macro-micro controllers were studied in [Madhani, 1998]), we found that an Inverse Jacobian type approach yielded good results in terms of stability and overall feel. The user could feel the inertia of the master, and an inertia due to movement of the micro axis, but the inertia of the macro axis was largely hidden. Contacts felt crisp, and using this simple system, remarkably small and soft objects such as erasers and tissue paper could be made to feel heavy and rigid using a force scaling of 50:1. Such behavior existed over the larger workspace of the macro axis. Friction and inertia, however, in the micro axis were also scaled up by a factor of 50. We found this to be barely noticeable using our system, but our friction (0.02 N at the tip of the micro axis) was miniscule by most robot standards. The level of force reflection and force scaling achievable will be directly limited by the end-effector friction levels when using this approach.

## 5. Acknowledgments

The authors gratefully acknowledge support of this work by ARPA under contract DAMD17-94-C-4123.

## References

[Buttolo and Hannaford, 1995] Buttolo, P. and Hannaford, B. (1995). Pen-based force display for precision manipulation in virtual environment. In *Proceedings IEEE Virtual Reality Annual International Symposium*, North Carolina.

[Hunter et al., 1989] Hunter, I. W., Lafontaine, S., Hunter, P. M. N. P. J., and Hollerbach, J. M. (1989). Manipulation and dynamic mechanical testing of microscopic objects using a tele-micro-robot system. In *Proc. IEEE International Conference on Robotics and Automation*, pages 1553–1557.

[Hwang and Hannaford, 1998] Hwang, D. and Hannaford, B. (1998). Teleoperation performance with a kinematically redundant slave robot. *International Journal of Robotics Research*, 17(16).

[Khatib, 1989] Khatib, O. (1989). Reduced effective inertia in macro/mini-manipulator systems. In *Robotics Research, Fifth International Symposium*, pages 279–284, Tokyo, Japan. MIT Press.

[Madhani, 1998] Madhani, A. J. (1998). *Design of Teleoperated Surgical Instruments for Minimally Invasive Surgery*. PhD thesis, Department of Mechanical Engineering, MIT.

[Madhani et al., 1998] Madhani, A. J., Niemeyer, G., and Salisbury, K. (1998). The black falcon: A teleoperated surgical instrument for minimally invasive surgery. In *Proceedings of the IEEE/RSJ International Conference on Intelligent Robots and Systems (IROS-98)*, Victoria, B.C.

[Salcudean and Yan, 1994] Salcudean, S. and Yan, J. (1994). Towards a force-reflecting motion-scaling system for microsurgery. In *Proc. IEEE International Conference on Robotics and Automation*.

[Sharon et al., 1988] Sharon, A., Hogan, N., and Hardt, D. (1988). High bandwidth force regulation and inertia reduction using a macro/micro manipulator system. In *IEEE*, pages 126–132.

# Haptic Surface Exploration

Allison M. Okamura, Michael A. Costa[1], Michael L. Turner,
Christopher Richard, and Mark R. Cutkosky
Dextrous Manipulation Laboratory
Stanford, CA 94305 USA
touch@cdr.stanford.edu

**Abstract:** We describe research at the Stanford Dextrous Manipulation Lab centered around haptic exploration of objects with robot hands. The research areas include object acquisition and manipulation and object exploration with robot fingers to measure surface features, textures and friction. We assume that the robot is semi-autonomous; it can receive guidance or supervision from humans regarding object selection and grasp choice, but is also equipped with algorithms for autonomous fine manipulation, surface exploration and feature identification. The applications of this work include object retrieval and identification in remote or hazardous environments.

## 1. Introduction

Haptic exploration is an important mechanism by which humans learn about the properties of unknown objects. Through the sense of touch, we are able to learn about characteristics such as object shape, surface texture, stiffness, and temperature. Unlike vision or audition, human tactile sensing involves direct interaction with objects being explored, often through a series of "exploratory procedures" [7]. Dextrous robotic hands are being developed to emulate exploratory procedures for the applications of remote planetary exploration, undersea salvage and repair, and other hazardous environment operations.

The challenges are formidable, including object detection and grasp planning, dextrous manipulation, tactile sensor development and algorithms for surface exploration and feature identification. As a consequence, fully autonomous object acquisition and exploration are not yet practical outside of carefully controlled laboratory experiments. At the other extreme, immersive telemanipulation and tele-exploration impose demands on bandwidth of force and tactile display that are not easily achieved, especially when the robot is located remotely. Our own approach therefore follows a middle road in which the robot is guided by humans but is also capable of some autonomous manipulation, local surface exploration and feature identification. Once the object is in hand, surface properties are identified and can also be replayed to human operators.

We begin by presenting an approach in which human operators can guide the process of grasping and manipulation using an instrumented glove, with

---

[1]also with NASA Ames Research Center, Intelligent Mechanisms Group, MS269-3, Moffett Field, CA 94035 USA

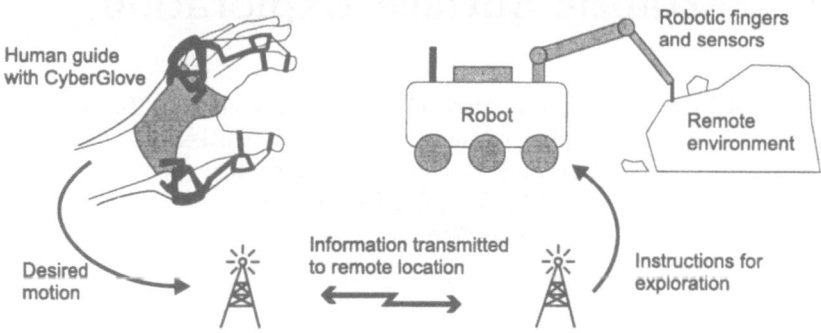

Figure 1. Guided haptic exploration of a remote surface.

or without force feedback (Figure 1). Once the object has been grasped, the robot can take over the task of surface exploration with occasional human intervention. We present several methods for detecting different surface properties. Features that are small (relative to the size of the robotic fingertip) are detected using intrinsic properties of the path traced by a spherical fingertip rolling or sliding over an object surface. Local texture profiles are recorded with a "fingernail" stylus, providing statistical measures of the surface roughness. Friction properties are estimated by measuring the relative velocities, accelerations and forces between the finger and the object during sliding. In addition, we show preliminary work on how models obtained via haptic exploration can be displayed through haptic feedback to users who may be located remotely.

## 2. Guiding Manipulation and Exploration

Haptic exploration requires stable manipulation and interaction with the object being explored. We are developing an approach that provides the flexibility of telemanipulation and takes advantage of developments in autonomous manipulation to reduce the required bandwidth for remote force and tactile display. A human "guide" instructs the robot through demonstration. While wearing an instrumented glove, the human reaches into a virtual environment (which may be updated with information from a real environment), grasps a virtual object and starts to manipulate it. The robot then determines the type of grasp used and the desired object motion and maps these to the kinematics of its own "hand".

There are several advantages to this approach. First, the interface is as humanly intuitive as possible and exploits the user's understanding of stable grasping and manipulation. The strategy is also flexible, because a person can quickly demonstrate a new maneuver. Meanwhile, the robot is performing the details of manipulation and force control autonomously, allowing it to optimize motions for its own capabilities and to respond immediately to information about the grasp state which it receives from its sensors. Our work indicates that when force information from the robot is displayed to the user, accurate positional resolution and force control can be achieved [17].

The most significant challenge is to obtain a best estimate of the intended

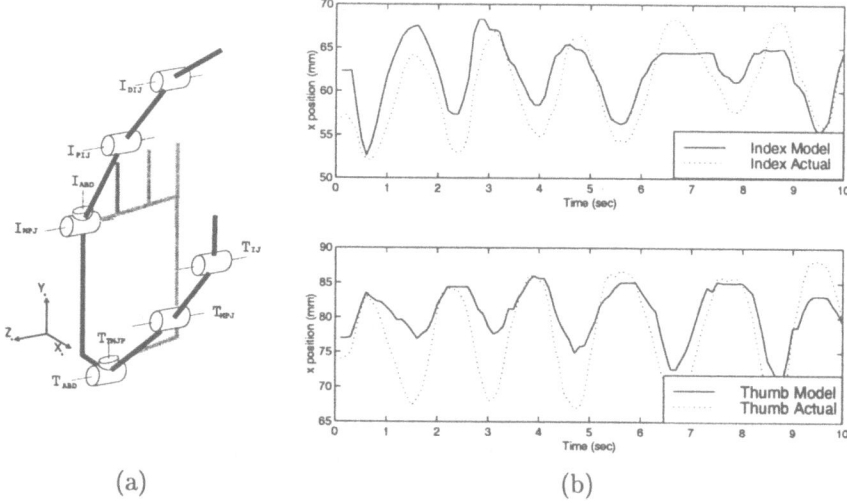

Figure 2. (a) Kinematic model of the thumb and index finger. (b) Modeled and actual fingertip trajectories.

object motion. Because tactile feedback from a virtual world is imperfect at best, and possibly non-existent, the human user may make finger motions that are not consistent with a rigid object or even with stable manipulation. However, during a maneuver we can assume that the user wishes to maintain a stable grasp. For a given motion of the user's fingers, the inferred motion of the virtual object is that which minimizes the error between the virtual fingers and measurements of the user's fingers, while satisfying the kinematic constraints of rolling, sliding and maintaining the grasp.

To assist in interpreting the motion of a human hand, we have developed a kinematic model and calibration routine. The human hand model used here (Figure 2a) is an adaptation of the models developed in [14] and [8]. The joints are approximated as pure rotations, with the base joint on the thumb modeled as a spherical joint. To calibrate the manipulation, the user wears a CyberGlove from Virtual Technologies, Inc. which records the angular position of each joint while manipulating an instrumented object. Knowledge of the object position is used to infer the user's actual fingertip position. The size of the user's hand and any errors in the glove measurements are calculated with a least squares fit [14].

Six users with disparate hand sizes were used to test the calibration procedure. Reliable calibration was achieved to within an average of 4mm of measured fingertip position. Figure 2b shows a typical post-calibration comparison of thumb and index tip trajectories. Further accuracy was limited by errors in localizing the hand in space. An improved calibration device has been developed, which will be presented in future work. The CyberGlove can also be augmented with CyberGrasp, a force-reflecting exoskeleton that adds resistive force-feedback to each finger. The addition of force-feedback will be used in future experiments with guided manipulation.

## 3. Surface Properties

### 3.1. Small Surface Features

Through guided manipulation, a robot can explore various properties of the remote environment. We consider the exploration three particular surface properties: small surface features, roughness, and friction. In this section, we present a method for detecting small surface features with a spherical robotic fingertip, with or without the use of tactile sensors. The approach can be used with guided or autonomous object exploration [13]. Although tactile sensing may be needed for control, it can be shown that surface modeling and feature detection can be accomplished without the use of tactile sensor data. We take a differential geometry approach to surface feature definition, comparing the principal curvatures of the object surface to the curvature of the fingertip.

Imagine a round fingertip rolling and sliding over the surface of an object. As the finger moves over the surface, the locus of points at the center point of the fingertip creates a parallel surface, whose principal curvatures are limited by the radius of the fingertip, $r_f$. For example, if the original surface has a cusp, the traced surface will have curvature $\frac{1}{r_f}$ around that point. In addition, there may be parts of the original surface that are unreachable by the fingertip, for example, a pit in the surface with radius of curvature less than $-\frac{1}{r_f}$, resulting in a discontinuity in the parallel surface. An estimate of the original surface can be calculated from the parallel surface by taking *its* parallel surface.

Features may then be extracted by comparing the curvature of either the parallel or estimated surfaces to the curvature of the fingertip. A curvature feature, as detected by a spherical robotic fingertip with radius $r_f$, is a region of a surface where one of the principle curvatures $k_i$ of the surface satisfies

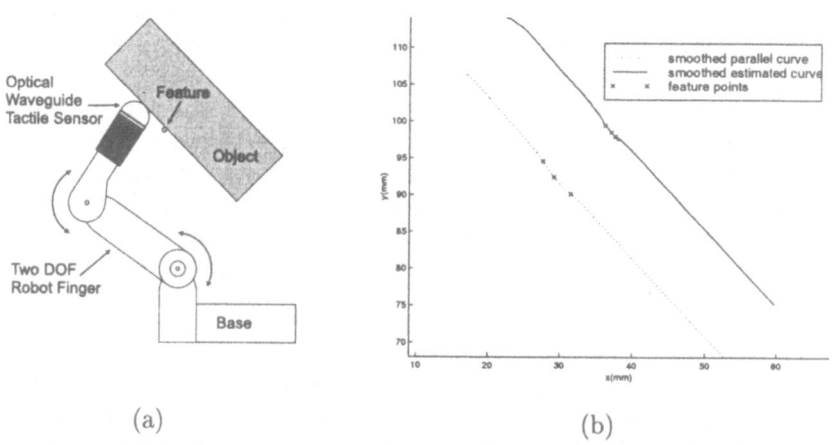

(a)          (b)

Figure 3. (a) Experimental apparatus for small surface feature detection. (b) Feature detection with a robotic finger rolling/sliding on a 45° surface with a 0.65 mm bump feature.

$k_i >= \frac{1}{r_f}$ or $k_i <= -\frac{1}{r_f}$. These are the positive curvature (convex) and negative curvature (concave) features, respectively. A discontinuity on the parallel surface is especially useful for detecting negative curvature features because it is a point with infinite curvature. The simple curvature feature definition can be extended to define macro features, which consist of patterns of curvature features. For example, a bump is defined as an area following the pattern {negative curvature feature}{positive curvature feature}{negative curvature feature} as the finger travels across the surface.

Experiments in using these definitions for feature detection were performed by rolling and sliding a hemispherical fingertip over a flat surface with small ridge features. The features were created by stretching a 0.65mm wire over the surface. The experiments were performed using two degree-of-freedom robotic fingers and 20mm diameter optical waveguide tactile sensors developed by Maekawa, et al. [9, 10] (Figure 3a). For a more detailed description of the experiments and apparatus, the reader is referred to [12].

As is typical of many robotic fingers, the workspace was limited and thus a combination of rolling and sliding was necessary to move the fingers over the surface of the object. A hybrid force/velocity control was used to obtain smooth sliding over the surface. Because the fingertip is spherical, the contact location on the finger gives the tangent and normal of the rigid surface. The velocity of the fingertip tangent to the surface and the force normal to the surface were controlled using a simple proportional law. The finger moved with an average speed of 0.03 m/sec and a normal force of $1 \pm 0.01$ N. Figure 3b shows the parallel and estimated surfaces and features detected for a bump on a flat surface angled at 45 degrees, with a 0.65 mm diameter wire stretched across the surface. The orientation of the object is the same as that in Figure 3a.

## 3.2. Surface Roughness

Below a certain scale, which depends on fingertip size and tactile sensor resolution, the perception of individual features gives way to texture. The texture of objects is an important part of their identification. Experiments with humans and robots alike have shown that the use of a "fingernail" or stylus can be a particularly effective way of characterizing textures [2]. An example application is the exploration of rock surfaces in remote planetary geology. In order to increase the sensations and science tools available to field geologists, NASA has been interested in incorporating haptics into its robotic rover field tests. Besides providing a new sensation available to geologists, it may also be possible to use haptics as a science tool to classify materials by measuring surface properties such as texture and roughness.

Surface height profiles are collected through a stylus or other sensor. In Figure 4a we compare profiles of test surface taken by a micro-machined stylus [5], Omron laser displacement sensor, and the integrated velocity signal of a phono cartridge stylus. The test surface was composed of five copper strips approximately 0.125 mm thick with adhesive backing laid onto a circuit board. While we were able to use the phono cartridge to capture the gross geometry of the test surface by integrating its velocity signal, a consistent integration error

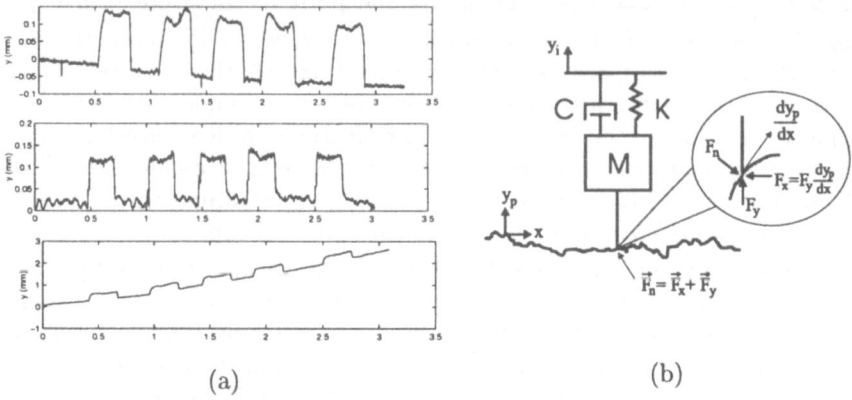

(a)                                                (b)

Figure 4. (a) Surface profiles of flat copper strips taken by a micro-machined stylus (top), laser displacement sensor (middle), and a phono cartridge stylus (bottom). (b) Virtual stylus model.

falsely indicates a large slant in the test surface.

The collected data may subsequently be explored through a haptic interface. Using a dynamic model of stylus to surface contact, forces are displayed to a user resulting from interaction with the collected surface profile data. One such model is shown in Figure 4b. We used this model to display virtual rock surfaces to planetary geologists during a NASA Ames field experiment in February 1999.

First a height profile as a function of position, $y_p(x)$, is collected by a sensor. The stylus is then modeled as a mass, spring, damper system connected to the vertical input of the user's haptic interface device, $y_i$. The horizontal position of the stylus is directly coupled to the horizontal position of the interface. While in contact with the surface, the stylus dynamics are computed by the following equation.

$$M\ddot{y}_p = K(y_i - y_p) + C(\dot{y}_i - \dot{y}_p) + F_y \qquad (1)$$

The constants were tuned for feel and stability to give virtual stylus parameters of $K = 0.973\frac{N}{mm}$, $C = 4.63 \times 10^{-5}\frac{N}{mm/s}$, and $M = 4.63 \times 10^{-5}g$.

The contact point of the stylus against the surface is modeled as frictionless contact point. As illustrated by the magnified portion of Figure 4b, the normal force $F_n$ is perpendicular to the tangent at the contact point. The tangent is computed by taking the derivative of the surface with respect to the $x$ coordinate system, $\frac{dy_p}{dx}$. In our application, this derivative is pre-computed, as we know the surface profile a priori.

The normal force is the sum of two $x$ and $y$ component forces. The $y$ component represents the reaction force supporting the stylus mass from going through the virtual surface vertically. The $x$ component represents the horizontal reaction force and is computed from the $\frac{dy_p}{dx}$ as shown in Figure 4b. While in contact with the surface, the user experiences $F_x$ and $F_y$.

In practice, the lateral force $F_x$ was scaled down to 5% of its full value. We found that using full lateral forces induced strong vibrations that detracted from the exploration of the surface. Not surprisingly, attempting to use a model with only lateral forces [11] was found to be unsatisfactory for rock surfaces. However, users who compared the models with and without lateral forces found that a small amount of lateral force improved the simulation. Interestingly, it has been claimed that surfaces modeled without friction feel glassy [4]. Although we did not use friction in this model, none of the geologists who used the system during the NASA field test or other users commented that surfaces felt glassy or slippery. We believe that a small amount of force opposing the user's horizontal motion, coupled with the irregularity of the surface, eliminates a slippery feel by creating an illusion of friction.

## 3.3. Friction

Although friction may not be necessary for the display of geological surfaces stroked by a hard stylus, it is an important component of the perception of surfaces in other applications. In addition, adequate levels of friction are necessary to ensure contact stability between an object and the fingers that are grasping it. Armstrong-Helouvry [1] is an overview of the issues involved in friction measurement. Shulteis, et al. [15] show how data collected from a telemanipulated robot can provide an estimate of the coefficient of friction between two blocks; one being manipulated by the robot and the other being at rest. Our own approach to the identification of friction involves force and acceleration measurements, in combination with a friction model.

The friction model used for the experiments presented below is a modified version of Karnopp's model [6]. This model includes both Coulomb and viscous friction and allows for asymmetric friction values for positive and negative velocities. It is distinct from the standard Coulomb + viscous friction model because it allows the static friction condition to exist at non-zero velocity if the magnitude of the velocity is less than a small, predefined value. The model is expressed as

$$
F_{friction}(\dot{x}, F_a) = \begin{cases} C_n sgn(\dot{x}) + b_n \dot{x} & : \quad \dot{x} < -\Delta v \\ \max\left(D_n, F_a\right) & : \quad -\Delta v < \dot{x} < 0 \\ \min\left(D_p, F_a\right) & : \quad 0 < \dot{x} < \Delta v \\ C_p sgn(\dot{x}) + b_p \dot{x} & : \quad \dot{x} > \Delta v \end{cases} \tag{2}
$$

where

$C_p$ and $C_n$ are the positive and negative values of the dynamic friction, $b_p$ and $b_n$ are the positive and negative values of the viscous friction, $\dot{x}$ is the relative velocity between the mating surfaces, $D_p$ and $D_n$ are the positive and negative values of the static friction, $\Delta v$ is the value below which the velocity is considered to be zero, and $F_a$ is the sum of non-frictional forces being applied to the system.

A one degree-of-freedom linear motion experiment was constructed in order to conduct friction identification experiments (See Figure 5). The friction and mass of an aluminum block sliding on a sheet of rubber were estimated. The

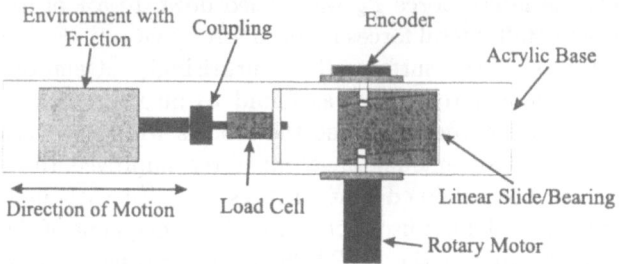

Figure 5. Schematic (top view) of the experimental apparatus.

procedure for friction identification can be summarized as follows: (1) Model the force/motion interaction of the system, (2) Move system over a range of velocities of interest, (3) Record force/motion variables included in the model, and (4) Solve for unknown parameters of the system model.

For each friction measurement experiment the aluminum block was connected to the apparatus and the apparatus was commanded to move with a periodic trajectory. Various periodic trajectories having frequencies ranging between of 0.5-3 Hz were explored. The trajectory presented here is a sinusoid with a frequency of 2 Hz and an amplitude of 0.01 meters. Trajectory amplitudes and frequency were selected to include the range of velocities for which friction estimates were desired.

The system was exercised for 10 seconds prior to collecting data for each experiment. This warm-up allowed the motion to come to steady state, and also eliminated the phenomenon of rising static friction because the system was not allowed to dwell at zero velocity for a significant period. After the 10 second warm-up was complete, the force applied to the aluminum block was recorded, along with the block's position, velocity and acceleration. Data were collected for 2 seconds, corresponding to four cycles and seven velocity reversals. The sample rate for data collection and motion control was 1 kHz.

By expressing the parameters of our model as linear coefficients of our inputs, the parameters can be estimated using least squares regression, or maximum likelihood methods. As a first step in expressing the model parameters in a linear fashion, we separate the data into two bins. One bin contains data for velocities of magnitude less than $\Delta v$. The second bin holds the remaining data. $\Delta v$ is selected as the smallest velocity range that fully encompasses the transition from static to dynamic friction. After the data points corresponding to low velocities are removed, the recorded velocity vector is split into two new velocity vectors. The velocity vector $vel_p$ is equal to the original vector $vel$ except that negative velocity values are replaced with zeros. The velocity vector $vel_n$ contains the negative portion of the original velocity vector and has zeros where there are positive values in $vel$.

Now, the measured force can be expressed as the sum of the inertia force, and the friction as

$$F_m = X\beta + \epsilon \qquad (3)$$

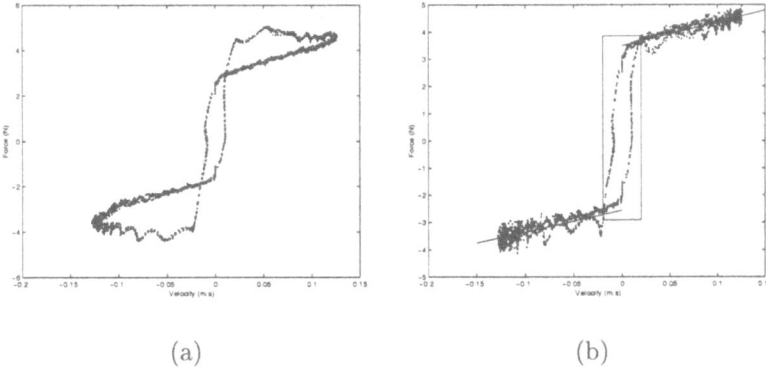

(a)                                         (b)

Figure 6. (a) Measured force versus velocity for aluminum on rubber (four motion cycles). (b) Measured force adjusted for mass and model of estimated friction for aluminum on rubber.

where

$F_m$ is the measured force,

$X$ is a matrix of measured accelerations and velocities,

$\beta$ is a vector of friction model parameters $[m, C_p, b_p, C_n, b_n]^T$, and

$\epsilon$ is the measurement error.

Least squares regression assumes that the measured variables, acceleration and velocity in our case, are free of measurement error. Fuller[3] presents several alternate estimators that account for errors in the input variables and provide unbiased estimates of $\beta$. Assuming that the errors in each of our measured variables are independent of each other, $\beta$ is estimated as

$$\tilde{\beta} = \left(N^{-1}X^T X - \ddot{S}_{uu}\right)^{-1}\left(N^{-1}X^T F_m\right) \qquad (4)$$

where $N$ is the number of samples,

$\ddot{S}_{uu} = diag(s_{\delta a}, 0, s_{\delta vel}, 0, s_{\delta vel})$,

$s_{\delta a}$ is the estimated variance of the acceleration error, and

$s_{\delta vel}$ is the estimated variance of the velocity error.

A friction estimate was obtained by using the method outlined to identify the friction and mass of an aluminum block sliding on rubber. The mass of the block was presumed unknown for each experiment; however, for verification purposes the block was weighed and found to be 0.419kg. Figure 6a shows the typical raw force data plotted against velocity. The effects of stick-slip vibration are evident. Figure 6b shows the raw force adjusted for the estimated mass. The solid lines in Figure 6b represent the predicted friction using the parameters in Equation 2. The rectangular box shows $\Delta v$, $D_n$ and $D_p$.

## 4. Conclusion and Future Work

In this work, we have outlined an approach for guided manipulation and haptic exploration using an instrumented glove, and presented various exploratory procedures utilizing different types of sensors.

In each of the sensing approaches, there is considerable future work in the development and refinement of detection schemes and haptic display algorithms to play back surface properties. At present, the different sensing methods have been individually tested; the eventual goal is to create an integrated system capable of using many different sensors and exploratory procedures. Further experiments in guided manipulation and haptic exploration are underway with a pair of two degree-of-freedom robotic fingers mounted on an Adept robot.

# References

[1] B. Armstrong-Helouvry, *Control of Machines with Friction*, Kluwer Academic Publishers, Norwell, MA, 1991.

[2] G. Buttazzo, et al., "Finger Based Exlorations," *Proceedings of SPIE - The International Society for Optical Enigneers*, Vol. 726, pp. 338-345, 1986.

[3] W. Fuller, *Measurement Error Models*, John Wiley and Sons, New York, 1987.

[4] D.F. Green, "Texture Sensing and Simulation Using the PHANToM: Towards Remote Sensing of Soil Porperties," *Second PHANToM User's Group Workshop*, Oct. 19-22, 1997.

[5] B. Kane, et al., "Force-Sensing Microprobe for Precise Stimulation of Mechanosensitive Tissues," *IEEE Trans. on Biomedical Engineering*, Vol. 42, No. 8, pp. 745-750, 1995.

[6] D. Karnopp, "Computer simulation of stick-slip friction in mechanical dynamic systems," *ASME Journal of Dynamic Systems, Measurement and Control*, Vol. 107, pp. 100-103, 1985.

[7] R. Klatzky and S. Lederman, "Intelligent Exploration by the Human Hand," Chapter 4, *Dextrous Robot Manipulation*, Springer-Verlag, 1990.

[8] J. Kramer, "Determination of Thumb Position Using Measurements of Abduction and Rotation," US Patent #5,482,056, 1996.

[9] H. Maekawa, et al., "Development of a Three-Fingered Robot Hand with Stiffness Control Capability," *Mechatronics*, Vol. 2, No. 5, pp. 483-494, 1992.

[10] H. Maekawa, et al., "A Finger-Shaped Tactile Sensor Using an Optical Waveguide," *IEEE Intl. Conf. on Systems, Man and Cybernetics*, pp. 403-408, 1993.

[11] M. Minsky, et al., "Feeling and Seeing: Issues in Force Display," *Computer Graphics (ACM)*, Vol. 24, No. 2, pp. 235-243.

[12] A. Okamura and M. Cutkosky, "Haptic Exploration of Fine Surface Features," *IEEE Intl. Conf. on Robotics and Automation*, in press, 1999.

[13] A. Okamura, et al., "Haptic Exploration of Objects with Rolling and Sliding," *IEEE Intl. Conf. on Robotics and Automation*, pp. 2485-2490, 1997.

[14] R. Rohling and J. Hollerbach "Calibrating the Human Hand for Haptic Interfaces," *Presence: Teleoperators and Virtual Environments*, Vol. 2, No. 4, pp. 281-296, 1993.

[15] T. Schulteis, et al., "Automatic Identification of Remote Environments," *Proc. of the ASME Dynamic Systems and Control Division*, Vol. 58, pp. 451-458, 1996.

[16] M. Tremblay and M. Cutkosky, "Estimating friction using incident slip sensing during a manipulation task," *IEEE Intl. Conf. on Robotics and Automation*, pp. 429-434, 1993.

[17] M. Turner, et al., "Preliminary Tests of an Arm-Grounded Haptic Feedback Device in Telemanipulation," *Proc. of the ASME Dynamic Systems and Control Division*, Vol. 64, pp. 145-149, 1998.

# Force and Vision Feedback for Robotic Manipulation of the Microworld

Bradley J. Nelson, Steve Ralis[1], Yu Zhou, Barmeshwar Vikramaditya
Department of Mechanical Engineering
University of Minnesota
Minneapolis, Minnesota USA
nelson@me.umn.edu

**Abstract**

Several emerging research areas related to microscale phenomenon, such as micromechatronics and biological cell manipulation, have the potential to significantly impact our society. Current manufacturing techniques, however, are incapable of automatically handling objects at these scales. The lack of manufacturing techniques for manipulating micron sized objects presents a technology barrier to the eventual commercial success of these fields. To overcome this technology barrier, new strategies in robotic control and robotic micromanipulation must be developed. These strategies which will enable the development of automatic micropart handling capabilities must address the following two issues: (1) extreme high relative positioning accuracy must be achieved, and (2) the vastly different microphysics that govern part interactions at micron scales must be compensated. In this paper, we present our work in visually servoed micropositioning and force sensor development for characterizing the mechanics of micromanipulation and controlling microforces throughout microassembly tasks.

## 1. Introduction

The eventual commercial success of micromechatronic technology, as well as other technologies dealing in microscales, requires that the handling of microparts be performed automatically in order to preserve potential economic benefits. However, economical assembly and packaging of micromechatronic devices of any complexity is currently not possible for two reasons. First, the manipulation of micron-sized objects is poorly understood [2][5]. Second, manufacturing facilities capable of quickly and cheaply assembling micromechatronic devices have not been developed, partly because research in microassembly has yet to demonstrate robust microassembly strategies. For example, precision alignment and bonding operations have already been identified as the single most costly manufacturing step in developing microoptoelectronic devices [3]. Though the future for microoptoelectronics is bright, the technology barrier that microassembly presents for the commercial success of this

---

1. Steve Ralis is with Electroglas Inspection Products, Inc., Corvallis, Oregon.

434

field is very real. This same barrier exists for the field of micromechatronics in general.

The two critical differences between macro and micro domains, i.e. the need for extremely high precision and the vastly different microphysics that govern part interactions at micron scales, indicate that new control and manipulation strategies must be developed for even the simplest micromanipulation strategies, such as alignment. The use of vision-based feedback has been identified as one of the more promising approaches for controlling the microassembly process, and the field of visual servoing is discussed. Applications of the field to micropositioning are demonstrated. The importance of force in assembly is obvious, and efforts at force sensing at the scales required are discussed as well.

## 2. Visually Servoed Micropositioning

### 2.1. Visual servoing with an optical microscope

We believe that visual servoing techniques must form an essential component of microassembly strategies. With appropriate vision tracking algorithms and control algorithms, submicron positioning precision can be obtained at mm/sec speeds. Fig. 1 shows a visual servoing system developed in our lab that uses an optical microscope.

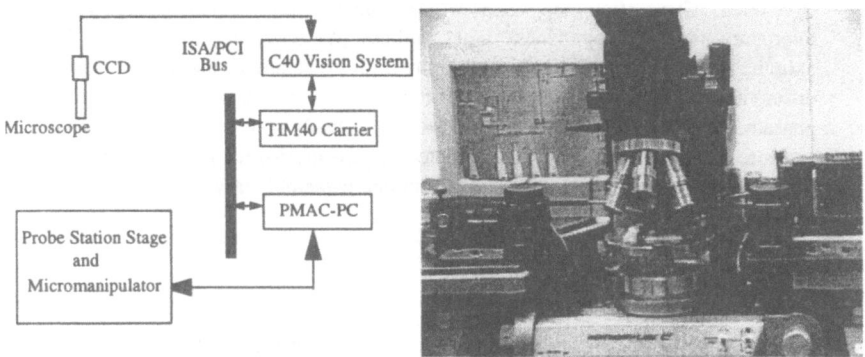

Fig. 1. Communication architecture and experimental
setup for visual servoing with an optical microscope.

Optimal control techniques are used to arrive at the following expression for the visual servoing control input

$$u(k) = -(TJ_v^T(k)QTJ_v(k) + L)^{-1}TJ_v^T(k)Q[x(k) - x_D(k+1)]$$

where $x(k) \in R^{2M}$ ($M$ is the number of features being tracked); $x_D(k)$ represents the desired image feature state; the weighting matrices $Q$ and $L$ allow the user to place more or less emphasis on the feature error and the control input; $T$ is the sampling period of the vision system; $u(k) = \begin{vmatrix} \dot{X}_T & \dot{Y}_T & \dot{Z}_T & \omega_X & \omega_Y & \omega_Z \end{vmatrix}$ is end-effector velocity; $J_v(k)$ is the image Jacobian matrix and is a function of the extrinsic and intrinsic parameters of the camera and the microscope optics as well as the number of features tracked and their locations on the image plane. The parameters of the image

Jacobian consist of the pixel dimensions on the CCD; the total linear magnification of the system; the image plane coordinates of the features being tracked; and the depth of the object from the CCD array. A derivation of the image Jacobian for an optical microscope can be found in [12]. Extensions to this control derivation that accounts for system delays, modeling and control inaccuracies, and measurement noise have been experimentally investigated [10]. The measurement of the motion of the features on the image plane is performed using an optical flow technique called Sum-of-Squares-Differences (SSD) [9]. The visual servoing loop is closed at 30Hz, and the performance of the system is demonstrated in Fig. 2. At low magnification (2X), mm/sec speeds are obtained. At 50X, 0.17μm precision is obtained without the use of tracking routines capable of subpixel sampling.

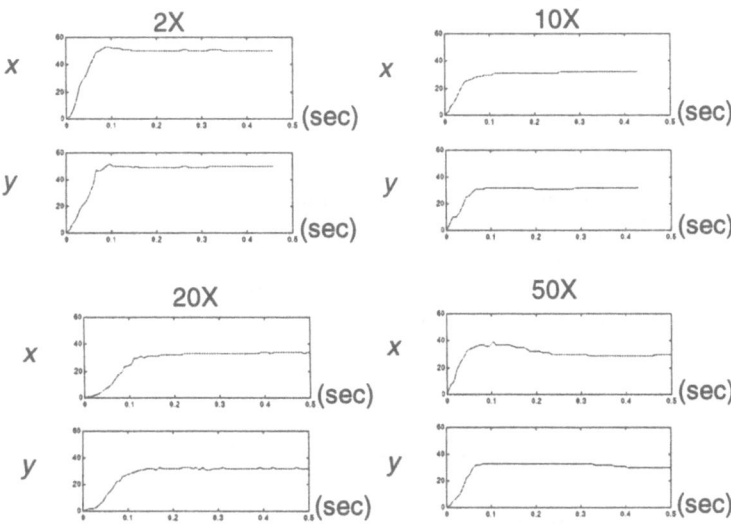

Fig. 2.    Step response of micro-visual servoing system at four different magnifications. The vertical axes are in units of pixels.

## 2.2. Supervisory Logic-based Controller

Multiple vision sensors that have different optical resolutions can be integrated to provide a large field-of-view with very high resolution for a microassembly visual servoing system. A logic-based supervisory controller can be used to accomplish this [6]. The appropriate sensor is determined by the controller and control system parameters are configured for the active sensor. The architecture is depicted in Fig. 3. and can be for-

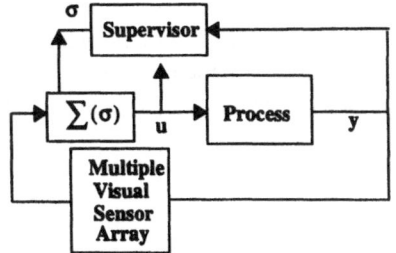

Fig. 3.   Supervisory logic-based control architecture.

mulated in state-space form as

$$\dot{x}_S = A_\sigma x_S + J_v u$$
$$u = F_\sigma x_S$$

(2)

where $\dot{x}_S$ is sensor space velocity; $x_S$ is the sensor space position; and $u$ is the input task space velocities. The supervisory signal, $\sigma$, is transmitted to the controller, $\sum(\sigma)$, that selects the controller input to the system based on the visual sensory feedback. The matrix $A$ is $I_2$; $J_v$ is the image Jacobian; the control input $u$ changes with respect to the signal output of the supervisor that switches the control gains, $F_\sigma$. This provides real time logic-based switching of multiple controllers via feedback of the process from multiple sensors. Fig. 4 shows a 3 DOF microassembly positioning stage and the low resolution and high resolution camera-lens systems used for controlling high precision probe insertions [11].

Fig. 5 shows the performance of the system during one complete insertion. From the plot and the images at the top of the figure, one can see the system being initially controlled using a "global" view of the 4x4 matrix of 254µm diameter holes and the tip of the 229µm diameter probe. As the stage is servoed in X and Y the error between the probe and hole reduces to a minimal threshold. The system then automatically switches to a high resolution "local" view of the insertion and begins moving the probe along Z until the probe enters the depth-of-field of the high resolution camera and appears in the image. At this point visual servoing in the XY plane occurs until the error reduces to a threshold, and the probe is then inserted along Z. The complete task takes less than a second for an insertion and has a precision of 2.2 µm in X and Y.

## 3. Force sensing for microassembly

Visual servoing experimental results with optical microscopes demonstrate large scales of controlled motion with mm/s speeds and submicron repeatability can be achieved using visual feedback alone [12]. It is also apparent that nanometric repeatability cannot be achieved. However, many microassembly tasks require this level of positioning repeatability, for example, alignment of microoptic components can require relative positioning in the range of 10nm. To improve the repeatability of the system, a second non-visual sensing modality must be used. The measurement of stress in micromachined components is an active area of research, and many different techniques have been investigated. This section discusses our approach to measuring microforces and the application of our nanonewton level force sensor for controlling impact forces during micropart manipulation.

### 3.1. Optical Beam Deflection for Force Sensing

For microforce sensing we believe optical techniques show a great deal of promise due to their electromagnetic immunity and the potential for high resolution [8][4]. Most commercially available atomic force microscopy systems use this technique, which further indicates that optical beam deflection is robust and fairly well characterized. A primary advantage of this technique is that it can be used in a non-contact

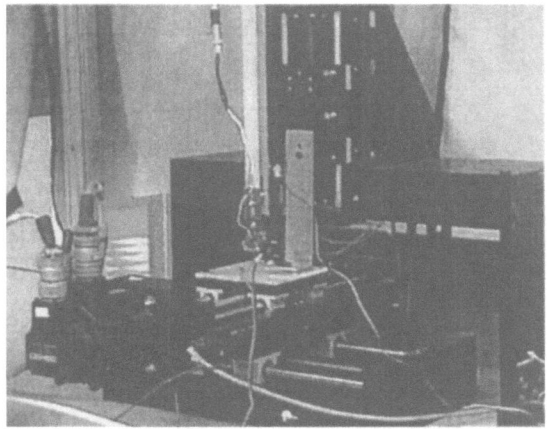

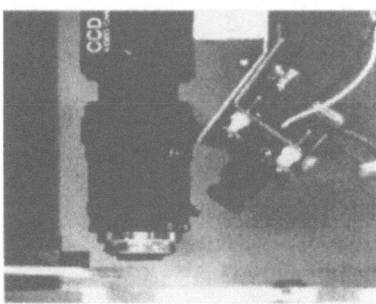

Fig. 4. Micro insertion workstation and close-up
view of the multiple visual sensors.

mode if certain structural members within a microdevice can be designed with appropriate stiffness so that they deflect a measurable amount without compromising the
structural integrity of the overall device. Fig. 6 shows our current optical deflection
strategy. Light from a laser diode is focused onto the tip of a Si or $Si_3Ni_4$ cantilever.
The laser beam is deflected by the cantilever onto a quadrature photodiode. The four
voltages output from the photodide are used to measure changes in the deflection of
the beam.

For optical beam deflection, the sensitivity of the sensor is inversely proportional
to the length of the cantilever. If cantilevers with a length on the order of 100μm are
used, an "optical lever" of 1cm can result in subangstrom resolution of cantilever
deflection using standard commercially available quadrature photodiodes. In this
case, environmental vibrations become one of the main factors limiting sensor resolution.

From simple beam theory, the measured deflection can be transformed into a
force applied to the cantilever from the equation

$$F = \frac{Ewt^3}{4l^3}\delta \qquad (3)$$

where $F$ is the applied force, $E$ is the modulus of elasticity of the cantilever, $w$, $t$, and

438

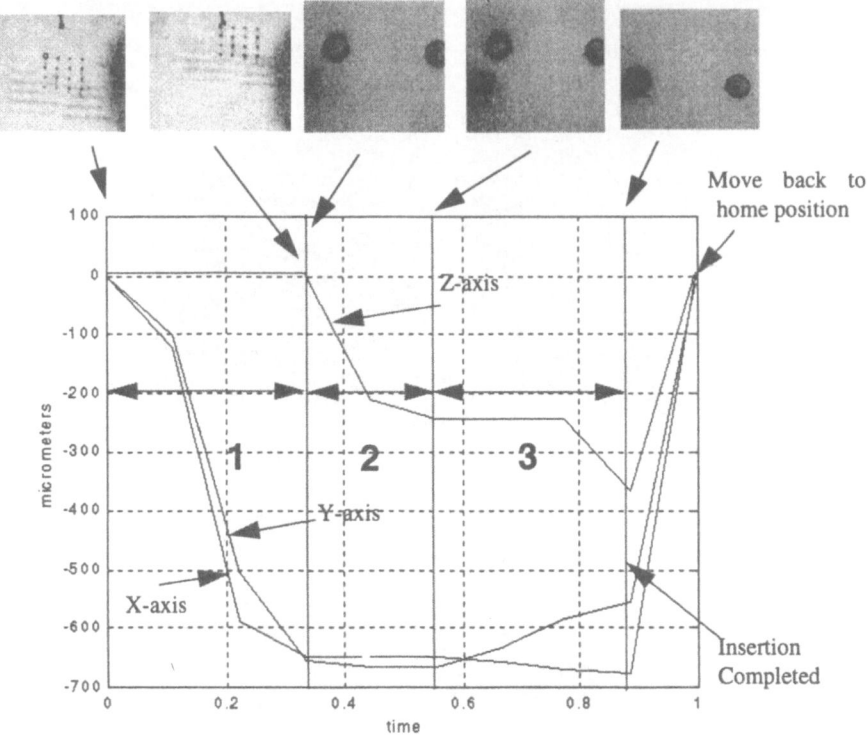

Fig. 5. Complete range of motion of one insertion with a
switching logic-based controller.

*l* are the width, thickness, and length, respectively, of the cantilever, and δ is the measured deflection.

Microfabrication techniques are used to make cantilevers [1] of either Si or $Si_3N_4$. For Si cantilevers made using bulk micromachining techniques, $E$ is a well known value derived from the literature.

Calibration of the optical beam sensor must be performed in order to determine how a given deflection maps into photodiode output. Calibration is performed using a piezoactuator that has been precisely calibrated using laser interferometric techniques. Advanced calibration techniques that result in position estimates with resolutions down to 0.05nm are commercially available [7].

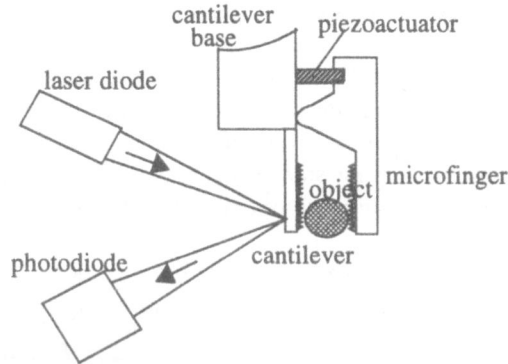

Fig. 6. Optical beam deflection technique.

As previously mentioned, a primary advantage of this technique is that it can be used in a non-contact mode if properly integrated with part handling devices. Fig. 6 shows a proposed device for handling controlling the gripping force of microobjects using optical beam deflection force sensing. Another advantage of optical beam deflection techniques is that the range and resolution of the sensor can be reconfigured simply by changing the mechanical properties of the cantilever. Ultra precise measurements require compliant cantilevers that are short, while less precise measurements with an increased range allow for the use of cantilevers with a higher stiffness and/or a greater length.

To test the performance of our OBD sensor (shown in Fig. 7), impact results were performed in which a piezoactuator was commanded to impact the Si cantilever. A/D output from the photodiodes in the OBD sensor was monitored and piezoactuator velocity was commanded at servo rates of 100Hz to generate a desired contact force of 2nN. Results from two trials are shown in Fig. 8. The top trial shows position and force as a function of time for a low gain force controller that generates a relatively small impact force of 9nN and very little position overshoot. The drawback to this controller is the slow approach velocity generated, approximately 7μm/s. The lower plot shows a much faster approach velocity, but an impact force over 15 times greater and a large position overshoot is generated. This illustrates a common problem with impact control in which approach velocity and impact force must be compromised. In the macroworld, inertial effects are mainly responsible for this behavior while in a microworld this trade-off is due to latencies in the system. One interesting aspect of these experiments is shown

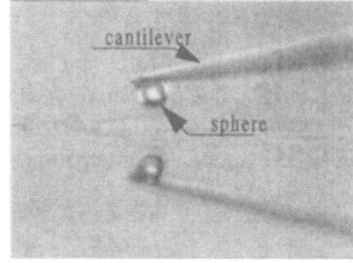

Fig. 7. OBD sensor and image of the cantilever with attached 10μm dia. sphere. The bottom sphere/cantilever is a reflection in the gold coated impact surface.

in the second plot in Fig. 8. Just before impact occurs at 1sec, a spike downward in the force plot demonstrates an adhesive force that begins pulling the cantilever toward the piezoactuator before impact actually occurs. This phenomenon also occurred in the second trial, however, the scale of the graph in the fourth plot precludes one from seeing the force. By integrating force and vision data within a single controller, impact performance can be significantly improved [13]. Results from an experimental trial are shown in Fig. 9. One can see a fast approach velocity of approximately 80μm/s; an adhesive force pulling the cantilever toward the surface; a small impact velocity of 9nN; and a quick convergence to the desired 2nN contact force. The lower plot combines the measured force from the trial in which force control alone was used with a high gain to increase the approach velocity, and from this

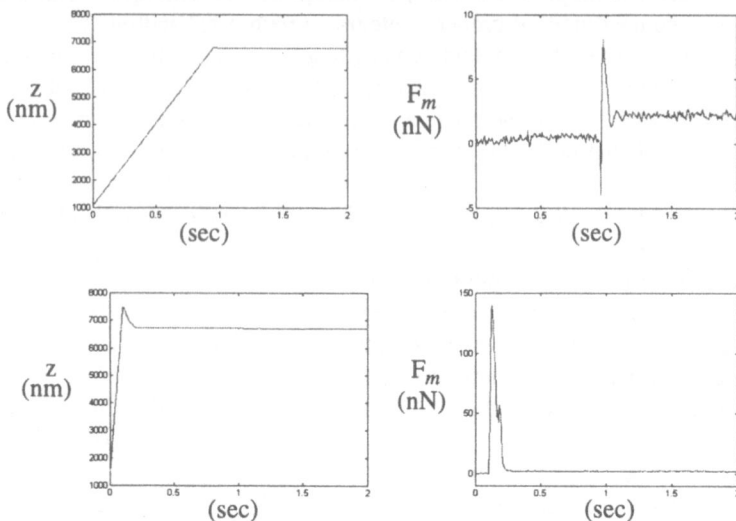

Fig. 8. Results from two trials of impact control using the OBD force sensor.

trial that combines force and vision feedback. It is apparent a much lower impact force results, however, there is an expense in a slower settling time to the 2nN contact force. This is due to the 30Hz limit in the visual servoing loop that was imposed by the CCD camera. A higher framerate camera will allow for better performance.

## 4. Conclusion

The technology barrier that micromanipulation presents to many emerging fields requires that control and assembly issues in ultra-high precision motion control be addressed. One control approach, visual servoing, applied in a microdomain shows promise in overcoming many of the problems that are unique to the microworld. As the wave properties of light limit the precision of visually controlled systems, new sensing modalities must be applied. Optical beam deflection sensing can be used to both characterize part interaction forces and control microassembly processes. Issues in integrating these sensing modalities are being addressed, with the goal being to very delicately control impact forces among microparts and direct assembly operations. Much work remains to be done in microassembly. For example, microgrippers that can overcome difficulties associated with part adhesion have yet to appear on the market, though, many research efforts are developing some very innovative gripping devices. One thing is certain for this research area; as microassembly matures and robust, economical micromanipulation strategies are developed, MEMS designers will envision a myriad of new hybrid MEMS devices previously thought impossible to manufacture.

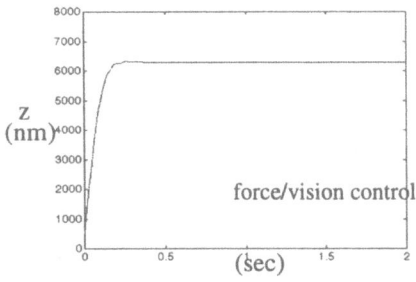 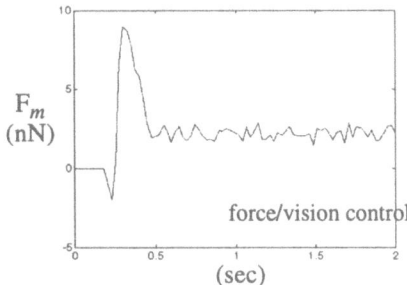

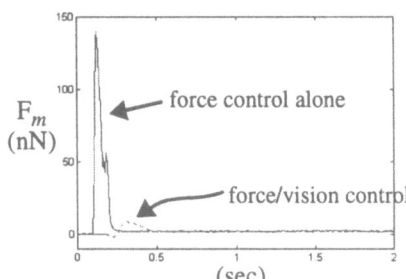

Fig. 9.  Results from an impact trial that integrates force and vision data and a combined plot (right) comparing the initial impact force resulting from force feedback alone and force and vision feedback.

## 5. Acknowledgments

This research was supported in part by the National Science Foundation through Grant Numbers IRI-9612329, CDA-9616757, and IRI-9702777, by the Office of Naval Research through Grant Number N00014-97-1-0668, and by DARPA through Grant Number N66001-97-1-8911.

## 6. References

[1]  T.R. Albrecht, S. Akamine, T.E. Carver, and C.F. Quate, "Microfabrication of cantilever styli for the atomic force microscope," *J. Vac. Sci. Technol.* **A 8(4)**, 3386-3396, 1990.

[2]  F. Arai, D. Ando, T. Fukuda, "Micro Manipulation Based on Micro Physics-Strategy Based on Attractive Force Reduction and Stress Measurement," *Proc. 1995 IEEE/RSJ Int. Conf. on Intelligent Robots and Sys. (IROS95)*, 2:236-241, Pittsburgh, August 5-9, 1995.

[3]  S.K. Ault and M.E. Lowry, "Automated Packaging: A Cure for Market Growing Pains," *Photonics Spectra*, pp. 120-124, September, 1996.

[4]  G. L. Cloud, *Optical Methods of Engineering Analysis*, New York: Cambridge University Press, Part VI &VII, 1995.

[5]  R.S. Fearing, "Survey of Sticking Effects for Micro Parts Handling," *Proc. 1995 IEEE/ RSJ Int. Conf. on Intelligent Robots and Sys. (IROS95)*, 2:212-217, Pittsburgh, August 5-9, 1995.

[6]  M. Fu, B. R. Barmish, "Adaptive stabilization of linear systems via switching controls", *IEEE Trans. on Automatic Control*; p.1097-1103, Dec. 1986.

[7]  T.R. Hicks and P.D. Atherton, *The NanoPositioning Book*, Queensgate Instruments Ltd.,

1997.

[8]  N. Kato, I. Suzuki, "Force-balancing force sensor with an optical lever", *Rev. Sci. Instrum.* pp. 5532-36, 66(12), Dec. 1995.

[9]  B. Nelson, N.P. Papanikolopoulos, and P.K. Khosla, "Visual servoing for robotic assembly," *Visual Servoing—Real-Time Control of Robot Manipulators Based on Visual Sensory Feedback.* ed. K. Hashimoto. River Edge, NJ:World Scientific Publishing Co. Pte. Ltd. pp. 139-164, 1993.

[10] N.P. Papanikolopoulos, Nelson, B. and Khosla, P.K., "Full 3-d tracking using the controlled active vision paradigm," *Proc. 1992 IEEE Int. Symp. on Intelligent Control (ISIC-92),* pp. 267-274, 1992.

[11] S. Ralis, B. Vikramaditya, and B.J. Nelson, "Visual Servoing Frameworks for Microassembly of Hybrid MEMS," *SPIE Vol. 3519 Microrobotics and Micromanipulation,* pp. 70-79, 1998.

[12] B. Vikramaditya and B.J. Nelson, "Visually Guided Microassembly Using Optical Microscopes and Active Vision Techniques," *Proc. 1997 IEEE Int. Conf. on Robotics and Automation,* pp. 3172-3177, Albuquerque, April 21-27, 1997.

[13] Y. Zhou, B.J. Nelson, and B. Vikramaditya, "Fusing Force and Vision Feedback for Micromanipulation," *Proc. 1998 IEEE Int. Conf. on Robotics and Automation,* Leuven, Belgium, May, 1998.

# Chapter 11

# Friction and Flexibility

Researchers are often noted for friction and for inflexibility. These papers threatened to provoke both at the conference since there are some widely differing points of view. However, robotics researchers need to confront both friction and flexibility to fully understand the performance limitations of machines.

Eftychios Christoforou and Christopher Damaren need to eliminate friction almost completely with sliding air pads to study the dynamics and control of flexible manipulators intended for space craft. Like many before, they are employing various nonlinear and adaptive control schemes. Friction in the motor drives is still a significant factor of course. Interestingly they exploit vibration suppression techniques to deal with unwanted modes and they also use the large payload mass as part of the controller design.

Gier Hovland, with his colleagues E. Berglund and O. Srdalen, present an interesting and very practical method for identifying the compliance (flexibility) of what ought to be rigid commercial robots. They need to do this without using many of the sensors we would take for granted since they are not fitted to industrial robots. The aim is to improve dynamic control which is valuable in high speed robot applications. While tracking accuracy is usually seen as the objective, the real practical issue is to ensure the robot reproduces identical movements at high and low speeds as near as possible. The operator can enter the safety cage to check the programmed movement, but safety rules dictate that only slow speeds are allowed. If the robot behaves differently at full speed when the operator is not allowed in the cage, debugging can be much more difficult.

Craig Tischler, with his colleagues Andrew Samuel and Stuart Lucas, presents a careful analysis of friction in his parallel wrist mechanism recently invented with Ken Hunt. Friction is usually difficult to quantify and handle in any but the simplest of mechanisms, but Craig has shown how highly complex kinematic chains can be analysed and has confirmed the theory with experimental results. The theory leads to considerable simplifications when designing a friction-compensating controller.

Masaru Uchiyama and Yoshihiro Tanno conclude this chapter with an exploration of dither techniques for reducing the effects of friction. They exploit the well known dithering technique used in hydraulic systems to measure friction and hence compensate for it without having to resort to delicate force-torque sensors at the end effectors of their robots. Using this method, they can allow electric drive motors to be driven backwards once they can measure and compensate for the friction which would otherwise make this impractical, even damaging for the motors and gear trains.

JPT

# Experiments on the Control of Flexible-Link Manipulators Carrying Large Payloads

Eftychios G. Christoforou & Christopher J. Damaren
Department of Mechanical Engineering
University of Canterbury
Private Bag 4800, Christchurch, New Zealand
christe@mech.canterbury.ac.nz, c.damaren@mech.canterbury.ac.nz

**Abstract:** The control problem for flexible multilink robots carrying large payloads is revisited and a globally stable controller together with its adaptive companion is considered. Both controllers were derived on the basis of a set of nonlinear approximate equations describing the payload dominated dynamics of the robot in conjunction with the passivity property of a suitably defined modified input/output pair for the system. Their ability to combine end–point tracking with simultaneous vibration suppression is demonstrated by a set of experimental results concerning a three degree–of–freedom planar robot with two flexible links.

## 1. Introduction

Flexible–link robots have been successfully involved in space operations and expanding human activity in space is expected to necessitate that they are extensively used. Space robots, which are characterized by long lightweight members, are able to perform a variety of tasks such as deploy and retrieve objects from the Shuttle's cargo bay, assist the assembly of large space assets, and the capture and despinning of satellites.

On the other hand, terrestrial–based robots have traditionally been designed as heavy and bulky structures so that flexibility effects are not encountered. Consequently, high capacity actuators are required, the cost of the manipulator is high, the operation is in general energy inefficient, and they are unable to perform rapid maneuvers. Lightweight flexible manipulators offer an attractive alternative that better suits the modern production requirements of reduced costs and high productivity rates, and great potential is seen for such robots to replace conventional models in many industrial applications.

Control of flexible robots is a most challenging engineering problem and will be a critical factor for their success. For rigid robots, controllers have traditionally been joint–based since the problem of tracking a prescribed task–space trajectory translates to having the joint degrees–of–freedom (DOF) follow the corresponding rigid inverse kinematics one. For a flexible–link robot though, the tip position depends on the elastic DOF as well and it is obvious that such schemes are in general not suitable.

A conceptually different approach to the control problem is examined here for the case of flexible robots carrying large payloads. The scheme takes into consideration the elastic nature of the system by effectively exploiting a non-linear approximate version of the motion dynamics, in conjunction with a suitably defined modified input and output for the plant which yields the passivity property. An adaptive version able to deal with the uncertainty of the mass properties is also examined. A set of experimental results is then presented, which clearly demonstrates the nature of the controllers and their ability to combine task space tracking with simultaneous active vibration suppression.

## 2. System Modeling

The system is modeled as a series of flexible and/or rigid bodies with single DOF rotational joints between them. The assumed modes method can be used for the discretization of the link deflections, and cantilevered shape functions were found to be an appropriate choice. Using the Lagrangian approach, the dynamic equations of motion were derived in a symbolic form suitable for simulating the behavior of the system and the software implementation of the controllers. The dynamic equations are of the following standard form:

$$M(\mathbf{q})\ddot{\mathbf{q}} + D\dot{\mathbf{q}} + K\mathbf{q} + C(\mathbf{q},\dot{\mathbf{q}})\dot{\mathbf{q}} = [\mathbf{1}\ \mathbf{O}]^T \tau(t)\,, \quad \mathbf{q} \triangleq col\{\theta, \mathbf{q}_e\} \qquad (1)$$

where $\theta$ is a column of joint rotations and $\mathbf{q}_e$ are the elastic coordinates. The mass matrix $M$, the structural damping matrix $D$, and the stiffness matrix $K$ can be partitioned consistent with the definition of $\mathbf{q}$ as:

$$M = \begin{bmatrix} M_{\theta\theta} & M_{\theta e} \\ M_{\theta e}^T & M_{ee} \end{bmatrix}, \quad K = \begin{bmatrix} \mathbf{O} & \mathbf{O} \\ \mathbf{O} & K_{ee} \end{bmatrix}, \quad D = \begin{bmatrix} \mathbf{O} & \mathbf{O} \\ \mathbf{O} & D_{ee} \end{bmatrix} \qquad (2)$$

with $M = M^T > \mathbf{O}$, $K_{ee} = K_{ee}^T > \mathbf{O}$, and $D_{ee} = D_{ee}^T > \mathbf{O}$. The vector $\tau$ contains the joint torques and $C$ is the matrix describing the nonlinear terms. The mass properties appearing in the dynamics were calculated from the solid model of each individual body in the chain created using CAD software.

The generalized position of the end–effector is described by a vector $\rho$, whose upper part consists of the position coordinates and at the bottom are three Euler angles parameterizing its orientation with respect to the base frame. As described in [1], the velocity of the end–effector is related to the generalized coordinate rates by:

$$\dot{\rho} = J_\theta(\theta, \mathbf{q}_e)\dot{\theta} + J_e(\theta, \mathbf{q}_e)\dot{\mathbf{q}}_e, \qquad (3)$$

where $J_\theta$ is the rigid Jacobian and $J_e$ the elastic one. For more details on the formulation of the above matrices refer to [1].

## 3. The Passivity Property and Robot Control

Passivity, which is a notion originally used in network theory, has become one of the fundamental concepts in control. On the basis of the passivity theorem

[2], the feedback interconnection of a passive and a strictly passive system is $L_2$-stable. In the case of rigid robots, which exhibit collocation of control inputs and outputs, it is well known that the mapping from joint torques to joint rates is passive. In the case of flexible–link robots the passivity property is preserved at the joint level, but it is not as strategic for controller design as in the rigid case, since the actual goal is end–point trajectory tracking.

A very important case concerning flexible–link robots arises when a large payload is attached to the tip of the manipulator. This commonly occurs in space manipulation scenarios where flexible robots are required to deal with payloads whose mass is much larger than their own. Given the desirability of the passivity property, Damaren [1] examined the multilink payload dominated case and defined the following modified input and output for the system:

$$\widehat{\tau}(t) \triangleq J_\theta^{-T}\tau, \quad \dot{\rho}_\mu \triangleq J_\theta\dot{\theta} + \mu J_e\dot{q}_e = \dot{\rho} - (1 - \mu)J_e(\theta, q_e)\dot{q}_e. \tag{4}$$

The modified output, $\dot{\rho}_\mu$, is called the $\mu$–tip rate, where $\mu$ is a real parameter. The true tip rates are captured by $\mu = 1$, while $\mu = 0$ considers only joint–induced motion. It was shown that the mapping from $\widehat{\tau}$ to $\dot{\rho}_\mu$ is passive for $\mu < 1$ when large payloads are involved. For $\mu = 1$ the mapping remains passive but the vibration modes become unobservable from the tip rates. The use of $\dot{\rho}_\mu$ was shown to be an effective way of introducing the elastic motion into the input of a suitable feedback controller and add damping to the vibration modes.

In [3], a Lagrangian approach was used to derive an approximate form of the motion dynamics, built upon the assumption that the total kinetic energy of the system can be adequately described by the kinetic energy residing within the large payload alone. That description of the dynamics was combined with the modified input/output idea to develop a globally stable controller for the task–space tracking problem. Its adaptive companion was derived in [4].

Further studies concerning the vibration modes of flexible robots were made by Damaren [5], and necessary and sufficient conditions for all vibration modes to exhibit a node at the manipulator's end–point were derived: $M_{\theta\theta}J_\theta^{-1}J_e = M_{\theta e}$. It was shown that this property can be closely achieved for large tip/link mass ratio and sufficiently small rotor inertias. The approach allowed an effective separation of the end–point dynamics from the elastic ones, and it was demonstrated that the nonlinear torque to end–effector motion dynamics become essentially equivalent to those of the corresponding rigid case:

$$M_{\rho\rho}\ddot{\rho} + C_\rho\dot{\rho} = J_\theta^{-T}\tau. \tag{5}$$

The corresponding rigid robot's mass, $M_{\rho\rho}$, and nonlinear terms, $C_\rho$, matrices are both evaluated for the configuration $\theta = \mathcal{F}_r^{-1}(\rho)$ and $q_e = 0$, where $\mathcal{F}_r(\cdot)$ is the rigid forward kinematics map. Equation (5) provides a better characterization for the task–space part of the approximate dynamics over the corresponding equation derived in [3], especially when the underlying large payload assumption is relaxed. The latter is equivalent to a version of the rigid task–space dynamics using only the mass properties of the large payload.

The elastic coordinates were shown to obey an equation of a standard Lagrangian form:

$$\widehat{M}_{ee}\ddot{q}_e + D_{ee}\dot{q}_e + K_{ee}q_e + c_e = -J_e^T J_\theta^{-T}\tau, \tag{6}$$

with $\widehat{M}_{ee}$ and $c_e$ the corresponding mass and nonlinear terms matrices respectively. Details concerning the formulation of the above equations and each individual component can be found in their original source.

In the present work, our experimental results use a modification of the two control schemes originally derived in [3] and [4]. The latest versions considered here were built upon Eqs. (5) and (6). The derivation and proofs of global asymptotic stability using Lyapunov arguments are similar to the ones for their original predecessors, and due to space limitations will be omitted. A summary of the schemes will only be presented. A complete reference describing the two updated versions is given by Christoforou & Damaren [6].

## 4. Rigid Inverse Dynamics and Passive Feedback Controller

The scheme consists of a model–based feedforward, $\tau_d$, combined with a feedback part, $\tilde{\tau}$. Extending the ideas presented in [3], a feedforward strategy was developed on the basis of Eq. (5):

$$\tau = \tau_d + \tilde{\tau}, \quad \tau_d = J_\theta^T(\theta, q_e)[M_{\rho\rho}(\rho)\ddot{\rho}_d + C_\rho(\rho, \dot{\rho})\dot{\rho}_d], \tag{7}$$

with $\rho_d$ the desired task–space trajectory. An estimate for the elastic coordinates, $q_{ed}$, can then be obtained from Eq. (6) produced by the application of $\tau_d$. Given $q_{ed}$ and the definition of $\dot{\rho}_\mu$ in (4), the desired trajectory $\dot{\rho}_{\mu d}$ is defined as:

$$\dot{\rho}_{\mu d} = \dot{\rho}_d - (1 - \mu)J_e(\theta, q_e)\dot{q}_{ed} \tag{8}$$

and the error trajectory for $\dot{\rho}_\mu$ becomes:

$$\dot{\tilde{\rho}}_\mu = \dot{\rho}_\mu - \dot{\rho}_{\mu d} = \dot{\tilde{\rho}} - (1 - \mu)J_e\dot{\tilde{q}}_e, \tag{9}$$

where $\tilde{\rho} \overset{\Delta}{=} \rho - \rho_d$ and $\tilde{q}_e \overset{\Delta}{=} q_e - q_{ed}$. The error dynamics were formulated and shown to preserve the passivity property for the modified input/output in (4), i.e., the mapping $\dot{\tilde{\rho}}_\mu = G(J_\theta^{-T}\tilde{\tau})$ is passive for $\mu < 1$. On the basis of the passivity theorem a simple PD feedback controller was selected to stabilize the system, so that the combined feedforward/feedback controller is given by:

$$\tau(t) = J_\theta^T(\theta, q_e)[M_{\rho\rho}(\rho)\ddot{\rho}_d + C_\rho(\rho, \dot{\rho})\dot{\rho}_d] - J_\theta^T(\theta, q_e)[K_d\dot{\tilde{\rho}}_\mu + K_p\tilde{\rho}_\mu]. \tag{10}$$

Given the passivity foundation of the scheme, a repertoire of different choices for the rate feedback part of the controller is available. Selecting a strictly positive real (SPR) controller is guaranteed to stabilize the passive system. Global asymptotic stability for the true tip tracking errors can be shown in a similar fashion to [4], using suitable Lyapunov arguments.

## 5. Adaptive Controller

In the case of rigid robots many different globally stable adaptive techniques have been proposed. A collection of some of the most important schemes is provided by Ortega & Spong [7] in a unified tutorial form. Important properties for their development are the linear dependence of a feedforward law on a set of suitably defined parameters, and the passivity of the mapping from joint torques to joint rates which is preserved in the error dynamics.

In the context of flexible-link robots, Damaren [4] has exploited the approximate form of the payload dominated dynamics derived in [3], in conjunction with the $\mu$-tip notion, to effectively realize the above properties for the modified input and output of the plant, and develop an adaptive scheme suitable for this class of manipulators. The adaptive scheme considered here, which is a natural extension of the original one, was constructed upon the motion dynamics as described by Eqs. (5) and (6).

The key property of linear parameter dependence of the dymamics can be realized on the basis of Eq. (5):

$$M_{\rho\rho}(\rho)\ddot{\rho}_d + C_\rho(\rho,\dot{\rho})\dot{\rho}_d = W(\ddot{\rho}_d,\dot{\rho}_d,\dot{\rho},\rho)\alpha, \tag{11}$$

where $W$ is called the *regressor matrix* and $\alpha$ is the vector of suitably selected parameters. The *filtered position rates*, $\dot{\rho}_r$, and a quantity which can be thought of as tracking accuracy, $s_\mu$, were defined as follows:

$$\dot{\rho}_r \overset{\Delta}{=} \dot{\rho}_d - \Lambda\tilde{\rho}_\mu , \quad s_\mu \overset{\Delta}{=} \dot{\tilde{\rho}}_\mu + \Lambda\tilde{\rho}_\mu , \quad \Lambda = \Lambda^T > O. \tag{12}$$

Based on Eq. (5) a feedforward strategy was developed and suitably modified using the above definitions to show that the mapping $s_\mu = G(J_\theta^{-T}\tilde{\tau})$ is passive for $\mu < 1$. As in [4], the control and parameter update laws were selected on the basis of the passivity theorem as follows:

$$\tau(t) = J_\theta^T(\theta,q_e)W(\ddot{\rho}_r,\dot{\rho}_r,\dot{\rho},\rho)\hat{\alpha}(t) - J_\theta^T(\theta,q_e)K_d s_\mu, \tag{13}$$

$$\dot{\hat{\alpha}} = -\Gamma W^T s_\mu , \quad \Gamma = \Gamma^T > O , \quad K_d = K_d^T > O, \tag{14}$$

with $\hat{\alpha}$ being the vector with the parameter estimates. In a similar fashion to [4], it was shown using Lyapunov arguments that the system can yield global asymptotic stability for the tracking errors $\tilde{\rho}$ and $\dot{\tilde{\rho}}$.

## 6. The Experimental Facility

Experimental results involving flexible-link robots are of great value since flimsy-link space robots can only undergo limited testing prior to launch. Unlike most of the relevant experimental work carried out in the field, which has focused on the one-link setup or two-links of which only one is flexible, our experiments involve the more realistic 3-link configuration with two flexible links.

Our results were obtained using an experimental facility which was designed and built in the Department of Mechanical Engineering at the University of Canterbury. The facility, which is shown in Figure 1, consists of two robotic arms possessing three rotational DOF and two flexible links each. The arms are planar, constrained to move in the horizontal plane so that gravity effects are not considered. A specially designed payload can be rigidly attached to the tip of each arm, and both its mass and moment of inertia can be varied. Each arm is supported on air–pads sliding on a glass–topped table in an almost frictionless fashion. For the arm under examination, the two flexible links are aluminium beams of cross section 6×30 mm and lengths 392 and 327 mm respectively. The length of the arm at the fully extended configuration is 1.08 m. The arm manipulates a 8.66 kg payload whose moment of inertia is 0.48 kg·m$^2$ with respect to a frame attached to the tip. The control of the system is based on a personal computer (PC) and a high–speed digital signal processor (DSP), which allows the execution of complex control algorithms at high sampling rates. The sampling frequency used for all our tests was 200 Hz. Our

Figure 1. The experimental facility

task–space controllers require feedback information from the joint rotations, which are measured using incremental encoders, and the Cartesian position and orientation of the end–point as well. End–point measurements are available from a 2–dimensional charge coupled device (CCD) camera located above the table. Although tip position feedback information can be provided by any suitable vision system, such solutions can sometimes be expensive and involve components external to the system. On the other hand, a method based on strain–gauge measurements can be advantageous in the sense that it can provide us with a low cost and more compact solution.

According to the method used for our tests, a number of $k$ strain gauges are attached at discrete positions along the length of each flexible link and then calibrated to provide the corresponding strain measurements. An $n$-th order polynomial is assumed to describe the deflected shape of each flexible link, the order of which is bounded by the number of strain gauges attached on the link $(n \leq k + 1)$. Given the success of cantilevered shape functions used in conjunction with the assumed modes method in modeling our system, polynomials satisfying cantilever boundary conditions at the inboard end were considered:

$$U_e(x,t) = q_{e,1}x^2 + q_{e,2}x^3 + \ldots + q_{e,n-1}x^n, \tag{15}$$

with $U_e$ the deflection, $x$ the distance along the length of the link, and the coefficients of the polynomial being a set of elastic DOF for the system. Mechanics of materials provide us with the strain–curvature relation for a flexible beam: $\varepsilon(x,t) = -(b/2)\ d^2U_e(x,t)/dx^2$, where $b$ is the thickness of the beam. Evaluating $\varepsilon$ at each of the $k$ measurement sites gives a linear system of simultaneous equations, the solution of which provides the elastic DOF of the arm. When $n < k+1$ the problem is overdetermined and a pseudoinverse least squares solution for the system can be obtained. Forward kinematics can then be applied to give the position of the end–point.

In our case three sets of strain gauges were attached on each flexible link, $(k = 3)$, and third order polynomials were considered, $(n = 3)$. The accuracy of the method was assessed for the planar case using the available measurements from the CCD camera.

## 7. Experimental Results – Discussion

For the implementation of the above class of controllers, direct measurements of the elastic coordinates $\mathbf{q}_e$ are required for the construction of the $\mu$–tip position and rate, and the Jacobian matrix $\boldsymbol{J}_\theta(\boldsymbol{\theta}, \mathbf{q}_e)$ as well. As in [3], in order to avoid the above problems and the need to calculate the trajectories for the elastic coordinates $\mathbf{q}_{ed}$ and $\dot{\mathbf{q}}_{ed}$, the following simplifications were incorporated: $\boldsymbol{J}_\theta(\boldsymbol{\theta}, \mathbf{q}_e) \doteq \boldsymbol{J}_\theta(\boldsymbol{\theta}, 0)$, $\rho_{\mu d} \doteq \rho_d$ so that $\dot{\tilde{\rho}}_\mu(t) \doteq [\mu\dot{\rho}(t) + (1-\mu)\boldsymbol{J}_\theta(\boldsymbol{\theta}, 0)\dot{\boldsymbol{\theta}}(t)] - \dot{\rho}_d(t)$ and $\tilde{\rho}_\mu(t) \doteq [\mu\rho(t) + (1-\mu)\mathcal{F}_r(\boldsymbol{\theta})] - \rho_d(t)$. On the basis of our numerical simulations and experiments, an appropriate value for $\mu$ is close to 1 which supports the validity of the above simplifications.

The desired end–effector trajectory, $\rho_d$, is taken to be such that all three end–effector DOF follow a quintic polynomial between an initial position, $\rho_i$, and a final position, $\rho_f$, in a certain length of time, $t_f$. Such trajectories, which are smooth both in velocity and acceleration, are commonly used in rigid robot motion planning:

$$\rho_d(t) = [10\left(\frac{t}{t_f}\right)^3 - 15\left(\frac{t}{t_f}\right)^4 + 6\left(\frac{t}{t_f}\right)^5](\rho_f - \rho_i) + \rho_i. \tag{16}$$

The initial position is taken to be the one corresponding to the rigid configuration $\boldsymbol{\theta}_i = [-\frac{\pi}{8}, \frac{\pi}{4}, 0]^T$ rad, the final one to $\boldsymbol{\theta}_f = [\frac{\pi}{8}, \frac{7\pi}{16}, \frac{\pi}{3}]^T$ rad, and $t_f = 3$ seconds. The deflections induced at the end of the first flexible link were found to exceed 23 mm. For the adaptive case, the same maneuver was used in a periodic fashion in order to allow more time for the adaptation to converge and better demonstrate the merits of adaptation.

## 7.1. Rigid inverse dynamics and passive feedback scheme

The most simple choice for the feedback gains would be a diagonal matrix. Here, the gains were selected as: $K_p = k_p \bar{J}_\theta^{-T} \bar{M}_{\theta\theta} \bar{J}_\theta^{-1}, K_d = k_d \bar{J}_\theta^{-T} \bar{M}_{\theta\theta} \bar{J}_\theta^{-1}$ with $k_p = 28$ and $k_d = 0.8$. The overbar notation, $(\bar{\cdot})$, denotes the matrix quantity evaluated at a constant rigid configuration, selected to be $\rho_d(t_f/2)$. Such a choice provides a natural weighting between the individual entries of the gain matrices corresponding to each one of the three task-space DOF.

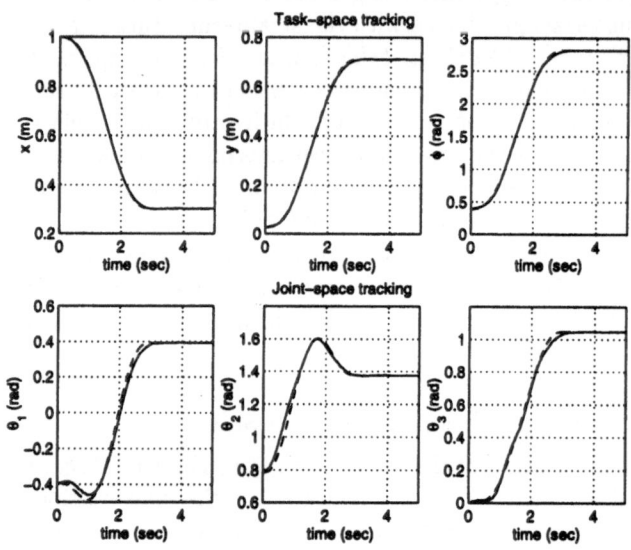

Figure 2. Tracking for the rigid inverse dynamics and passive feedback scheme (– – – desired/rigid, ——— actual)

In Figure 2, the three graphs at the top show the tip position tracking of each one of the three controlled task-space coordinates for $\mu = 0.92$. Both the actual and desired trajectories are shown on the same plots for comparison. It is clear that very good tracking is achieved in all three task-space coordinates, without any residual vibrations after the completion of the useful motion. The three graphs at the bottom of the same figure show the actual joint-space motion, which is drawn in solid line, and the dashed one is the trajectory the corresponding rigid robot would have to follow in order to yield the desired task-space motion. It is not surprising that the two trajectories are different, given that the end-point position of a flexible robot not only depends on the joint rotations, but on the link deflections as well.

## 7.2. Adaptive scheme

For the experiments involving the adaptive scheme, 10 parameters of the robot and the manipulated payload were updated on-line, i.e., $\alpha = [m_2, m_3, I_1, I_2, I_3, c_{x2}, c_{x3}, I_{r1}, I_{r2}, I_{r3}]^T$, where $m_i$ is the mass, $c_{xi}$ is the first moment of inertia, and $I_i$ is the second moment of inertia of the $i$-th link of the corresponding 3-link rigid robot. Parameter $I_{ri}$ is the inertia of rotor $i$. All parameters in $\boldsymbol{\alpha}$ were considered to be completely unknown and their initial estimates were set to zero.

The feedback gains matrix was selected in a fashion similar to the non-adaptive case as $\boldsymbol{K_d} = k_d \bar{\boldsymbol{J}}_\theta^{-T} \bar{\boldsymbol{M}}_{\theta\theta} \bar{\boldsymbol{J}}_\theta^{-1}$ with $k_d = 1.2$, and the weighting matrix $\boldsymbol{\Lambda} = 1$. The adaptation gains matrix was taken to be diagonal and tuned experimentally: $\boldsymbol{\Gamma} = diag\{1, 10, 0.4, 0.7, 0.05, 15, 0.3, 0.03, 0.0005, 0.2\}$. The relative size of its entries is directly related to the persistency of excitation (PE) property concerning the desired trajectory, *i.e.*, parameters difficult to identify require that the corresponding entry in the adaptation gains matrix is large.

Figure 3 shows the tracking for each task–space coordinate for $\mu = 0.92$. Comparing each subsequent cycle, it becomes clear how tracking performance improves with time due to the adaptation. The trajectories for some of the adaptively updated parameters are shown in Figure 4, all of which converge to the vicinity of some constant values.

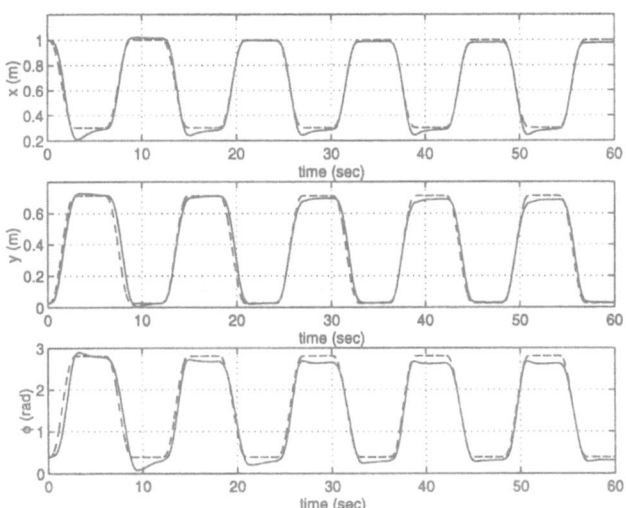

Figure 3. Task–space tracking for the adaptive scheme ($---$ desired, —— actual)

## 8. Conclusions

The paper has indicated that the dynamics of flexible robots carrying large payloads possess some special characteristics which can be effectively exploited for control design. The updated versions of two controllers derived on that assumption were summarized and the actual hardware implementation has demonstrated their applicability, and their ability to provide good task–space tracking with simultaneous active vibration suppression.

The most difficult problem related to the implementation of the controllers was found to be the quality of the velocity measurements, which restricts the

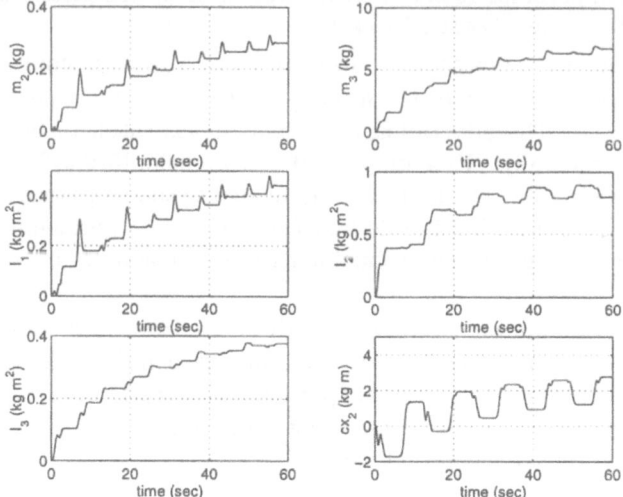

Figure 4. Parameter estimates for the adaptive scheme

achievable performance. All our velocity feedback information was obtained by differencing the available position measurements, and low order Butterworth filters were used to considerably improve the quality of these signals.

Extensive experimental tests have been performed which explore the controller design problem and various implementation issues. The robustness of the schemes with respect to uncertainty related to the size of the manipulated payload has also been demonstrated. Future work will involve more theoretical and experimental investigations concerning the two schemes. Extensions to the closed–loop configuration, representing the case of two cooperating flexible arms manipulating the same payload, are currently under study.

## References

[1] C. J. Damaren. Passivity analysis for flexible multilink space manipulators. *AIAA J. of Guidance, Control, and Dynamics*, 18(2):272–279, 1995.

[2] C. A. Desoer and M. Vidyasagar. *Feedback Systems: Input-Output Properties*. Academic Press, 1975.

[3] C. J. Damaren. Approximate inverse dynamics and passive feedback for flexible manipulators with large payloads. *IEEE Trans. on Robotics and Automation*, 12(1):131–138, 1996.

[4] C. J. Damaren. Adaptive control of flexible manipulators carrying large uncertain payloads. *J. of Robotic Systems*, 13(4):219–228, 1996.

[5] C. J. Damaren. Modal properties and control system design for two-link flexible manipulators. *Int. J. Robotics Research*, 17(6):667–678, 1998.

[6] E. G. Christoforou and C. J. Damaren. Passivity–based controllers for flexible–link manipulators carrying large payloads. In *Proc. of IPENZ Annual Conf.*, volume 2, pages 249–254, Auckland, New Zealand, 1998.

[7] R. Ortega and M. W. Spong. Adaptive motion control of rigid robots: a tutorial. *Automatica*, 25(6):877–888, 1989.

# Identification of Joint Elasticity of Industrial Robots

G.E. Hovland[†], E. Berglund[*] and O.J. Sørdalen[†]

[†]ABB Corporate Research,
Information Technology and Control Systems Division,
Bergerveien 12, N-1375 Billingstad, Norway.
E-mail: {geir.hovland,ole.sordalen}@nocrc.abb.no

[*]Department of Engineering Cybernetics,
The Norwegian University of Science and Technology,
N-7034 Trondheim, Norway.
E-mail: Einar.Berglund@itk.ntnu.no

**Abstract:** *In this paper we present a new method for the identification of joint elasticities of industrial robots. The presented method is able to identify the elasticity parameters in presence of coupled dynamics between two joints. An accurate model description of the joint elasticities is absolutely necessary to achieve accurate path tracking of industrial robots at high speeds. The industrial requirements for path accuracy in applications such as spray painting, waterjet cutting and assembly are so high that elasticity effects have to be accounted for by the control system.*

## 1. Introduction

The modelling of modern robot systems as rigid mechanical systems is an unrealistic simplification. The industrial requirements for path accuracy in applications such as spray painting, waterjet cutting, plasma cutting and assembly are so high that elasticity effects have to be accounted for by the control system. There are two sources of vibration in robot manipulators: 1) joint elasticity, due to the elasticity of motion transmission elements such as harmonic drives, gear-boxes, belts or long shafts [8], and 2) link elasticity, introduced by a long reach and slender/lightweight construction of the arm, [1, 5].

Many industrial robot systems use elastic transmission elements which introduce joint elasticity. Given a dynamic model description including the rigid-body dynamics and joint elasticity, several control algorithms have been proposed in the literature, see for example [2] for a survey. To be able to implement such control algorithms and achieve the strict path tracking requirements from industry, one needs very accurate parameter estimates of the stiffness and damping constants of the transmission elements.

In this paper we present a new approach to the identification problem of joint elasticity. The proposed solution assumes that the rigid-body dynamics are known. For a rigid-body dynamics identification method, see for example the work by [9]. We show that when the rigid-body dynamics are known the joint elasticity identification becomes a linear problem. The linear formulation avoids the problems associated with non-linear estimation techniques, such as proper selection of initial values and convergence. Another advantage of the method presented in this paper is the little amount of measurement data required. In principle, only one sinoid measurement series is required to identify four elasticity parameters (spring and damper constants for two coupled joints). To increase the accuracy of the estimated parameters, additional measurement series can easily be utilised.

Most industrial robots do not have any velocity or acceleration sensors. Instead, the velocity and acceleration measurements are normally generated by numerical differentiation. In this paper we present a different approach where the position measurements are approximated by Fourier coefficients and the velocity and acceleration data are found analytically. The approach used in our paper avoids the troublesome noise problems associated with numerical differentiation.

The problem of controlling robots with joint elasticities has received a significant amount of attention in the literature, see for example [3, 4, 6, 10]. The identification problem, however, has received less attention. The work by De Luca, for example, assumes that the joint elasticity parameters are known and the elasticity parameters are used directly in the design of the control laws. Hence, the methods and results presented in our paper will be directly beneficial to several of the existing control strategies in the literature.

## 2. Model of Joint Dynamics

Figure 1 presents a simplified illustration of a robot with two elastic joints with coupled dynamics. The elasticity model of these joints with coupled inertia terms is shown in Figure 2. The goal of the identification algorithm is to identify $K_1$, $D_1$, $K_2$ and $D_2$ when the rigid-body inertia parameters $J_{m1}$, $J_{m2}$, $J_{a1}$, $J_{a2}$, $J_{12}$ and $J_{21}$ are known. The equations of motion are given by

$$J_{m1}\ddot{\theta}_{m1} + D_1(\dot{\theta}_{m1} - \dot{\theta}_{a1}) + K_1(\theta_{m1} - \theta_{a1}) = \tau_{m1}$$
$$J_{a1}\ddot{\theta}_{a1} + J_{12}\ddot{\theta}_{a2} + D_1(\dot{\theta}_{a1} - \dot{\theta}_{m1}) + K_1(\theta_{a1} - \theta_{m1}) = 0$$
$$J_{m2}\ddot{\theta}_{m2} + D_2(\dot{\theta}_{m2} - \dot{\theta}_{a2}) + K_2(\theta_{m2} - \theta_{a2}) = \tau_{m2}$$
$$J_{a2}\ddot{\theta}_{a2} + J_{21}\ddot{\theta}_{a1} + D_2(\dot{\theta}_{a2} - \dot{\theta}_{m2}) + K_2(\theta_{a2} - \theta_{m2}) = 0$$

$$(1)$$

Note that the set of equations (1) do not contain any velocity dependent coupling terms, such as Coriolis or centripetal forces. The main reason why we have excluded these terms, is to make the system linear in all state variables $(\theta_{m1}, \dot{\theta}_{m1})$, $(\theta_{a1}, \dot{\theta}_{a1})$, $(\theta_{m2}, \dot{\theta}_{m2})$ and $(\theta_{a2}, \dot{\theta}_{a2})$. The Coriolis and Centripetal forces contain nonlinear terms such as $\dot{\theta}_{a1}\theta_{a2}$ and $\dot{\theta}_{a1}^2$.

A linear model significantly simplifies the identification algorithm and the model parameters can be found directly from a frequency response analysis.

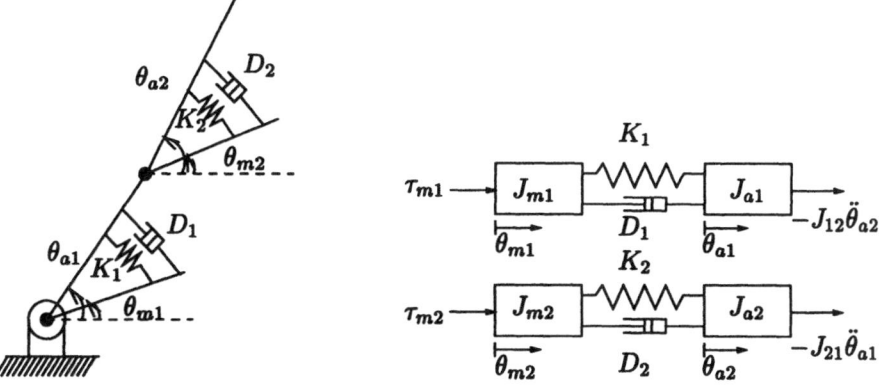

Figure 1. *Visualisation of robot with two dynamically coupled joints and joint elasticities. The differences between the motor variables $\theta_m$ and the arm variables $\theta_a$ are exaggerated to illustrate the elasticity parameters.*

Figure 2. *Two-mass elastic model of two joints with coupled-inertia dynamics.*

However, care should be taken when the motions of the robot joints are generated during the identification experiments to keep the velocity terms as low as possible.

## 3. Identification Algorithm

In this section we present the three steps of the identification algorithm; in 3.1 the generation of Fourier coefficients from a sinoid excitation of the robot joints and in 3.2 the identification of the model parameters.

### 3.1. Fourier Coefficients

Let $f(t)$ be a measurement signal, either joint position or joint torque. In the experiments these signals are generated as sinoid functions. Hence, to remove the effects from measurements noise, friction and other nonlinear terms, we approximate the measurement $f(t)$ by a pure sinoid function using Fourier coefficients given by

$$
a = \frac{1}{L/2} \int_0^L (f(t) \sin(\omega t)) \, dt
$$

$$
b = \frac{1}{L/2} \int_0^L (f(t) \cos(\omega t)) \, dt \tag{2}
$$

where $\omega$ is the natural frequency of the generated measurement series and $L$ is an integer number times the period of sinoid. $L$ should be chosen larger than one period to achieve time averaging of the data and further reduce the effects of noise and nonlinearities.

Finally, a complex number representation of the sinoid measurement is given by

$$y = a + ib \tag{3}$$

The complex number $y$ contains both amplitude and phase information and will be used in the next section to find the transfer function matrix. Given the complex number $y$ for the position measurements, the velocities and accelerations are found analytically, ie.

$$
\begin{aligned}
\theta &= a + ib \\
\dot{\theta} &= \theta(i\omega) \\
\ddot{\theta} &= \theta(i\omega)^2
\end{aligned}
\tag{4}
$$

where $\omega$ is the known excitation frequency in $(rad/sec)$. In the experiments, the Fourier coefficients are found for several different frequencies $\omega$ in the robot's bandwidth. The analytical generation of the velocity and acceleration signals is a major advantage. For most robot systems, only joint position measurements are available from encoder data. By differentiation the encoder data twice, a large amount of quantization noise is introduced in the acceleration data. The analytical generation of velocity and acceleration from sinoid functions avoids the quantization noise problem. By time-averaging over a large number of periods (L large), the influence from quantization noise in the velocity and acceleration estimates is neglectable.

### 3.2. Identification of Model Parameters

First, we write the system equations (1) on vector form.

$$Ms^2q + Ds(q - q_a) + K(q - q_a) = \tau \tag{5}$$
$$M_2s^2q_a + Ds(q_a - q) + K(q_a - q) = 0 \tag{6}$$

where $s$ is the Laplace transform. The matrices and vectors are given by

$$
\begin{aligned}
q &= \begin{bmatrix} \theta_{m1} \\ \theta_{m2} \end{bmatrix}, \quad q_a = \begin{bmatrix} \theta_{a1} \\ \theta_{a2} \end{bmatrix}, \quad \tau = \begin{bmatrix} \tau_{m1} \\ \tau_{m2} \end{bmatrix} \\
M &= \begin{bmatrix} J_{m1} & 0 \\ 0 & J_{m2} \end{bmatrix}, \quad M_2 = \begin{bmatrix} J_{a1} & J_{12} \\ J_{21} & J_{a2} \end{bmatrix} \\
D &= \begin{bmatrix} D_1 & 0 \\ 0 & D_2 \end{bmatrix}, \quad K = \begin{bmatrix} K_1 & 0 \\ 0 & K_2 \end{bmatrix}
\end{aligned}
\tag{7}
$$

By substitution of equation (6) into equation (5) the motor dynamics can be written on the following linear form in $D$ and $K$.

$$D[(I + M_2^{-1}M)s^3q - M_2^{-1}s\tau] + K[(I + M_2^{-1}M)s^2q - M_2^{-1}\tau] = s^2\tau - Ms^4q \tag{8}$$

From the Fourier coefficients of $\mathbf{q}$ and $\tau$, we generate the following vectors.

$$\begin{bmatrix} a_1 \\ a_2 \end{bmatrix} = (\mathbf{I} + \mathbf{M}_2^{-1}\mathbf{M})s^2\mathbf{q} - \mathbf{M}_2^{-1}\tau \tag{9}$$

$$\begin{bmatrix} a_3 \\ a_4 \end{bmatrix} = (\mathbf{I} + \mathbf{M}_2^{-1}\mathbf{M})s^3\mathbf{q} - \mathbf{M}_2^{-1}s\tau = i\omega \begin{bmatrix} a_1 \\ a_2 \end{bmatrix} \tag{10}$$

$$\begin{bmatrix} b_1 \\ b_2 \end{bmatrix} = s^2\tau - \mathbf{M}s^4\mathbf{q} \tag{11}$$

We then have

$$\begin{aligned}
\mathbf{A} &= \begin{bmatrix} \Re(i\omega a_1) & \Re(a_1) & 0 & 0 \\ \Im(i\omega a_1) & \Im(a_1) & 0 & 0 \\ 0 & 0 & \Re(i\omega a_2) & \Re(a_2) \\ 0 & 0 & \Im(i\omega a_2) & \Im(a_2) \end{bmatrix} \\[2mm]
&= \begin{bmatrix} -\omega\Im(a_1) & \Re(a_1) & 0 & 0 \\ \omega\Re(a_1) & \Im(a_1) & 0 & 0 \\ 0 & 0 & -\omega\Im(a_2) & \Re(a_2) \\ 0 & 0 & \omega\Re(a_2) & \Im(a_2) \end{bmatrix} \\[2mm]
\mathbf{b} &= \begin{bmatrix} \Re(b_1) \\ \Im(b_1) \\ \Re(b_2) \\ \Im(b_2) \end{bmatrix}, \quad \mathbf{p} = \begin{bmatrix} D_1 \\ K_1 \\ D_2 \\ K_2 \end{bmatrix}
\end{aligned}$$

$$\mathbf{A}\mathbf{p} = \mathbf{b} \tag{12}$$

The parameter vector $\mathbf{p}$ can be identified from only one sinoid measurement series when we use both the real and imaginary parts of the Fourier coefficients. To achieve a robust identification result, it is advisable to use several sinoid measurement series with different frequencies. The $\mathbf{A}$ matrix and the $\mathbf{b}$ vector will then have $4n$ rows, where $n$ is the number of different frequencies. The parameter vector $\mathbf{p}$ is then simply found from the pseudo-inverse, ie.

$$\mathbf{p} = \mathbf{A}^+\mathbf{b} \tag{13}$$

$$\mathbf{A}^+ = (\mathbf{A}^T\mathbf{A})^{-1}\mathbf{A}^T \tag{14}$$

The pseudo-inverse is equal to a least-squares estimation method, see for example [7]. The explicit solutions for $D_1$, $K_1$, $D_2$ and $K_2$ found by solving equations (13)-(14) are given below.

$$D_1 = -\frac{\sum_{i=1}^{n} \Im(a_1)\Re(b_1) - \Re(a_1)\Im(b_1)}{\sum_{i=1}^{n} \omega_i(\Re(a_1)^2 + \Im(a_1)^2)}$$

$$K_1 = \frac{\sum_{i=1}^{n} \Re(a_1)\Re(b_1) + \Im(a_1)\Im(b_1)}{\sum_{i=1}^{n} \Re(a_1)^2 + \Im(a_1)^2}$$

$$D_2 = -\frac{\sum_{i=1}^{n} \Im(a_2)\Re(b_2) - \Re(a_2)\Im(b_2)}{\sum_{i=1}^{n} \omega_i(\Re(a_2)^2 + \Im(a_2)^2)}$$

$$K_2 = \frac{\sum_{i=1}^{n} \Re(a_2)\Re(b_2) + \Im(a_2)\Im(b_2)}{\sum_{i=1}^{n} \Re(a_2)^2 + \Im(a_2)^2} \tag{15}$$

where $n$ is the number of frequencies.

## 4. Experiments

In this section we present the experimental results from the elasticity identification of an ABB robot used for waterjet cutting at a customer site in Sweden. We also present the direct improvements achieved in path tracking performance when using the identified elasticity parameters in the control structure.

Section 4.1 describes the non-parametric identification of the multivariable transfer function matrix. The elements in this matrix are used to verify the identification results. In section 4.2 we present the experimental results from the identification algorithm developed in section 3. Finally, in section 4.3 we present the improvements in path tracking performance for a waterjet cutting application.

### 4.1. Identification of Transfer Function Matrix

Only the motor state variables $(\theta_{m1}, \dot{\theta}_{m1})$ and $(\theta_{m2}, \dot{\theta}_{m2})$ can be measured and approximated by Fourier coefficients. Hence, for the identification of the model parameters we only use the $2 \times 2$ subsystem of equation (1) corresponding to the motor variables.

$$\begin{bmatrix} \ddot{\theta}_{m1} \\ \ddot{\theta}_{m2} \end{bmatrix} = \begin{bmatrix} H_{11} & H_{12} \\ H_{21} & H_{22} \end{bmatrix} \begin{bmatrix} \tau_{m1} \\ \tau_{m2} \end{bmatrix} \tag{16}$$

Let $y_1$ and $y_3$ be the complex numbers of $\ddot{\theta}_{m1}$ and $\ddot{\theta}_{m2}$ for one measurement series. Let $\tau_1$ and $\tau_3$ be the corresponding complex numbers for the joint torques $\tau_{m1}$ and $\tau_{m2}$, respectively. Similarly, $y_2, y_4, \tau_2$ and $\tau_4$ describe a second measurement series. We then have

$$\begin{bmatrix} y_1 \\ y_2 \\ y_3 \\ y_4 \end{bmatrix} = \begin{bmatrix} \tau_1 & \tau_3 & 0 & 0 \\ \tau_2 & \tau_4 & 0 & 0 \\ 0 & 0 & \tau_1 & \tau_3 \\ 0 & 0 & \tau_2 & \tau_4 \end{bmatrix} \begin{bmatrix} H_{11}(i\omega) \\ H_{12}(i\omega) \\ H_{21}(i\omega) \\ H_{22}(i\omega) \end{bmatrix} \tag{17}$$

For each frequency $\omega$, the four complex transfer function elements are simply found by inversion of the $\tau$ matrix, ie.

$$\begin{bmatrix} H_{11}(i\omega) \\ H_{12}(i\omega) \\ H_{21}(i\omega) \\ H_{22}(i\omega) \end{bmatrix} = \begin{bmatrix} \tau_1 & \tau_3 & 0 & 0 \\ \tau_2 & \tau_4 & 0 & 0 \\ 0 & 0 & \tau_1 & \tau_3 \\ 0 & 0 & \tau_2 & \tau_4 \end{bmatrix}^{-1} \begin{bmatrix} y_1 \\ y_2 \\ y_3 \\ y_4 \end{bmatrix} \tag{18}$$

To identify the entire transfer function matrix $H$, two measurement series generating sinoid functions in $\ddot{\theta}_{m1}$ and $\ddot{\theta}_{m2}$ are required for several frequencies $\omega$ within the joint bandwidths.

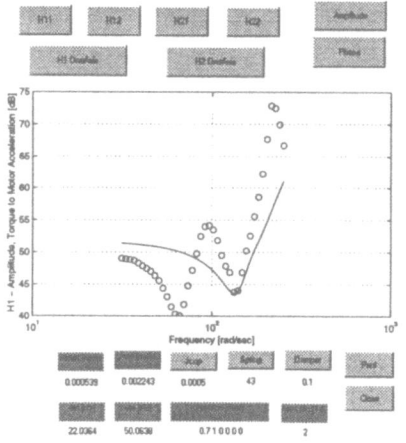

**Figure 3.** *Measured transfer function* $|H_{11}(i\omega) + H_{12}(i\omega)\frac{\tau_{m2}}{\tau_{m1}}|$ *(dotted) and model* $|\frac{\ddot{\theta}_{m1}}{\tau_{m1}}|$ *(solid) with estimated elasticity parameters* $D_1 = 0.10$, $K_1 = 43.0$, $D_2 = 0.015$ *and* $K_2 = 7.5$.

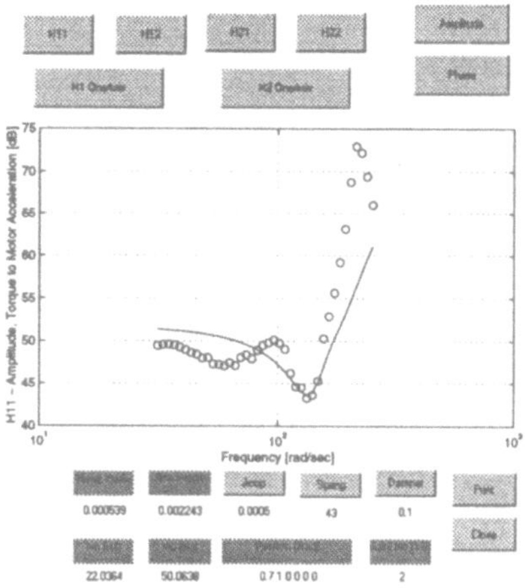

**Figure 4.** *Estimated transfer function* $|H_{11}(i\omega)|$ *(dotted) and model* $|\frac{\ddot{\theta}_{m1}}{\tau_{m1}}|$ *(solid) with estimated elasticity parameters* $D_1 = 0.10$, $K_1 = 43.0$, $D_2 = 0.015$ *and* $K_2 = 7.5$.

## 4.2. Identification of Elasticity Parameters

In Figure 3 the dotted curve shows the measured transfer function from motor torque to motor acceleration. During the experiments, the controller loops on all the joints are activated which is necessary to prevent the manipulator from collapsing due to the gravity torque. Note that two resonance modes are visible in the transfer function in Figure 3. One of these resonance modes is caused by the dynamic inertia coupling between the joints and the controller loop of the coupled joint. Hence, to identify the spring and damper constants directly from the measurements of Figure 3 is difficult.

The dotted curves in the Figures 4 and 5 show the estimated transfer functions $H_{11}$ and $H_{22}$ found using equation (18). Notice that these two transfer functions both have only one dominating resonance mode. The four elasticity parameters $K_1 = 43.0$, $K_2 = 7.5$, $D_1 = 0.10$ and $D_2 = 0.015$ were estimated using the method described in equations (5)–(15) when all the inertia parameters $\mathbf{M}$ and $\mathbf{M}_2$ were known.

To verify that the estimated results correspond to the estimated transfer functions $H_{11}$ and $H_{22}$ from section 4.3, we have plotted the model (solid curve) in all the Figures 3–5. We see from Figure 3 that the resonance mode caused by the joint elasticity, corresponds to the second elasticity mode in the measured transfer function. Figures 4 and 5 show that the identified model given $K_1$, $K_2$, $D_1$ and $D_2$ match the estimated transfer functions $H_{11}$ and $H_{22}$ well. We believe that the differences in the Figures are caused by unmodelled dynamics, such as the nonlinear velocity terms or joint friction.

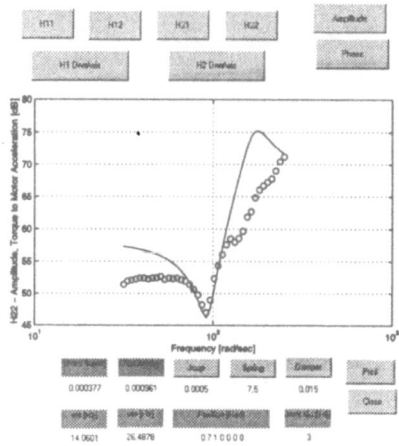

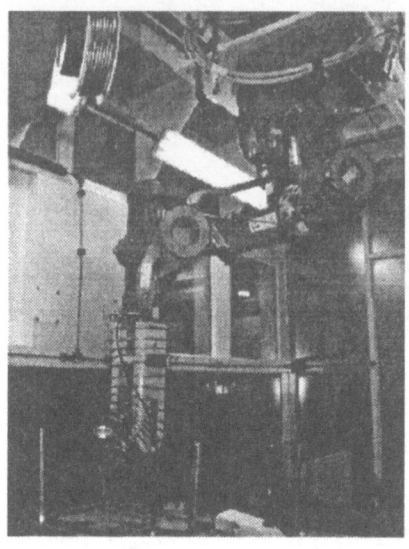

Figure 5. *Estimated transfer function* $|H_{22}(i\omega)|$ *(dotted) and model* $|\frac{\ddot{\theta}_{m2}}{\tau_{m2}}|$ *(solid) with estimated elasticity parameters* $D_1 = 0.10$, $K_1 = 43.0$, $D_2 = 0.015$ *and* $K_2 = 7.5$.

Figure 6. *Experimental setup consisting of a Waterjet cutting cell and an ABB robot.*

### 4.3. Waterjet Cutting Application

To demonstrate the importance of accurate estimates of the elasticity parameters, we present results from tests in a waterjet cutting cell from ABB I-R Waterjet Systems in Sweden. ABB I-R is a global leader in sophisticated three-dimensional robotized waterjet cutting systems for trimming and finishing components in general manufacturing and the automobile industry. The company is also a leading manufacturer of cutting tables and associated systems for two-dimensional waterjet cutting. Waterjet cutting is used in a wide range of industries: interior finish on vehicles, aerospace manufacture, metal fabrication, stone cutting, textile industries, automotive and truck manufacture, sheet metal industries, plastic processing, architectural metal and glass cutting and food processing. The path tracking requirements for finished products are extremely high. Using manipulators with elastic joints, the requirements from industry can only be satisfied with accurate estimates of the elasticity parameters.

The cutting cell is shown in Figure 6, where an ABB robot is mounted in the ceiling and the cutting is performed in the horizontal plane in a $1mm$ thick plastic material. Figure 7 shows the outlines of the cutting paths. The squares have dimension $28mm \times 45mm$. The Cartesian cutting speed was chosen $300mm/sec$. The three shapes at the top of the figure were cut with a non-optimal choice of elasticity parameters in the servo controllers. The right-most figure has the largest corner zones and an oscillatory behaviour is clearly visible. In fact, the largest path tracking error for this particular test was several millimeters. The three shapes at the bottom of the figure show the same tests when using the identified elasticity parameters in the control

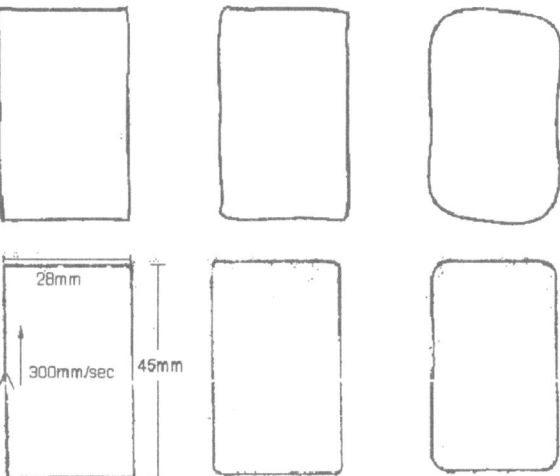

Figure 7. *Top: Cutting path with non-optimised elasticity parameters. Bottom: Cutting path with identified elasticity parameters.*

structure. This time the oscillatory behaviour is significantly reduced. With the identified elasticity parameters, the path tracking performance of the ABB I-R Waterjet cell is well inside the requirements for the industrial waterjet cutting applications mentioned above.

## 5. Conclusions

In this paper we have presented a new approach to the identification of joint elasticities between two robot joints with coupled dynamics. The main advantage of the method is the linear problem formulation in the elasticity parameters (joint stiffness and damping). The linear formulation allows for a straightforward estimation of the parameters using a weighted least squares approach. Hence, typical problems associated with non-linear estimation techniques, such as convergence problems and proper selection of initial parameter values are avoided.

The results show that the identification method is able to compute accurate parameter estimates with a good match between the identified model and measured transfer function responses. The identified elasticity parameters can be used directly in the control structure to achieve accurate path tracking performance at high speeds of industrial robots.

The method has been tested in a waterjet cutting application. When using the identified elasticity parameters, the waterjet cutting cell is able to meet strict industrial requirements with large cutting velocities.

# References

[1] Book W.J. (1988): "Modelling, design, and control of flexible manipulator arms: A tutorial review", *Proceedings of the 29th IEEE Conference on Decision and Control*, Honolulu, HI, pp.500-506.

[2] De Luca A. (1998a): "Trajectory Control of Flexible Manipulators", *Control Problems in Robotics and Automation*, pp. 83-104, Editors: B. Siciliano and K.P. Valavanis, Springer.

[3] De Luca A. (1988b): "Dynamic Control of Robots with Joint Elasticity", *Proceedings of the 29th IEEE Conference on Decision and Control*, Honolulu, HI, pp.500-506.

[4] De Luca A. (1988c): "Control Properties of Robot Arms with Joint Elasticity", In: Byrnes C.I., Martin C.F., Saeks R.E. (eds), *Analysis and Control of Nonlinear Systems*, North-Holland, Amsterdam, The Netherlands, pp. 61-70.

[5] Fraser A.R. and R.W. Daniel (1991): *Perturbation Techniques for Flexible Manipulators*, Kluwer, Boston, MA.

[6] Kelly R. and V. Santibanez (1998): "Global Regulation of Elastic Joint Robots Based on Energy Shaping", *IEEE Transactions on Automatic Control*, Vol. 43, No. 10, pp.1451-1455.

[7] Strang G. (1988): *Linear Algebra and its Applications*, Harcourt Brace Jovanovich, Inc., San Diego.

[8] Sweet L.M. and M.C. Good (1985): "Redefinition of the Robot Motion Control Problem", *IEEE Control System Magazine*, Vol. 5, No. 3, pp. 18-24.

[9] Swevers J., et.al. (1997): "Optimal Robot Excitation and Identification", *IEEE transactions on robotics and automation*, Vol. 13, No. 5, pp. 730-740.

[10] Wilson, G.A. and G.W. Irwin (1994): "Robust Tracking of Elastic Joint Manipulators Using Sliding Mode Control", *Transactions of the Institute of Measurement and Control*, Vol. 16, No. 2, pp.99-107.

# Modelling Friction in Multi-Loop Linkages

C.R. Tischler, S.R. Lucas and A.E. Samuel
Department of Mechanical and Manufacturing Engineering
The University of Melbourne
Parkville, Victoria, Australia
craigt@, stuart@, aes@mame.mu.oz.au

**Abstract:** For a serial robot, the friction in each joint is overcome by the actuator driving that joint. This is not the case, however, for a parallel or multi-loop robot. These linkages contain unactuated joints whose movement is controlled by the geometry of the linkage and the motion of the actuated joints. The actuators must work to overcome the frictional forces in the passive joints, and the question of how this work is apportioned amongst them is the topic of this paper. We also explore the degradation, by joint friction, of the linkage's ability to apply forces. A technique for constructing a force model of the linkage is briefly described before being applied to the Melbourne Dextrous Finger, which has a multi-loop structure. Experimental results show the benefit of the predictions.

## 1. Introduction

Design procedures that generate robot linkages capable of applying precise forces have received less attention than procedures that deliver robots with precise control of position. In some applications, however, precise control of forces is perhaps the more important. Robot grasping, for example, requires good resolution of the grasp forces to ensure both equilibrium and stability of the grasped object, and this sets criteria for the design of robot hands.

This paper explores the governing role of joint friction in the performance of multi-loop kinematic chains, when they are used to apply forces. Joint friction in a serial linkage is overcome by the direct actuation of the joint. In contrast, multi-loop linkages have some unactuated joints and the work done against joint friction in these passive joints is performed by actuators located elsewhere in the linkage. The effort to overcome friction is shared between the actuators in a way which depends upon the configuration of the robot and the load being exerted.

Here we model the transmission of forces through a linkage using Davies' method of virtual power in mechanical networks [1]. This method is particularly well suited for tracking the action of applied forces and torques, and finding where each manifests itself as reactions in the joints. Knowing the joint reactions allows us to infer the joint friction using Coulomb's law or some other suitable friction model. In turn, these friction forces influence the joint reactions, and an iterative process must be used to arrive at a solution.

The Melbourne Dextrous Finger was analysed in this way, and compared with experimental results, obtained by statically measuring the actuator forces and finger-tip loads. Because there was no motion of the finger, the senses of frictional forces were assumed to act in opposition to the actuator forces. Frictional forces can also act to assist the actuator forces if the finger is being back-driven, and since both situations are likely to occur in the operation of the hand, the model predicts a range in which the measured forces are expected to lie. Ideally, this range should be narrow. We present results for the performance of the Melbourne Finger and identify design refinements that flow from our friction predictions.

## 2. Related Research

Armstrong-Hélouvry [2] is a careful blending of tribology and control theory, and its application to the first three joints of a PUMA 560. This work shows, however, that the frictional behaviour of the PUMA 560 is relatively unaffected by changes in the loading of the robot, implying that most of the friction is caused by the preloading of bearings and drive-chains, which is required to reduce the amount of slop or backlash to acceptable levels. Under such conditions, and at low speeds, the joint reactions are effectively constant. For the Melbourne Dextrous Finger, few of the joints require preloading to reduce the slop in the location of the finger-tip [3]. The joint reactions are then functions of forces being applied to the finger.

Techniques for identifying the parameters of friction models that accurately describe the behaviour of robot drives have been investigated [4, 5, 6] and this is a necessary step in the design and implementation of appropriate compensation schemes for robot controllers [7, 8]. In compensator design the aim is to get the best response from a given device but, unfortunately, this does not aid the design of new, and possibly better, robot linkages, which is our primary concern. In the work presented here, we aim for a procedure which will provide guidance for designers as to how best to minimise the intrusion of friction into the performance of a proposed linkage. After all, even the best control scheme cannot extract good responses from a poorly designed robot which is prone to jamming or clamping.

Methods for the analysis of forces in linkages are well established. Hain [9] gives a good account for planar linkages while Paul [10] gives a method for the analysis of friction in planar linkages. Force analysis in spatial linkages is more difficult and we use Davies' virtual power method, which is described in Section 3, because the format of the solution is well suited to our multi-loop linkage. Nevertheless, we admit that a simple application of static equilibrium, body by body, should also give an adequate although less direct result.

## 3. A Model for Distributing Frictional Forces

Davies' virtual power method [1, 11] for the analysis of loads in mechanical networks draws an analogy with current in an electrical network. The linkage can be represented as a graph, where the bodies are nodes and the joints are arcs. A set of independent loops in the linkage are identified and a circuit

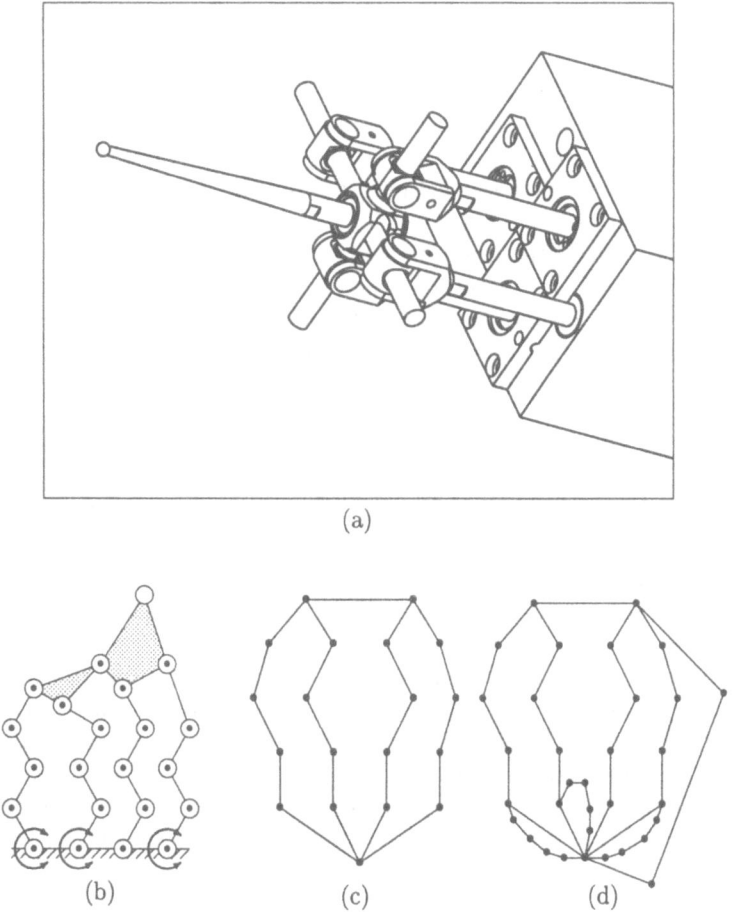

(a)

(b)        (c)        (d)

*Figure 1. The Melbourne Dextrous Finger, (a) a CAD model, (b) a schematic of the kinematic chain, and (c) its graph, which is extended in (d) to account for loads applied at the finger-tip and actuators.*

wrench is assigned to each. The circuit wrenches are unknown but must satisfy Kirchhoff's nodal law, which is essentially the condition of static equilibrium. The reaction in any joint is then the sum of the circuit wrenches passing through the corresponding arc in the graph.

Davies' work should be consulted for the details of the method, but for brevity here we give only an application of it to the Melbourne finger, which is shown in Figure 1(a). The kinematic chain of this device is shown in Figure 1(b) and this is an intermediate step towards the graph shown in Figure 1(c). The details of the drive have been omitted for simplicity.

For each force or load applied to the linkage another path is added to the graph between the bodies acted on by the force, thus increasing the number of loops in the graph by one. The additional loops formed when finger-tip and actuators loads are included is shown in Figure 1(d). The number of freedoms in the added paths is governed by the dimension of the space spanned by the applied force. For the actuators where a simple force is to be applied, the space is one dimensional and there must be $(6 - 1 =)5$ freedoms in the added path. Assuming a simple frictional point contact is made between the finger-tip and the grasped object, any pure force acting through the centre of the finger-tip is admissible and these forces span a three system. Consequently there are $(6 - 3 =)3$ freedoms in the added path around the finger-tip. There are seven independent loops in Figure 1(d).

Each freedom in the graph is assigned an arbitrary direction and, using these, a loop matrix $\mathbf{B}$ can be constructed with one column for each joint and one row for each loop (i.e. $\mathbf{B}$ is a $7 \times 39$ matrix). The elements of the matrix indicate the freedoms traversed by each loop and whether the sense of the joint corresponds to the sense of the loop.

We next need to form the motion screw matrix $\mathbf{M}$ which describes the joint freedoms. The screw coordinates of the freedoms are expressed in a global coordinate frame and stacked to form $\mathbf{M}$ with one set of screw coordinates per row, (i.e. $\mathbf{M}$ is a $39 \times 6$ matrix). For the Melbourne Finger it is easy to write the screw coordinates symbolically so that they follow the movement of the linkage [12]. The freedoms in the added action loops about the actuators and finger-tip must be linearly independent and reciprocal [13] to the wrenches which are to be applied to the linkage at that point.

The freedom matrix $\mathbf{F}$ of the linkage can be assembled by taking each row of the loop matrix $\mathbf{B}$, writing it as a diagonal matrix, and multiplying it by the motion screw matrix $\mathbf{M}$. The freedom matrix $\mathbf{F}$ is the concatenation of the resulting matrices. For the case being considered $\mathbf{F}$ is a $39 \times 42$ matrix.

An action screw matrix $\mathbf{A}$ is formed from the unknown circuit wrench coordinates by writing them all in one column in the appropriate order. Kirchhoff's nodal law is then expressed as:

$$\mathbf{FA} = 0. \tag{1}$$

For our case, this gives 39 expressions to be satisfied by the 42 unknowns, and the three independent components of the circuit wrenches are directly related to the actuator forces. Reducing $\mathbf{F}$ to row echelon form gives expressions of which each has one dependent component of the circuit wrenches in terms of the independent ones.

The reactions in each of the joints can then be found after forming a matrix $\mathbf{A}^*$ which has the coordinates of one circuit screw on each row. The joint reactions $\mathbf{R}$ are then found by:

$$\mathbf{R} = \mathbf{B}^T \mathbf{A}^* \tag{2}$$

into which expressions for the dependent components of the circuit wrenches can be substituted to give $\mathbf{R}$ in terms of only the actuator efforts.

The work so far has not considered the effects of friction, but to include it provides no conceptual difficulty, despite the increase in the size of the matrices involved. For each joint where the friction is to be considered another path must be added to the graph to form a loop around the joint of interest. The path contains screws that are reciprocal to the friction wrench in the joint, and their number depends on the number of freedoms permitted by the joint. A cylindrical joint requires four and a simple hinge needs five.

The magnitude of the friction wrench is considered to be an independent quantity in the first stages, and we are able to calculate its effect on the joint reactions throughout the linkage. The reaction at a joint can be decomposed into pure force components which are orthogonal and aligned with the extent of the bearings. The magnitudes of these components are multiplied by a friction coefficient, determined from manufacturers' data sheets or tables of approximate coefficients. From this, the magnitude of the friction wrench can be determined and this is substituted into Equation 2. As a result, the joint reactions change and we need to iterate until the solution converges.

One problem not yet addressed is the question of the sense that the friction wrench takes. When the linkage is in motion the sense of movement is determined and the sign to apply is clear. However, for the static case the direction of impending motion is needed, and here two alternative assumptions are possible. The first is that the friction wrenches oppose the forces applied by the actuators. In the experiments that follow, the finger-tip load is applied by a passive force/torque sensor, and we believe this first case to be appropriate. The other alternative is that the friction acts to assist the actuators in supporting the load. This would be common in the normal operation of the hand, when the grasped object is being pushed, by some fingers, into another and thus back-driving that finger. Interestingly, these two cases are not symmetrical about the frictionless results. The actual performance of the finger will lie between these two extremes and, for 'good' performance a narrow range is desirable. We assume that if free the finger-tip would move in a direction that is parallel to the finger-tip force, and hence presume this to be the impending finger-tip velocity. From the presumed finger-tip velocity the corresponding joint velocities are determined.

## 4. Experimental Procedure

The experiment observed the relationship between the static finger-tip forces and the actuated forces. The finger-tip forces were measured by constraining the spherical finger-tip in a specially designed ball and socket clamp fixed to a six-axis ATI FT3031 force/torque sensor. Since the connection between the finger-tip and sensor is equivalent to a ball and socket joint, only pure forces acting through the centre of the finger-tip were measured, see Figure 2. When connected in this manner, the combined finger and sensor have a mobility of zero, and consequently the actuator forces produce no motion and provide internal loading of the linkage only. It is the reactions to this internal loading that are measured by the force/torque sensor.

Owing to its inherent complexity, the drive system was seen to distract from the specific questions of this work. Therefore, the drives of the finger were removed, and the finger and sensor were mounted vertically so that actuator loads could be applied by suspended weights hanging directly from the sliders

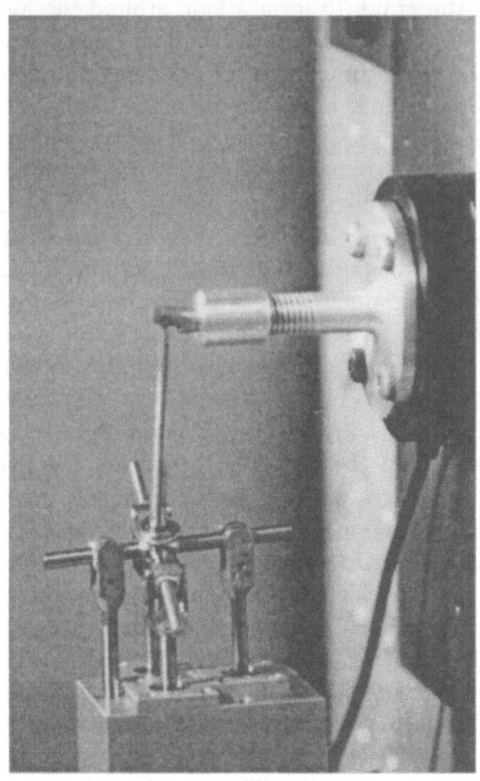

*Figure 2. The experimental setup showing the finger-tip clasped to the force/torque sensor.*

where they protrude from the back of the finger. This arrangement provided a direct measure of the forces applied to the linkage.

The finger was set in a known configuration and the gravity loading of the finger linkage was zeroed out of the force/torque sensor signal. Known actuating forces were applied to the linkage and the resulting finger-tip loads were measured for several combinations of actuator loads. The finger-tip load was determined by sampling the force/torque sensor at 800Hz and averaging one second's worth of data. Three configurations were tested and, in all, 85 sets of measurements were taken.

## 5. Results

The results collected are shown in Figure 3(a) and (b) for the $x$-, $y$- and $z$-directions. The coordinate system is arranged with the positive $z$-direction vertically up in Figure 2, the positive $x$-direction is, more or less, towards the viewer, and the positive $y$-direction is to the right of Figure 2 so that a force in this direction presses straight into the force torque sensor without applying a moment.

The graphs compare components of the measured finger-tip forces with the

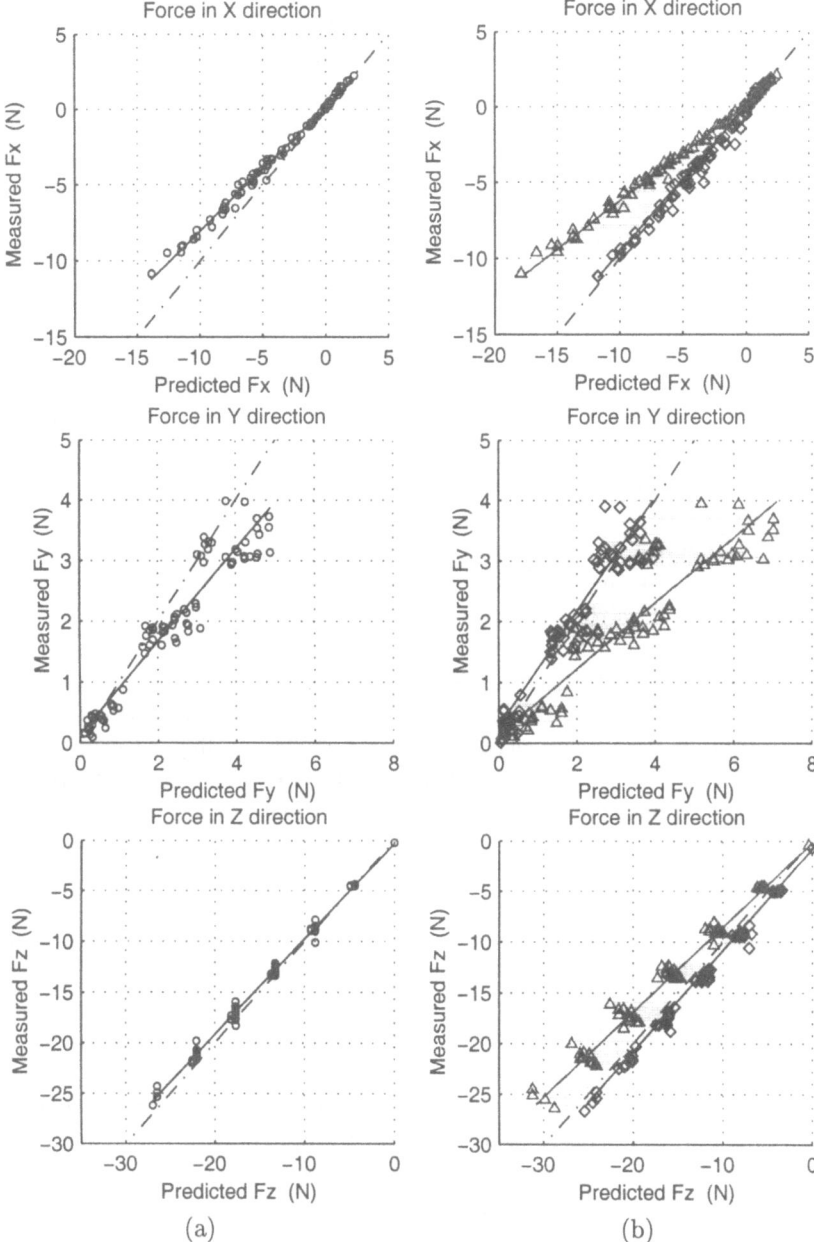

*Figure 3. A comparison of the measured and predicted finger-tip forces over a range of actuator forces: (a) the predictions of the model when friction is neglected, and (b) the range of operation of the finger when friction is included in the model. The cases where friction opposes the actuator efforts are marked with a ◇, while a △ marks the cases where friction assists the actuators in supporting the finger-tip load.*

predictions of the model for the three configurations. The circles in Figure 3(a) indicate the predicted forces when no joint friction is modelled. The dashed line shows where predicted and measured finger-tip data are equal and it can be seen that the model overestimates the finger-tip force, which is as expected. The lines of best fit to the data are found and these have gradients of 0.82, 0.76, and 0.94, for the $x$-, $y$- and $z$-directions. From these we can infer an efficiency for the linkage of about 80% and, although the actual efficiency is dependent of the configuration, we see that it is reasonable consistent over the three configurations shown. The $r^2$ values for the regressions give an estimate of the spread from a strict linear relationship; they are 0.993, 0.927, and 0.995 which shows that there is a good linear relationship.

With a passive force/torque sensor connected to load the finger-tip, one would expect the joint friction to reduce the predicted loads. Including friction in the model alters the predicted finger-tip loads as shown in Figure 3(b). Here the predictions with friction opposing the actuator efforts are marked with an '$\diamond$' and those with friction assisting the actuator efforts to support the finger-tip load are marked with a '$\triangle$'. The dotted lines between the two sets indicate the predicted range in which the performance of the finger lies; on one edge the finger is driving the grasp and on the other it is being back-driven by the force applied through the grasped object.

Figure 3(b) shows that the data with friction opposing the actuator forces is very close to the measured finger-tip forces. The gradients of the lines of best fits are 0.961, 0.945, and 1.000 for the $x$-, $y$- and $z$-directions, showing that the friction model now accounts for the measured finger-tip loads. The corresponding $r^2$ values are 0.983, 0.919, and 0.994 which indicates that the quality of the fit has dropped slightly but is still very good.

The triangles in Figure 3(b) show the difference between the cases when friction opposes the actuators and when it assists the actuators in supporting the finger-tip load. The size of the difference determines the responsiveness of the finger when applying forces. If during the manipulation of an object, the finger-tip changes the direction of its motion from towards the object to away from it while always maintaining a squeezing force then a step change in the actuator effort is required to maintain the same grasping force. The model predicts that this effect is about 30% of the finger-tip load in the $x$- and $y$-directions and, as such, this will be of some importance in strategies for synthesising and controlling the grasp. Design changes which reduce the disparity between the two cases will make the control task easier.

For the graphs in Figure 3 there was no special tuning of the coefficients of friction to achieve the good fit shown. Thus these figures have been produced without information that is unavailable to the designer or cannot be reasonably estimated. Nevertheless, the results are sensitive to those coefficients which contribute most significantly to the reduction in the loads measured at the finger-tip.

For one particular configuration and loading we can breakdown the contribution of the friction in each joint to the overall lowering of performance. Figure 4 shows a diagram of the finger with the approximate proportion of dif-

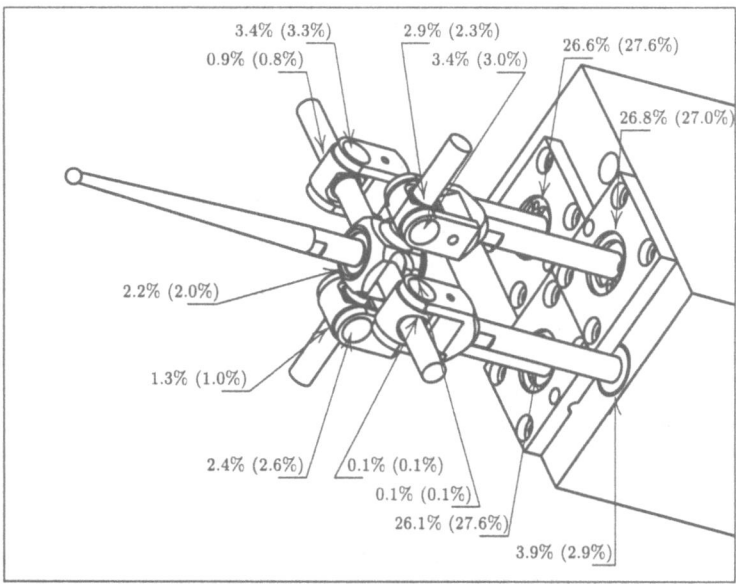

*Figure 4.* The proportional contribution of the linkage joints to friction losses measured at the finger-tip. The first percentages are for the friction opposing the actuators while the numbers in brackets are for the back-driven case.

ference between the frictionless and friction model. The bracketed percentages apply to the case where friction loads assist the actuator forces in supporting the finger-tip load. It can be seen that three joints account for about 80% of the performance loss owing to friction. An assessment of the reactions in these joints shows that the moment reactions lead normal forces that are an order of magnitude greater that the direct reactions, and this suggests one way in which the design of the linkage can be improved. Any effort to replace other journal bearings in the linkage with rolling element bearings appears to be of little benefit.

## 6. Conclusions

We have applied Davies' virtual power method to analyse the role of joint friction in a multi-loop linkage. We found that static and finger-tip loads applied by a passive device are well described by the model when the frictionless values are adjusted with friction opposing the actuator loads. Moreover, the predictions are based only on information known by the designer before the linkage is constructed and no fine tuning of the friction coefficients has occurred. Therefore the model seems a useful tool for design. Comparing the frictionless predicted and measured finger-tip loads, an efficiency of the linkage can be inferred; for the Melbourne Dextrous Finger this efficiency is about 80%.

An important property of the finger is its ability to be back-driven. In this case friction acts to assist the actuators in supporting the load on the finger-tip. The model can be used to predict this property, but as yet we have not been able to measure this effect and it is part of our on going work.

The normal operation of the hand is then expected to lie in a range between the case where the finger is driving the grasp and the finger is being driven by the grasp. The model predicts that the dead-band so formed is of the order of perhaps 30% of the finger-tip load, although the exact amount is dependent on the configuration of the finger. We believe this has serious consequences for the way in which grasps are planned and executed by dextrous robot hands. To this point in time we are unaware of any attempts in the grasping literature that takes proper account of such an effect.

The analysis of individual configurations and loading cases shows how the friction in individual joints contributes to the overall behaviour of the finger. We show that most of the friction is attributed to just three joints and the analysis suggests a means to improve the design of the finger. Furthermore, for control purposes it seems sufficient to model only the friction in these three joints and this is a substantial saving in effort.

# References

[1] T. H. Davies. Circuit actions attributable to active couplings. *Mechanism and Machine Theory*, 30(7):1001–1012, 1995.

[2] B. Armstrong-Hélouvry. *Control of Machines with Friction*. Kluwer Academic Press, Boston, 1991.

[3] C. R. Tischler and A. E. Samuel. Predicting the slop of in-series/parallel manipulators caused by joint clearances. In *Advances in Robot Kinematics: Analysis and Control*, pages 227–236, Dordrecht, 29th June–4th July 1998. Kluwer Academic Publishers.

[4] J. Swevers, C. Ganseman, D. B. Tükel, J. De Schutter, and H. Van Brussel. Optimal robot excitation and identification. *IEEE Transactions on Robotics and Automation*, 13(5):730–740, 1997.

[5] M. R. Elhami and D. J. Brookfield. Identification of Coulomb and viscous friction in robot drives: An experimental comparison of methods. *Proc. IMechE C, Journal of Mechanical Engineering Science*, 210(6):529–540, 1996.

[6] M. R. Elhami and D. J. Brookfield. Sequential identification of Coulomb and viscous friction in robot drives. *Automatica*, 33(3):393–401, 1997.

[7] C. Canudas de Wit. Experimental results on adaptive friction compensation in robot manipulators: low velocities. In *Experimental Robotics I*, pages 196–214, Montreal, June 19-21 1989. Springer-Verlag.

[8] B. Armstrong and B. Amin. PID control in the presence of static friction: A comparison of algebraic and describing function analysis. *Automatica*, 32(5):679–692, 1996.

[9] K. Hain. *Applied Kinematics*. McGraw-Hill, New York, 2nd edition, 1967.

[10] B. Paul. *Kinematics and Dynamics of Planar Machinery*. Prentice Hall, Eaglewood Cliffs NJ, 1979.

[11] T. H. Davies. Mechanical networks—III: Wrenches on circuit screws. *Mechanism and Machine Theory*, 18(2):107–112, 1983.

[12] C. R. Tischler. *Alternative Structures for Robot Hands*. Ph.D. thesis, University of Melbourne, 1995.

[13] K. H. Hunt. *Kinematic Geometry of Mechanisms*. Clarendon Press, Oxford, 2nd edition, 1990.

# Implementation of Dual-Arm Cooperative Control by Exploiting Actuator Back-drivability

Masaru Uchiyama and Yoshihiro Tanno [1]
Department of Aeronautics and Space Engineering, Tohoku University
Aoba-yama 01, Sendai 980-8579, Japan
uchiyama@space.mech.tohoku.ac.jp

Kunio Miyawaki
Technical Research Institute, Hitachi Zosen Corporation
2-11, Funamachi 2-chome, Taisho-ku, Osaka 551, Japan
miyawaki@omlab.lab.hitachizosen.co.jp

**Abstract:** This paper presents hybrid position/force control for dual-arm cooperation without using any force/torque sensor. The control uses only motor current as force/torque information and exploits motor back-drivability. A key technique in this control is compensation of friction at the motors, for which this paper presents a new method named dithered measurement. Experiments in the paper show that this method works well in order to implement dependable force control.

## 1. Introduction

It was not late after the emergence of robotics technologies that multi-arm robot systems began to be interested in by robotics researchers. There have been proposed quite a few different control schemes, for example [1]–[3], that require force-related control such as hybrid position/force control. A survey by the first author of this paper is presented in [4]. Those control schemes have been proven to be effective for cooperative tasks. But the proof was done mainly in laboratories using force/torque sensors. In practical implementation in industry, however, such sophisticated equipments as force/torque sensors tend to be avoided by many reasons: unreliability, expensiveness, etc.

An alternative method to implement the force control is to exploit actuator back-drivability; rebirth of the early methods [2], [5], [6] should be attractive for people in industry. Inoue [5] demonstrated effectiveness of force control with using servo errors at joints which correspond to motor currents. Takase and others [6] designed a torque-controlled manipulator for force control. Fujii and Kurono [2] proposed compliance control for dual-arm cooperation without using any force/torque sensor but exploiting only actuator back-drivability.

---

[1] Currently with Robot Laboratory, FANUC, Ltd.

This paper presents hybrid position/force control for dual-arm cooperation that does not use any force/torque sensor. The control uses only motor current as force/torque information and exploits motor back-drivability. Its structure is based on our previous work presented, for example, in [7]. A key technique in this control is compensation of friction at the motors, for which this paper presents a new method named dithered measurement. Experiments in the paper show that this method works well in order to implement dependable force control.

## 2. Experimental Setup

The robot, named C-ARM (Cooperative-ARM), used in the experiment is drawn in Figure 1. It is designed for laboratory experiment of dual-arm robotics based on which building a dual-arm robot to be applied to tasks in shipyard is aimed at. Each arm has three degrees of freedom in the same vertical plane. Each joint is actuated by an AC servo motor. Table 1 lists specifications of the motors. M1 to M3 denote the motors from the base, shoulder and the wrist of each arm, respectively. It is noted that the reduction ratio for M1 has a unit of Nm/N since the output of the joint is linear.

An idea regarding the weight of the robot can be obtained through the numbers in Table 2 where forces and torques due to arm weights when the arms are extended horizontally are presented. Also, the values of parameters $a_2$ and $a_3$ in Figure 1, that are 800 mm and 300 mm, respectively, will give an idea regarding the size.

Table 1. Motor specifications.

| Motor | M1 | M2 | M3 |
|---|---|---|---|
| Maximum speed [rpm] | 3,000 | 3,000 | 3,000 |
| Rated torque [Nm] | 0.98 | 0.98 | 0.2 |
| Maximum torque [Nm] | 2.94 | 2.16 | 0.54 |
| Reduction ratio | 1/200 [Nm/N] | 1/100 | 1/100 |

Table 2. Forces/torques due to arm weights when extended horizontally.

| Motor | M1 | M2 | M3 |
|---|---|---|---|
| Arm 1 | 0 [N] | 37.7138 [Nm] | 8.7870 [Nm] |
| Arm 2 | 0 [N] | 32.3092 [Nm] | 9.0894 [Nm] |

## 3. Force/Moment Measurement by Motor Current

### 3.1. Relation between Force/Moment and Motor Current

The torque $\tau_{mi}$ at the $i$th motor is proportional to the motor current $I_{ai}$, that is:

$$\tau_{mi} = K_{ti}I_{ai} \tag{1}$$

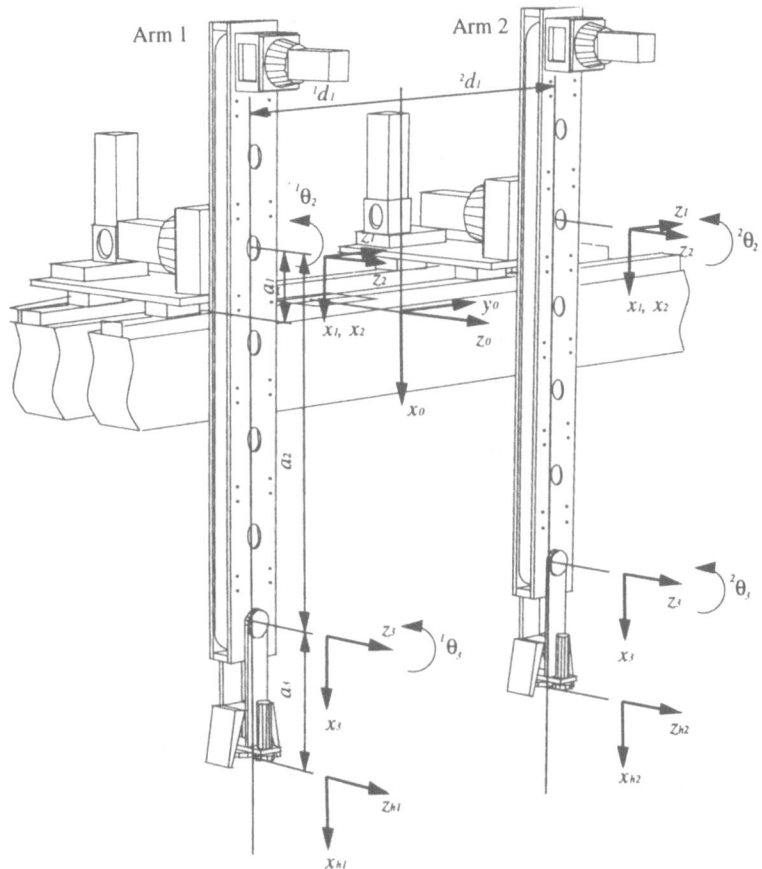

Figure 1. An experimental two-arm robot.

where $K_{ti}$ is a constant of proportionality for the $i$th motor. Generally, the motor torques at each joint are amplified by the reduction gears. The reduction gear ratios are shown in Table 1. Therefore, the amplified torques for the robot are represented in a vector form as $K_r^{-1}\tau_m$ where $K_r$ is a diagonal matrix with reduction gear ratios as its diagonal elements and $\tau_m$ is a motor torque vector consisting of motor torques to drive the robot. It is noted that the M1 components of $K_r^{-1}\tau_m$ are forces.

The force/torque vector $K_r^{-1}\tau_m$ is to move the robot and to balance the external forces and moments imparted to the end-effectors. This relation is written by

$$K_r^{-1}\tau_m = \tau_p + \tau_f \tag{2}$$

where $\tau_p$ is a force/torque vector to move the robot and $\tau_f$ is a force/torque vector caused by the external forces and moments at the end-effectors.

The term $K_r^{-1}\tau_m$, that is the left-hand side of the equation, is obtained from the motor currents $I_{ai}$. Then, a question is how to extract $\tau_f$ from the right-hand side of the equation. To answer this, we need to derive equations of motion of the robot after considering even dynamics of the motors and reduction gears, strictly speaking. Here, however, we neglect terms due to the dynamics of the robot in the equations of motion assuming that the robot does not move fast. Then, we have

$$\tau_p = \tau_v + \tau_g \tag{3}$$

where $\tau_v$ and $\tau_g$ are force/torque vectors due to frictions and the gravity, respectively. Then, $\tau_p$ is deducted from the joint force/torque $K_r^{-1}\tau_m$, to yield $\tau_f$, which in turn is used to calculate the external forces and moments being applied to the end-effectors.

### 3.2. Friction Model

The friction model used for the calculation of $\tau_v$ is shown in Figure 2. This model includes both Coulomb and viscous frictions. In Figure 2, $\theta_i$ and $R_{si}$ are the joint angle and the maximum static friction, respectively, while $\varepsilon_i$ is a constant parameter which is a threshold for the approximation of $\dot{\theta}_i$ to be zero.

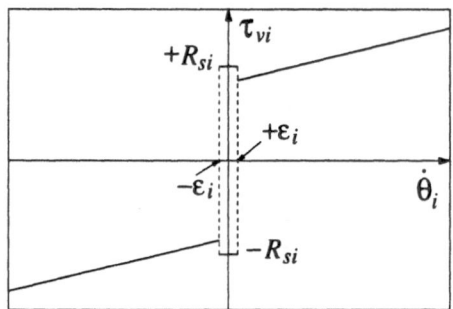

Figure 2. A friction model.

When $\dot{\theta}_i$ is outside the threshold, that is, $|\dot{\theta}_i| > \varepsilon_i$, the $i$th frictional torque (or force for M1) is given by

$$\tau_{vi} = V_i\dot{\theta}_i + R_{ci}\,\mathrm{sgn}(\dot{\theta}_i) \tag{4}$$

where $V_i$ and $R_{ci}$ are the coefficients of viscous and Coulomb frictions, respectively, and are determined experimentally. Tables 3 and 4 list the determined values.

### 3.3. Proposal of Dithered Measurement

The real static friction exists between the maximum static frictions $\pm R_{si}$. It is difficult to decide the direction and magnitude of the static friction when $\theta_i$ is stationary. Therefore, it makes sense to assume that the static friction is zero when $\dot{\theta}_i$ is inside the threshold. With this assumption, $\tau_v$ and consequently $\tau_p$ is obtained. The value may be used for the calculation of $\tau_f$ from the measured value of $K_r^{-1}\tau_m$.

Table 3. Viscous friction coefficients.

| | $V_1$ | $V_2$ | $V_3$ |
|---|---|---|---|
| Arm 1 | 182.5796 [N/(m/s)] | 42.4204 [Nm/(rad/s)] | 8.6715 [Nm/(rad/s)] |
| Arm 2 | 146.5936 [N/(m/s)] | 32.2094 [Nm/(rad/s)] | 8.6218 [Nm/(rad/s)] |

Table 4. Coulomb friction coefficients.

| | $R_{c1}$ | $R_{c2}$ | $R_{c3}$ |
|---|---|---|---|
| Arm 1 | 59.4631 [N] | 14.5059 [Nm] | 3.9694 [Nm] |
| Arm 2 | 56.6499 [N] | 19.8123 [Nm] | 4.4954 [Nm] |

However, this method does not work well when the static frictions $\pm R_{si}$ are large. This is explained by the experimental results presented in Figure 3. The results are for the joint 2 (M2) of the arm 2. The arm was loaded at the wrist by three known weights 0 kg, 1.4 kg and 3.4 kg. The top, middle and the bottom figures show the joint motion, measured torques, and the measurement errors, respectively. The angle in the joint motion is measured from the vertical line in the $x_0$ direction (see Figure 1). It is seen that the measurement errors are large when the joint is stationary although they are satisfactorily small when the joint is moving. Tables 5 and 6 give a summary of the measurement errors for the all joints of the robot. The percentages in the parentheses after the values of the errors are ratios of the errors to the maximum forces or torques of the motors. Large measurement errors exist when the robot is stationary. Those errors are apparently due to that the static friction is assumed to be zero.

To solve this problem, we propose a new method in which the static friction for compensation is switched alternately between $+R_{si}$ and $-R_{si}$ at each sampling period $\Delta t$ when $\dot{\theta}_i$ is inside the threshold. In this case, the friction model is written as follows:

$$
\tau_{vi} = \begin{cases} +R_{si} \overset{\Delta t}{\leftrightarrow} -R_{si} & -\varepsilon_i \leq \dot{\theta}_i \leq \varepsilon_i \\ V_i \dot{\theta}_i + R_{ci} \operatorname{sgn}(\dot{\theta}_i) & \dot{\theta}_i < -\varepsilon_i \quad \varepsilon_i < \dot{\theta}_i \end{cases} . \tag{5}
$$

As to the rest, the procedure is the same as the above-mentioned method in which the static friction is assumed to be zero. The model of Equation (5) is in a sense the same as the previous one because in the both cases the average static friction is zero. However, they are different from the viewpoint of dynamics. The latter model causes a dither effect on motors when the measurement is in a feedback loop. This is an important point of difference because the dither effect may reduce the joint friction. This effect is well known in the field of hydraulic power technology. With the latter model the measurement error does not decrease but becomes rather noisy due to chattering coming from the switching in the friction model. But, the chattering will deduce the friction at the joints when it appears on the control signal to the motors after going around the feedback loop. This effect eventually increases back-drivability of the joints

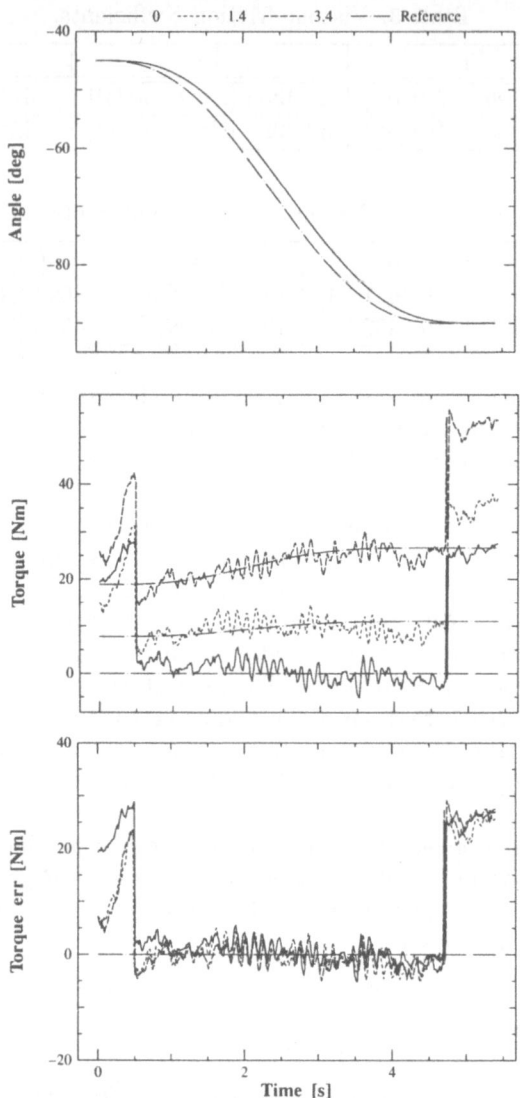

Figure 3. Experimental results.

and is, of course, preferable for force control. The experiment presented in the next section will verify this.

## 4. Experiment of Dual-Arm Cooperative Control

### 4.1. Control Scheme

The control scheme employed in the experiment is the one presented in [3]. It has been applied to C-ARM in [7] where force/torque sensors at the joints are

Table 5. Measurement errors when motor is moving.

| Motor | M1 | M2 | M3 |
|-------|-----|-----|-----|
| Arm 1 | ±15 [N] (±2.70%) | ±6 [Nm] (±1.80%) | ±1 [Nm] (±1.22%) |
| Arm 2 | ±15 [N] (±2.70%) | ±5 [Nm] (±1.50%) | ±1 [Nm] (±1.22%) |

Table 6. Measurement errors when motor is stationary.

| Motor | M1 | M2 | M3 |
|-------|-----|-----|-----|
| Arm 1 | ±33 [N] (±5.95%) | ±26 [Nm] (±7.81%) | ±5 [Nm] (±6.10%) |
| Arm 2 | ±25 [N] (±4.51%) | ±25 [Nm] (±7.51%) | ±5 [Nm] (±6.10%) |

used. In this paper we implement the control scheme with using the motor currents instead of the force/torque sensors. A block diagram of the control scheme is presented in Figure 4. The task vectors are the same as those in [7].

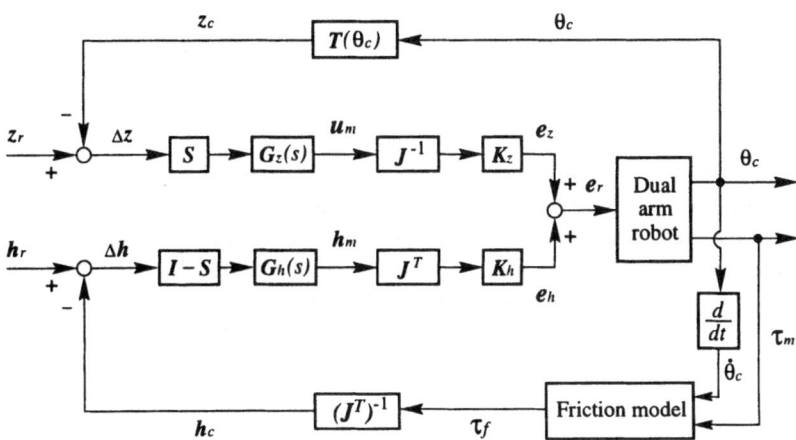

Figure 4. A block diagram of the control scheme.

The upper loop in the figure is for position control while the lower one is for force control. The vectors $z$ and $h$ represent task vectors of position and force, respectively. The both vectors consist of external and internal task vectors. The subscripts $r$ and $c$ denote reference and current, respectively. The matrix $S$ is for selecting position and force control components and is named a selection matrix. The matrices $G_z(s)$ and $G_h(s)$ are operators representing control laws for position and force, respectively. The matrix $J$ is a Jacobian matrix. The matrices $K_z$ and $K_h$ are diagonal matrices to convert velocity and force command into motor commands. The vector $\theta$ is a joint angle vector.

## 4.2. Experiment

The task chosen for the experiment is to carry an object up and down by the two arms. Figure 5 shows this task. The object is a metal bar with mass of 5.4 kg,

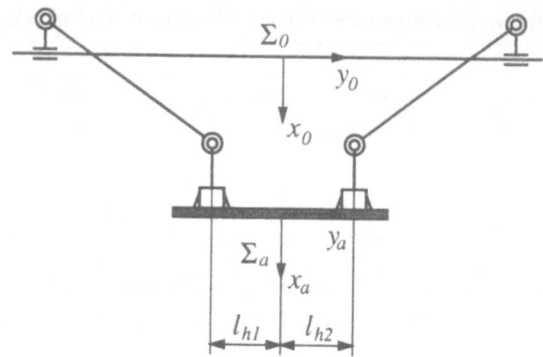

Figure 5. Task tested in the experiment.

at the middle of which a force/torque sensor is embedded to measure internal forces and torques. The length of the bar is 900 mm and $l_{h1} = l_{h2} = 300$ mm.

On applying the hybrid position/force control scheme to this task, it is found that the robot has three external and three internal degrees of freedom. In the experiment, the external and internal coordinates are controlled for position and force, respectively. This is decided by choosing the selection matrix appropriately.

For reference positions/orientations, those at the object center expressed with respect to the base frame $\Sigma_0$ are used; these are components of $z$ and are denoted by $^0p_a$. For reference forces/moments, the internal forces/moments on the object expressed with respect to the object frame $\Sigma_a$ are taken. These are denoted by $^af_r$. The components of $^0p_a$ and $^af_r$ are written as follows:

$$^0p_a = \begin{bmatrix} ^0x_a & ^0y_a & ^0\gamma_a \end{bmatrix}^T \tag{6}$$

$$^af_r = \begin{bmatrix} ^a\Delta F_{rx} & ^a\Delta F_{ry} & ^a\Delta N_{rz} \end{bmatrix}^T. \tag{7}$$

Experimental results are shown in Figure 6. The reference position in the $x_0$ direction is given as a cosine curve and the rest of the components of $^0p_a$ are set constant. The reference internal forces/moments are all set to be constant. To see if the internal forces/moments calculated from the motor currents give good estimation, we compared those with the internal forces/moments measured by the force/torque sensor embedded in the object. The "Sample" in Figure 6 (c) and (d) means that the force data are obtained by the force/torque sensor.

It is observed that large vibrations in the measured values of the internal forces appear when the object velocity approaches to zero. These vibrations are caused by the employed method, that is, the method of dithered measurement in which the friction model determined by Equation (5) is used. The vibrations go into the motor commands eventually in order to contribute to reduction of the static friction, although experimental data to show this are not included in the paper. The forces measured by the force/torque sensor are held close to the

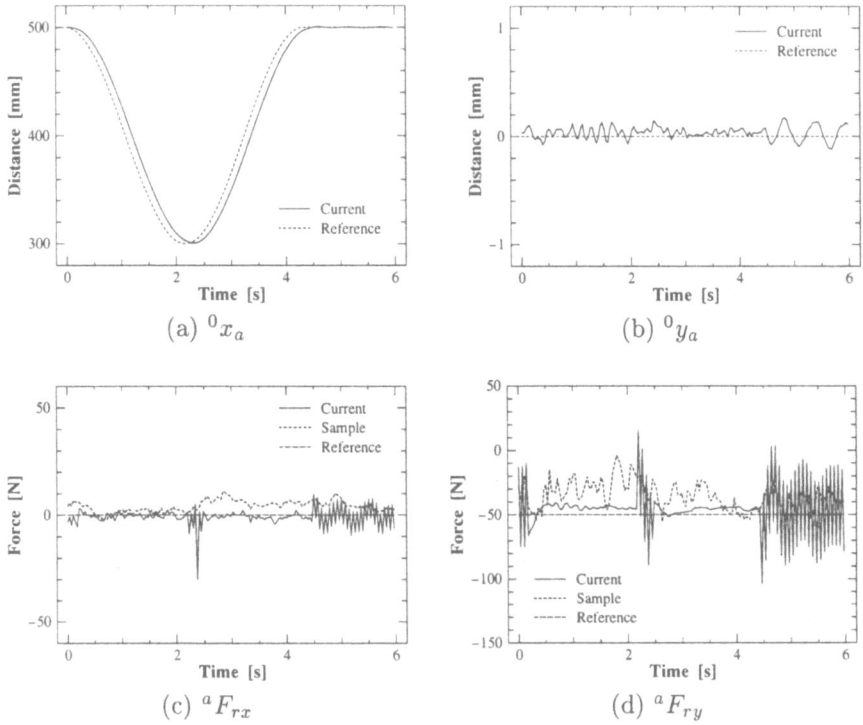

Figure 6. Experimental results.

reference values during the whole task. This shows that our method worked well.

## 5. Conclusions

This paper has presented experiments on hybrid position/force control for dual-arm cooperation. The control is implemented without using any force/torque sensor. The experiments have been conducted with a dual-arm robot, named C-ARM, which is designed for laboratory experiments toward applications to tasks in shipyard. The control uses only motor currents as force/torque information and exploits motor back-drivability. A key technique in this method is compensation of friction at the motors, for which this paper presents a new method called dithered measurement. The method uses a unique friction model in which the static friction is estimated as a dithered signal generated by switching between plus and minus of the maximum static friction at each sampling period. Experiments in the paper have shown that the method works well. Future research, therefore, will be directed to designing practical dual-arm robots for tasks in shipyard based on the results presented in this paper. Optimal or adaptive load distribution for better performance of the robot may be studied interestedly within the framework of cooperative control presented in this paper.

# References

[1] Nakano E, Ozaki S, Ishida T, Kato I 1974 Cooperational control of the anthropo-morphous manipulator 'MELARM.' In: *Proc. 4th Int. Symp. Industrial Robots.* Tokyo, Japan, November 1974, pp 251–260

[2] Fujii S, Kurono S 1975 Coordinated computer control of a pair of manipulators. In: *Proc. 4th IFToMM World Congress.* Newcastle upon Tyne, England, September 1975, pp 411–417

[3] Uchiyama M, Dauchez P 1993 Symmetric kinematic formulation and non-master/slave coordinated control of two-arm robots. *Advanced Robotics: The International Journal of the Robotics Society of Japan.* 7(4):361–383

[4] Uchiyama M 1998 Multi-arm robot systems: a survey. In: Chiacchio P, Chiaverini S (Eds) 1998 *Complex Robotic Systems.* Springer-Verlag, London (Lecture Notes in Control and Information Sciences 233), pp 1–31

[5] Inoue H 1971 Computer controlled bilateral manipulator. *Bul. JSME.* 14(69):199–207

[6] Takase K, Inoue H, Sato K, Hagiwara S 1974 The design of an articulated manipulator with torque control ability. In: *Proc. 4th Int. Symp. Industrial Robots.* Tokyo, Japan, November 1974, pp 261–270

[7] Uchiyama M, Kitano T, Tanno Y, Miyawaki K 1996 Cooperative multiple robots to be applied to industries. In: *Proc. World Automation Congress (WAC '96), vol. 3.* Montpellier, France, May 1996, pp 759–764

# Chapter 12

# Humanoid and Human Interaction

Personally I have had misgivings about the effort being devoted to humanoid robot development, but these are eased when I realise the calibre of some of the results. Indeed, the challenge of making robots is likely to inspire many new and elegant solutions which will be applied to solve much more mundane problems.

Alex Zelinsky and his colleagues report an elegant method for a computer vision system to work out what a person is looking at, just by measuring head and eye movements. The driver for this is the need to record driver gaze to evaluate the ergonomics of new car designs. Earlier ways to do this have necessitated special head gear which carries small infra red projectors and cameras to monitor eye movements. This equipment often bothers the human subjects and makes them aware of the monitoring process. The new method requires no special equipment other than a video camera (and the specialised image processing hardware of course). However, there are many potential applications ranging from computer screen pointers to new ways to command robots. In this paper they discuss the potential for this, using the Barrett whole-arm-manipulator. The arm is much safer than a conventional rigid robot because it is force-controlled and back-drivable. The eventual aim is happy and safe coexistence between humans and robots who, for the time being, have to be kept safely apart. In his presentation, Zelinsky reported that the reliability of the manipulator needs considerable improvement before really close contact could be contemplated.

Satoshi Kagami and his colleagues continue this theme. They have been inspired by the need to emulate human skin which is so effective when our arms and hands contact other bodies. They describe experiments with different artificial skins and sensor arrangements and their video showed a trusting researcher allowing his head to be cuddled by the arm of a robot with the skin attached.

Yoshihiko Nakamura, Masafumi Okada and Shin-ichirou Hoshino describe a novel mechanism which mimics the operation of a human shoulder which has been notoriously difficult to replicate mechanically. The intriguing part of their design is the way in which controlled damping and compliance can be simply introduced with tailor made components in the mechanism. The motion of the mechanism is difficult to understand without seeing the videotape.

The last paper continues the human interaction theme developed by the first two papers, but this time in the context of surveillance. The intention is to detect automatically the one person in a million who will try to steal a valuable object from an art gallery or museum. Paulo Peixoto and his colleagues exploit the geometry of an enclosed space to keep track of several people moving around a room, even through they often occlude each other as they pass. Both active vision (camera following a person's movement) and passive fixed cameras are used with a high degree of success.

JPT

# Towards Human Friendly Robots: Vision-based Interfaces and Safe Mechanisms

A. Zelinsky, Y. Matsumoto, J. Heinzmann and R. Newman
Research School of Information Sciences and Engineering
The Australian National University
Canberra, ACT 0200, Australia
URL: http://wwwsyseng.anu.edu.au/rsl/

**Abstract:** To develop human friendly robots we required two key components; smart interfaces and safe mechanisms. Smart interfaces facilitate natural and easy interfaces for human-robot interaction. Facial gestures can be a natural way to control a robot. In this paper, we report on a vision-based interface that in real-time tracks a user's facial features and gaze point. Human friendly robots must also have high integrity safety systems that ensure that people are never harmed. To guarantee human safety we require manipulator mechanisms in which all actuators are force controlled in a manner that prevents dangerous impacts with people and the environment. In this paper we present a control scheme for the Barrett-MIT whole arm manipulator (WAM) which allows people to safely interact with the robot.

## 1. Introduction

If robotics technology is to be introduced into the everyday human world, the technology must not only operate efficiently executing complex tasks such house cleaning or putting out the garbage, the technology must be safe and easy for people to use. Executing complex tasks in unstructured and dynamic worlds is an immensely challenging problem. Researchers have realized that to create robotic technology for the human world, the most efficient form is the humanoid robot form, as people have surrounded themselves with artifacts and structures that suit dual-armed bipeds e.g. a staircase. However, merely giving a robot the human shape does not make it friendly to humans. Present day humanoid robots instead of stepping over a person lying on the ground would most probably walk on a person without falling over

What constitutes a human friendly robot? Firstly, human friendly robots must be easy to use and have natural communication interfaces. People naturally express themselves through language, facial gestures and expressions. Speech recognition in controlled situations using a limited vocabulary with minimal background noise interference is now possible and could be included in human-robot interfaces. However, vision-based human-computer interfaces are only recently began to attract considerable attention (e.g.[1, 3, 4]). With a visual interface a robot could recognise facial gestures such as "yes" or "no", as well being able to determine the user's gaze point i.e. where the person is looking. The ability to estimate a person's gaze point is most important for a human-friendly robot. For example a robot assisting the disabled may need to pick up items that attract the user's gaze. Our goal is to build "smart" interface for use in human-robot applications. In this paper we describe our most recent results in building real-time vision based interfaces.

The second requirement for a human-friendly robot is that it must possess a high integrity human safety system that ensures the robot's actions never result in physical harm. One way to solve the human-safety problem is to build robots that could never harm people, by making the mechanisms, small, light-weight and slow moving. However, this will result in systems that can't work in the human world i.e. won't be able to lift or carry objects of any significance. The only alternative is to build systems that are strong and fast, which are human-friendly. Current commercial robot manipulator technology is human-unfriendly. This is because these robots only use point to point control. Point to point control is quite dangerous to use in dynamic environments where situations and conditions can unexpectedly change. If an object unexpectedly blocked a planned point to point path, either the object would be smashed out of the robot's path or the robot would sustain considerable damage. Clearly a compliant force based controller is needed in such situations. However, the technology force-based control of robots is not yet readily available. The current practice is to add force sensors to existing robot manipulators, usually at the tool plate. This allows force control of only the robot's end point. This means the robot could still easily collide with unexpected objects with other parts of the manipulator. To ensure safety a Whole Arm Manipulation (WAM) approach must be adopted, in which all the robot's joints are force controlled. Another important aspect that must be guarded against are software failures. A framework is needed to prevent a robot from continuing to work if a computer hangs or sends erroneous commands. In this paper we describe our recent results to build a safe control architecture and scheme for our Barrett WAM robot arm manipulator.

## 2. A Vision-based Human-Robot Interface

We have implemented a visual interface that tracks a person's facial features in real time (30Hz). Our system does not require special illumination nor facial makeup. The system is user independent and insensitive to fluctuations in lighting conditions. We use the IP5000 vision processing system to implement our visual interface. This vision processing system is manufactured by Hitachi and is designed to implement real-time vision processing such as smoothing, erosion, template correlation in each frame of a colour NTSC video stream. The hardware consists of a single PCI-bus card, running on an Pentium processor running Linux. Currently, we are interfacing a stereo vision system by using the video interlace on a single channel to transmit alternating images from the left and right cameras [6].

Our vision process for the visual interface consists three parts as follows:

- 3D Face model acquisition
- 3D Face tracking
- 3D Gaze point estimation

## 2.1. 3D Face Model Acquisition

Our three dimensional face model is currently composed of six template images of facial features (which are located in each side of two eyes and a mouth ) and its corresponding coordinates (x, y and z). In order to acquire such models, facial features should be detected as the first stage.

Figure 1. Facial feature detection process

Figure 1 illustrates the processing steps to detect facial features. The first step is skin color detection, using color information in HSI space to eliminate the effect of the fluctuations in lighting conditions. However, this segmentation is not sufficient to clearly detect the facial features. This is because the threshold value for skin color classification has to be relaxed to allow detection of faces of different races.

In the second stage, the largest skin color region is selected, and black blobs inside the area are filled in to build a "facial mask". This mask is used to select a gray-scale facial image from the original input, thus eliminating the backgound images other than the facial region. The gray-scale facial image is binarized to produce black areas based on a variable threshold of image intensity in the facial region. Finally the black regions inside the skin region are investigated and the features that satisfy certain geometrical constraints are extracted. Figure 2 shows the results of detection of facial features for two different persons. Once facial features have been detected, local images of the corners of the features are recorded as template images from the right camera image (shown in the right image of Figure 3). To calculate the 3D

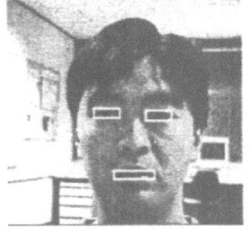

Figure 2. Detected facial features

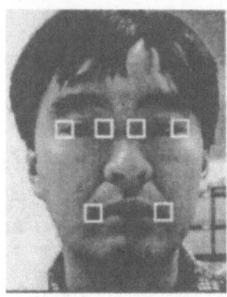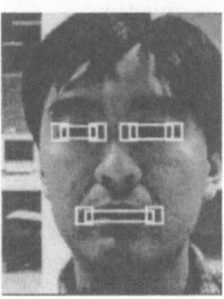

Figure 3. Right: Extract templates of each feature. Left: Use stereo matching of feature templates to build a 3D model.

coordinates of each feature, stereo matching of the extracted templates with images from the left camera (shown in the left image of Figure 3) is carried out. The template images and corresponding coordinates are stored as a 3D facial model. The model acquisition process is currently performed at 10Hz.

## 2.2. 3D Face Tracking

In each video frame, all feature templates are correlated with the current image, and the 3D position of every feature is calculated. The correlation function of IP5000 uses normalized correlation matching, and therefore is unaffected by fluctations in lighting.

Next, the 3D measurements from stereo matching are fitted to a the 3D model. This is done by using the results of the tracking from the previous frame, and then performing gradient descent using a spring connection model, where each feature correlation value is treated as the stiffness in a spring between the model and the measurements. In this approach, features that are tracking well have strong springs which result in helping estimate the positions of features that are tracking poorly. As a result of model fitting, six parameters representing the position and orientation of the face are acquired. The best fitting 3D model is then back projected into 2D, and the search areas used to track features in the next frame are updated. The entire tracking process runs at 30Hz.

Figure 4 shows the results of face tracking for various situations. Even when some features are deformed or occluded, the tracking process performes quite well.

## 2.3. 3D Gaze Point Estimation

Our system outputs the position and posture of the face, as well as the position of the pupils. Pupil detection is done using the Circular Hough Transform [5], the results of which are also shown in Figure 4.

There are ten parameters in total that describe our 3D face model. Six parameters represent the face position and orientation, and two parameters are used to represent the vertical and horizontal positions of each eye. Figure 5, shows an facial image and a facial model that has been animated with the ten parameters.

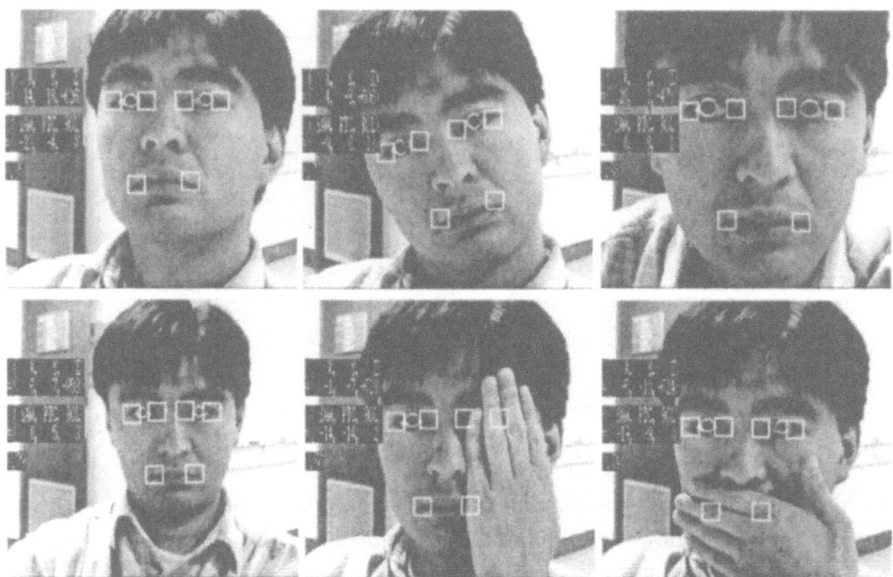

Figure 4. 3D Face Tracking

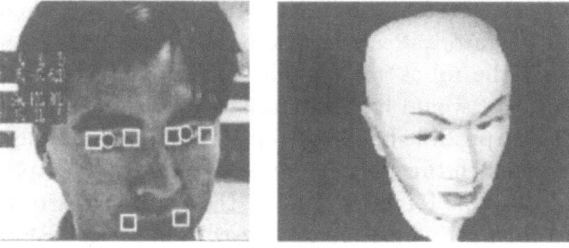

Figure 5. A facial image and the corresponding animated 3D Facial Model

Gaze point estimation is done by fusing the 3D face position and posture with the 2D positions of the pupils. The eyeball is modelled as a sphere, and using the corners of the eye the center of the eyeball can be determined. The eyeball center is connected to the detected pupil's center to form a gaze line for each eye. The gaze lines are then projected into 3D space, and the intersection point becomes the estimated gaze point. Quite often due to noisy measurements the gaze lines do not intersect, in this case the mid point of the shortest line segment joining the gaze lines is used. The gaze direction is considered to be the line joining the estimated gaze point to a point located between the eyes. Figure 6 shows a stereo image pair of a face, and and the corresponding animated facial model shows the estimated gaze direction. Currently, the gaze detection process runs at 15Hz.

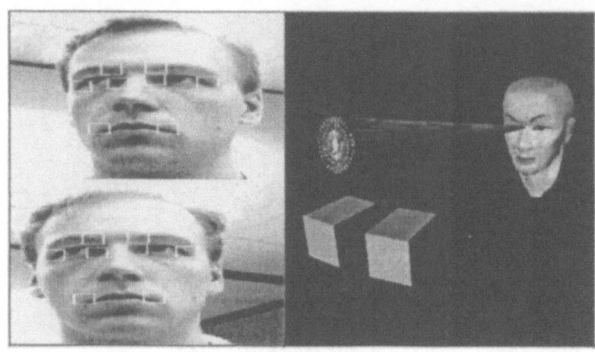

Figure 6. Gaze Point Estimation.

## 3. A Safe Mechanism

In this section we will describe what we believe are requirements for a human-friendly robot and how we are planning to implement a real system which will meet these requirements.

### 3.1. A Philosophy for Safe Robots

Designing the control architecture for a human-friendly robot involves decisions about the distribution of responsibilities between the robot's control algorithms and the human operator. People must be viewed as intelligent agents that can interfere with the robot accidentally, intentionally and even maliciously. A perfectly safe robot would have to react to all events in an appropriate way which minimises the possible harm to people, itself and other objects. However, there is a limit to which this goal can be achieved. For example, limiting both the velocity and possible forces applied by the robot seems to be a reasonable safety strategy. However, when interacting with humans both restrictions can not be guaranteed by the control system. If a person applies forces to the robot arm that cause velocities beyond the preset velocity limit, the control system can react in two ways. It may choose an active strategy and restrain the motion using higher torques than the ones defined as safe and thus violate the force restriction, or it may choose a rather passive strategy using forces only inside the restrictions to slow down the arm, and thereby allowing arm velocities that exceed the preset limits.

A dilemma arises from the uncertainty in the robot's perception of the current situation. The most appropriate action in one context may be unsafe in a different context. If the context can not be determined correctly, it may be unsafe to take an action, or not to take an action! Thus, the robot should not be charged with the task of correctly interpreting and reacting to all events possibly created by humans. A better strategy is to charge the robot with the task of behaving in a way that is generally safe, but more importantly, in a way that is *easily understood and predictable*. Human-friendly robots should rely on human intelligence for safe operations, just the same as when other dangerous tools like knifes, power saws and cars are used.

Since the robot may have no idea of when a collision may occur, it has to limit its kinetic energy to safe levels when moving autonomously. It also has to limit its bandwidth of motion to allow a human observer to reasonably predict its motion. Both issues are strongly interrelated. The kinetic energy of a robot at a low bandwidth is less dangerous than the same amount of kinetic energy of a robot when it is operating at a higher bandwidth.

## 3.2. Safe Mechanical Design

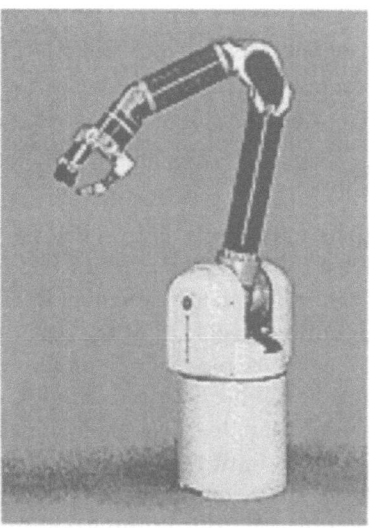

Figure 7. Barrett WAM robot

The hardware platform for the project is a recently acquired Barrett Technology WAM arm (Whole Arm Manipulator), shown in Figure 7. The arm is a commercial version of the original MIT arm built by Townsend and Salisbury [7]. The WAM robot not only allows manipulation with its end-effector, but also manipulation with the upper and lower links of the arm. Also, since the joints are driven by cables transmissions instead of gears there is no backlash present. More importantly, the friction in the WAM transmissions is low. This has two important implications:

- Each joint of the robot is easily backdrivable. Thus, humans interacting with the robot can change the pose of the robot simply by pushing the links or the endtip of the robot. Forces applied to the endtip or any other part of the robot get reflected back to the motors with very low loss in the transmission.

- Torques applied to the motor get transmitted to the joint with negligible loss. This means that the forces the robot is applying to it's environment can be derived with a high accuracy from the motor commands.

This means that the usually strict distinction between actuators and sensors becomes blurred.

Rotational joints can only sense and comply to forces which are tangential to its axis. Forces parallel or radial to the axis can not be transmitted to the motor. For this reason the WAM is a seven axis robot, and so theoretically sense forces in all directions.

In terms of safety the cable transmission has some important advantages. The two major threats to persons entering a robots workspace to interact with the robot are

- the kinetic energy of the robot, which directly determines its impact energy, and

- the forces applied by the robot after contact between the robot and the person is established.

The only way to limit the first threat is to move the robot slowly. Since the impact energy transmitted to the person can't be instantaneously reduced even if the robot can detect the collision. By designing light weight robots equipped with padding the maximum safe velocity of the robot could be increased. The WAM is a light weight design. Almost all of the parts in the arm are made of aluminium and magnesium, significantly reducing the weight of the arm. The weight of the robot's components from shoulder joint is only 15kg.

The classical approach to limit the second threat is to use a force/torque (FT) sensor attached to the wrist of the robot. The forces applied at the ende-effector can be measured precisely, this is important for dextrous manipulation. However, as a safety feature such a FT sensor gives only limited protection. If for example the robot's elbow makes accidental contact with a person the robots control program can not react to the situation. The WAM arm with its low friction transmission and its ability to directly measure and control motor currents is well suited to safe human-robot interaction.

### 3.3. Safe Hardware

Apart from the mechanical design of the WAM, there is a need for electronic safety measures. The amplifiers for the joint actuators of the WAM are digitally controlled and can be setup to limit the output current to the motors. The output currents that are needed to safely move the robot are sufficiently large enough to enable it to reach velocities that may not be considered safe. This is particularly true if the motion is exaggerated due to gravitational effects. However, limiting the current does provide protection from extreme robot motions. This is particularly important during the development of control software. Also we monitor the joint velocities continuously with an analog circuit. If any of the joint velocities exceed preset limits, the robot is shut down. The emergency shutdown procedure can also be issued by software. It disables all amplifiers and shortcuts the motor power wires with a solid state contactor. This procedure also works as an electric brake for the motors. When a shut down occurs, the robot gently falls down without posing any real danger to people inside the robots work space. However, this safety measure is meant as a last resort measure.

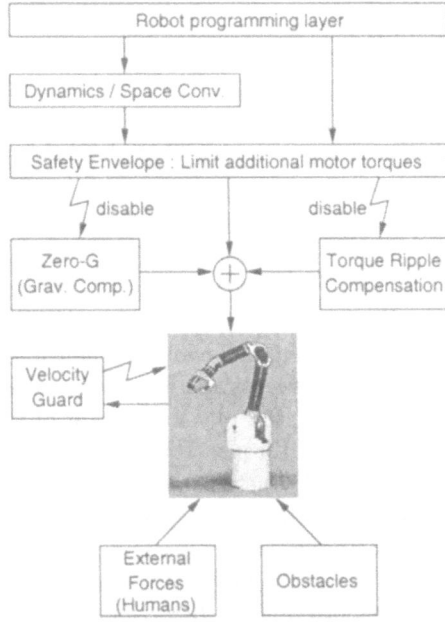

Figure 8. Software architecture

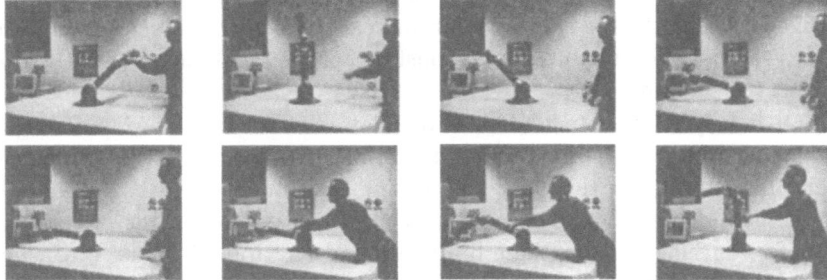

Figure 9. Image sequence from a Zero-Gravity experiment

### 3.4. Safe Software

The control software for the robot is structured in a hierarchical way, shown in Figure 8. At the lowest layer is a passive module which monitors the velocities of the elbow and the hand in cartesian space. The hardware velocity limit can only monitor raw joint velocities. However, the same joint velocity can result in significantly different velocities of the links of the robot depending on the kinematic configuration of the robot. In order not to limit the usability of the robot by rigorous hardware limits, the limits have to be set to values that may result in cartesian space velocities that violate the safety limits. Thus, the cartesian space velocities must also be monitored in order to shut the robot down if a dangerous situation arises. Since the maximum cartesian space velocity of a link is always reached at one of its end points, it is sufficient to monitor the position of the elbow joint and position of the hand. If the velocity of any one of these points exceeds the preset safety margin the hardware safety circuit is instructed to shut down the robot.

On the next level two modules are responsible for eliminating the effects of gravity and the torque ripple of the motors. The Zero-G module generates motor torques based on the current position together with the robot's kinematic model to counteract the gravitational forces acting on the links. The robot becomes effectively weightless and floats freely when pushed by a person. It is only slowed down by the friction in the joints which is low and thus; even small pushes can cause the robot travel quite long distances before the kinetic energy is consumed by the friction.

To generate a smooth floating behaviour of the robot another effect must be compensated for which in conventional robot controllers can be ignored. A torque ripple (TR) is caused by the physical design of the motors. It occurs because of the variation in the torques that are generated by the motor at different positions of the shaft at constant motor currents. For brushless motors of the type used in the WAM the effect is sufficiently strong enough to give the impression that the motors are snapping into equally spaced positions. By adding extra torques to the output of the Zero-G module we can compensate for the TR effect and a smooth floating behaviour can be generated.

Above the Zero-G level the Safety Envelope provides the lowest level programming interface. Its purpose is to limit the torques that can be added to the outputs of the float level by any controller. These limits are not static but change according to the current joint and cartesian space velocities. Since the robot is already floating in Zero-G, constant torques could accelerate the robot beyond the safe limits.

Obviously the robot should still be allowed to brake even when it is going faster than the preset limits. The definition of the functions generating the appropriate limits for the current situation are bounded by the torques that can accelerate the robot to the defined safe velocity limits in the cycle time of the safety envelope. Thus, these limits are linear reciprocal values of the cycle time. However, allowing the robot to accelerate to its standard safety velocity in one cycle of the safety envelope is not necessary for interaction tasks and the resulting jerky motions can gives users an 'unsafe feeling'. By chosing conservative limits, gives more natural floating motions.

Above the Safety Envelope any control strategy may be implemented to control the robot. The limits in the safety envelope will always guarantee safe operation conditions. For convenience a module may be implemented which offers a programming interface with modified robot dynamics or cartesian space control rather than the joint space control offered by the Safety Envelope. Since we wish to permit the robot to move slowly, the control program may ignore dynamic effects like centrifugal and coriolis forces. However, when moving slowly the effects of stiction and friction become significant and have to be considered either in a conversion module hiding the effects or in the controller itself.

The control algorithm to be used on the top level can be any algorithm used for conventional robots. The Safety Envelope is meant to will always guarantee safe and human-friendly operation. To achieve a point-to-point motion a PID controller can be used. If the robot is in a configuration far away from the goal configuration, the PID controller will output huge torques. Those will be clipped in the Safety Envelope to levels that will accelerate the robot slowly towards its goal configuration. At any time instance a person can interfere with the arm, change its configuration, block its motion, but eventually the robot will reach its goal if it's not permanently blocked.

Figure 9 shows a sequence of frames from a 5 second sequence showing the robot in floating mode. Only the velocity guard, the TR compensation and the Zero-G module are active. In the first image the robot is being pushed by a person and keeps floating until it bumps into the foam rubber tabletop. When the person is pulling the robot back by its upper link, note how despite the low velocity the inertial effects flex the lower arm.

### 3.5. Heavy Object Manipulation

The restrictions that guarantee safety under normal conditions disable the robot from lifting heavy objects. When heavy objects are to be manipulated and moved in a safe manner the Zero-G module must adapt to the new robot configuration. The robot with the payload has to be gravity compensated for before the controller can be allowed to move the object safely. To keep the same level of safety, the maximum speed of the robot has to be adjusted since the kinetic energy at the same maximum speed will be considerably higher. There is also the risk of the robot loosing its grip of the object. In this case the Zero-G model changes abruptly, this needs to be detected by the controller.

## 4. Human-Robot Co-Existence

Almost all industrial robot arms are set up inside cages that prevent people from entering the workspace. We see the main application area of human-friendly robots as helping hands in complex tasks that require more than two hands and for the elderly or disabled. For such applications areas, people and robots must share the workspace. While the robot needs a certain degree of autonomy, all its actions are continuously supervised by the human opearator.

As a first step in this direction we are currently implementing a system which combines the perceptual abilities of the gaze tracking application with the dexterity of the WAM arm. Our face tracking system can be utilized as an interface for human-robot interaction. Presently, we are considering two approaches to indicate a person's intention to a robot: Gestures and Gaze point.

Gesture recognition such as "yes" or "no" can easily be realized once accurate facial position and posture is detected in real-time. Our group has already developed such a system using previous face tracking system with monocular camera[2]. Human-Robot interaction using gaze point could be realized in such a way that allows the selection of an object of interest by fixing one's eyes on the object. The robot could then pick up the object and pass it to the user. This is an intuitive way for people to interact with robots.

## 5. Conclusions

From the human perspective a robot is a machine characterised by its geometry, reach, weight, speed, bandwidth, strength and behaviour. These characteristics should be non-threatening if users are to feel comfortable working with such a robot.

The user should also be able to easily communicate with the robot. If communication is human-like using speech, touch and gesture, the acceptability of the technology is likely to be high. As speech technology is rapidly becoming a well understood technology, our interests are directed to advancing vision and tactile interfaces.

The vision-based interface that we have developed can determine in real-time and in an inobtrusive manner through facial feature tracking determine the user's gaze point. This can be done independent of the user. Our interface represents a considerable advance in the face tracking technology.

The mechanical design of the WAM robot is fundamental to the realisation of a human-friendly robot. Additional hardware and software safety measures provide a high level of safety. The software architecture allows for the implementation of different safety strategies and a safe development environment for new robot control algorithms. The encapsulation of the Zero-G module by the safety envelope guarantees minimum restrictions to achieve efficient robot motions inside predefined safety constraints. Thus providing the user with the abstraction of a light and gentle arm that only meekly resists when manipulated by a person.

Human-friendly robots are now within technological reach. This paper presented substantial first steps towards the realisation of this goal.

## References

[1] A.Azarbayejani, T.Starner, B.Horowitz, and A.Pentland. Visually controlled graphics. *IEEE Trans. on Pattern Analysis and Machine Intelligence*, 15(6):602–605, 1993.

[2] A.Zelinsky and J.Heinzmann. Real-time Visual Recognition of Facial Gestures for Human Computer Interaction. In *Proc. of the Int. Conf. on Automatic Face and Gesture Recognition*, pages 351–356, 1996.

[3] Black and Yaccob. Tracking and Recognizing Rigid and Non-rigid Facial Motions Using Parametric Models of Image Motion. In *Proc. of Int. Conf. on Computer Vision (ICCV'95)*, pages 374–381, 1995.

[4] J.Heinzmann and A.Zelinsky. 3-D Facial Pose and Gaze Poing Estimation using a Robust Real-Time Tracking Paradigm. In *Proc. of the Int. Conf. on Automatic Face and Gesture Recognition*, pages 142–147, 1998.

[5] Par Keirkegaard. A Method for Detection of Circular Arcs Based on the Hough Transform. *Machine Vision and Applications*, 5:249–263, 1992.

[6] Y. Matsutmoto, T. Shibata, K. Sakai, M. Inaba, and H. Inoue. Real-time Color Stereo Vision System for a Mobile Robot based on Field Multiplexing. In *Proc. of IEEE Int. Conf. on Robotics and Automation*, pages 1934–1939, 1997.

[7] W.T.Townsend and J.K.Salisbury. Mechanical Design for Whole-Arm Manipulation. In *Robots and Biological Systems: Toward a New Bionics?*, pages 153–164, 1993.

# Development of a Humanoid H4 with Soft and Distributed Tactile Sensor Skin

Satoshi KAGAMI[†]    Atsushi KONNO[‡]    Ryosuke KAGEYAMA[†]
Masayuki INABA[†]    Hirochika INOUE[†]

† Dept. of Mechano-Informatics, University of Tokyo.
7-3-1, Hongo, Bunkyo-ku, Tokyo, 113-8656, Japan.
{kagami,kageyama,inaba,inoue}@jsk.t.u-tokyo.ac.jp

‡ Dept. of Aeronautics and Space Engineering, Tohoku University.
01 Aoba-yama, Sendai, Miyagi, 980-8579, Japan.
konno@space.mech.tohoku.ac.jp

**Abstract:** A wheel type humanoid robot "H4" is developed as a platform for the research on vision-tactile-action coupling in intelligent behavior of robots. The H4 has the features as follows : 1) upper-body consists of 19DOF (3DOF waist, 6DOF arm, 4DOF head), light weight and enough space to attach skin, 2) multi-value tactile switch and matrix gel tactile sensor, two types soft and distributed sensor skin covers upperbody, 3) each tactile sensor module has microprocessor module which is connected via $I^2C$ serial bus and make sensor network, 4) PC/AT compatible onboard computer which is controlled by RT-linux and connected to a network via radio ethernet, 5) 3D vision facility and 6) self-contained. This paper describe the design, implementation and experiment of H4, which is expected to be a common test-bed in experiment and discussion for tactile based aspects of intelligent robotics.

## 1. Introduction

A stiff robot which behaves in human world may injure human being or may break surrounding and itself. For example, without softness, no robot can be safe in an occasion of contacting to a unconscious human being. Passive and active softness are required to make such a robot safe. The former is achieved by skin, soft structural material, soft mechanisms and so on. The latter is achieved by soft actuator control from tactile/force sensory inputs. Indeed, a living being has both passive and active softness functions and keeps its body/surrounding safe by their combination.

So far, there are many researches on elemental functions for robot softness, such as mechanism(suspension or other mechanical compliance), control(compliance and inpeedance control), soft tactile sensor(ex. [1]), and so on. But there are very few research on integrating these elemental soft functions and on solving the problem which is caused in a real-world robot behavior.

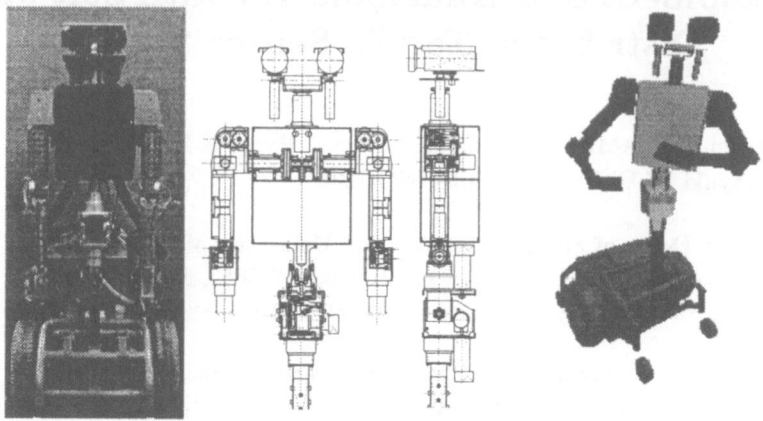

Figure 1. Humanoid "H4"(left), H4 mechanical components (center) and 3D geometric model in Euslisp (right)

Therefore a wheel type humanoid robot "H4" is developed as a platform for the research on vision-tactile-action coupling in intelligent behavior of robots. Because soft skin which covers its entire body and tactile sensors which are implanted within skins, are fundamentally important for a robot which behaves in human world, H4 is designed to be attached skin. Also, H4 is designed to have a wide work area from floor to shelf, and it is self-contained. Multi-value touch switch and conductive gel matrix sensor are developed and implemented on H4 in order to cover upper body. This paper describes design, implementation and experiment of H4 tactile sensor side.

## 2. Requirements and Design of Vision-Tactile-Action Coupling Behavior Research Platform

### 2.1. Requirements for Vision-Tactile-Action Coupling Behavior Research Platform

For a robot that behaves in real-world, robot may contact human/object with or without expectation. In both cases, tactile sensor which covers whole body is important to detect such a contact. Also robot surface should be soft in both cases to absorb impact force and to contact by face to face. In case of unexpected contact, no system can handle sensor processing or actuator control latency without soft skin. Therefore, soft skins which covers whole body is required.

So far, there are many remarkable tactile sensor researches have been proposed to designate high density resolution(ex. [2]), simple signal processing (ex. [3]), force tensor data acquisition(ex. [4]), and slip vibration data acquisition(ex. [5]).

As for implement a sensor which covers all over the body, not only its sensor function, but also there are five requirements as follows.

1. Flexible enough to cover curved surface
2. Simple circuit
3. Small number of wiring
4. Simple processing for sensor signal
5. Soft and deform enough

### 2.2. Basic Design of H4

1. Upper-body is humanoid to interact human being, has smooth surface and has enough space in order to attach skin,
2. Wheel base is adopted which has enough capability in a office-like environment,
3. 2/3 size with an adult to not frighten, and make a body as light as possible in order to safety,
4. 4DOF for head, 6DOF for each arms, 3DOF for waist and 2DOF for wheel base,
5. Harmonic drive gear is adopted to every joint in order to augment back-driverbility
6. Self-contained including PC/AT clone and battery. Computer is connected to a network via wireless ethernet,
7. RT-Linux controls from low-level software servo to high level intelligent software (ex. model calculation),
8. Onbody microprocessor(H8) network for sensor data transmission,
9. Synchronous stereo camera and field mixing circuit makes binocular vision processing possible.

### 2.3. Evaluation and Improvement of Sensor Distributed Artificial Skin

A tactile sense function is considered to be classified as follows: 1) touch switch, 2) multi-value touch switch, 3) analog force/position sensor, 4) force tensor sensor. Tactile sensor which is implanted within a soft skin should also be soft. We have developed these three sensors, 1) Multi-value tactile sensor[6], 2) Robohip(Pressure sensor with silicon rubber)[7], 3) Conductive gel matrix sensor[8], and we will evaluate them through real-world examination.

Combining soft tactile sensor skin with stereo vision, H4 has a capability to work in human world. Following sections, hardware, software components of H4, conductive gel matrix sensor and real-time depth map generation method are denoted.

## 3. H4 Hardware and Software System

### 3.1. Hardware System based on PC/AT Clone

CPU board of H4 is PC/AT clone for FA use. It has PentiumMMX-233MHz, 128MBytes EDO-RAM, two serial and one parallel port, IDE, floppy, keyboard, mouse, VGA and Ethernet. A 810MBytes 2.5inch IDE harddisk is connected. This board has PISA connector to connect to a backplane.

Backplane board has one PISA connector for CPU board, two ISA connector and three PCI connector. Fujitsu tracking vision board (TRV-CPD6),

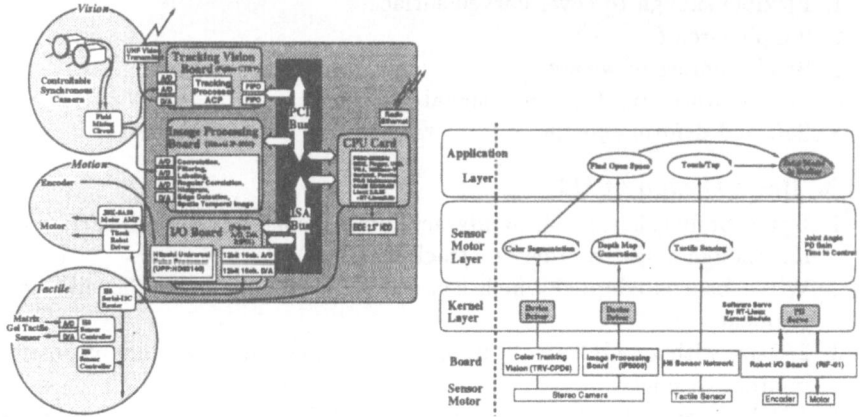

Figure 2. H4 hardware and software components

Hitachi vision board (IP5000) are attached to PCI bus. Two Fujitsu robot I/O boards are attached to ISA bus (Fig.2(left)).

### 3.2. Software System on RT-Linux

In general, servoing is executed on a kind of real-time monitor, but operation system (OS) of a robot that behaves in real-world should handle both servoing and high-level intelligent software. Therefore OS for a robot should have functions such as; software servoing, multi-processing, inter-process communication, networking, development environment and so on.

### 3.3. Motor Servo

Motor servo for 21DOF actuator is implemented as one RT-Linux hard real-time module (Motor Servo in Fig.2(right)). There are two modes input for this module, one is position of each actuator, and the other is the end point of each arm. In the latter mode, the module calculates inverse kinematics to acquire the current position of each actuator. The module also gets the gain for PD control of each actuator, so that the module counts encoder of each actuator and decides output force using PD gain and interpolation of designated position, and outputs through D/A. The cycle is 10msec.

### 3.4. Model based control by Euslisp

Since the robot has many degrees of freedom, it is hard to generate stable body motion in a complex environment. To solve this problem, 3D geometric model is implemented in Euslisp[9] (Fig.1). This special lisp is developed for robot software, and has the features as follows: 1) object oriented, 2) 3D solid modeler, 3) multi-thread programming functions, 4) original X-window toolkit, 5) OpenGL support, 6) efficient memory management using Fibonacci and Mark & Sweep type GC, 7) Unix system-call functions, 8) interface to other programming languages, 9) multiple platforms support (Solaris, SunOS, OSF-1, Irix, Linux, Windows95).

So far, under Remote-Brained approach, robot modeling library has been

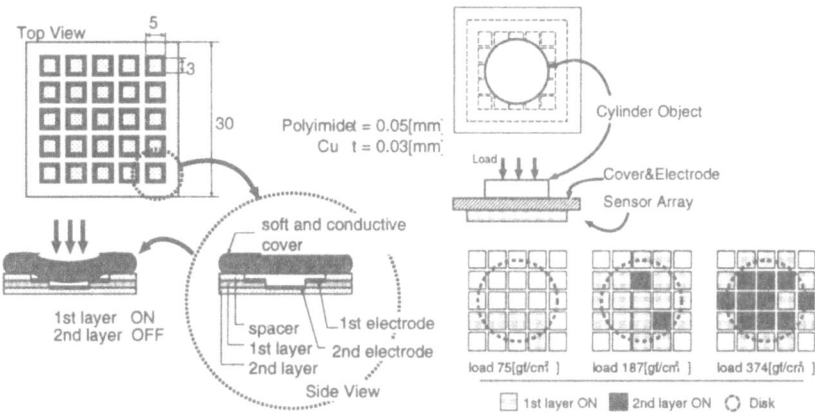

Figure 3. Multivalue touch sensor (left) and experiment (right)

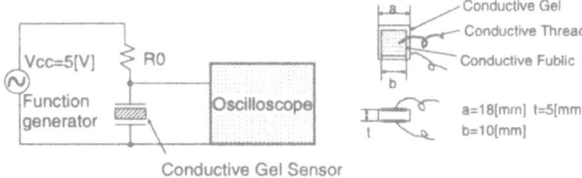

Figure 4. Conductive gel impedance examination circuit

developed in Euslisp[10]. We applied this result to generate H4 postures in various environment.

## 4. Soft and Distributed Tactile Sensor Skin

### 4.1. Multi-value Tactile Switch

Multi-value tactile switch which satisfies the requirements above is developed. It has soft cover and flexible to attach cylindrical surface. Its circuit and sensor signal processing are simple enough because it is digital switch. In this paper, we developed two value touch switch and Fig.3 shows its components. In this switch, sensing pressure threshold can be changed by the spacer which divides electrodes. Basic experiment using cylinder is shown in Fig.3. This sensor is very simple and has possibility to make very small system using MEMS technology. But since its sensor itself is not soft, it has a limitation to attach curved surface. Therefore we implemented this sensor to a robot body.

### 4.2. Conductive Gel Matrix Sensor

Soft tactile sensor which covers all over the body is required for a robot that behaves in human world. So far several research on soft tactile sensor that can be implanted in a soft skin (ex. [4]). We developed a matrix tactile sensor which consists of conductive gel that is originally developed as for a living body electrode. This conductive gel is made from monomer in ionized solu-

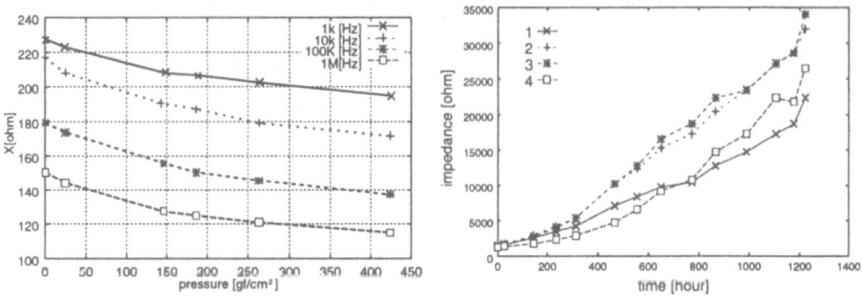

Figure 5. Conductive gel sensor basic examination circuit (left) and time change of conductive gel impedance(right)

Figure 6. Matrix gel tactile sensor module (left) and H4 arm with gel tactile sensor module (right)

tion by polymerizing. Deformation of a gel made a change of impedance so that input force can be measured. Because direct current will cause electrochemical polarization or electrolysis of metal electrode, pulse current are used for measuring[8]. Fig.4 shows its basic experiment by measuring pressure and impedance. Gel impedance is measured by alternating current (AC).

Because most of conductive gel is made by water, time changing of the impedance causes a problem. We measured time changing of gel impedance (Fig.5), and solve this problem by packing inside the polyethylene film.

### 4.3. Onbody Tactile Sensor Network System

Using multiplexor circuit and Hitachi micro-processor H8-3334YF, matrix tactile sensor and onbody $I^2C$ network are developed(Fig.7(right)), in order to reduce the number of wires. For an electrode, conductive fabric and conductive string are adopted(Fig.6(left)).

Multi-value tactile switch modules which is divided into $12 \times 8$ patches with three values, are attached for H4 body. Four conductive gel tactile sensor modules which is divided into $8 \times 8$ patches, are attached for H4 arms(Fig.6(right) shows a lower arm). Six H8 micro-processors are connected through $I^2C$ bus (five of which are for sensors and one is for data transmission), and send the

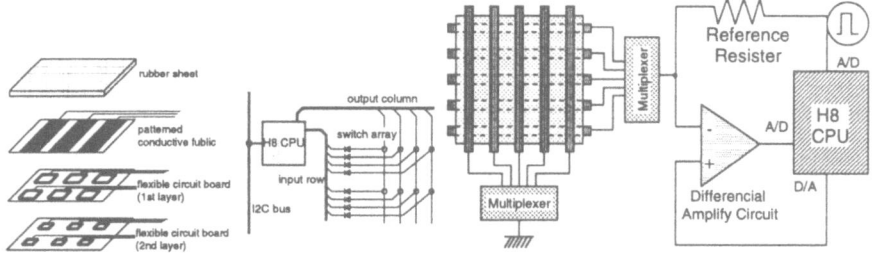

Figure 7. Multivalue touch sensor circuit (left) and gel tactile sensor module (right)

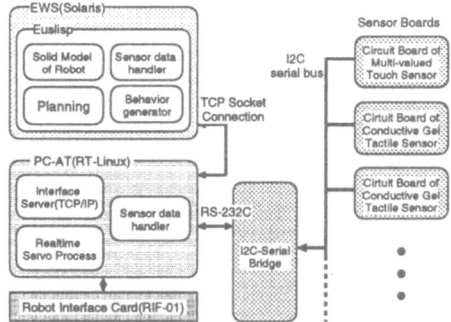

Figure 8. Tactile sensor skin software components

data to PC/AT clone CPU which controls actuatorsFig.8.

### 4.4. Experiments

Fig.9 shows experiments of reflection sensor measurement results to 3D geometric model in Euslisp. Fig.9 (left) is multi-value tactile switch experiment and Fig.9 (right) is matrix gel sensor experiment.

Fig.10 shows experiments of matrix gel sensor. H4 looks the point where tactile sensor output is changed (Fig.10 (left)) and looks the point where tactile sensor output is the biggest (Fig.10 (right)).

Finally we've succeeded to touch to human being(Fig. 11, 12).

## 5. Conclusion

A wheel type humanoid robot "H4" is developed as a platform for the research on tactile-action coupling in intelligent behavior of robots. The H4 has the features as follows : 1) upper-body consists of 19DOF (3DOF waist, 6DOF arm, 4DOF head), light weight and enough space to attach skin, 2) multi-value tactile switch and matrix gel tactile sensor, two types soft and distributed sensor skin covers upperbody, 3) each tactile sensor module has microprocessor module which is connected via $I^2C$ serial bus and make sensor network, 4) PC/AT compatible onboard computer which is controlled by RT-linux and connected to a network via radio ethernet, 5) 3D vision facility and 6) self-

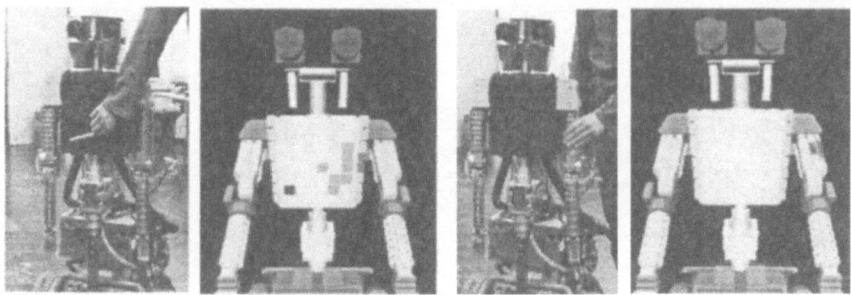

Figure 9. Tactile display on the robot solid model.

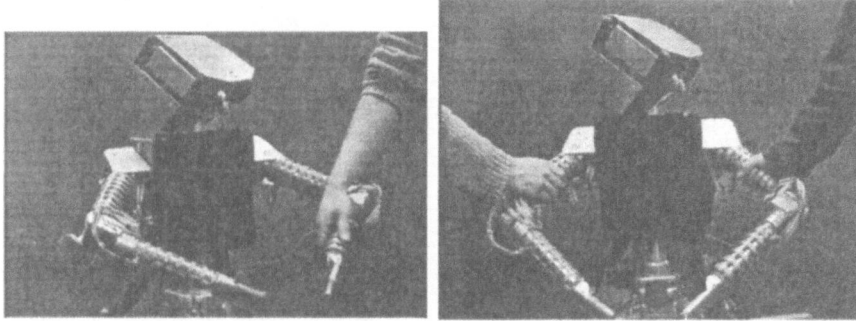

Figure 10. Experiment of measuring a force

contained.

This paper describe the design, implementation and experiment of H4. Then development of two types of soft and distributed tactile sensors are denoted. These sensors are soft enough so that robot can touch to a human without injuring, and a) matrix sensing enables circuit simple, b) serial bus enables number of wires small, c) flexible enough to attach robot surface. Furthermore, experiment of touch to a human being is denoted.

## Acknowledgments

This research has been supported by Grant-in-Aid for Research for the Future Program of the Japan Society for the Promotion of Science, "Research on Micro and Soft-Mechanics Integration for Bio-mimetic Machines (JSPS-RFTF96P00801)" project.

# References

[1] H. R. Nicholls and M. H. Lee. A Survey of Robot Tactile Sensing Technology. *Journal of Robotics Research*, Vol. 8, No. 3, pp. 3–30, 1989.

[2] M. Raibert and J.E. Tanner. A VLSI Tactile Array Sensor. In *Proc. of 12th ISIR*, 1982.

[3] M. Shimojo, M. Ishikawa, and K. Kanaya. A flexible high resolution tactile imager with video signal output. In *Proc. of 1991 IEEE/International Conference on Robotics and Automation*, pp. 384–391, 1991.

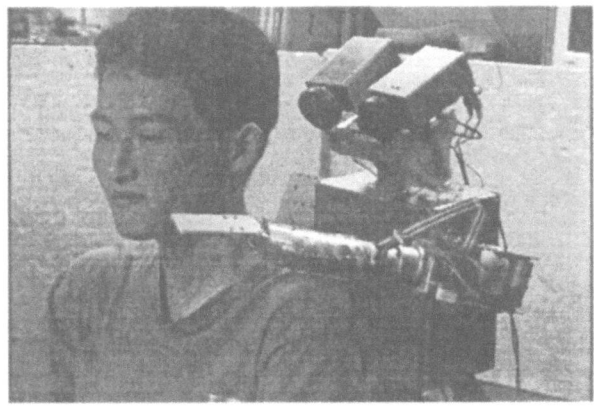

Figure 11. Experiment of tapping a human shoulder

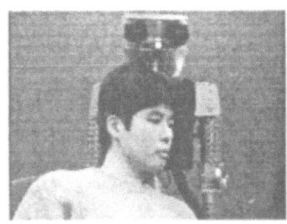 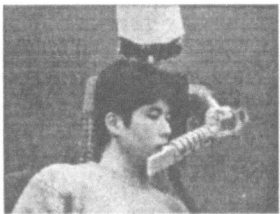 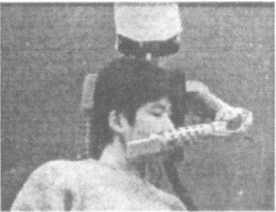

Figure 12. Experiment of grabbing a human head

[4] H. Shinoda, K. Matsumoto, and S. Ando. Acoustic Resonat Tensor Cell for Tactile Sensing. In *Proc. of IEEE Int. Conf. on Robotics and Automation*, pp. 3087 – 3092, 1997.

[5] H. Morita, Y. Yamada, and Y. Umetani. Vibrotactile Sensor Elements with Surface Ridges for Robot Hands (in Japanese). In *Proc. of the Third Robotics Symposia*, pp. 89–94, 1998.

[6] R. Kageyama, K. Nagashima, A. Konno, M. Inaba, and H. Inoue. Development and applications of multi-value soft tactile sensor that can covers robot surface. In *Proc. of '98 Annual Symposium of Robotics-Mechatronics*, pp. 1CI1-2, 1998.

[7] K. Ogawara, N. Futai, F. Kanehiro, M. Inaba, and H. Inoue. Sitting Down Behavior of Humanoid Robot using Soft Tactile Sensors Implanted in Skin. In *Proc. of 16th Annual Conference of Robotics Society of Japan*, Vol. 3, pp. 1361–1362, 1998.

[8] R. Kageyama, S. Kagami, M. Inaba, and H. Inoue. Development and Applications of Soft Tactile Sensor Made of Conductive Gel. In *Proc. of 16th Annual Conference of Robotics Society of Japan*, Vol. 2, pp. 873–874, 1998.

[9] T. Matsui and S. Sekiguchi. Design and Implementation of Parallel EusLisp Using Multithread (in Japanese). *Information Processing Society*, Vol. 36, No. 8, pp. 1885–1896, 1995.

[10] M. Inaba. Remote-Brained Robotics: Interfacing AI with Real World Behaviors. In *Robotics Research: The Sixth International Symposium*, pp. 335–344. International Foundation for Robotics Research, 1993.

# Development of the Torso Robot – Design of the New Shoulder Mechanism 'Cybernetic Shoulder'

Yoshihiko NAKAMURA      Masafumi OKADA      Shin-ichirou HOSHINO

University of Tokyo, Department of Mechano-Informatics

7-3-1 Hongo Bunkyo-ku Tokyo 113, Japan

nakamura, okada, hoshino@ynl.t.u-tokyo.ac.jp

**Abstract:** In this paper, we develop a three DOF mechanism for humanoid robots, which we call the cybernetic shoulder. This mechanism imitates the motion of the human shoulder and does not have a fixed center of rotation, which enables unique human-like motion in contrast to the conventional design of anthropomorphic seven DOF manipulators that have base three joint axes intersecting at a fixed point. Taking advantage of the cybernetic shoulder's closed-link structure, we can easily introduce the programmable passive compliance adopting the actuation redundancy and elastic members. This is important for the integrated safety of humanoid robots that are inherently required to physically interact with the human.

## 1. Introduction

The humanoid robot is similar to a human, has same degree-of-freedom as a human, and can do same works as a human. Many researches have been made for it [1]~[4]. The most important demands for humanoid robots are as follows:

1. **Human-Like Mobility** To have human-like geometry would be a requirement for a humanoid robot working among the humans since it provides with a geometric sufficiency for the robot working in the structured environment by the humans. The same reasoning applies to the mobility of humanoid robots. Moving, namely reaching and walking, like the humans is important from the geometric functionality point of view as well as the psychological point of view.

2. **Human-Like Compliance** Working and moving among humans requires special concerns on the safety issues. A humanoid robot should weigh not significantly more than a human. Mechanical compliance of the surfaces and joints is also a necessity.

In this paper, we propose a new shoulder mechanism for humanoid robots. A closed kinematic chain with three degree-of-freedom structured mechanism, which meets the first requirement and potentially enables the implementation of the second requirement. We call this mechanism the cybernetic shoulder.

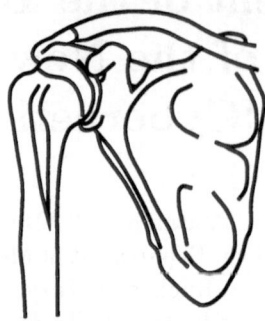

Figure 1. Human shoulder mechanism

## 2. Human Shoulder Mechanism

Figure 1 shows the human shoulder mechanism. The human shoulder is composed by 5 joints so that it can move smoothly, which is caused by a collarbone[5]. Figure 2 shows the motion of the human shoulder. This figure means that the human shoulder's motion does not have a fixed center of rotation. We consider that causes the human-like motion. The conventional design

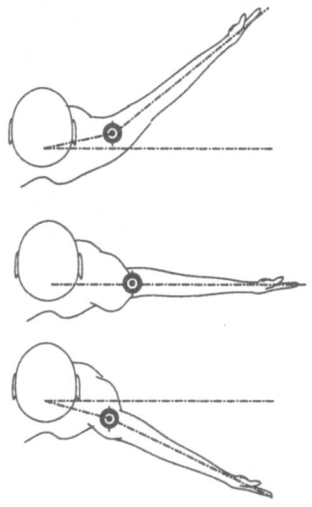

Figure 2. Motion of the human shoulder

of anthropomorphic seven degree-of-freedom manipulators cannot realize this motion.

## 3. Cybernetic Shoulder

### 3.1. A Mechanism of the Cybernetic Shoulder

Figure 3 shows the model of the cybernetic shoulder. This mechanism imitates the human shoulder geometry and has a human-like motion. $\beta$ and $\delta$ are two degree-of-freedom gimbal mechanisms, $d$ is a ball joint, $b$ is a two degree-of-

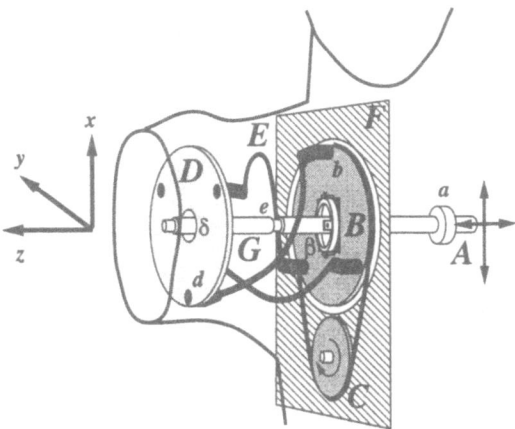

Figure 3. Cybernetic shoulder

freedom universal joint, $a$ is a four degree-of-freedom spherical and prismatic joint and $e$ is a prismatic joint. Moving point $A$ within vertical plane alters the pointing direction of the main shaft $G$, which determines, along with the constraints due to the free curved links $E$ between points $b$ and $d$, the direction of the normal vector of $D$. The rotation about the normal of $D$ is mainly determined by the rotation of $C$ through $B$ and $E$. Note that the rotation of $C$ affects the pointing direction of $D$ when $B$ and $D$ are no longer parallel. The advantages of this mechanism are summarized as follows.

**Compactness** Since the cybernetic shoulder can locate its actuators inside the chest, the shoulder geometry occupies rather small volume and shows a smooth shape, compared with the conventional designs of shoulder joints of manipulators.

**Large mobile area** In this mechanism, the angle magnification ratio can be wide, which causes the large mobile area. Figure 4 shows the motion of the cybernetic shoulder, as of the reduced two dimensional model. When the link $G$ rotates $\psi$ within $\pm 45$ degree, the normal vector of $D$ rotates within $\pm 90$ degree.

**Human like motion** As we mentioned, this mechanism imitates the human shoulder motion and does not have a fixed center of rotation. Figure 5 shows a locus of centers of rotation when it moves along $y$ axis.

From these figures we can understand that the cybernetic shoulder does not have a fixed center of rotation, which is same as human shoulder motion shown in Fig.2. It is impossible for conventional design of anthropomorphic three degree-of-freedom mechanisms.

**Singularity Free** In theory, link $G$ can rotate within $\pm 90$ degrees, namely $\psi$ can rotate in the range of $|\psi| < 90$ degrees. When $\psi$ equals to $\pm 90$ degrees, point $A$ should move an infinite distance even for a small required motion

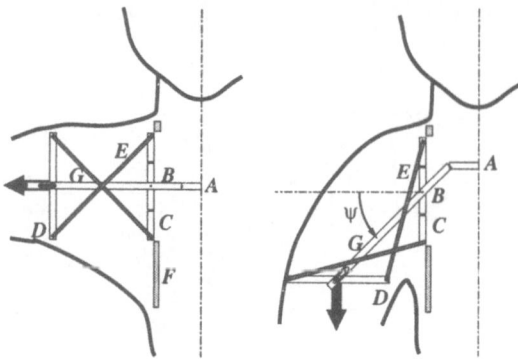

Figure 4. Motion of the Cybernetic Shoulder

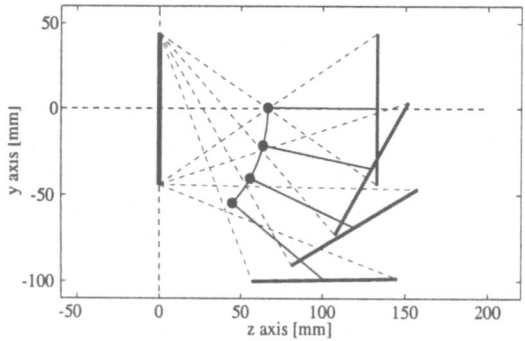

Figure 5. Center of rotation

of plate $D$, namely it is a singular point. In practice, mechanical design may limit the range of $\psi$ a bit smaller. In the two dimensional model, $\psi$ can rotate to 45 degrees, and the normal vector of plate $D$ can rotate to 90 degrees. Although the upper limit of $\psi$ depends on the rotation about the $z$ axis in the three dimensional model, whose effect is not much. In our prototype, the range of $\psi$ takes $\pm 45$ degrees at the minimum. This range of $\psi$, along with the coupled change of the center of rotation of plate $D$, keeps the work space of arms larger than the conventional design. Note that there is no shoulder singularity within the workspace.

**Small Backlash** The cybernetic shoulder has a double universal-joint structure which yields a human shoulder geometry and has a human like motion[3]. Many types of double universal-joint structures were developed [3, 6, 7], where the double universal-joint mechanisms were driven by gears, and resulted in large backlash. On the other hand, the kinematic constraint of the cybernetic shoulder is provided by the closed kinematic chains, which realizes small backlash.

Figure 6. Photographs of the cybernetic shoulder

Figure 6 and 7 show photographs of a prototype of the cybernetic shoulder. The height of the body is about 400mm and the width between the end of the left shoulder and right shoulder is about 600mm. The diameter of $D$ is about 110mm. Three actuators are DC motors of 90W. The planner motion of point $A$ is made by two perpendicular ball-screw axes assembled in series.

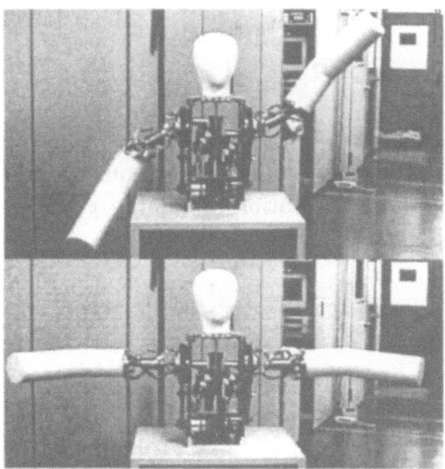

Figure 7. Motion of the experimental system

## 3.2. Kinematics of the cybernetic shoulder

Due to the complexity of a closed kinematic chain, it is difficult to get a closed form solution of the inverse or the forward kinematics. Therefore we develop a numerical method to solve the kinematics.

We define parameters and coordinate systems as shown in Fig.8. $x_0 y_0 z_0$ is the absolute coordinate system which has the origin at the $\beta$, center of $B$. $x_1 y_1 z_1$ is the end plate coordinate system which has the origin at the $\delta$, center

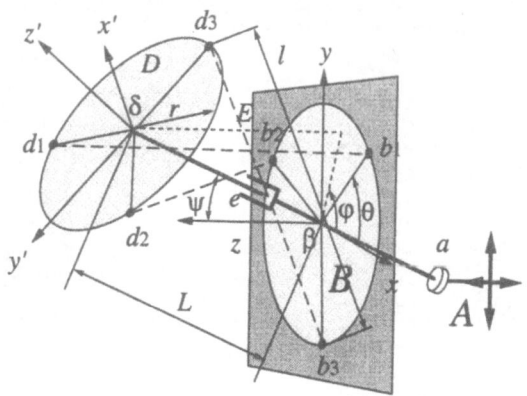

Figure 8. Model of the cybernetic shoulder

of $D$. In the initial condition, $b_1$ has the same direction as $x_0$. In Fig.8, $x_0 y_0 z_0$ is rotated $\phi$ around the $z_0$ axis. In the following, $[ \cdot ]^i$ means a vector in the $x_i y_i z_i$ coordinate system and $R_\theta^\xi$ implies the rotation $\theta$ around the $\xi$ axis. Accordingly Fig.8

$$b_i^0 = R_\phi^{z_0} R_{\frac{2}{3}(i-1)\pi}^{z_0} [ \; r \quad 0 \quad 0 \; ]^T \tag{1}$$

$$d_i^1 = R_{\frac{2}{3}(i-1)\pi}^{z_0} [ \; r \quad 0 \quad 0 \; ]^T \tag{2}$$

are satisfied. Since $\beta$ and $\delta$ have a gimbal mechanism, we can consider a $\theta_{x0}$ rotation about the $x_0$ axis, a $\theta_{y0}$ rotation about the $y_0$ axis, a $\theta_{x1}$ rotation about the $x_1$ axis and a $\theta_{y1}$ rotation about the $y_1$ axis. We now obtain the following equations:

$$d_i^0 = \delta + R d_i^1 \quad (i = 1, 2, 3) \tag{3}$$

$$\delta := L R_A e_z \tag{4}$$

$$R := R_A R_B \tag{5}$$

$$R_A := R_{\theta_{x0}}^{x_0} R_{\theta_{y0}}^{y_0} \tag{6}$$

$$R_B := R_{\theta_{y1}}^{y_1} R_{\theta_{x1}}^{x_1} \tag{7}$$

$$e_z = [ \; 0 \quad 0 \quad 1 \; ]^T \tag{8}$$

where $\theta_{x0}, \theta_{y0}$ and $\phi$ are given by the position of $A$ and the rotation angle of $C$. Any parameters such as $\theta_{x1}, \theta_{y1}$ and $L$ are unknown. The kinematic constraints of the mechanism are represented by that the lengths of link $E$ are constant. Namely,

$$\left| d_i^0 - b_i^0 \right|^2 = \ell^2 \tag{9}$$

From equation (3), (9) is written as follows:

$$\left| L R_A e_z + R d_i^1 - b_i^0 \right|^2 = \ell_i^2 \tag{10}$$

which is then expanded in the following form:

$$2Le_z^T R_A^T \left( R_A R_B d_i^1 - b_i^0 \right) - 2b_i^{0^T} R_A R_B d_i^1$$
$$+L^2 + 2r^2 - \ell_i^2 = 0 \quad (i = 1, 2, 3) \tag{11}$$

We apply numerical computation to solve equation (11).

## 4. Design of the passive compliance
### 4.1. Design of a compliant link

A humanoid robot should have a mechanical softness so as not to injure the people sharing the space. Since the cybernetic shoulder is composed of a closed kinematic chain, we can design the softness of mechanism.

Figure 9. Compliant link with damper

Table 1. Spring constant of links

|  | Material | Spring constant (N/m) |
|---|---|---|
| 1. | $\phi$3mm carbon fiber | 1456.7 |
| 2. | $\phi$5mm carbon fiber | 5845.5 |
| 3. | $\phi$4.31mm SMA | 2295.5 |
| 4. | Aluminum alloy | 15775.0 |

Figure 9 shows the compliant link of a $\phi$5mm carbon fiber rod fiber and a damper, which is designed as seen in Figure 10. Assembled two thick squared rings in chain form three chambers, which are filled with Temper Foam®[†]. By replacing this material with alternatives one, we can design the elasticity and viscosity of a link. We prototyped four types of elastic link. The configuration and the spring constant of these links are shown in Table 4.1. While the aluminum alloy link shows rather rigid property, links designed by using carbon fiber rod and a shape memory alloy possesses flexibility of different levels.

---

[†]Temper Foam : Produced by EAR SPECIALITY COMPOSITES Corp. shows frequency dependent characteristics. Namely, it shows high stiffness for high frequency, high viscosity for low frequency.

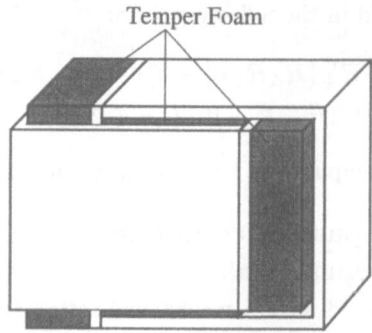

Figure 10. Design of a damper

Figure 11. The cybernetic shoulder with $\phi$3mm carbon fiber rod

## 4.2. Compliance of the cybernetic shoulder

Figure 11 shows the cybernetic shoulder with $\phi$3mm carbon fiber rod link, and we measured the viscosity of the cybernetic shoulder. The configuration of the experimental setup is shown in Figure 12. We measured the oscillation of the end grip when an impulsive disturbance is added. Figure 13 shows the experimental result using $\phi$5mm carbon fiber rod link with damper (dashed line) and without damper (solid line). This figure shows the effectiveness of the damper. In order to evaluate the effect of viscosity on damping, the maximum amplitude of swings are normalized in the figure. Regarding the temper memory alloy, the elasticity and viscosity vary depending on the temperature[8]. By the choice of the phase transformation temperature, we can design the property of links, which is shown in Figure 14 using $\phi$4.31mm SMA links (cool and heated).

## 5. Conclusions

In this paper, we proposed, designed and fabricated a new shoulder mechanism (the cybernetic shoulder) for humanoid robots. The followings and the

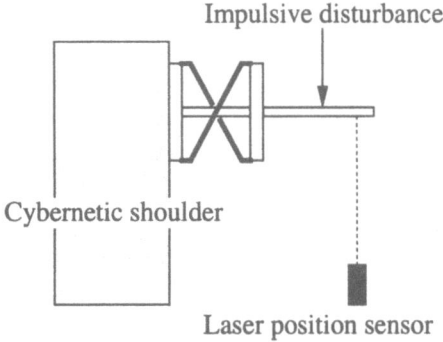

Figure 12. Experimental setup

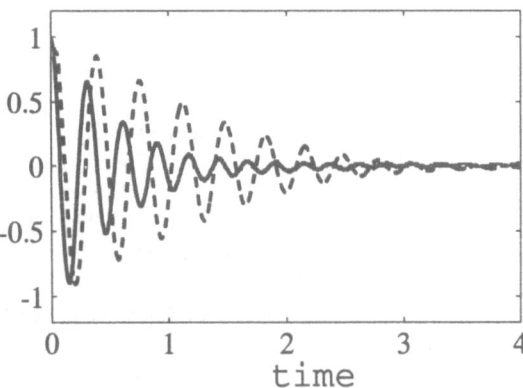

Figure 13. Disturbance responses ($\phi$5mm carbon fiber link with a damper and without damper)

summary of outcomes:

1. The advantages of the cybernetic shoulder are compactness, large and singularity free work space, human-like mobility and small backlash.

2. Taking an advantage of closed kinematic chain structure, the cybernetic shoulder allows mechanical compliance and damping design by appropriate choice of materials.

3. We design the viscosity of link $E$ using temper foam® and shape memory alloy.

The authors would like to thank Mr. N. Mizushima for his help with the experimental setup. This research was supported through the Research for the Future Program, the Japan Society for the Promotion of Science (Project No. JSPS-RFTF96P00801).

518

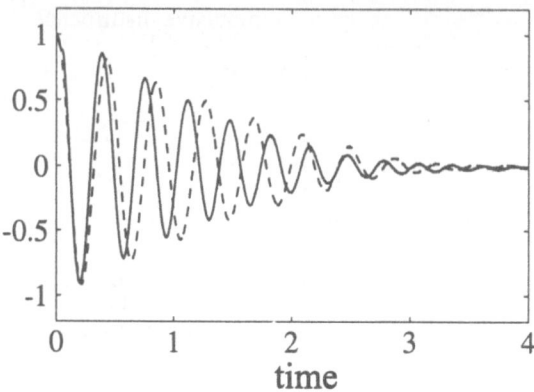

Figure 14. Disturbance responses of link 3(solid line) and heated link 3(dashed line)

## References

[1] J.Yamaguchi and A.Takanishi: Development of a Biped Walking Robot Having Antagonistic Driven Joints Using Nonlinear Spring Mechanism; Proc. of the IEEE International Conference on Robotics and Automation, pp. 185-192 (1997)

[2] K.Hirai, M.Hirose, Y.Haikawa and T.Takenaka: The Development of Honda Humanoid Robot; Proc. of the IEEE International Conference on Robotics and Automation, pp. 1321-1326 (1997)

[3] M.E.Rosheim: Robot Evolution, The Development of Anthrobotics; JOHN & SONS, INC., (1994)

[4] M.Inaba, T.Ninomiya, Y.Hoshino, K.Nagasaka, S.Kagami and H.Inoue: A Remort-Brain Full-Body Humanoid with Multisensor Imaging System of Binocular Viewer, Ears, Wrist Force ant Tactile Sensor Suit; Proc. of the International Conference on Robotics and Automation, pp. 2497-2502 (1997)

[5] I.A.Kapandji: Physiologie Articulaire; Maloine S.A. Editeur, (1980)

[6] W. Pollard: Position Controlling Apparatus; U.S. Patent 2,286,571, (1942)

[7] J.P. Trevelyan: Skills for a Shearing Robot : Dexterity and Sensing; Robotics Research, pp. 273-280 (1985)

[8] A.Terui, T.Tateishi, H.Miyamoto and Y.Suzuki: Study on Stress Relaxation Properties of NiTi Shape Memory Alloy; Trans. of the Japan Society of Mechanical Engineering (A), Vol.51, No.462,(1985) (in Japanese)

# Combination of Several Vision Sensors for Interpretation of Human Actions

Paulo Peixoto, Jorge Batista, Helder Araújo and A. T. de Almeida
ISR - Institute of Systems and Robotics
Dept. of Electrical Engineering - University of Coimbra
3030 COIMBRA - PORTUGAL
peixoto,batista,helder,aalmeida@isr.uc.pt

**Abstract:** In this paper we describe how the combination of several vision sensors can be used to track multiple human targets in a typical surveillance situation. The experimental system devised is made up of a fixed camera and a binocular active vision system. The fixed camera is used to provide the system with peripheral motion detection capabilities. The human activity is monitored by extracting several parameters that are useful for their classification. The system enables the creation of a record based on the type of activity. These logs can be selectively accessed and provide images of the humans in specific areas.

## 1. Introduction

Surveillance applications have specific features that depend upon the type of environment they are designed to operate on. In most surveillance applications the main focus of interest is human activity [1, 2, 3, 4]. Automatic interpretation of the data is very difficult and most systems in use require human attention to interpret the data.

Images of the environment are acquired either with static cameras with wide-angle lenses (to cover all the space), or with cameras mounted on pan and tilt devices (so that all the space is covered by using good resolution images) [5, 6, 7, 8]. In some cases both types of images are acquired but the selection of the region to be imaged by the pan and tilt device depends on the action of a human operator. The combination of both types of images to exploit their complementarity and redundancy is not common. This is an important and essential step towards the full automation of high-security applications in man-made environments.

The system described in this paper tries to explore the combination of several vision sensors in order to cope with the proposed goal of autonomously detect and track human intruders in man-made environments.

In the current setup the system is made up of a static camera and a binocular active system (see fig. 1). The static camera acquires wide-angle images of the environment enabling the early detection of intrusion. The system associated to the fixed camera can track several moving objects simultaneously. This system behaves as a peripheral motion detector. If one of the intruders approaches a critical area the active system starts its tracking. By using optical

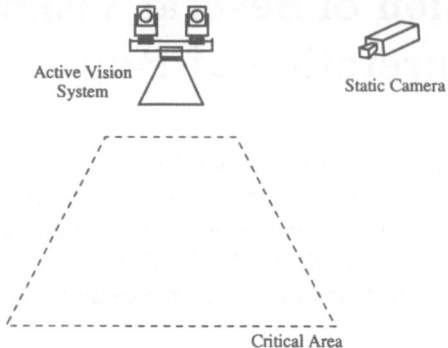

Figure 1. Integrated Visual Surveillance System

flow the system is able to track binocularly non-rigid objects in real-time. Simultaneously motion segmentation is performed, which enables the extraction of high quality target images (since they are foveated images). These are useful for further analysis and action understanding. The 3D trajectory of the target is also recovered by using the proprioceptive data.

## 2. Peripheral Vision

The Peripheral Vision system (the wide-angle static camera) is responsible for the detection and tracking of all targets visible in the scene. It is also responsible for the selection of the target that is going to be tracked by the binocular active vision system.

### 2.1. Target Tracking and Initialization

Target segmentation is achieved using a simple motion detection scheme. In each frame a segmentation procedure based on optical flow allows the detection of the possible available targets. A Kalman filter is attached to each detected target and the information returned by the filter is used to predict the location of the target in the next frame. This target predicted position is used to determine the actual target position in the next frame. If the uncertainty in position becomes too large over a significant amount of time then the target is considered to be lost and the associated tracking process is terminated. This can occur when the target walks out of the image, is heavily occluded or stops.

A priority level is dynamically assigned to each detected target according to some pre-established, application-dependent criteria. This allows the system to sort the several targets available in the image according to their priority level. The top most priority target will be the one followed by the active vision system. If a new high priority target is detected then the active vision system should direct its attention to this new target. This is achieved using the fixation ability of the active vision system.

The image sequence depicted in Figure 2 exemplifies the tracking process. In this example, three intruders move in front of the active vision system. The peripheral vision system segments and sets up a track process for each intruder. These images of the sequence are separated by time intervals of approximately

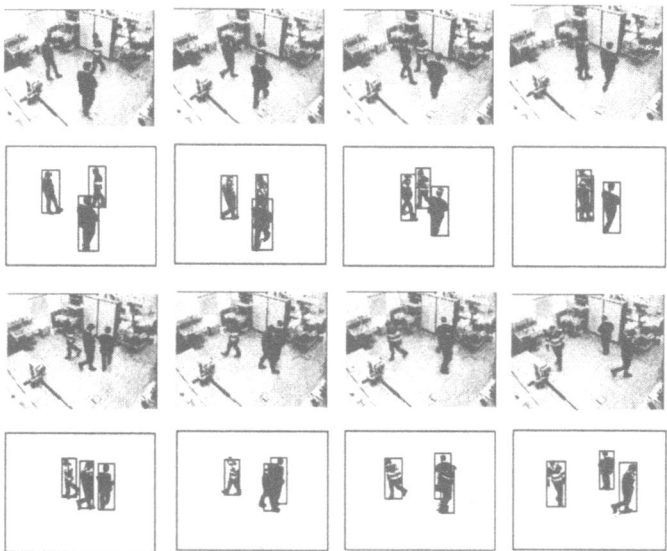

Figure 2. Eight pictures taken from an image sequence.

500 *ms.*

## 2.2. Ground Plane Correspondence

In order to redirect the attention to a new target the active vision system should know where to look for it. Since the position of the target is known in the static camera image, we will need to map that position in terms of rotation angles of the neck (pan and tilt) and vergence (we are assuming that both vergence angles are equal). The goal would be to fixate the active vision system on the target head.

Assuming that all target points considered in the static camera image lie in the ground plane then any point in this plane can be mapped to a point in the image plane of the static camera using an homography [9].

The pair of corresponding points $p_i$ and $P_i$ in the two planes is related projectively by an homography $kP = Ap$ where $k$ is a scale factor of the point pair $(p,P)$ which accounts for the fact that the representation in homogeneous coordinates is not unique. $A$ is an invariant 3x3 matrix with only eight degrees of freedom because scale is arbitrary under projective equivalence. This matrix $A$ can be recovered by establishing the correspondence between at least four points.

$$k_i \begin{pmatrix} X_i \\ Y_i \\ 1 \end{pmatrix} = \begin{bmatrix} s_1 & s_2 & s_3 \\ s_4 & s_5 & s_6 \\ s_7 & s_8 & 1 \end{bmatrix} \begin{pmatrix} x_i \\ y_i \\ 1 \end{pmatrix}$$

For each detected target in the image plane $p(x,y)$ we can compute the correspondent point in the ground plane and then the relationship between the point $P(X,Y)$ in the plane and the joint angles can be derived directly from the geometry of the active vision system (see Fig. 3):

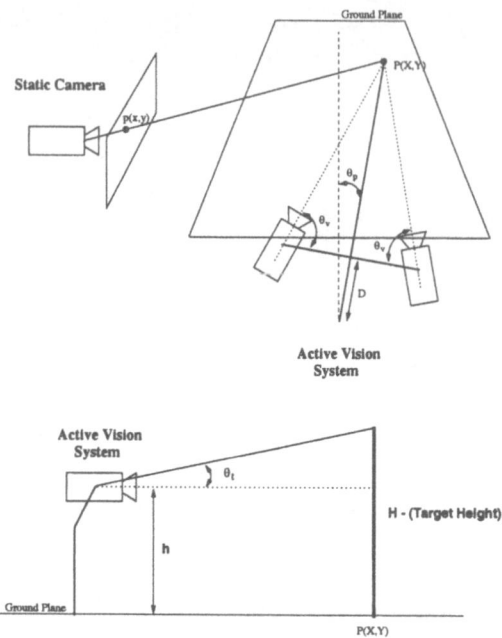

Figure 3. Correspondence between image points and ground plane points

$$\theta_p = \arctan \frac{X}{Y}, \; \theta_v = \arctan \frac{\frac{B}{2}}{\sqrt{X^2+Y^2}-D} \text{ and } \theta_t = \arctan \frac{H-h}{\sqrt{X^2+Y^2}}$$

with $B$ the baseline distance.

To compute the tilt angle we must know the target height which can be easily computed since we can obtain the projection of the targets head and feet points (detected in the image plane) in the ground plane. Assuming that the static camera height and position in the ground plane referential is known we can very easily compute the targets height (see Fig. 4).

Since each target corresponds to a blob and not all the blob pixels are in the ground plane one has to define which pixels should be used to compute the target position in the ground plane. In our case (due to our specific geometric configuration) we use the blob pixels closest to the lower part of the image. This assumption is valid as long as the target is not occluded.

## 2.3. Mutual Cross-Checking

Situations of partial target occlusion, and others can be dealt with by using the information available to both systems. If partial occlusion occurs then there will be an error in the mapping of the target position in the ground plane. If the active vision system is directed to track that specific target it may not find it (or it may find it beyond its expected position). In that case the peripheral vision system will be informed of such an occurrence and a new estimate of the position can be computed. Other typical situations are the cases when the active vision system starts tracking a specific target and changes to a different one (due to, for example, a frontal crossing of two intruders). This situation and others that are similar can be accounted for by cross-checking the location

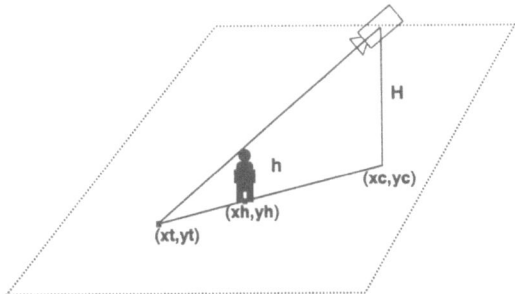

Figure 4. Target Height Computation

of the target on both systems.

At each frame the active vision system reports the coordinates of the actual target to the peripheral system (using inverse mapping). This position is then cross-checked against the detected target position in the peripheral system. If the positional error is above a certain threshold then a new fixation is forced. Cross-checking is possible in our case because both systems are fully synchronized. Since the images are synchronized and time-stamped all the events are referenced in the same time base.

## 3. Active Vision System Visual Routines

The active vision system is responsible for pursuit of a specific target in the scene. This task is achieved using two different steps: fixation and smooth pursuit. In the first one the attention of the active vision system is directed to the target and in the second one the target is tracked [10].

### 3.1. Fixation

The fixation process is realized using gross fixation solutions, defined as saccades, followed by a fine adjustments in both eyes in order to achieve vergence.

Having the target information given by the peripheral vision system, redirecting gaze can be regarded as controlling the geometry of the head so that the images of the target will fall into the fovea after the fixation. Because of the rapid movement of the MDOF head joints during saccades, visual processing cannot be used. Therefore, saccades must be accomplished at the maximum speed or in the minimum possible time. The solution adopted was moving neck and eyes in order to achieve symmetric vergence and having the cyclopean eye looking forward to the target. Saccade motion is performed by means of position control of all degrees of freedom involved.

In order to achieve perfect gaze of both eyes in the moving target, and since the center of mass is probably not the same in both retinas for non rigid objects, after the saccade a fine fixation adjustment is performed. A grey level cross-correlation tracker is used to achieve perfect fixation of both eyes.

Some tests have been made in order to measure the ability of the active vision system to precisely fixate on the requested target. Fig. 5 represents the mapping error in terms of pan and vergence angles in function of the position

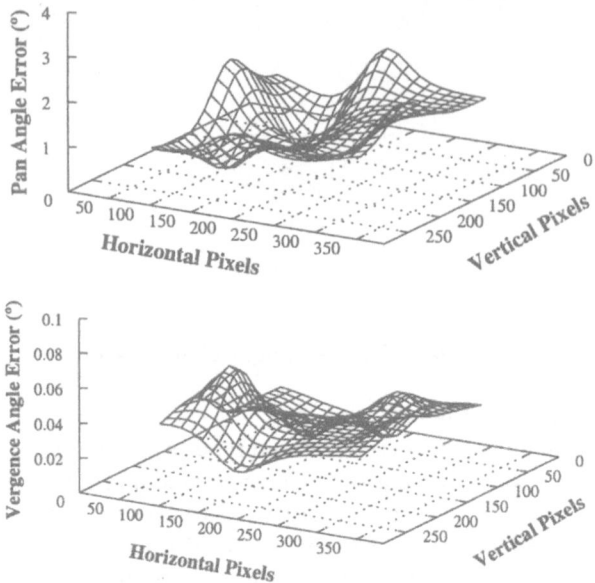

Figure 5. Pan and Vergence Angle Error.

of the target in the peripheral camera image. Only points in the critical area (a rectangle with 4m x 5m) were considered. This errors justify the use of a fine fixation adjustment to guarantee vergence necessary for the smooth pursuit process.

## 3.2. Smooth Pursuit Using Optical Flow

After fixating on the target the pursuit process is started by computing the optical flow. During the pursuit process velocity control of the degrees of freedom is used instead of position control as in the case of the saccade.

Optical flow is used to estimate the target velocity in both left and right images. Assuming that the moving target is inside the fovea after a saccade, the estimated target velocity in the cyclopean eye is used to control the movement of the pan in order to guarantee that the cyclopean eye is always pointing straight to the target. The target velocity disparity between both left and right images is used to control vergence [11].

The smooth pursuit process starts a Kalman filter estimator, which takes the estimated image motion velocity of the target as an input. Using this approach, the smooth pursuit controller generates a new prediction of the current image target velocity, and this information is sent to the motion servo controller every 10ms.

Fig. 6 shows an example of a frame taken during the smooth pursuit process. Both the static camera image and the images taken by the active vison system can be seen.

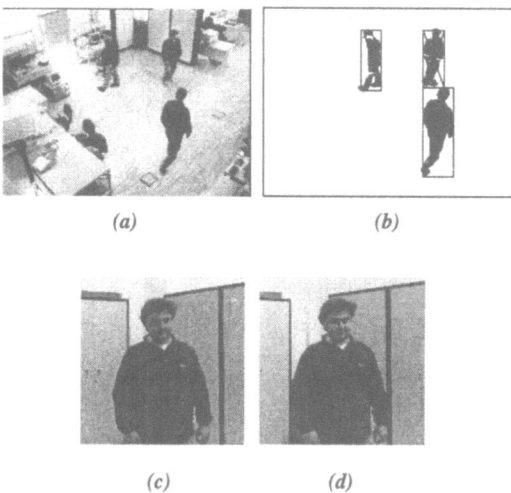

Figure 6. Target detection and tracking (a) Image acquired with the peripheral vision system. (b) Target segmentation (c),(d) Active vision system left and right images

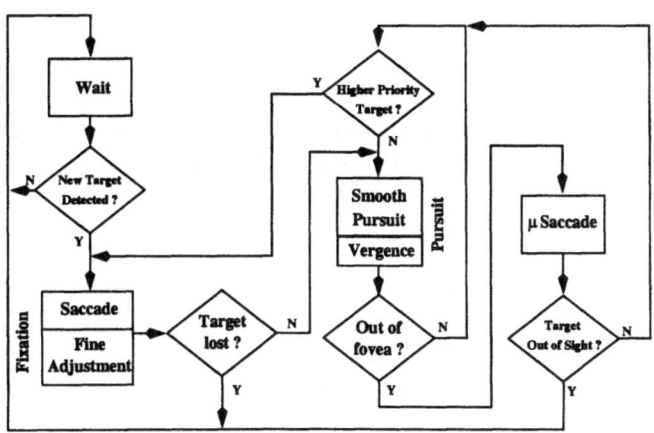

Figure 7. State Transition System

## 4. The Global Controller Strategy

The strategy adopted by the gaze controller to combine saccade, smooth pursuit and vergence to track moving objects using the active vision system was based on a *State Transition System*. This controller defines four different states: *Waiting, Fixation (Saccade), Smooth Pursuit and μSaccade*. Each of these states receives control commands from the previous state, and triggers the next state in a closed loop transition system. The peripheral vision system is used as the supervisor of the active vision system, triggering the transition between the different states. The global configuration of this state transition system is show in figure 7.

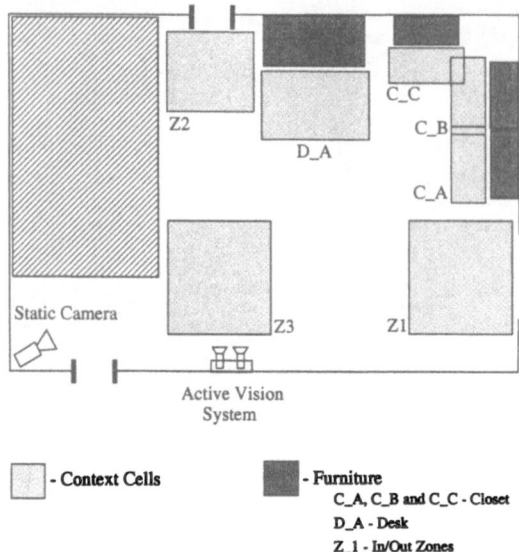

Figure 8. Context Cells Definition

## 5. Human Activity Logging

A fundamental problem to be addressed in any surveillance or human activity monitoring scenario is that of information filtering: how to decide whether a scene contains an activity or behavior worth analyzing. Our approach to detection and monitoring of such situations is based on the fact that typically actions are somehow conditioned by the context in which they are produced. For instance the action of opening a closet only makes sense in the near vicinity of a closet.

We assumed the concept of "*context cells*" to discriminate portions of the scene where any behavior can be important in terms of the application. It is assumed that a set of rules is known "*a priori*" and that these rules have the adequate relevance for the purpose of the monitoring task. It is also assumed that these context cells have enough information to trigger a logging event in order to register the actions of the human subjects.

Since in this first approach we only have as an input the position of the target in the plane and its height we defined a very simple set of context cells in our lab (see fig. 8). Three different context cells were defined: closets, desks and in/out zones that correspond to areas where targets can enter/exit the scene. The rule used to describe the possible actions in the Desk context cell is shown Fig. 9. Two actions are logged in this case: "Near Desk" and "Seated at Desk".

An advantage of this concept based on the context is that it can be used to predict the expected appearance and location of the target in the image, to predict occlusion and initiate processes to deal with that occlusion.

The system creates log files that describe the actions that occurred during a certain period of time. A picture is taken by the active vision system for

Figure 9. An example of a rule used to describe an action, in this case the use of a desk

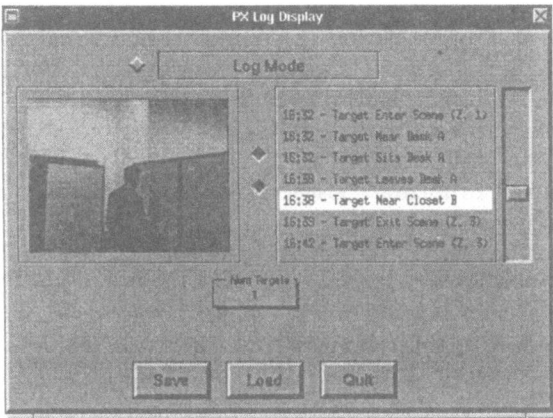

Figure 10. This picture shows an example of a typical Human Activity log output

each recorded action. These images can then be used for posterior analysis and processing, for instance for identification purposes. Figure 10 shows an example of a portion of a log file recorded by the system.

Different actions require different views in order to understand what is going on. For instance if the target is near a closet, then his hands and not his head, should have the attention of the system. The advantage of the use of the active vision system is that if the "best" view needed for a particular action understanding has not been captured then the active vision system can be instructed to redirect its attention to the right place. Once again the definition of "best" view is context dependent and if this context is known then the search space for the best view can be substantially reduced.

Another aspect of this work (not yet in real-time) is the modeling of more complex behaviors using the same underlaying principle. The logging ability can be extended the to the detection of more elaborate actions like the detection of opening or closing of closets, and other typical indoor actions.

# 6. Conclusions

In this paper we described a system aimed at detecting and tracking intruders in man-made environments. This system is based on a peripheral motion detector and a foveated tracking system. The specific features of the active system enable the tracking of non-rigid objects while handling some degree of occlusion. The degree of occlusion that can be tolerated depends upon the target distance. Behavior modeling can advantageously use the 3D trajectories reconstructed with the proprioceptive data. Experimental results suggest that, by integrating both systems, a higher degree of robustness is achieved. Logging of human activity is performed in real time and by analyzing changes in that data (changes in height and position) some limited interpretation of action is performed. In addition, the redundancy of the system enables cross-checking of some types of information, enabling more robustness.

# References

[1] B. Rao and H. Durrant-Whyte. A decentralized bayesian algorithm for identification of tracked targets. *IEEE Trans. on Systems, Man and Cybernetics*, 23(6):1683–1698, November/December 1993.

[2] K. Rao. Shape description of curved 3d objects for aerial surveillance. In *ARPA96*, pages 1065–1076, 1996.

[3] R. Howarth and H. Buxton. Visual surveillance monitoring and watching. In *ECCV96-II*, pages 321–334, 1996.

[4] L. Davis, R. Chellapa, Y. Yacoob, and Q. Zheng. Visual surveillance and monitoring of human and vehicular activity. In *DARPA97*, pages 19–27, 1997.

[5] Y. Cui, S. Samarasekera, Q. Huang, and M. Greiffenhagen. Indoor monitoring via the collaboration between a peripheral sensor and a foveal sensor. In *Proc. of the IEEE Workshop on Visual Surveillance*, pages 2–9. IEEE Computer Society, January 1998.

[6] T. Kanade, R. T. Collins, A.J. Lipton, P. Anandan, P. Burt, and L. Wixson. Cooperative multisensor video surveillance. In *Proc. of the DARPA Image Understanding Workshop*, pages 3–10, 1997.

[7] M. Yedanapudi, Y. Bar-Shalom, and K. R. Pattipati. Imm estimation for nultitarget-multisensor air-traffic surveillance. *Proc. IEEE*, 1:80–94, January 1997.

[8] E. Grimson, P. Viola, O. Faugeras, T. Lozano-Perez, T. Poggio, and S. Teller. A forest of sensors. In *Proc. of the DARPA Image Understanding Workshop*, volume 1, pages 45–50, New Orleans, LA, 1997.

[9] K. J. Bradshaw, I. Reid, and D. Murray. The active recovery of 3d motion trajectories and their use in prediction. *IEEE Trans. on PAMI*, 19(3):219–234, March 1997.

[10] J. Batista, P. Peixoto, and H. Araújo. Real-time vergence and binocular gaze control. In *IROS97–IEEE/RSJ Int. Conf. on Intelligent Robots and Systems*, Grenoble, France, September 1997.

[11] J. Batista, P. Peixoto, and H. Araujo. Real-time visual surveillance by integrating peripheral motion detection with foveated tracking. In *Proc. of the IEEE Workshop on Visual Surveillance*, pages 18–25. IEEE Computer Society, January 1998.

# Lecture Notes in Control and Information Sciences

**Edited by M. Thoma**

**1993–1999 Published Titles:**

**Vol. 204:** Takahashi, S.; Takahara, Y.
Logical Approach to Systems Theory
192 pp. 1995 [3-540-19956-X]

**Vol. 205:** Kotta, U.
Inversion Method in the Discrete-time
Nonlinear Control Systems Synthesis
Problems
168 pp. 1995 [3-540-19966-7]

**Vol. 206:** Aganovic, Z.; Gajic, Z.
Linear Optimal Control of Bilinear Systems
with Applications to Singular Perturbations
and Weak Coupling
133 pp. 1995 [3-540-19976-4]

**Vol. 207:** Gabasov, R.; Kirillova, F.M.;
Prischepova, S.V.
Optimal Feedback Control
224 pp. 1995 [3-540-19991-8]

**Vol. 208:** Khalil, H.K.; Chow, J.H.;
Ioannou, P.A. (Eds)
Proceedings of Workshop on Advances
inControl and its Applications
300 pp. 1995 [3-540-19993-4]

**Vol. 209:** Foias, C.; Özbay, H.;
Tannenbaum, A.
Robust Control of Infinite Dimensional
Systems: Frequency Domain Methods
230 pp. 1995 [3-540-19994-2]

**Vol. 210:** De Wilde, P.
Neural Network Models: An Analysis
164 pp. 1996 [3-540-19995-0]

**Vol. 211:** Gawronski, W.
Balanced Control of Flexible Structures
280 pp. 1996 [3-540-76017-2]

**Vol. 212:** Sanchez, A.
Formal Specification and Synthesis of
Procedural Controllers for Process Systems
248 pp. 1996 [3-540-76021-0]

**Vol. 213:** Patra, A.; Rao, G.P.
General Hybrid Orthogonal Functions and
their Applications in Systems and Control
144 pp. 1996 [3-540-76039-3]

**Vol. 214:** Yin, G.; Zhang, Q. (Eds)
Recent Advances in Control and Optimization
of Manufacturing Systems
240 pp. 1996 [3-540-76055-5]

**Vol. 215:** Bonivento, C.; Marro, G.;
Zanasi, R. (Eds)
Colloquium on Automatic Control
240 pp. 1996 [3-540-76060-1]

**Vol. 216:** Kulhavý, R.
Recursive Nonlinear Estimation: A Geometric
Approach
244 pp. 1996 [3-540-76063-6]

**Vol. 217:** Garofalo, F.; Glielmo, L. (Eds)
Robust Control via Variable Structure and
Lyapunov Techniques
336 pp. 1996 [3-540-76067-9]

**Vol. 218:** van der Schaft, A.
$L_2$ Gain and Passivity Techniques in Nonlinear
Control
176 pp. 1996 [3-540-76074-1]

**Vol. 219:** Berger, M.-O.; Deriche, R.;
Herlin, I.; Jaffré, J.; Morel, J.-M. (Eds)
ICAOS '96: 12th International Conference on
Analysis and Optimization of Systems -
Images, Wavelets and PDEs:
Paris, June 26-28 1996
378 pp. 1996 [3-540-76076-8]

**Vol. 220:** Brogliato, B.
Nonsmooth Impact Mechanics: Models,
Dynamics and Control
420 pp. 1996 [3-540-76079-2]

**Vol. 221:** Kelkar, A.; Joshi, S.
Control of Nonlinear Multibody Flexible Space
Structures
160 pp. 1996 [3-540-76093-8]

**Vol. 222:** Morse, A.S.
Control Using Logic-Based Switching
288 pp. 1997 [3-540-76097-0]

**Vol. 223:** Khatib, O.; Salisbury, J.K.
Experimental Robotics IV: The 4th
International Symposium, Stanford, California,
June 30 - July 2, 1995
596 pp. 1997 [3-540-76133-0]

**Vol. 224:** Magni, J.-F.; Bennani, S.;
Terlouw, J. (Eds)
Robust Flight Control: A Design Challenge
664 pp. 1997 [3-540-76151-9]

**Vol. 248:** Chen, Y.; Wen C.
Iterative Learning Control
216pp: 1999 [1-85233-190-9]

**Vol. 249:** Cooperman, G.; Jessen, E.; Michler,
G. (Eds)
Workshop on Wide Area Networks and High
Performance Computing
352pp: 1999 [1-85233-642-0]